AF388657

Wearable Movement Sensors for Rehabilitation: From Technology to Clinical Practice

Wearable Movement Sensors for Rehabilitation: From Technology to Clinical Practice

Editors

Gerrit Ruben Hendrik Regterschot
Gerard M. Ribbers
Johannes B. J. Bussmann

MDPI • Basel • Beijing • Wuhan • Barcelona • Belgrade • Manchester • Tokyo • Cluj • Tianjin

Editors
Gerrit Ruben Hendrik Regterschot
Erasmus MC University Medical
Centre Rotterdam
The Netherlands

Gerard M. Ribbers
Erasmus MC University
Medical Centre Rotterdam
The Netherlands

Johannes B. J. Bussmann
Erasmus MC University
Medical Centre Rotterdam
The Netherlands

Editorial Office
MDPI
St. Alban-Anlage 66
4052 Basel, Switzerland

This is a reprint of articles from the Special Issue published online in the open access journal *Sensors* (ISSN 1424-8220) (available at: https://www.mdpi.com/journal/sensors/special_issues/clinical_practice).

For citation purposes, cite each article independently as indicated on the article page online and as indicated below:

LastName, A.A.; LastName, B.B.; LastName, C.C. Article Title. *Journal Name* **Year**, *Volume Number*, Page Range.

ISBN 978-3-0365-2063-6 (Hbk)
ISBN 978-3-0365-2064-3 (PDF)

Contents

About the Editors

Gerrit Ruben Hendrik Regterschot

(Ph.D.) is a specialist in the development and application of wearable movement sensors for measuring the quantity and quality of physical behaviour in motor rehabilitation populations. Currently (2021) he is the coordinator of the national ArmCoach4Stroke project and other consortium research projects at the Department of Rehabilitation Medicine of the Erasmus Medical Center. At the same time he works as a Research Fellow at the University of Twente where he develops new and innovative sensor-based methods and machine learning models for the measurement of the quality of arm use after stroke. He also has extensive experience working in industry (Philips Research), and holds a PhD degree from the University of Groningen.

Gerard M. Ribbers

(Prof. Dr.) started as resident neurosurgery in 1989 and switched to rehabilitation medicine in 1991. As resident he chaired the 'kerngroep revalidaticum' and participated in the 'concilium revalidaticum'. Since 1995 he has worked at Rijndam rehabilitation in Rotterdam. His clinical tasks have focused on patients with severe cognitive, behavioral and linguistic disorders due to acquired brain injury. Until 2007 he managed the inpatient division with 120 beds. In 2007 he founded the Rotterdam Neurorehabilitation Research (RoNeRes) consortium, a collaboration between Rijndam rehabilitation and the Erasmus University MC. He was appointed professor Neurorehabilitation in 2012 and head of the department of Rehabilitation Medicine Erasmus MC in 2018.

Johannes B. J. Bussmann

(Ph.D.) his research aims at understanding and improving physical behaviour (PB) of people with chronic conditions, and to apply this in rehabilitation care. In this research the focus is on the use of technology in the "at home" situation. Although the theme of research is not population-specific, his research focuses on people with acquired neurological conditions such as stroke. Studies include topics as the development, validation and application of objective measurement and feedback devices. Examples of this are posture and motion monitors, a wheelchair monitor, an objective stress monitor, and an upper limb use monitor for stroke patients. In 2008 Hans Bussmann initiated the ICAMPAM conference (International Conference for Ambulatory Measurement of Physical Activity and Movement) and in 2015 the International Society for the Measurement of Physical Behaviour (ISMPB).

Editorial

Wearable Movement Sensors for Rehabilitation: From Technology to Clinical Practice

Gerrit Ruben Hendrik Regterschot [1,*], Gerard M. Ribbers [1,2] and Johannes B. J. Bussmann [1]

[1] Department of Rehabilitation Medicine, Erasmus University Medical Center Rotterdam, P.O. Box 2040, 3000 CA Rotterdam, The Netherlands; g.ribbers@erasmusmc.nl (G.M.R.); j.b.j.bussmann@erasmusmc.nl (J.B.J.B.)
[2] Rijndam Rehabilitation, Westersingel 300, 3015 LJ Rotterdam, The Netherlands
* Correspondence: g.r.h.regterschot@erasmusmc.nl

Citation: Regterschot, G.R.H.; Ribbers, G.M.; Bussmann, J.B.J. Wearable Movement Sensors for Rehabilitation: From Technology to Clinical Practice. *Sensors* **2021**, *21*, 4744. https://doi.org/10.3390/s21144744

Received: 30 June 2021
Accepted: 6 July 2021
Published: 12 July 2021

Publisher's Note: MDPI stays neutral with regard to jurisdictional claims in published maps and institutional affiliations.

Motor disorders are a common and age-related problem in the general community. Therefore, medical rehabilitation often focuses on reducing the burden of motor disorders. With an aging society, the burden of motor disorders is expected to grow, while the healthcare capacity is not expected to match this growth. For this reason, there is an urgent need to optimize medical rehabilitation of motor disorders both in terms of effectiveness and efficiency.

The rapid innovations in wearable movement sensors in recent years may provide an opportunity to translate these innovations into the field of motor rehabilitation. Wearable movement sensors can provide objective and precise measurements of the quantity and quality of physical activities, body postures, and movements in clinical as well as normal daily life environments, thereby providing clinicians with data that can be used to guide, personalize, and optimize therapy. Since wearable sensors are portable, inexpensive, unobtrusive, and also have the ability to provide information that is unique and cannot be obtained otherwise (e.g., by standardized clinical tests or questionnaires), they have an enormous potential for the tracking of patient functioning and recovery during motor rehabilitation. In addition, wearables can play a crucial role in the existing tendency towards at-home monitoring and treatment, and in substituting more complex measurement devices, such as camera systems.

Despite their potential to optimize motor rehabilitation, wearable movement sensors are relatively scarcely applied in rehabilitation of motor disorders. Important challenges remain, such as the development of reliable and valid wearable movement sensors in clinical populations and free-living environments, barriers in the deployment of wearable movement sensors in clinical care, development and optimization of innovative sensor configurations and data analysis techniques (such as machine learning-based algorithms that enable detection of specific activities and movements in free-living conditions), and the development of disease-specific sensor-based outcome measures that are relevant and interpretable by patients and clinicians.

This Special Issue, entitled "Wearable Movement Sensors for Rehabilitation: From Technology to Clinical Practice", aims to facilitate the application of wearable movement sensors in clinical practice. It intends to explore the opportunities for the application of wearable movement sensors in motor rehabilitation.

A total of 17 papers are published in this Special Issue. These papers mainly focus on the following topics: algorithm development, technical validation, clinical validation, monitoring of physical behavior in daily life conditions, and implementation in motor rehabilitation. Hereafter, we provide a brief overview of each paper.

Yang et al. (2020) [1] developed an online gait-planning algorithm based on sensing signals to enable balance control during exo-skeleton assisted walking with crutches in spinal cord patients. Results from this pilot study in healthy adults indicate that the developed online gait-planning algorithm can plan the landing point of the swing leg to improve balance control during exo-skeleton assisted walking. The algorithm may be useful to improve balance control during exoskeleton assisted walking in spinal cord injury patients and reduce the need of using crutches.

Sy et al. (2020) [2] present a novel Lie group constrained extended Kalman filter to estimate lower limb kinematics with a minimal sensing solution during different physical activities (e.g., walking). Healthy adults performed an activity protocol with three inertial measurement unit (IMU) sensors, placed on the pelvis and on both ankles. Results showed relatively small errors for the knee and hip joint angles compared to an optimal motion capture system, indicating the validity of the algorithm. This paper contributes to the development of a sensor-based method that enables comfortable and long-term monitoring of lower limb kinematics in rehabilitation patient populations.

Fadel et al. (2020) [3] propose a new algorithm that quantifies the characteristics of walking based on a hip-worn accelerometer in a more detailed manner than activity counts. The algorithm uses the fast Fourier Transform to obtain periodic characteristics of walking, and it reduces the dimensionality of the raw sensor data into a form that retains details of the original signal while enabling existing statistical methods for analyses. The paper serves as a proof of concept for how researchers can extract the walking characteristics from sensor data and investigate the association with relevant health-related outcomes in rehabilitation. An example is provided of a study that investigates the associations of walking spectra obtained from the fast-paced 400 m walk with age and BMI in older adults.

Roossien et al. (2021) [4] developed a sensor-based method for the measurement of lumbar load. The method consists of six IMU sensors on the upper and lower arms, sternum, and pelvis. Lumbar load is quantified as the net moment around the L5/S1 intervertebral body, estimated using a method that is based on artificial neural networks. The validity of the sensor-based method was supported in healthy adults since the differences in the estimated lumbar load were consistent with the perceived intensity levels and the character of the work tasks. The method may be used to monitor lumbar load in people with musculoskeletal disorders such as lower back pain, to assess muscular overload during rehabilitation, and to help clinicians to tailor treatments.

Bravi et al. (2021) [5] investigated the validity of a sensor-based system for shoulder range of motion assessment in cervical spinal cord injury patients. The sensor-system consists of two IMU sensors placed on the wrist and upper arm. Patients and healthy controls performed four shoulder movements. Every movement was evaluated with a goniometer and with the IMU system at the same time. The validity of IMU system was partially confirmed, since relative agreement between the IMU system and goniometer was high but absolute agreement was relatively low. The proposed IMU system may be a potential tool for monitoring shoulder range of motion in patients with cervical spinal cord injury.

Tak et al. (2020) [6] presented a new sensor-based method for the measurement of knee, hip and spine joint angles during the single leg squat. The sensor system consists of four IMU sensors placed on the trunk, pelvis, upper leg, and lower leg. In this study, healthy adults performed single leg squats. Results showed high correlations between the joint angles measured with the sensor system and an optical reference system, indicating the validity of the sensor system. The sensor method is potentially relevant for monitoring and optimizing lower extremity kinematics during rehabilitation interventions.

Ahmadi et al. (2020) [7] investigated the accuracy of group, group-personalized, and fully-personalized machine learning physical activity classification models in children with cerebral palsy. Models were trained and tested using accelerometer data from the hip, wrist, and ankle. To assess the validity, the classification accuracy was evaluated and compared in a laboratory trial and a simulated free-living trial with 38 children while wearing a wrist-worn accelerometer. Results showed that group-personalized and fully-personalized Random Forest activity classification models provide a more accurate recognition of physical activity in children with CP than "one-size-fits-all" group models.

Labarierre et al. (2020) [8] performed a systematic review to identify and summarize studies in which motion sensors and machine learning algorithms have been used to adapt the behavior of orthotic/prosthetic devices to user locomotion mode (e.g., stair ascent/descent, walking on flat floor). Results showed that classification accuracies were, in general, very high in healthy people and people with unilateral transtibial and transfemoral amputation. These findings support the validity of sensor methods and machine learning algorithms to recognize locomotion mode.

Yang et al. (2020) [9] developed a sensor-based method to estimate relative 3D orientations between finger tips and the dorsal side of the hand with inertial motion sensors but without magnetometers to avoid magnetic disturbance. The method consists of one sensor on the dorsal side of the hand, and one on the most distal finger segment. Results in three healthy adults show that errors in relative orientation between fingers and hand are relatively small during hand movements and during a functional water-drinking task. The sensor method is potentially useful for clinical assessments during stroke rehabilitation.

Prasanth et al. (2021) [10] performed a systematic review of sensor-based methods applied for real-time gait analysis. Inertial measurement units on the shank and foot are most often used for gait analysis in combination with threshold or peak identification methods for gait detection. Less than one third of the sensor-based methods for gait analysis were validated on pathological gait data. For clinical gait assessments, a combination of inertial measurement units and rule-based methods are recommended as an optimal solution.

Regterschot et al. (2021) [11] investigated to what extent arm use measurements with wrist-worn accelerometers in stroke patients are affected by whole-body movements, such as walking. Wrist-worn accelerometers are often applied to measure arm use after stroke. They measure arm use by recording all arm movements, including non-functional arm movements due to whole-body movements. Results of the study show that whole-body movements substantially increase cross-sectional arm use outcomes when not correcting arm use data for whole-body movements, thereby threatening the validity of arm use outcomes and measured arm use changes.

Zhou et al. (2020) [12] evaluated a sensor-based method to classify fallers from non-fallers based on spatial-temporal gait characteristics. Wearable sensors were placed on both ankles and the lower back. A partial least square discriminant analysis was used to classify fallers and non-fallers based on gait features derived from the sensor data. Results showed that fallers differed from non-fallers in gait patterns. The presented sensor-based method may be useful in rehabilitation to identify persons with a high fall risk and to monitor the effects of interventions on fall risk.

Mazzarella et al. (2020) [13] investigated in this pilot study whether a 3D motion capture system can detect changes over time in pre-reaching and reaching behaviors in infants with perinatal stroke and cerebral palsy. Results showed that spatiotemporal characteristics of upper extremity movements measured with a 3D motion capture system change over time in infants with typical development, cerebral palsy and perinatal stroke, with potential differences between infants with typical development and cerebral palsy. This study shows the potential of wearable sensors for measuring characteristics of upper extremity movements in infants with perinatal stroke and cerebral palsy.

Fleiner et al. (2021) [14] investigated the association between physical behavior and subjectively-rated circadian chronotypes in older adults. Physical activity was measured in 81 older adults during one week with a motion sensor on the lower back and the wrist. Results showed that the timing of mobility-related activity is associated with subjectively-rated chronotypes in older adults. The presented sensor-based method may provide a useful approach for early detecting and tailoring the treatment of circadian disruptions in rehabilitation populations.

Hofstad et al. (2020) [15] measured the number of consecutive steps and walking bouts in persons with a lower limb amputation using three accelerometers: one in each trouser pocket and one on the sternum. Measurements were performed for two consecutive days in 20 persons with a lower limb amputation and 10 age-matched controls. Results showed that objectively measured mobility was highly affected in persons with an amputation and that self-reported mobility did not match with the objective sensor-based measurements. This study recommends the use of accelerometers to measure mobility in persons with a lower limb amputation.

Lang et al. (2020) [16] discussed the major barriers for the application of wearable movement sensors in motor rehabilitation and proposed benchmarks for the implementation of sensors in clinical practice. Barriers in the clinic are the busy clinical environment and the lack of realization of the value of the information that can be obtained with sensors. Technology-related barriers include: (1) sensor systems that are inaccurate for many patient populations; (2) sensor systems that are not user-friendly for clinicians and/or patients; (3) the lack of published data regarding reliability and clinical validity of sensor systems.

Braakhuis et al. (2021) [17] explored the use, perspectives, and barriers to wearable activity monitoring in day-to-day stroke care routines amongst physical therapists. Results of the online survey showed that 27% of the respondents were using activity monitoring, and the concept of remote activity monitoring was perceived as useful. The identified barriers to clinical implementation were lack of skills and knowledge of patients, financial constraints, and not knowing what type of monitor to apply.

This Special Issue shows a range of potential opportunities for the application of wearable movement sensors in motor rehabilitation. However, the papers surely do not cover the whole field of physical behavior monitoring in motor rehabilitation. Most studies in this Special Issue focused on the technical validation of wearable sensors and the development of algorithms. Clinical validation studies, studies applying wearable sensors for the monitoring of physical behavior in daily life conditions, and papers about the implementation of wearable sensors in motor rehabilitation are under-represented in this Special Issue. Studies investigating the usability and feasibility of wearable movement sensors in clinical populations were lacking. We encourage researchers to investigate the usability, acceptance, feasibility, reliability, and clinical validity of wearable sensors in clinical populations to facilitate the application of wearable movement sensors in motor rehabilitation.

Author Contributions: Conceptualization, G.R.H.R., G.M.R. and J.B.J.B.; writing—original draft preparation, G.R.H.R.; writing—review and editing, G.R.H.R., J.B.J.B. and G.M.R. All authors have read and agreed to the published version of the manuscript.

Funding: This research received no external funding.

Acknowledgments: We would like to thank all authors who contributed to this Special Issue.

Conflicts of Interest: The authors declare no conflict of interest.

References

1. Yang, W.; Zhang, J.; Zhang, S.; Yang, C. Lower Limb Exoskeleton Gait Planning Based on Crutch and Human-Machine Foot Combined Center of Pressure. *Sensors* **2020**, *20*, 7216. [CrossRef] [PubMed]
2. Sy, L.W.F.; Lovell, N.H.; Redmond, S.J. Estimating Lower Limb Kinematics Using a Lie Group Constrained Extended Kalman Filter with a Reduced Wearable IMU Count and Distance Measurements. *Sensors* **2020**, *20*, 6829. [CrossRef] [PubMed]
3. Fadel, W.F.; Urbanek, J.K.; Glynn, N.W.; Harezlak, J. Use of Functional Linear Models to Detect Associations between Characteristics of Walking and Continuous Responses Using Accelerometry Data. *Sensors* **2020**, *20*, 6394. [CrossRef] [PubMed]
4. Roossien, C.C.; Baten, C.T.M.; van der Waard, M.W.P.; Reneman, M.F.; Verkerke, G.J. Automatically Determining Lumbar Load during Physically Demanding Work: A Validation Study. *Sensors* **2021**, *21*, 2476. [CrossRef] [PubMed]
5. Bravi, R.; Caputo, S.; Jayousi, S.; Martinelli, A.; Biotti, L.; Nannini, I.; Cohen, E.J.; Quarta, E.; Grasso, S.; Lucchesi, G.; et al. An Inertial Measurement Unit-Based Wireless System for Shoulder Motion Assessment in Patients with Cervical Spinal Cord Injury: A Validation Pilot Study in a Clinical Setting. *Sensors* **2021**, *21*, 1057. [CrossRef]
6. Tak, I.; Wiertz, W.P.; Barendrecht, M.; Langhout, R. Validity of a New 3-D Motion Analysis Tool for the Assessment of Knee, Hip and Spine Joint Angles during the Single Leg Squat. *Sensors* **2020**, *20*, 4539. [CrossRef]
7. Ahmadi, M.N.; O'Neil, M.E.; Baque, E.; Boyd, R.N.; Trost, S.G. Machine Learning to Quantify Physical Activity in Children with Cerebral Palsy: Comparison of Group, Group-Personalized, and Fully-Personalized Activity Classification Models. *Sensors* **2020**, *20*, 3976. [CrossRef]

8. Labarrière, F.; Thomas, E.; Calistri, L.; Optasanu, V.; Gueugnon, M.; Ornetti, P.; Laroche, D. Machine Learning Approaches for Activity Recognition and/or Activity Prediction in Locomotion Assistive Devices—A Systematic Review. *Sensors* **2020**, *20*, 6345. [CrossRef] [PubMed]

9. Yang, Z.; van Beijnum, B.J.F.; Li, B.; Yan, S.; Veltink, P.H. Estimation of Relative Hand-Finger Orientation Using a Small IMU Configuration. *Sensors* **2020**, *20*, 4008. [CrossRef] [PubMed]

10. Prasanth, H.; Caban, M.; Keller, U.; Courtine, G.; Ijspeert, A.; Vallery, H.; von Zitzewitz, J. Wearable Sensor-Based Real-Time Gait Detection: A Systematic Review. *Sensors* **2021**, *21*, 2727. [CrossRef] [PubMed]

11. Regterschot, G.R.H.; Selles, R.W.; Ribbers, G.M.; Bussmann, J.B.J. Whole-Body Movements Increase Arm Use Outcomes of Wrist-Worn Accelerometers in Stroke Patients. *Sensors* **2021**, *21*, 4353. [CrossRef] [PubMed]

12. Zhou, Y.; Rehman, R.Z.U.; Hansen, C.; Maetzler, W.; Din, S.D.; Rochester, L.; Hortobágyi, T.; Lamoth, C.J.C. Classification of Neurological Patients to Identify Fallers Based on Spatial-Temporal Gait Characteristics Measured by a Wearable Device. *Sensors* **2020**, *20*, 4098. [CrossRef] [PubMed]

13. Mazzarella, J.; McNally, M.; Richie, D.; Chaudhare, A.M.W.; Buford, J.A.; Pan, X.; Heathcock, J.C. 3D Motion Capture May Detect Spatiotemporal Changes in Pre-Reaching Upper Extremity Movements with and without a Real-Time Constraint Condition in Infants with Perinatal Stroke and Cerebral Palsy: A Longitudinal Case Series. *Sensors* **2020**, *20*, 7312. [CrossRef] [PubMed]

14. Fleiner, T.; Trumpf, R.; Hollinger, A.; Haussermann, P.; Zijlstra, W. Quantifying Circadian Aspects of Mobility-Related Behavior in Older Adults by Body-Worn Sensors—An "Active Period Analysis". *Sensors* **2021**, *21*, 2121. [CrossRef] [PubMed]

15. Hofstad, C.J.; Bongers, K.T.J.; Didden, M.; van Ee, R.F.; Keijsers, N.L.W. Maximal Walking Distance in Persons with a Lower Limb Amputation. *Sensors* **2020**, *20*, 6770. [CrossRef] [PubMed]

16. Lang, C.E.; Barth, J.; Holleran, C.L.; Konrad, J.D.; Bland, M.D. Implementation of Wearable Sensing Technology for Movement: Pushing Forward into the Routine Physical Rehabilitation Care Field. *Sensors* **2020**, *20*, 5744. [CrossRef] [PubMed]

17. Braakhuis, H.E.M.; Bussmann, J.B.J.; Ribbers, G.M.; Berger, M.A.M. Wearable Activity Monitoring in Day-to-Day Stroke Care: A Promising Tool but Not Widely Used. *Sensors* **2021**, *21*, 4066. [CrossRef] [PubMed]

Article

Lower Limb Exoskeleton Gait Planning Based on Crutch and Human-Machine Foot Combined Center of Pressure

Wei Yang [1,2], Jiyu Zhang [3], Sheng Zhang [1,2,*] and Canjun Yang [1,2]

[1]   Ningbo Research Institute, Zhejiang University, Ningbo 315100, China; simpleway@zju.edu.cn (W.Y.); ycj@zju.edu.cn (C.Y.)
[2]   School of Mechanical Engineering, Zhejiang University, Hangzhou 310027, China
[3]   School of Instrumentation Science and Engineering, Harbin Institute of Technology, Harbin 150001, China; 19b901013@stu.hit.edu.cn
*   Correspondence: szhang1984@nit.zju.edu.cn

Received: 3 November 2020; Accepted: 14 December 2020; Published: 16 December 2020

Abstract: With the help of wearable robotics, the lower limb exoskeleton becomes a promising solution for spinal cord injury (SCI) patients to recover lower body locomotion ability. However, fewer exoskeleton gait planning methods can meet the needs of patient in real time, e.g., stride length or step width, etc., which may lead to human-machine incoordination, limit comfort, and increase the risk of falling. This work presents a human-exoskeleton-crutch system with the center of pressure (CoP)-based gait planning method to enable the balance control during the exoskeleton-assisted walking with crutches. The CoP generated by crutches and human-machine feet makes it possible to obtain the overall stability conditions of the system in the process of exoskeleton-assisted quasi-static walking, and therefore, to determine the next stride length and ensure the balance of the next step. Thus, the exoskeleton gait is planned with the guidance of stride length. It is worth emphasizing that the nominal reference gait is adopted as a reference to ensure that the trajectory of the swing ankle mimics the reference one well. This gait planning method enables the patient to adaptively interact with the exoskeleton gait. The online gait planning walking tests with five healthy volunteers proved the method's feasibility. Experimental results indicate that the algorithm can deal with the sensed signals and plan the landing point of the swing leg to ensure balanced and smooth walking. The results suggest that the method is an effective means to improve human–machine interaction. Additionally, it is meaningful for the further training of independent walking stability control in exoskeletons for SCI patients with less assistance of crutches.

Keywords: gait planning; stride length; center of pressure; human–machine interaction

---

1. Introduction

More than 250,000 individuals annually sustain spinal cord injuries worldwide, mainly due to traffic accidents and fall from heights [1]. Spinal cord injury (SCI) patients can become paraplegic due to lesion characteristics. Moreover, long-term sitting and lying may cause poor health conditions and complications, e.g., muscular atrophy, pressure sores, constipation, and osteoporosis. Therefore, helping SCI patients to stand, walk, and engage in self-care may mitigate a crucial social problem.

With the development of wearable robotics and sensors, increased attention has been paid to the research of the lower-limb exoskeleton for the locomotion impaired, such as SCI and stroke patients. Zhang et al., Kawamoto et al., and Nilsson et al. developed exoskeleton systems for stroke rehabilitation through referencing normal walking gait guidance or reference gait-based impedance control [2–4]. Husain et al., Fineberg et al., and Jung et al. are more concerned about various functions of the exoskeleton to help SCI patients walk independently [5–7]. The former series focused on topics, such as patient walking intention recognition and joint torque control, and the latter on issues, such as human-machine dynamic balance control, walking mode switch for various terrains, and gait trajectory planning. Still, a significant amount of research and development is required to assist SCI patients to walk naturally and in balance, especially regarding exoskeleton gait planning. Besides efficacy, lack of community involvement, etc., the performance of self-controlled walking balance inhibits the wide application of commercial medical lower-limb exoskeletons on the market, such as Rewalk [8], HAL [9], and Ekso [10].

The simplest method to perform a "standard" gait is to predefine the normal gait trajectory. The desired joint trajectory is either recorded from several healthy volunteers or extracted from a gait analysis database. To improve the adaptation to various subject heights, the predefined gait trajectory is usually parameterized by lower limb sizes. Suzuki et al. prerecorded hip and knee joint angles from a healthy subject and divided a gait cycle into stance and swing phases, which are triggered by plantar reaction forces and torso tilt angles [11]. Similarly, Mina is controlled by a predefined gait trajectory [12], and gait-related parameters, including single- or continuous-step mode selection, walking speed, and step transition duration, can be tuned. Wang et al. proposed the MINDWALKER exoskeleton for gait assistance in the coronal and sagittal planes [13]. The gait is prerecorded by a healthy subject walking in MINDWALKER in zero-assistance mode. The predefined standard gait in the sagittal plane guarantees a natural walking posture, and step width control realized by online adjustment of exoskeleton hip abduction/adduction joints in the coronal plane ensures walking stability. Therefore, gait planning online or offline with additional measured information has been studied in recent years. Jeon et al. presented a fast wearable sensor-based gait phase classification method with the help of a convolutional neural network, which represents human-machine walking intention and is useful for exoskeleton motion control [14]. The polymer optical fiber sensors reported by Leal-Junior et al. show promising application scenarios for soft and wearable gait measurement for exoskeleton online gait planning [15].

Jung et al. presented the online computation of centroidal momentum (CM) [16], i.e., linear and angular momenta at the center of mass (CoM), in the exoskeleton-supported walking, which is regarded as a stability index to estimate the actual state of balance. Preliminary trials confirmed the assumption, and further research is in the progress of using CM to trigger a controller of the exoskeleton to maintain or recover the balance. Aphiratsakun et al. proposed a leg exoskeleton balancing control using a zero-moment point (ZMP) and a fuzzy logic controller [17]. The ground contact points on each foot were measured by a load cell and compared with the target ZMP, and input the differences into the controller, which generates the compensating angles of the left and right (L/R) ankle joints to position ZMP in the convex hull of the support area. Similarly, the center of pressure (CoP) of a human-machine system was investigated by Kim et al. for walking balance validation [18]. With measurements of both human gravity and exoskeleton support force, the CoP is calculated, and the stability condition is judged. Chen et al. calculated the CoP of a human-machine-crutch system and controlled an exoskeleton with an offline designed gait [19]. During walking, the gait is modified online if the calculated CoP exceeds a predefined stable area. Deng et al. used the capture point theory for biped robot balance control to guide the target landing point of a human and exoskeleton swing foot [20]. The instantaneous capture point is obtained by modeling the human-machine system, and the gait trajectory is corrected to solve the instability problem caused by random forward/backward leaning of the subject's upper body. Most of the above work is focused on human-machine dynamic stability during walking with the help of gait planning. Although CoP, ZMP, CoM, etc. are key facts to guide gait planning, the influence of crutches is not always considered. The subject's strength on crutches somehow determines the human-machine walking balance state. We focus on the support distribution of crutches and exoskeleton soles, to guide gait planning. This paper designs a gait planning algorithm, aiming to improve human-machine coordination and gradually improve a subject's active stability control during walking. The gait planning uses CoP calculation based on crutch reaction force and human-machine plantar force, and stride length mapping determined by the calculated combined CoP.

The remainder of this paper is organized as follows. Section 2 introduces the exoskeleton prototype, mathematical method, and related simulation. Section 3 presents the experimental design for gait planning algorithm validation. Results are analyzed and the effects of the algorithm are discussed in Section 4. Section 5 relates our conclusions, along with future research directions.

2. Materials and Methods

In this work, a commercial lower-limb exoskeleton (UGO, RoboCT, Inc., Hangzhou, China) is adopted as a testbed for gait planning algorithm validation. Besides, customized crutches and exoskeleton soles are designed as the accessory equipment for the crutch endpoint pressure and human-machine foot pressure measurement, respectively. Thus, the combined CoP can be calculated based on the pressure of the crutch endpoint and human-machine foot. By establishing the relationship between stride length and combined CoP, the stride length can be obtained. Finally, due to the determination of the stride length, the target gait will be planned by forward and inverse kinematic models. For the forward and inverse kinematics models' validation, the gait planning algorithm is simulated by MATLAB (Mathworks). The balance control of the human-machine system in the sagittal plane is the primary concern, since the hip and knee flexion/extension (f/e) joints with the developed lower-limb exoskeleton are motor driven. The human-exoskeleton-crutch system in the sagittal plane is shown in Figure 1a. The local coordinate system is established, in which X- and Y-axes indicate the vertical and horizontal moving directions, respectively. The origin of the local coordinate system is located at the ankle joint of the exoskeleton support leg. Figure 1b shows the four-DoF kinematic model of the lower-limb exoskeleton. Since the torso motion does not affect forward/inverse kinematic modeling for gait planning, the torso is simplified as one rigid body.

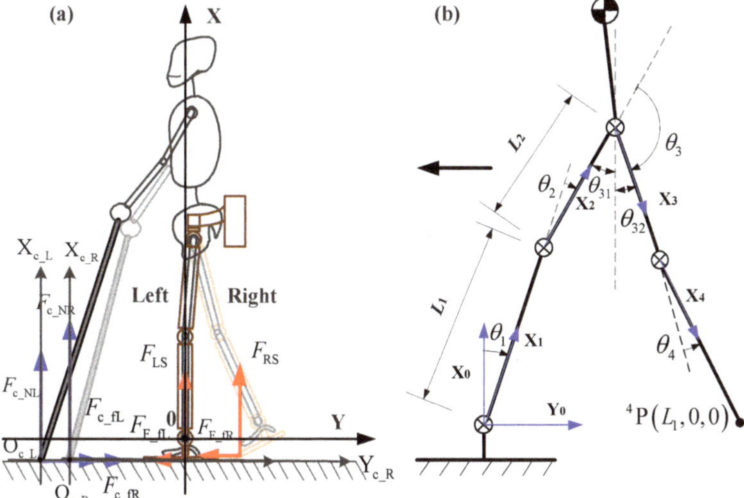

Figure 1. Human-exoskeleton-crutch system in the sagittal plane. (**a**) The local coordinate system is located at the rotation center of support of the leg ankle joint. The L/R crutch coordinate is located at the endpoint of the crutch. (**b**) Four-DoF kinematic model of the lower-limb exoskeleton. Denavit-Hartenberg parameters are tagged, including axes on joints and rotation angles of adjacent links.

2.1. Exoskeleton

The powered lower-limb exoskeleton system UGO (Figure 2) is developed to assist the balanced walk of patients with SCI or stroke. It is designed for subjects with heights between 1.5 and 1.9 m, and weigh below 100 kg. The hip and knee f/e joints are driven by servo motors and harmonic reducers coupled with a maximum rated torque of 100 Nm, while the ankle dorsi-flexion/ plantar-flexion joint is fully passive. All joints in the sagittal plane are limited to a certain range of motion, i.e., hip flexion (+110°) and extension (−40°), knee flexion (+95°) and extension (0°), ankle dorsi-flexion (+20°), and plantar-flexion (−30°). It should be emphasized that the passive ankle joint will not affect the gait planning since we focus on the ankle joint coordinates during forwarding/inversing kinematics instead of the footplate coordinate. Each exoskeleton motor-driven joint is equipped with a magnetic rotary encoder (AS5048A, AMS, Inc., Graz, Austria) with 14-bit resolution, and a torque sensor for joint angle and torque measurement, respectively. The torque sensor is customized with a range of 150 Nm, resolution of 0.05 Nm, and an accuracy of 0.3% FS. Both angle and torque signals are recorded with a sampling rate of 100 Hz.

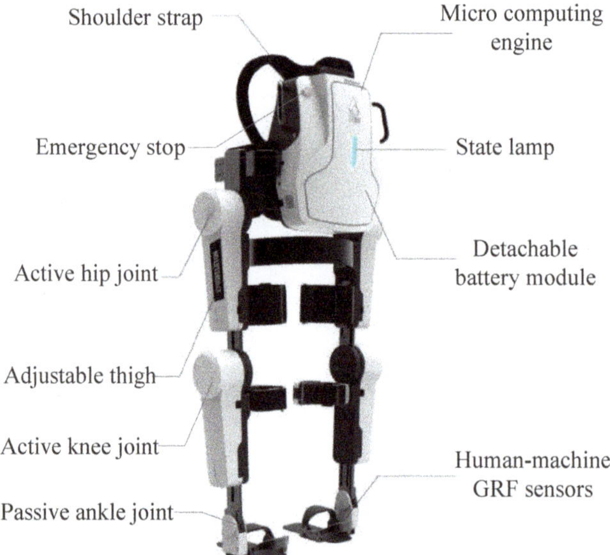

Figure 2. Lower limb exoskeleton system UGO developed by RoboCT, Inc.

2.2. Foot and Crutch Ground Reaction Force (GRF) Measurements

2.2.1. Foot GRF Measurement

Force-sensing resistor (FSR) sensors mounted on exoskeleton soles and crutches determine the GRFs and the CoP of the human-exoskeleton-crutch system. Figure 3a shows a human-machine GRF measurement sole, including seven FSR sensors, a rubber baseboard, and a processing circuit. FSR sensors were calibrated first with a designed force loading test bench, as shown in Figure 3b. All results are fitted by fifth-order polynomials. The calibration results are shown in Figure 3c. The maximum RMS error of all calibration results is 3.601 N.

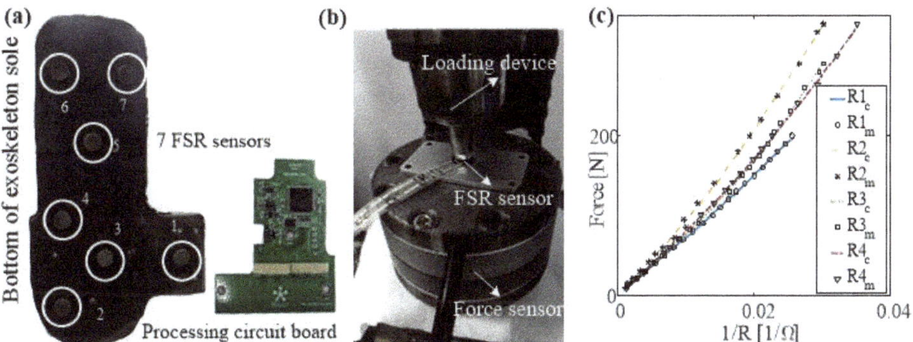

Figure 3. Calibration of accessory equipment. (**a**) Seven force-sensing resistor (FSR) sensors were mounted at the bottom of the exoskeleton sole to measure human-machine coupled ground reaction force (GRF). (**b**) Each FSR sensor was calibrated with a loading test bench by a force sensor (M4325K, Sun Rise Instrument, Inc., Nanning, China) with a range of 400 N and an accuracy of 0.5% FS. (**c**) Fitted results of calibration. Maximum RMS error of all calibration results was 3.601 N.

2.2.2. Crutch GRF Measurement

Chen et al. considered the influence of a crutch on human-machine balance control and designed crutches with FSR sensors at the bottom [19]. On the one hand, these sensors are helpful to measure GRF directly and conveniently. On the other hand, the interference caused by the change of the pitch angle of the crutch is an uncertain point in the application of GRF. To solve this problem, an indirect measurement of GRF for the crutch is proposed in this work. Since only the palm and forearm contact the crutch, FSR sensors are mounted on both contact areas, as shown in Figure 4a. One inertial measurement unit (IMU, JY901, Wit-Motion, Inc., Shenzhen, China) is mounted on the crutch to measure the crutch roll/pitch/yaw angles. With the above sensors, GRF on the crutch can be calculated by simple force analysis, as shown in Figure 4b. Compared to the direct measurement method in [19], this indirect measurement method is: (i) Easily acquires the contacting force data of the palm and forearm, and (ii) has the ability to remove the disturbance of the crutch pitch angle. The disadvantage is that some yaw, pitch, roll angle errors, palm, and forearm contacting force errors may lead to large vertical GRF errors. Therefore, calibration was performed in the following part. For this model, we assume the crutch endpoint is almost landed in the sagittal plane when the crutch yaw angle is within a range of −10~15 degree. Here, −10 degree denotes the inner side yaw angle boundary and 15 degree is the outside yaw angle boundary. F_N and F_f are the normal and tangential components, respectively, of GRF; θ_c and θ_a are the crutch pitch angle and intrinsic geometry angle, respectively; and T_h, T_a are the palm and forearm pressure loaded on the crutch when the subject uses it for balance control:

$$\begin{cases} F_N = T_a \cos(\theta_c - \theta_a) + T_h \sin\theta_c \\ F_f = -T_a \sin(\theta_c - \theta_a) + T_h \cos\theta_c \end{cases} . \tag{1}$$

By using Equation (1), the normal component of GRF was calculated for each crutch, since both the palm and forearm pressures were measured as well as the pitch angles of crutches. To ensure accurate GRF measurement, further calibration was done with the commercial load cell. The real vertical component of GRF loaded by the crutch, denoted as F_Z, was measured by a load cell. The subject was asked to use the crutches to maintain balance while the upper body was leaning slightly forward. After data recording, the subject was asked to adjust his posture slightly and stand still to record data again. Figure 5a shows the calibration results of L/R crutches. Since the subject was accustomed to holding the crutches in a certain posture for balance control, most normal components of GRF were less than 100 N. Figure 5b is the calibration results of L/R soles. RMS errors of L/R soles are 10.36 and 14.89 N, respectively.

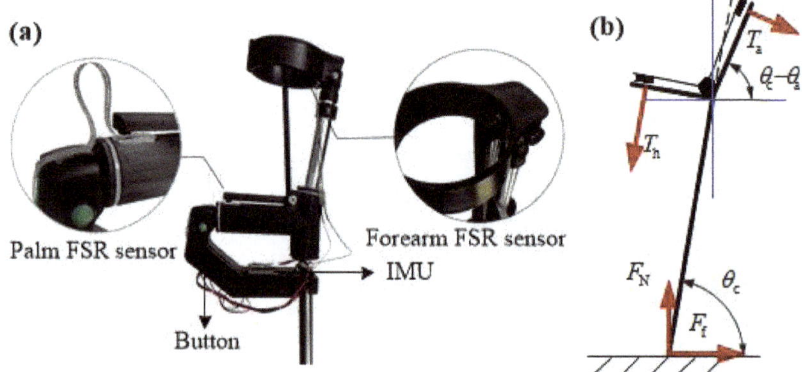

Figure 4. Customized crutch for GRF measurement. (**a**) Locations of palm and forearm FSR sensors, as well as the mounting position of inertia measurement unit (IMU) for crutch pitch angle measurement. (**b**) With force analysis, the crutch, normal and tangential components of GRF are obtained.

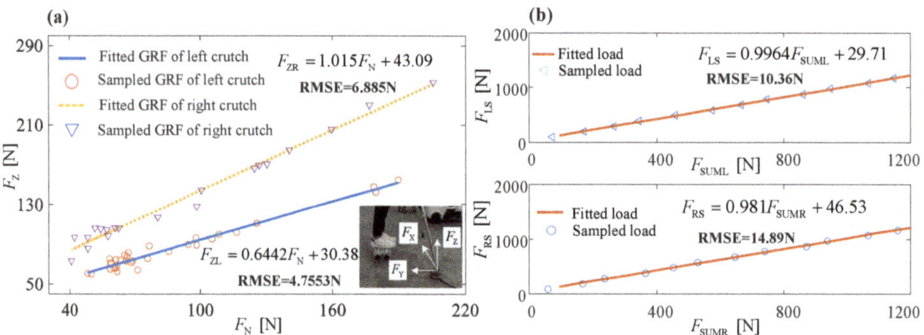

Figure 5. Calibration results of L/R crutches and L/R soles. (**a**) Calibration results of L/R crutches. F_{ZL} and F_{ZR} denote vertical components of L/R crutch endpoint GRF, respectively, and F_N is the calculated vertical components of GRF based on measured palm and forearm pressure. (**b**) Calibration results of L/R soles. F_{LS} and F_{RS} are L/R gravities loaded on the soles, respectively. F_{SUML} and F_{SUMR} are the sum of gravities measured by 7 FSR sensors on L/R soles, respectively.

2.3. Combined CoP and Stride Length Calculation

The main idea of gait planning comes from the combined CoP based on the crutch GRF and human-machine GRF, which differs from the conventional CoP that is acquired from human plantar force interacting with the exoskeleton soleplate. We are concerned not only with subject gravity but with the GRF from the crutch, as well as that caused by the exoskeleton's own gravity. With the force analysis shown in Figure 1, this combined CoP in the sagittal plane can be determined as:

$$CoP_y = \frac{\sum_{i=1}^{4}(y_i \cdot F_i)}{\sum_{i=1}^{4} F_i},$$ (2)

where CoP_y is the y-coordinate of the combined CoP in the local coordinate system. $y_i(i = 1, 2, 3, 4)$ are the y-coordinates in the support ankle frame, i.e., X0-Y0 coordinate, as shown in Figure 1b, which denote the L/R crutch and L/R foot, respectively. $F_i(i = 1, 2, 3, 4$, which denote F_{LS}, F_{RS}, F_{c_NL}, F_{c_NR}) are the corresponding GRFs. With this in mind, the combined CoP can be obtained online as long as the crutch and foot coordinates and GRFs are determined.

Figure 6 shows the algorithm for gait planning based on combined CoP, which is obtained from the measured and calculated results: L/R crutch coordinates, L/R crutch GRF, L/R foot coordinates, and L/R foot GRFs. The L/R foot coordinates can be acquired directly by the four-DoF forward kinematics in support ankle coordinate (see Figure 1b), while the L/R crutch coordinates are calculated based on the forward kinematics of the support shank, thigh, torso, and upper limb-crutch serial-links model. The torso angle in the sagittal plane is measure by IMU placed on the torso. The upper limb-crutch coupled link pitch angle is measured by IMU mounted on the crutch. The stride length is determined by the combined CoP through a mapping function. The reference stride length is acquired from forward kinematics based on the reference gait. With the stride length and reference stride length, the target ankle joint coordinate can be calculated, and is converted into the joint space by inverse kinematics. After joint space gait sequence interpolation, the target gait trajectory is determined for the current gait semi-cycle. After gait planning before each swing, the human exoskeleton system swing step is triggered by button 2 and the planned semi-cycle gait will be performed during the swing phase. That is to say, Figure 6 denotes the procedure of each single swing step.

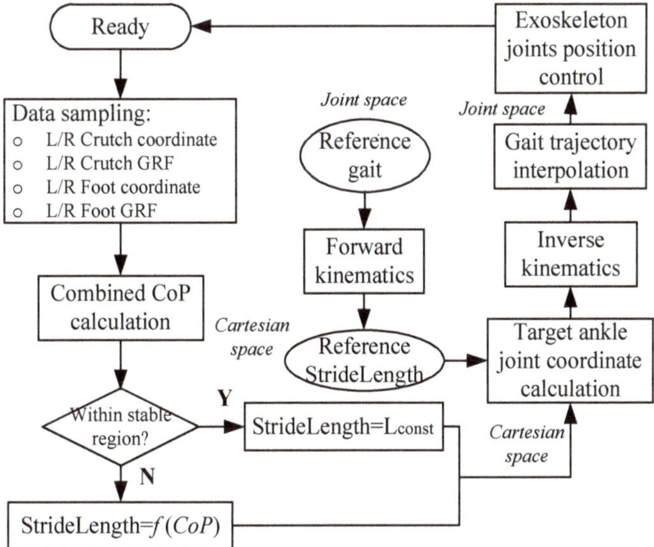

Figure 6. Algorithm for gait planning based on combined CoP. Once combined CoP is calculated, the stride length is obtained with a predefined mapping function. Compared with the reference stride length, the target gait trajectory is determined by converting the inverse kinematics. L_{const} is a constant stride length. When the combined CoP locates within a stable region (between L/R legs), the next stride length is determined to be L_{const}.

To establish a mapping relationship between the combined CoP and stride length, the subject's walking behavior with the exoskeleton was studied. Based on walking tests with nominal reference gait (Kirtley, 2013) on healthy subjects and SCI patients with a height from 1.5 to 1.9 m, the stride length is mostly between 600 and 1200 mm, which coincides well with experimental data from Mendoza et al. [21]. Since a further relationship between the combined CoP and stride length is still unclear and may depend on human behavioral analysis, we simplified it to a linear mapping:

$$L_{stride} = \begin{cases} a \cdot CoP_y + b & -500 \leq CoP_y \leq 0 \\ L_{const} & else \end{cases}, \tag{3}$$

where L_{stride} is the stride length, denoting the Y-axis length of the swing ankle joint from the beginning of swing to the termination of swing during one step, a and b are linear mapping function coefficients, and L_{const} is constant. With the condition that:

(1) The stride length of a subject walking in an exoskeleton is mostly between 600 to 1200 mm based on trials with nominal reference gait [22];
(2) The relationship between the combined CoP and the desired stride length is linear;
(3) CoP_y is bounded within [−500, 0] mm when it is located in the negative Y-axis. We did a pre-test for a total of 100 steps on 5 healthy subjects, and the CoP_y was always within [−500, 200] mm.

Thus, a = −1.2, b = 600. The constant L_{const} = 600 reflects that the stride length, when a subject walks in a stable region, is 600 mm.

2.4. Forward and Inverse Kinematic Modeling

Consider a simplified lower limb exoskeleton with four DoF, as shown in Figure 1b. Only pitch angles in the sagittal plane are of concern. Hip L/R f/e joints are considered to coincide in the sagittal plane. The ankle joint of the support leg is set as the origin of frame 0. The coordinate of the swing ankle joint in frame 4 is ^{4}P. With a forward kinematic model, we obtain its coordinate in the local coordinate system (frame 0). The Denavit–Hartenberg (DH) table of the Craig version [23] is shown in Table 1.

Table 1. **Denavit-Hartenberg** (DH) parameters of the lower-limb exoskeleton.

i	$\alpha_{i\text{-}1}$	$a_{i\text{-}1}$	d_i	θ_i
1	0	0	0	θ_1
2	0	L_1	0	θ_2
3	0	L_2	0	θ_3
4	0	L_2	0	θ_4

With the DH parameters from Table 1, the transformation matrix from the i-1th to ith coordinate can be obtained as:

$$
{}^{i-1}_{i}\mathbf{T} =
\begin{bmatrix}
c\theta_i & -s\theta_i & 0 & a_{i-1} \\
s\theta_i c\alpha_{i-1} & c\theta_i c\alpha_{i-1} & -s\alpha_{i-1} & -s\alpha_{i-1}d_i \\
s\theta_i s\alpha_{i-1} & c\theta_i s\alpha_{i-1} & c\alpha_{i-1} & c\alpha_{i-1}d_i \\
0 & 0 & 0 & 1
\end{bmatrix}.
\tag{4}
$$

Thus, the transformation matrix from the fourth frame to the base frame can be represented as:

$$
{}^{0}_{4}\mathbf{T} = {}^{0}_{1}\mathbf{T}\cdot{}^{1}_{2}\mathbf{T}\cdot{}^{2}_{3}\mathbf{T}\cdot{}^{3}_{4}\mathbf{T}.
\tag{5}
$$

The swing ankle joint coordinate is then calculated as:

$$
{}^{0}\mathbf{P} = {}^{0}_{4}\mathbf{T}\cdot{}^{4}\mathbf{P}.
\tag{6}
$$

With the forward kinematic model, the swing ankle joint coordinate is obtained when the gait trajectory is known. We adopt the gait trajectory from Kirtley (2013) as the reference gait. Hence, the reference swing ankle joint coordinate in Cartesian space can be calculated, which is denoted as (y_{a_r}, x_{a_r}). Similarly, the reference hip joint coordinate in Cartesian space (y_{h_r}, x_{h_r}) is calculated. Since the target gait planning is a scaling operation based on the reference gait (the reference ankle joint and hip joint coordinates) in Cartesian space, the target swing ankle joint and hip joint coordinates are calculated by:

$$
k_a = \frac{L_{\text{stride}}(k) + L_{\text{stride}}(k-1)}{2L_{\text{stride\_r}}}
\tag{7}
$$

$$\begin{cases} y_{a\_t} = k_a \cdot y_{a\_r} \\ x_{a\_t} = \lambda_a \cdot k_a \cdot x_{a\_r} \end{cases} \tag{8}$$

$$\begin{cases} y_{h\_t} = k_a \cdot y_{h\_r} \\ x_{h\_t} = \lambda_h \cdot k_a \cdot x_{h\_r} \end{cases}, \tag{9}$$

where k_a, λ_a, and λ_h are the stride length scale, ankle, and hip optimization factors, respectively. The scale factor k_a is determined by the target and reference stride lengths. The larger the target stride length is, the larger the scale factor k_a will be, leading to magnified hip and ankle joint coordinates in Cartesian space compared to reference ones. Both λ_a and λ_h are within [0.9, 1.1] to prevent singular solutions. L_{stride} and L_{stride_r} are the target and reference stride lengths, respectively. k denotes the kth semi-cycle of walking. (y_{a_t}, x_{a_t}) and (y_{h_t}, x_{h_t}) are the planned swing ankle joint and hip joint coordinates, respectively.

Once the target swing ankle joint and hip joint coordinates are obtained, gait planning becomes an inverse kinematic calculation. The geometric solution of the inverse kinematic of the above 4-DoF kinematic model is:

$$\begin{cases} \theta_4 = a\cos\frac{L_{\text{HA\_SW}}^2 - L_1^2 - L_2^2}{2L_1 L_2} \\ \theta_2 = a\cos\frac{L_{\text{HA\_ST}}^2 - L_1^2 - L_2^2}{2L_1 L_2} - 180 \\ L_{\text{HA\_SW}}^2 = (y_{a\_t} - y_{h\_t})^2 + (x_{a\_t} - x_{h\_t})^2 \\ L_{\text{HA\_SW}}^2 = y_{h\_t}^2 + x_{h\_t}^2 \end{cases} \tag{10}$$

$$\begin{cases} \theta_{31} = 90 + \theta_2 - \text{atan2}(x_{h\_t} + \sqrt{L^2_{\text{HA\_ST}} - k_2^2}, k_2 + y_{h\_t}) \\ k_2 = L_1 + L_2 \cos\theta_2 \end{cases}. \tag{11}$$

$$\begin{cases} \theta_{32} = \gamma_4 - \text{atan2}(y_{a\_t} - y_{h\_t}, x_{a\_t} - x_{h\_t}) \\ \gamma_4 = a\sin(\frac{L_1 \sin\theta_4}{L_{\text{HA\_SW}}}) \end{cases}. \tag{12}$$

$$\theta_1 = -\theta_2 - \theta_{31}, \tag{13}$$

where $L_{\text{HA\_SW}}$ and $L_{\text{HA\_ST}}$ are the respective distances from the hip joint to the swing and support ankle joints. Both k_2 and γ_4 are intermediate variables. θ_{31} and θ_{32} are hip joint angles for the support and swing leg, respectively. For exoskeleton hip and knee f/e joint control, the target gait is defined by:

$$\theta = [\theta_2, \theta_{31}, \theta_{32}, \theta_4]. \tag{14}$$

2.5. Gait Planning Algorithm Simulation

The reference Cartesian position coordinates of hip, knee, and ankle joints can be obtained through forward kinematics. Figure 7 shows the resultant reference trajectories in Cartesian space. An example of the target ankle joint trajectory is plotted for comparison. Having assumed $L_{\text{stride}}(k)$ and $L_{\text{stride}}(k-1)$, k_a is obtained through Equation (6). Thus, the target ankle and hip joint trajectories are planned.

Since the lower limb exoskeleton is powered by motor-driven joints, including the hip and knee f/e, the planned Cartesian position coordinates of the ankle and hip joints should be converted to the joint space through inverse kinematics as shown in Equations (10)–(14). Consecutive steps of walking are simulated with a sequence of stride lengths, $L_{\text{stride}} = [1245, 1150, 1050, 950, 850, 750, 650]$. Figure 8 shows the planned left ankle joint and hip joint trajectories derived from the planned gait. Simulated stride lengths are obtained by subtracting the first and last Y-coordinate of ankle joint trajectories during each step. Compared to the input stride lengths, the RMS error of seven left leg steps is 2.315 mm.

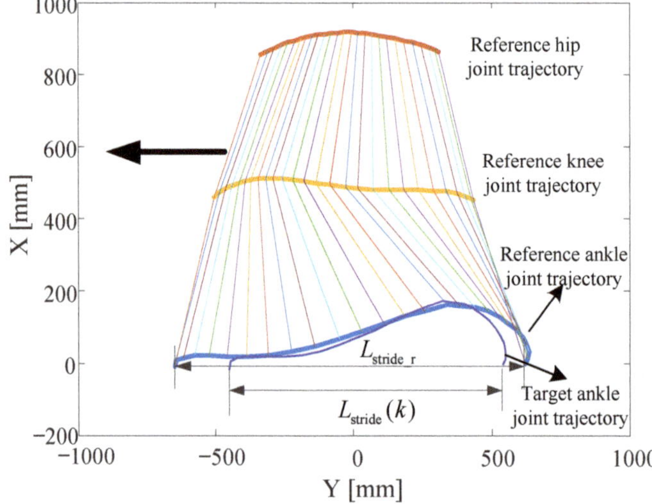

Figure 7. Resultant reference trajectories in Cartesian space and an example of a target ankle joint trajectory. Lower limb sizes are L_1 = 490 mm, L_2 = 430 mm, and L_{stride_r} = 1290.52 mm is the resultant stride length with the reference gait.

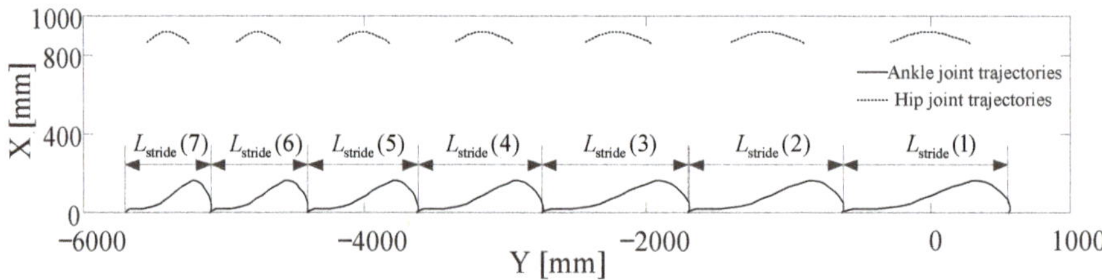

Figure 8. Left ankle joint and hip joint trajectories derived from planned gait with forward kinematics. Stride length can be obtained with simulation results, and the RMS error of seven left leg steps is 2.315 mm.

2.6. Online Gait Planning Walking

The procedures of exoskeleton-assisted walking are shown in Figure 9. For the first step, the gait was predefined to transfer from the standing posture to the double support phase with the left leg in front. At this moment, the subject would adjust L/R crutches to a certain posture to assure stability. Button 2 was pressed once the subject felt stable, and the corresponding signals were sampled to calculate the combined CoP and stride length, followed by gait planning in the micro-computing engine within several milliseconds. When the subject heard buzz 2, indicating the completion of the next step in gait planning, the subject would walk one step with an exoskeleton-assisted gait once button 1 was pressed. Buzz 1 reminded the subject when the current step was completed. The procedure continued until the end of the cyclic step. The walking test was ended when the last step was transferred from the double support phase to the standing posture.

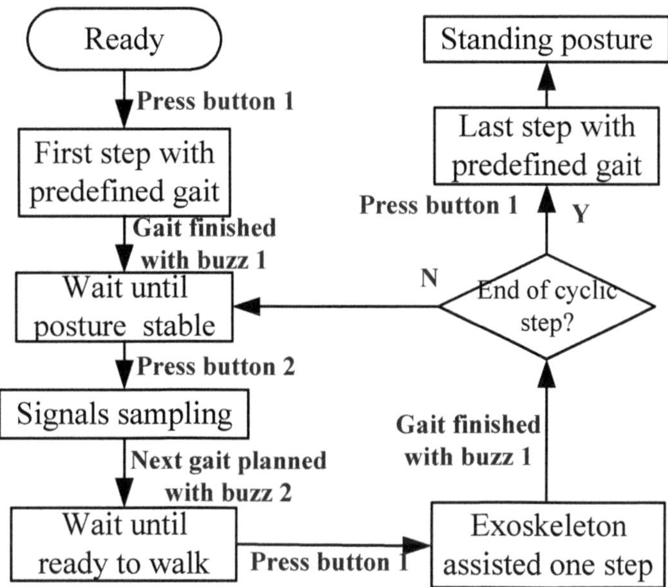

Figure 9. Exoskeleton-assisted walking test procedure. The subject was told to press buttons 1 and 2 before walking and signal sampling, respectively. The subject was reminded with buzz 1 that the current gait was finished, and with buzz 2 that signal sampling was completed. The gaits of the first and last steps were predefined and independent of CoP.

3. Experiment

Three experiments were conducted to respectively verify (1) the advantage of selected linear mapping, (2) feasibility of the gait planning algorithm, and (3) how balanced walking is ensured. One healthy subject (male, 28 years of age, 70 kg, 1.76 m) volunteered for the second experiment, and 5 healthy volunteers (5 males, with 29.20 ± 3.90 (SD) year of age, 74.4 ± 4.39 (SD) kg, 1.74 ± 2.95 (SD) m) were recruited for the first and third experiments. Informed consent was obtained from the volunteers.

3.1. Stride Length Mapping Function Verification Test

The comparison tests were designed to prove the suitability of linear mapping for the subject to learn how to control the crutch landing point. Another two quadratic mapping functions (denoted as Quad1 and Quad2) were selected for comparison. L_{stride} and CoP_y of both quadratic functions were bounded with [600, 1200] and [−500, 200], respectively, which is the same as the linear function. Five volunteers were recruited and each one repeated 3 sets of walking trials with 27 steps, respectively. For the first set, a linear mapping function (denoted as Line1) was applied to determine stride length while for the other two sets, Quad1 and Quad2 were applied, respectively. The subjects were asked to control their weight distribution between the legs and crutches to realize the target stride length. For the first 9 steps, the target stride length was 650 mm. For the next 9 steps, the target stride length was 800 mm, and for the last 9 steps, the target stride length was 1000 mm. After each step, the realized stride length and deviation from the target one were told to the subject as feedback for adjustment of the next step. Deviations of the last 6 steps among every 9 steps were recorded.

3.2. Online Gait Planning Algorithm-Based Walking Test

The exoskeleton-assisted walking test was motion captured by an OptiTrack motion capture system (Natural Point, Inc., Corvallis, OR, USA) as shown in Figure 10. Four trackable points (each trackable combines three markers) were built for motion tracking of the right crutch endpoint, exoskeleton right ankle joint, hip joint, and trunk. With the first trackable, the landing coordinate and the pitch angle of the right crutch endpoint were recorded. With the other three trackable points, the right hip and knee f/e angles and trunk tilt angles were recorded. The right ankle joint coordinate was also recorded by trackable 2. The exoskeleton system stored the sampling signals and the calculation results, including L/R hip and knee angles, torques, GRFs from L/R human-machine soles, crutches, palm and forearm pressures loaded on L/R crutches, calculated CoP, and planned gait of each step. Six trials were conducted on one volunteer, and seven steps were performed for each trial because of the limited range of the motion capture system.

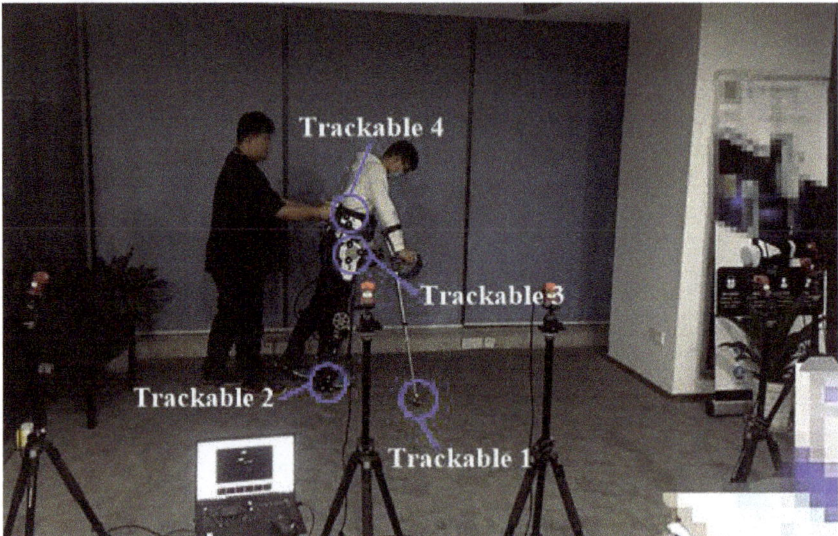

Figure 10. Exoskeleton-assisted walking test. Six cameras were grouped for motion capture of four trackable points. The exoskeleton was programmed to perform seven steps, including the first and last predefined transfer steps. The gaits of the other five steps were planned online based on CoP results.

3.3. Balanced Walking Verification Test

The comparison tests were designed for the verification of balance control that was realized by online gait planning. The same 5 volunteers were recruited for exoskeleton-assisted walking tests. Each one was asked to finish 10 steps of walking for 2 trials: one trial was conducted with the aforementioned online gait planning algorithm while the other was conducted with the fixed reference gait. Torques of both hip and knee f/e joints were recorded during the whole walking tests. The combined CoP was calculated and recorded before each swing phase.

4. Results and Discussion

The results of the first experiment are shown in Figure 11. In addition, two quadratic mapping functions were built for comparison, which are shown in Figure 11a. For each test, the mapping function determined the relationship between the combined CoP and stride length. The location of the resultant stride lengths in Figure 11a indicates that the linear mapping function helps the subject adjust to the target stride length easier. Figure 11b shows the stride length error of each test. For all three target stride length trials, errors with the linear function were smaller, which means that, with linear characteristics, the subjects could adjust their weight distribution easier to map the target stride length. Therefore, the linear mapping function was selected at this stage since it built a more transparent relationship between the combined CoP and stride length.

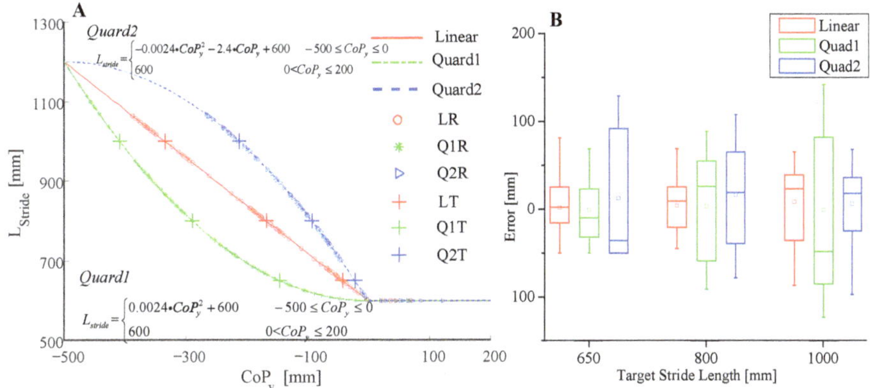

Figure 11. Stride length mapping function verification tests. (**A**) Walking test results with target stride lengths. "Linear" denotes the linear mapping function as shown in Equation (2), "Quad1" and "Quad2" are two quadratic functions for comparison. "LR", "Q1R", and "Q2R" are test results (the combined CoP and stride length) with the Linear, Quad1, and Quad2 mapping functions, respectively. "LT", "Q1T", and "Q2T" are the target stride lengths for the test with the Linear, Quad1, and Quad2 mapping functions, respectively. (**B**) Errors toward the target stride length of each test. For each mapping function of each target stride length, all 5 volunteers' results were concerned.

Figure 12 shows resultant GRFs, the combined CoP locations, and trajectories of ankle position in the sagittal plane. Since the stride length is determined by the mapping function, Equation (2), at the end of the swing phase, the combined CoP (coordinate on Y-axis) is always located between L/R legs. The results in Figure 12c confirm this conclusion. Therefore, a balanced walking state is ensured at this stage and less strength is applied to the crutches at the moment. For the analysis of the online gait planning algorithm, intermediate variables sampled and processed before each step are shown in Figure 12a. F_L Crutch and F_R Crutch are the respective normal components of the L/R crutch GRFs. F_L Foot and F_R Foot are the respective exoskeletons L/R sole GRFs. During step 2, the L/R crutches-related components for CoP were lower because of a smaller forward-leaning range of the subject's upper body, i.e., the subject was less dependent on the support of the crutches, leading to a positive CoP (see Figure 12b). According to the local coordinate system shown in Figure 1, this positive value represented that the combined CoP was between the L/R legs (blue hollow circle on the right side in Figure 12c). Thus, the corresponding stride length was set to a minimum value of 600 mm for smooth walking. According to steps 3 and 4, as shown in Figure 12a, the L/R crutch GRFs dominated, leading to negative values of CoP (see Figure 12b). Consequently, the combined CoPs were located ahead of the L/R legs (the black hollow circle near −1000 and the blue hollow circle near −1500 in Figure 12c).

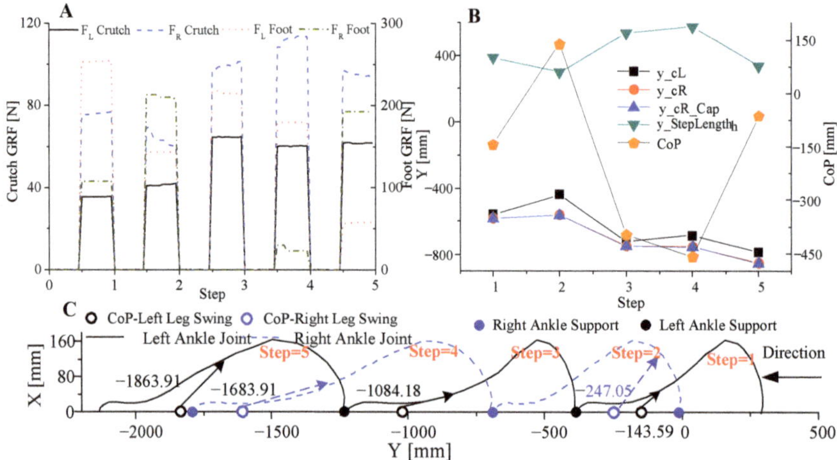

Figure 12. Intermediate variables and L/R ankle joint trajectories. (**A**) Sampled and preprocessed L/R crutches and exoskeleton sole GRFs before the swing phase, representing contributions for human-machine support. GRFs during the other phases were set to 0 here for clarity. (**B**) Calculation results of L/R crutch endpoint coordinates (y_cL and y_cR), step length, combined CoP coordinates, and right crutch endpoint coordinate from the motion capture system (y_cR_Cap), representing bases of the bgait planning algorithm. (**C**) L/R ankle joint Cartesian position coordinates and corresponding combined CoP (hollow circle) and support ankle joint (solid circle). The arrows indicate the corresponding relationship between CoP and ankle joint Cartesian position. Except for step 2, each CoP was outside L/R legs before the swing, and finally located within L/R legs after the swing as shown here.

With the third experiment, the driven torques of the hip and knee f/e joints, as well as L/R crutch GRFs, were measured and compared between test set 1 (walking trials with online gait planning based on the combined CoP) and test set 2 (walking trials with fixed reference gait). Figure 13a shows the driven torques of 4 joints during 10 steps walking trial. The grey area selected parts to denote joint torques of the double support phase, and the remaining parts denote the joint torques of the swing phase. During the swing phase, the subject's lower limbs are dragged or pulled to walk by exoskeleton joint torques, while during the double support phase, the subject's lower limbs are maintained as a changeless posture with the help of exoskeleton joint torques. Thus, joint torques are generated within both phases for motion control. Figure 13b is a box plot of the normalized sum of four joint torques during the swing phase based on the five volunteers' trials. The normalized torque is acquired with:

$$\Gamma_k = \left(\sum_{i=1}^{10} \sum_{j=1}^{4} \tau_{kij} \right) / (10 m_k), \tag{15}$$

where Γ_k denotes the normalized torque of the kth volunteer, $k = 1, 2, \ldots 5$, τ_{kij} is the jth joint torque of the ith step for the kth volunteer, $j = 1, 2, \ldots 4$, and $i = 1, 2, \ldots 10$, m_k is the weight of the kth volunteer. Normalized torques of S1 and S2 are closed, which indicates that the summed driven torques of four joints during the swing phase are almost the same between the walking test with the online gait planning algorithm and the walking test with the fixed reference gait. However, compared to the walking test with the fixed reference gait, the driven torques of four joints were reduced during the double support phase with the online gait planning algorithm (Figure 13c). The average reduction of joint torques for 5 volunteers is 18.5%. This reduction, on the one hand, may indicate that when the stride length is determined based on the combined CoP (to ensure that it is located between L/R legs), fewer torques are needed to maintain a double support posture. While, on the other hand, it may result from that the subjects using their own body forces and moments to maintain balance since the subjects were able-bodied volunteers. Furthermore, L/R crutches GRFs were measured after each swing phase. The average values of the two aforementioned test sets are shown in Figure 13d. There is an average reduction of 32.4% of crutch GRF when the online gait planning algorithm was applied. This mainly results from a better balance state of the human-machine system during the double support phase.

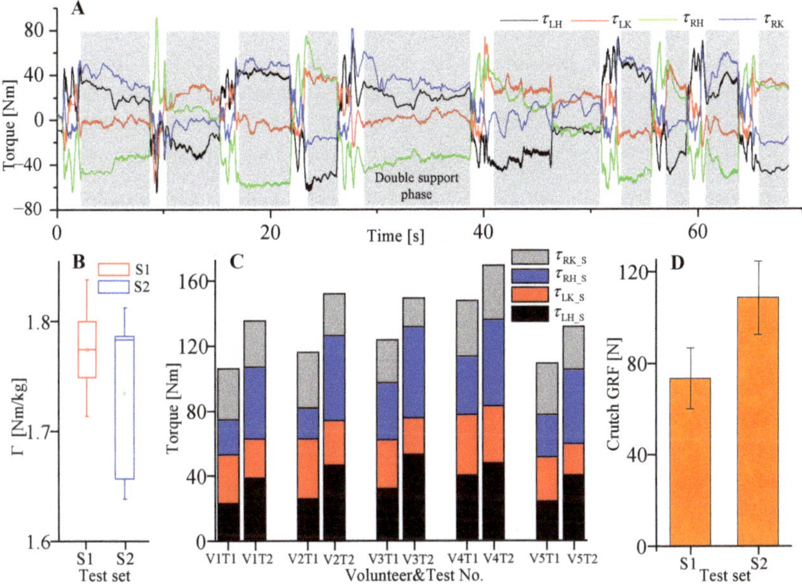

Figure 13. Experimental results of 2 walking test sets. (**A**) Driven torques of 4 joints during one 10-step walking trial. τ_{LH}, τ_{LK}, τ_{RH}, and τ_{RK} are the torques of the left hip, left knee, right hip, and right knee, respectively. (**B**) Box plot of the normalized sum of 4 joint torques during the swing phase based on 5 volunteers' trials. S1 and S2 denote test set 1 (walking trials with online gait planning based on the combined CoP) and test set 2 (walking trials with fixed reference gait), respectively. (**C**) Comparison of joint torques during the double support phase. τ_{LH_S}, τ_{LK_S}, τ_{RH_S}, and τ_{RK_S} are torques of the left hip, left knee, right hip, and right knee during the double support phase, respectively. (**D**) Average L/R crutches GRFs measured after each swing phase. For each test set, L/R crutches GRFs of each step were measured and averaged based on 5 volunteers' walking tests.

The present work mainly focuses on the validation and verification of an online gait planning algorithm based on the combined CoP of human-machine and crutch GRFs. The target of gait planning in this research is to gradually reduce the subject's dependence on crutches for stability control. However, the relationship between the combined CoP and stride length has not been studied in detail. Our plans for future work will focus on behavior analysis of the SCI patient when walking with crutches, and further investigate the relationship between the combined CoP and stride length.

5. Conclusions

In this work, an online gait planning algorithm was presented, which is adaptive to combined CoP based on both human-machine and crutch GRFs. The mapping relationship between the combined CoP and stride length was preliminarily constructed with a linear function and validated by a comparison experiment. According to the experimental results, the algorithm can process the sensing signals and plan the landing point of the swing leg to ensure balance and stable walking. The performance of the gait planning algorithm was validated by the data analysis of the results, i.e., an 18.5% reduction of joint torques and a 32.4% reduction of crutch GRF during the double support phase compared to the fixed reference gait. This is of great significance for the further training of independent walking stability control in exoskeletons for SCI patients with less assistance of crutches. With this algorithm, the SCI patients can try to adjust their posture with a self-determined range by adjusting the weight distribution between the exoskeleton soles and crutches, because the corresponding gait with walking stability is planned accordingly. Thus, the SCI patient can be guided gradually by a physical therapist based on professional experience, or automatically by the exoskeleton system. The work in this paper forms the basis of a new solution for exoskeleton-assisted walking stability control with less help of crutches, i.e., gradually training the patient to be compatible with the exoskeleton for balanced walking. In the future work, the crutch model should be improved to consider it in the X-Y-Z coordinate, with which the real vertical component of GRF could be obtained instead of performing calibration. Meanwhile, paraplegic patients need to be tested to further verify the improvement of walking balance.

Author Contributions: This work was carried out in collaboration among all authors. W.Y., S.Z. and C.Y. developed the concept. W.Y. designed the experiments. Data process and analysis were performed by W.Y. and J.Z. C.Y. and S.Z. reviewed the draft and made substantial comments. All authors have read and agreed to the published version of the manuscript.

Funding: This research was partially funded by both National natural Science Foundation of China (No. 51805469) and Science and technology plan project of drug regulatory system of Zhejiang province (No. 2020016).

Acknowledgments: We appreciate Tian Wang and Qingyu Zhao for helping design of the experiments. Many thanks to Zhenting Hu and Qinglin Sun for assisting with the experiments, and to Xiangyu Sun, Quanquan Fan, Guangxing Ye and Kun Zhou for aiding in programming and mechanical design work.

Conflicts of Interest: The authors declare no conflict of interest.

References

1. Biering-Sørensen, F.; Bickenbach, J.E.; El Masry, W.S.; Officer, A.; Von Groote, P.M. ISCoS–WHO collaboration. International Perspectives of Spinal Cord Injury (IPSCI) report. *Spinal Cord* **2011**, *49*, 679–683. [CrossRef] [PubMed]

2. Kawamoto, H.; Kamibayashi, K.; Nakata, Y.; Yamawaki, K.; Ariyasu, R.; Sankai, Y.; Sakane, M.; Eguchi, K.; Ochiai, N. Pilot study of locomotion improvement using hybrid assistive limb in chronic stroke patients. *BMC Neurol.* **2013**, *13*, 141. [CrossRef] [PubMed]

3. Nilsson, A.; Vreede, K.S.; Häglund, V.; Kawamoto, H.; Sankai, Y.; Borg, J. Gait training early after stroke with a new exoskeleton—The hybrid assistive limb: A study of safety and feasibility. *J. Neuroeng. Rehabil.* **2014**, *11*, 92. [CrossRef] [PubMed]

4. Zhang, J.-F.; Dong, Y.-M.; Yang, C.-J.; Geng, Y.; Chen, Y.; Yang, Y. 5-Link model based gait trajectory adaption control strategies of the gait rehabilitation exoskeleton for post-stroke patients. *Mechatronics* **2010**, *20*, 368–376. [CrossRef]

5. Fineberg, D.B.; Asselin, P.; Harel, N.Y.; Agranova-Breyter, I.; Kornfeld, S.D.; Bauman, W.A.; Spungen, A.M. Vertical ground reaction force-based analysis of powered exoskeleton-assisted walking in persons with motor-complete paraplegia. *J. Spinal Cord Med.* **2013**, *36*, 313–321. [CrossRef]

6. Husain, S.; Ramanujam, A.; Momeni, K.; Garbarini, E.; Augustine, J.; Forrest, G. Effects of Exoskeleton Training Intervention on Net Loading Force Profile for SCI: A Case Study. *Arch. Phys. Med. Rehabil.* **2017**, *98*, e62–e63. [CrossRef]

7. Jung, J.; Jang, I.; Riener, R.; Park, H. Walking intent detection algorithm for paraplegic patients using a robotic exoskeleton walking assistant with crutches. *Int. J. Control Autom. Syst.* **2012**, *10*, 954–962. [CrossRef]

8. Raab, K.; Krakow, K.; Tripp, F.; Jung, M. Effects of training with the ReWalk exoskeleton on quality of life in incomplete spinal cord injury: A single case study. *Spinal Cord Ser. Cases* **2016**, *2*, 15025. [CrossRef]

9. Watanabe, H.; Marushima, A.; Kawamoto, H.; Kadone, H.; Ueno, T.; Shimizu, Y.; Endo, A.; Hada, Y.; Saotome, K.; Abe, T.; et al. Intensive Gait Treatment Using a Robot Suit Hybrid Assistive Limb in Acute Spinal Cord Infarction: Report of Two Cases. *J. Spinal Cord Med.* **2017**, *42*, 395–401. [CrossRef]

10. Bach Baunsgaard, C.; Vig Nissen, U.; Katrin Brust, A.; Frotzler, A.; Ribeill, C.; Kalke, Y.-B.; León, N.; Gómez, B.; Samuelsson, K.; Antepohl, W.; et al. Gait training after spinal cord injury: Safety, feasibility and gait function following 8 weeks of training with the exoskeletons from Ekso Bionics article. *Spinal Cord* **2018**, *56*, 106–116. [CrossRef]

11. Suzuki, K.; Mito, G.; Kawamoto, H.; Hasegawa, Y.; Sankai, Y. Intention-based walking support for paraplegia patients with Robot Suit HAL. *Adv. Robot.* **2007**, *21*, 1441–1469. [CrossRef]

12. Neuhaus, P.; Noorden, J.H.; Craig, T.J.; Torres, T.; Kirschbaum, J.; E Pratt, J. Design and evaluation of Mina: A robotic orthosis for paraplegics. *IEEE Trans. Neural Syst. Rehabil. Eng.* **2011**, *2011*, 5975468.

13. Wang, S.S.; Wang, L.L.; Meijneke, C.C.; Van Asseldonk, E.E.; Hoellinger, T.; Cheron, G.; Ivanenko, Y.Y.; La Scaleia, V.V.; Sylos-Labini, F.; Molinari, M.M.; et al. Design and Control of the MINDWALKER Exoskeleton. *IEEE Trans. Neural Syst. Rehabil. Eng.* **2015**, *23*, 277–286. [CrossRef] [PubMed]

14. Jeon, H.; Kim, S.L.; Kim, S.; Lee, D. Fast Wearable Sensor–Based Foot–Ground Contact Phase Classification Using a Convolutional Neural Network with Sliding-Window Label Overlapping. *Sensors* **2020**, *20*, 4996. [CrossRef] [PubMed]

15. Leal-Junior, A.G.; Avellar, L.; Jaimes, J.; Diaz, C.A.R.; Dos Santos, W.M.; Siqueira, A.A.G.; Pontes, M.J.; Marques, C.; Neto, A.F.F. Polymer Optical Fiber-Based Integrated Instrumentation in a Robot-Assisted Rehabilitation Smart Environment: A Proof of Concept. *Sensors* **2020**, *20*, 3199. [CrossRef] [PubMed]

16. Jung, J.H.; Veneman, J.F. Real time computation of Centroidal Momentum while human walking in the lower limbs rehabilitation exoskeleton: Preliminary trials. In Proceedings of the 2019 IEEE 16th International Conference on Rehabilitation Robotics (ICORR), Toronto, ON, Canada, 24–28 June 2019; pp. 721–726.

17. Aphiratsakun, N.; Parnichkun, M. Balancing control of leg exoskeleton using zmp-based jacobian compensation. *Int. J. Robot. Autom.* **2010**, *25*, 359–371. [CrossRef]

18. Kim, J.-H.; Han, J.W.; Kim, D.Y.; Baek, Y.S. Design of a Walking Assistance Lower Limb Exoskeleton for Paraplegic Patients and Hardware Validation Using CoP. *Int. J. Adv. Robot. Syst.* **2013**, *10*, 113. [CrossRef]

19. Chen, B.; Zhong, C.-H.; Zhao, X.; Ma, H.; Qin, L.; Liao, W.-H. Reference Joint Trajectories Generation of CUHK-EXO Exoskeleton for System Balance in Walking Assistance. *IEEE Access* **2019**, *7*, 33809–33821. [CrossRef]

20. Deng, M.-Y.; Ma, Z.-Y.; Wang, Y.-N.; Wang, H.-S.; Zhao, Y.-B.; Wei, Q.-X.; Yang, W.; Yang, C. Fall preventive gait trajectory planning of a lower limb rehabilitation exoskeleton based on capture point theory. *Front. Inf. Technol. Electron. Eng.* **2019**, *20*, 1322–1330. [CrossRef]

21. Mendoza-Crespo, R.; Torricelli, D.; Huegel, J.C.; Gordillo, J.L.; Pons, J.L.; Soto, R. An Adaptable Human-Like Gait Pattern Generator Derived From a Lower Limb Exoskeleton. *Front. Robot. AI* **2019**, *6*, 6. [CrossRef]

22. Kirtley, C. Clinical Gait Analysis. *IEEE Eng. Med. Biol. Mag.* **2013**, *7*, 35–40.

23. Craig, J.J. *Introduction to Robotics: Mechanics and Control*; Pearson/Prentice Hall: New York City, NY, USA, 2005.

Article

Estimating Lower Limb Kinematics Using a Lie Group Constrained Extended Kalman Filter with a Reduced Wearable IMU Count and Distance Measurements [†]

Luke Wicent F. Sy [1,*], Nigel H. Lovell [1] and Stephen J. Redmond [2]

[1]   Graduate School of Biomedical Engineering, UNSW Sydney, Sydney 2052, Australia; n.lovell@unsw.edu.au
[2]   UCD School of Electrical and Electronic Engineering, University College Dublin, Belfield, 4 Dublin, Ireland; stephen.redmond@ucd.ie
*   Correspondence: l.sy@unsw.edu.au
†   This paper is an extended version of our paper published in Estimating Lower Limb Kinematics using a Lie Group Constrained EKF and a Reduced Wearable IMU Count. In Proceedings of the 2020 8th IEEE RAS/EMBS International Conference for Biomedical Robotics and Biomechatronics (BioRob) held at New York City, USA on 29th November–1st December.

Received: 21 October 2020; Accepted: 17 November 2020; Published: 29 November 2020

Abstract: Tracking the kinematics of human movement usually requires the use of equipment that constrains the user within a room (e.g., optical motion capture systems), or requires the use of a conspicuous body-worn measurement system (e.g., inertial measurement units (IMUs) attached to each body segment). This paper presents a novel Lie group constrained extended Kalman filter to estimate lower limb kinematics using IMU and inter-IMU distance measurements in a reduced sensor count configuration. The algorithm iterates through the prediction (kinematic equations), measurement (pelvis height assumption/inter-IMU distance measurements, zero velocity update for feet/ankles, flat-floor assumption for feet/ankles, and covariance limiter), and constraint update (formulation of hinged knee joints and ball-and-socket hip joints). The knee and hip joint angle root-mean-square errors in the sagittal plane for straight walking were $7.6 \pm 2.6°$ and $6.6 \pm 2.7°$, respectively, while the correlation coefficients were 0.95 ± 0.03 and 0.87 ± 0.16, respectively. Furthermore, experiments using simulated inter-IMU distance measurements show that performance improved substantially for dynamic movements, even at large noise levels ($\sigma = 0.2$ m). However, further validation is recommended with actual distance measurement sensors, such as ultra-wideband ranging sensors.

Keywords: Lie group; constrained extended Kalman filter; gait analysis; motion capture; pose estimation; wearable devices; IMU; distance measurement

---

1. Introduction

Human pose estimation is the tracking of position and orientation (i.e., pose) of body segments. Joint angles can then be calculated from the relative pose of linked body segments. Applications exist in robotics, virtual reality, animation, and healthcare (e.g., gait analysis). Traditionally, human pose is captured within a laboratory setting using optical motion capture (OMC) systems with up to millimeter level position accuracy [1] when properly configured and calibrated. However, the setup for OMC systems is time consuming and inconvenient (e.g., multiple markers are taped to the body) and requires considerable expertise. Recent miniaturization of inertial measurements units (IMUs) has paved the path toward inertial motion capture (IMC) systems suitable for prolonged use outside of the laboratory. Some examples of clinical applications in the current literature include determining level of spinal damage [2] and Parkinson's disease diagnosis [3]. Furthermore, developing a comfortable IMC for routine daily use may facilitate interactive rehabilitation [4,5], and allow the study of movement disorder progression to enable predictive diagnostics.

Commercial IMCs attach one sensor per body segment (OSPS) [6], which may be considered too cumbersome and expensive for routine daily use by a consumer due to the number of IMUs required. Similar works also exist in the literature using quaternions [7], Kalman filters (KF) [8], and particle filters [9]. Each IMU typically tracks the orientation of the attached body segment using an orientation estimation algorithm (e.g., [10,11]), which is then connected via linked kinematic chain, usually rooted at the pelvis. A reduced-sensor-count (RSC) configuration, where IMUs are placed on a subset of body segments, can improve user comfort, reduce setup time and system cost. However, using fewer wearable sensor units necessarily reduces the kinematic information available, which must otherwise be inferred from (i) our knowledge of human movement (e.g., enforcing mechanical joint constraints or making dynamic balance assumptions), or (ii) by using additional sensing modalities within each wearable sensor unit. Each approach will be described in the next subsections.

1.1. Existing Algorithms for Human Motion Tracking Using Fewer IMUs

The performance of algorithms with RSC configuration depends (i) on how the kinematic information of uninstrumented body segments is inferred and (ii) on how body pose is represented.

The kinematic information of body segments which do not have sensors attached to them may be inferred by the algorithm, either through data obtained in the past (i.e., observed correlations between co-movement of different body segments or data-driven approaches) or by using a simplified kinematic model of the human body (i.e., model-based approaches). Data-driven approaches (e.g., nearest-neighbor search [12] and bi-directional recurrent neural network [13]) are able to recreate realistic motion suitable for animation-related applications. However, these approaches are normally biased toward motions already contained in the database, which may limit their use in monitoring pathological gait. Model-based approaches reconstruct human movement using kinematic and biomechanical models (e.g., linear regression [14], constrained KF [15], extended KF (EKF) [16], particle filter [9], and window-based optimization [17]). The use of optimization-based estimators is sometimes favoured due to its relative ease to setup and ease of understanding. However, the algorithm can be very inefficient when tracking a larger number of dimensions (e.g., when tracking body pose over a long duration). To significantly increase the efficiency of the algorithm when estimating body pose across time, a recursive estimator can take advantage of the state substructure and reduce the state dimension being tracked [18].

Body poses are usually represented using Euler angles or quaternions [9,16]. However, recent work on pose estimation has shown that using a Lie group to represent the states of recursive estimator is a promising approach. Such algorithms typically represent the body pose as a chain of linked segments using matrix Lie groups to represent the orientation or pose of each body segment; specifically the special orthogonal group, $SO(n)$, and special Euclidean group, $SE(n)$, where $n = 2, 3$, are the spatial dimensions for humam body kinematics problems. The early works of Wang et al. [19] and Barfoot et al. [20] investigated representations and propagation of pose uncertainty, the former in the context of manipulator kinematics and the latter focused on $SE(3)$, followed by the formulation of recursive estimators using Lie group representation (e.g., EKF [21] and unscented KF (UKF) [22]). Recent literature has reported the use of Lie group based recursive estimators to estimate human movement. Cesic et al. estimated the full body pose from OMC marker measurements and achieved significant improvements compared to an Euler angle representation [23]; and even supplemented the approach with an observability analysis [24]. Joukov et al. represented the human body using $SO(n)$, tracking the pose using measurements from IMUs under an OSPS configuration [25]. Joukov et al.'s algorithm was tested using an arm tracking experiment, where the results improved especially during arm poses that cause a singularity when using an Euler angle representation (i.e., Lie group representation is singularity free).

1.2. Improving Human Pose Estimation Using Additional Sensor Measurements

Another approach is to supplement kinematic information from the IMU with another kind of sensor, which inherently increases cost and reduces battery life. Note that we will focus on systems that supplement pose estimation, not on the global position estimation of the subject (e.g., [26]). For example, IMCs can be supplemented with standard video cameras (e.g., fused using an optimization-based algorithm [27], and deep neural networks [28]) or depth cameras [29] at fixed locations in the capture environment, external to the subject. The combination of IMCs and portable cameras solves a weakness of OMCs (i.e., marker or body segment occlusion) and a weakness of IMCs (i.e., global position drift). However, the system still requires an external sensor that is carried by another person or requires some quick setup. IMCs can also be supplemented by distance measurements (using ultrasonic devices and KF in OSPS configuration [30], using constrained KF in RSC configuration [31]), removing dependence on any external non-body-worn sensor.

1.3. Novelty

This paper describes a novel human pose estimator that represents the state using Lie groups with the state propagated iteratively using a constrained EKF (CEKF) to estimate lower body kinematics for an RSC configuration of IMUs and inter-IMU distance measurements; the Lie group framework and inclusion of inter-IMU distance measurements, along with the exploration of its effect on pose estimation accuracy, are the major advancements made in this paper. It extends the work of [32] and builds on prior work of [15,31], but instead represents the state variables as elements of Lie groups, specifically $SE(3)$, to track both position and orientation (whereas [15] only tracks position and assumes orientation measurements are noise-free). Furthermore, this paper presents in detail a novel Lie group formulation for assumptions typically used in human pose estimation (e.g., zero velocity update, constant body segment lengths, and a hinged knee joint). While not our focus here, our algorithm, with its $SE(3)$ representation, is able to track the global position of body segments, taking into account IMU measurements during the prediction step, and a simpler implementation of certain measurement assumptions (e.g., zero velocity update), though at the expense of having an additional constraint step to ensure satisfaction of biomechanical constraints. The algorithm design was motivated by the need for a body pose representation that more closely models the human biomechanical system (without a dramatic increase in the dimensions of the tracked state) from which the missing kinematic information of uninstrumented body segments are inferred. The contributions of this paper advance the development of gait assessment tools for comfortable and long-term monitoring of lower body movement.

2. Mathematical Background: Lie Group and Lie Algebra

The matrix Lie group G is a group of $n \times n$ matrices (e.g., $SE(3)$). Mathematically speaking, it is also a smooth manifold with smooth group composition and inversion (i.e., matrix multiplication and inversion). The Lie algebra $\mathfrak{g}$ can be represented in the vector space and is closely related to Lie group G. It represents a tangent space of a group at the identity element [33]. The elegance of Lie theory lies in its ability to represent pose using a vector space (e.g., Lie group G is represented by $\mathfrak{g}$) without additional constraints (e.g., without requiring $\mathbf{R}^T\mathbf{R} = \mathbf{I}$ which is using a rotation matrix representation of orientation, or $\|q\| = 1$ which is using a quaternion representation of orientation) [34].

The matrix exponential $\exp_G : \mathfrak{g} \rightarrow G$ (Equation (1)) and matrix logarithm $\log_G : G \rightarrow \mathfrak{g}$ relates (i.e., local diffeomorphism) the Lie group G and its Lie algebra $\mathfrak{g}$. The Lie algebra $\mathfrak{g}$ is a $n \times n$ matrix that can be represented compactly in an n-dimensional vector space. A linear isomorphism (i.e., one-to-one mapping) between $\mathfrak{g}$ and $\mathbb{R}^n$ is given by operators $[\ ]_G^{\vee} : \mathfrak{g} \rightarrow \mathbb{R}^n$ and $[\ ]_G^{\wedge} : \mathbb{R}^n \rightarrow \mathfrak{g}$, which map between the compact and matrix representation of the Lie algebra $\mathfrak{g}$. Figure 1 shows an illustration of the said mappings.

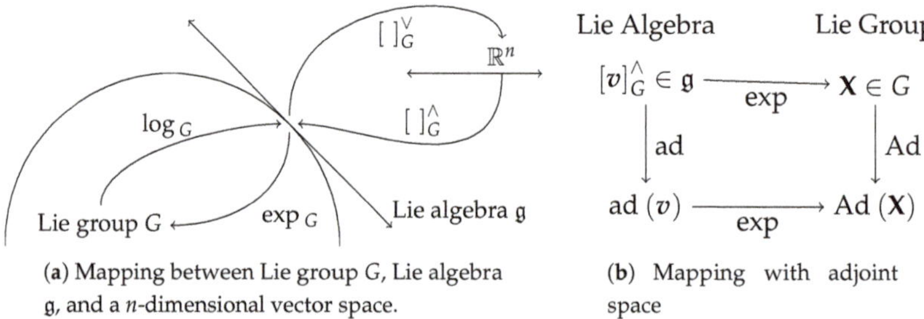

(a) Mapping between Lie group G, Lie algebra $\mathfrak{g}$, and a n-dimensional vector space.

(b) Mapping with adjoint space

Figure 1. Overview of Lie group theory mappings. When $G = SE(3)$, Lie group $\mathbf{X} = \mathbf{T}$ is a 4×4 transformation matrix representing pose (i.e., 3D rotation and translation). Similarly, $v = \boldsymbol{\xi}$ where Lie algebra $[\boldsymbol{\xi}]^{\wedge}_{SE(3)} \in \mathfrak{se}(3)$ and the vector $\boldsymbol{\xi} \in \mathbb{R}^n$ with $n = 6$.

Furthermore, the adjoint operator of a Lie group, $\mathrm{Ad}_G(\mathbf{X})$, the adjoint operator of a Lie algebra, $\mathrm{ad}_G(v)$, and the right jacobian, $\Phi_G(v)$ (Equation (2)), where $\mathbf{X} \in G$ and $[v]^{\wedge}_G \in \mathfrak{g}$ will be used in later sections. Multiplying an n-dimensional vector representation of a Lie algebra with $\mathrm{Ad}_G(\mathbf{X}) \in \mathbb{R}^{n \times n}$ (i.e., the product $\mathrm{Ad}_G(\mathbf{X})v$) transforms the vector from one coordinate frame to another, similar to how rotation matrices transform points from one frame to another. $\mathrm{ad}_G(v)$ is the Lie algebra of $\mathrm{Ad}_G(\mathbf{X})$. A summary of the operators for Lie groups $SO(3)$, $SE(3)$, and R^n will be explained in the next subsections. They will serve as building blocks to implement the algorithm being described by this paper. For a more detailed introduction to Lie groups refer to [18,34,35].

$$\exp([v]^{\wedge}_G) = \sum_{n=0}^{\infty} \frac{1}{n!}([v]^{\wedge}_G)^n \tag{1}$$

$$\Phi_G(v) = \sum_{i=0}^{\infty} \frac{(-1)^i}{(i+1)!} \mathrm{ad}_G(v)^i \, , \, v \in \mathbb{R}^n \tag{2}$$

2.1. Special Orthogonal Group $SO(3)$

The special orthogonal group, $SO(3) := \{\mathbf{R} \in \mathbb{R}^{3 \times 3} | \mathbf{R}\mathbf{R}^T = 1, \det \mathbf{R} = 1\}$, represents orientation, where $\mathbf{R}$ is the typical 3×3 rotation matrix whose column vectors represent the x, y, and z basis vectors. The operations for $SO(3)$ are listed below, and will serve as building blocks for $SE(3)$, which will be described in the next subsection. Note that $[a]^{\wedge}_{SO(3)} b$ is equivalent to the cross product of a and b. See Chapter 7 of [18] for details.

$$[\boldsymbol{\phi}]^{\wedge}_{SO(3)} = \begin{bmatrix} \phi_1 \\ \phi_2 \\ \phi_3 \end{bmatrix}^{\wedge}_{SO(3)} = \begin{bmatrix} 0 & -\phi_3 & \phi_2 \\ \phi_3 & 0 & -\phi_1 \\ -\phi_2 & \phi_1 & 0 \end{bmatrix} , \begin{bmatrix} 0 & -\phi_3 & \phi_2 \\ \phi_3 & 0 & -\phi_1 \\ -\phi_2 & \phi_1 & 0 \end{bmatrix}^{\vee}_{SO(3)} = \begin{bmatrix} \phi_1 \\ \phi_2 \\ \phi_3 \end{bmatrix} = \boldsymbol{\phi} \tag{3}$$

If $\boldsymbol{\phi}/|\boldsymbol{\phi}|$ represents a unit vector axis we wish to rotate around, and $|\boldsymbol{\phi}|$ is the angle by which we wish to rotate, then the rotation matrix, $\mathbf{R}$, which will implement this rotation is given by Equation (4), which is also known as the Rodrigues' axis-angle rotation formula. When $\boldsymbol{\phi}$ is very small, $\mathbf{R} \approx \mathbf{I}_{3 \times 3} + [\boldsymbol{\phi}]^{\wedge}_{SO(3)}$.

$$\mathbf{R} = \exp\left([\boldsymbol{\phi}]^{\wedge}_{SO(3)}\right) = \cos\left(|\boldsymbol{\phi}|\right) \mathbf{I}_{3 \times 3} + (1 - \cos\left(|\boldsymbol{\phi}|\right)) \frac{\boldsymbol{\phi}\boldsymbol{\phi}^T}{|\boldsymbol{\phi}|^2} + \sin\left(|\boldsymbol{\phi}|\right) \left[\frac{\boldsymbol{\phi}}{|\boldsymbol{\phi}|}\right]^{\wedge}_{SO(3)} \tag{4}$$

Furthermore, the Lie algebra adjoint, Lie group adjoint, and inverse operators are listed in Equation (5).

$$\mathrm{ad}_{SO(3)}(\boldsymbol{\phi}) = [\boldsymbol{\phi}]^{\wedge}_{SO(3)} , \quad \mathrm{Ad}_{SO(3)}(\mathbf{R}) = \mathbf{R}, \quad \mathbf{R}^{-1} = \mathbf{R}^T \tag{5}$$

Lastly, to approximate the compound rotations, $\mathbf{R}_1 \mathbf{R}_2$, in the Lie algebra space where $\mathbf{R}_1 = \exp([\boldsymbol{\phi}_1]^{\wedge}_{SO(3)})$ and $\mathbf{R}_2 = \exp([\boldsymbol{\phi}_2]^{\wedge}_{SO(3)})$, we can use Equation (6). The right Jacobian, $\Phi_{SO(3)}(\boldsymbol{\phi}) \in \mathbb{R}^{3 \times 3}$, is obtained using Equation (7).

$$[\log(\mathbf{R}_1 \mathbf{R}_2)]^{\vee}_{SO(3)} \approx \boldsymbol{\phi}_1 + \Phi_{SO(3)}(\boldsymbol{\phi}_1)^{-1} \boldsymbol{\phi}_2 \in \mathfrak{so}(3) \tag{6}$$

$$\Phi_{SO(3)}(\boldsymbol{\phi}) = \frac{\sin(|\boldsymbol{\phi}|)}{|\boldsymbol{\phi}|} \mathbf{I}_{3\times 3} + \left(1 - \frac{\sin(|\boldsymbol{\phi}|)}{|\boldsymbol{\phi}|}\right) \frac{\boldsymbol{\phi}\boldsymbol{\phi}^T}{|\boldsymbol{\phi}|^2} - \frac{1-\cos(|\boldsymbol{\phi}|)}{|\boldsymbol{\phi}|} \left[\frac{\boldsymbol{\phi}}{|\boldsymbol{\phi}|}\right]^{\wedge}_{SO(3)} \in \mathbb{R}^{3\times 3} \tag{7}$$

2.2. Special Euclidean Group, SE(3)

The special Euclidean group, $SE(3) := \left\{ \mathbf{T} = \begin{bmatrix} \mathbf{R} & t \\ 0^T & 1 \end{bmatrix} \in \mathbb{R}^{4\times 4} \mid \{\mathbf{R}, t\} \in SO(3) \times \mathbb{R}^3 \right\}$, represents orientation and translation, where $\mathbf{T}$ is the typical 4×4 transformation matrix, $\mathbf{R}$ is the rotation matrix, and t represents a coordinate point in Euclidean space. The operations for $SE(3)$ are listed below. $\mathbf{I}_{i \times i}$ and $\mathbf{0}_{i \times j}$ denote $i \times i$ identity and $i \times j$ zero matrices. See Chapter 7 of [18] for details.

$$[\boldsymbol{\xi}]^{\wedge}_{SE(3)} = \begin{bmatrix} \rho \\ \boldsymbol{\phi} \end{bmatrix}^{\wedge}_{SE(3)} = \begin{bmatrix} [\boldsymbol{\phi}]^{\wedge}_{SO(3)} & \rho \\ \mathbf{0}_{1\times 3} & 0 \end{bmatrix}, \quad \begin{bmatrix} [\boldsymbol{\phi}]^{\wedge}_{SO(3)} & \rho \\ \mathbf{0}_{1\times 3} & 0 \end{bmatrix}^{\vee}_{SE(3)} = \begin{bmatrix} \rho \\ \boldsymbol{\phi} \end{bmatrix} \tag{8}$$

$$\mathbf{T} = \exp([\boldsymbol{\xi}]^{\wedge}_{SE(3)}) = \begin{bmatrix} \exp([\boldsymbol{\phi}]^{\wedge}_{SO(3)}) & \Phi_{SO(3)}(-\boldsymbol{\phi})\rho \\ \mathbf{0}_{1\times 3} & 1 \end{bmatrix} = \begin{bmatrix} \mathbf{R} & t \\ \mathbf{0}_{1\times 3} & 1 \end{bmatrix} \tag{9}$$

$$\mathrm{ad}_{SE(3)}(\boldsymbol{\xi}) = \begin{bmatrix} [\boldsymbol{\phi}]^{\wedge}_{SO(3)} & [\rho]^{\wedge}_{SO(3)} \\ \mathbf{0}_{3\times 3} & [\boldsymbol{\phi}]^{\wedge}_{SO(3)} \end{bmatrix}, \quad \mathrm{Ad}_{SE(3)}(\mathbf{T}) = \begin{bmatrix} \mathbf{R} & [\rho]^{\wedge}_{SO(3)}\mathbf{R} \\ \mathbf{0}_{3\times 3} & \mathbf{R} \end{bmatrix}, \quad \mathbf{T}^{-1} = \begin{bmatrix} \mathbf{R}^T & -\mathbf{R}^T\rho \\ \mathbf{0}_{1\times 3} & 1 \end{bmatrix} \tag{10}$$

Lastly, we note the useful identity defined in Equation (11) where $[a]^{\wedge}_{SE(3)}, [b]^{\wedge}_{SE(3)} \in \mathfrak{se}(3)$ which is the Lie algebra of the Lie Group $SE(3)$ ([18], Equation (72)), which will be used to compute the Jacobians of our model later.

$$[a]^{\wedge}_{SE(3)} b = [b]^{\odot}_{SE(3)} a, \text{ where } b = \begin{bmatrix} \epsilon \\ \eta \end{bmatrix}, [b]^{\odot} = \begin{bmatrix} \eta \mathbf{I}_{3\times 3} & -[\epsilon]^{\wedge}_{SO(3)} \\ \mathbf{0}_{1\times 3} & \mathbf{0}_{1\times 3} \end{bmatrix}, \epsilon \in \mathbb{R}^3, \eta \in \mathbb{R} \tag{11}$$

2.3. Real Numbers $\mathbb{R}^n$

In order to represent vector state variables (e.g., translation, velocity, and acceleration) and be consistent with how we used $SE(3)$ to represent pose, we represented the real numbers $s \in \mathbb{R}^n$ as $SE(n)$ poses with position and no rotation. The operations for $\mathbb{R}^n$ are listed below.

$$[s]^{\wedge}_{\mathbb{R}^n} = \begin{bmatrix} \mathbf{0}_{n\times n} & s \\ \mathbf{0}_{1\times n} & 0 \end{bmatrix}, \quad \begin{bmatrix} \mathbf{0}_{n\times n} & s \\ \mathbf{0}_{1\times n} & 0 \end{bmatrix}^{\vee}_{\mathbb{R}^n} = s \tag{12}$$

$$\mathbf{S} = \exp([s]^{\wedge}_{\mathbb{R}^n}) = \begin{bmatrix} \mathbf{I}_{n\times n} & s \\ \mathbf{0}_{1\times n} & 1 \end{bmatrix}, \quad \left[\log\left(\begin{bmatrix} \mathbf{I}_{n\times n} & s \\ \mathbf{0}_{1\times n} & 1 \end{bmatrix}\right)\right]^{\vee}_{\mathbb{R}^n} = s, \quad \exp([s]^{\wedge}_{\mathbb{R}^n})^{-1} = \begin{bmatrix} \mathbf{I}_{n\times n} & -s \\ \mathbf{0}_{1\times n} & 1 \end{bmatrix} \tag{13}$$

$$\mathrm{ad}_{\mathbb{R}^n}(s) = \mathbf{0}_{n\times n}, \quad \mathrm{Ad}_{\mathbb{R}^n}(\mathbf{S}) = \mathbf{I}_{n\times n}, \quad \Phi_{\mathbb{R}^n}(s) = \mathbf{0}_{n\times n} \tag{14}$$

Note that the multiplication of two elements of the Lie group (i.e., the exponential of s_1 and s_2) is equivalent to the vector addition of two elements of the Lie algebra (i.e., $s_1 + s_2$).

$$[\log(\exp([s_1]^{\wedge}_{\mathbb{R}^n}) \exp([s_2]^{\wedge}_{\mathbb{R}^n}))]^{\vee}_{\mathbb{R}^n} = s_1 + s_2 \tag{15}$$

3. Algorithm Description

The proposed algorithm, L5S-3IMU, uses a similar model and assumptions to our prior works in [15,31], denoted as CKF-3IMU and CKF-3IMU+D, albeit expressed in Lie group representation. The algorithm uses measurements from three IMUs attached at the pelvis (sacrum) and shanks (slightly above the ankles), and the inter-IMU distance measurements to estimate the orientation of five body segments (i.e., pelvis, thighs, and shanks) with respect the world frame, W (Figure 2). The Lie group representation enables the tracking of both position and orientation (note that CKF-3IMU only tracked position and assumed orientation measurements were noise-free). Figure 3 shows an overview of the proposed algorithm. L5S-3IMU predicts the ankle and pelvis positions through double integration of their linear 3D acceleration, and predicts the shank and pelvis orientation through integration of their linear 3D angular velocity. The orientation estimates are further updated using a third-party orientation estimation algorithm. Drift in the position estimates due to sensor noise, which accumulates quadratically in the double integration of acceleration, was mitigated through the following assumptions: (1) the pelvis position is approximated as the height of the pelvis when the leg(s) are unbent, or as informed by inter-IMU distance measurements, when available; and (2) the ankle velocity and height above the floor are zeroed whenever a footstep is detected. Furthermore, a pseudo-measurement equal to the current global position estimate of the pelvis and ankles is made with a fixed covariance to contain the ever-growing error covariance of the said states. Lastly, constant body segment lengths and hinged knee joints (one degree of freedom (DOF)) with limited range of motion (ROM) were enforced as biomechanical constraints. The pre- and post-processing parts remain exactly the same as the CKF-3IMU algorithm (e.g., acceleration due to body movement was calculated by expressing the acceleration of the instrumented body segment in the world frame using the orientation estimate and then subtracting acceleration due to gravity (i.e., $g = [0\ 0\ 9.81]^T$) [6]).

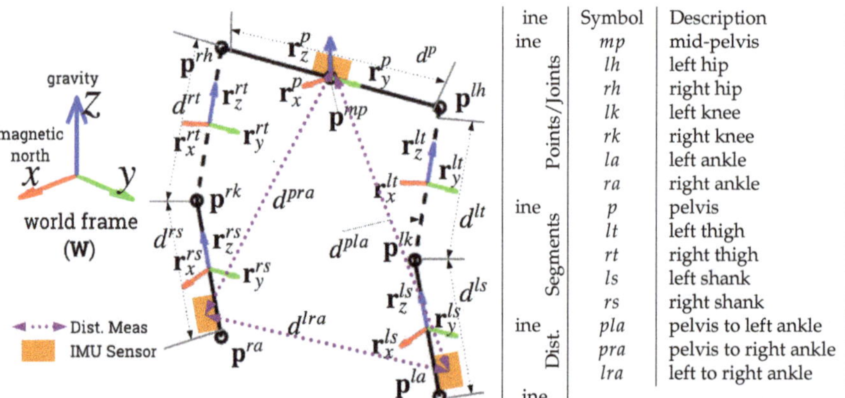

	Symbol	Description
Points/Joints	mp	mid-pelvis
	lh	left hip
	rh	right hip
	lk	left knee
	rk	right knee
	la	left ankle
	ra	right ankle
Segments	p	pelvis
	lt	left thigh
	rt	right thigh
	ls	left shank
	rs	right shank
Dist.	pla	pelvis to left ankle
	pra	pelvis to right ankle
	lra	left to right ankle

Figure 2. Model of the lower body used by LGKF-3IMU. The circles denote joint positions, the solid lines denote instrumented body segments, whilst the dashed lines denote segments without IMUs attached (i.e., thighs). Dotted lines denote inter-IMU measurements.

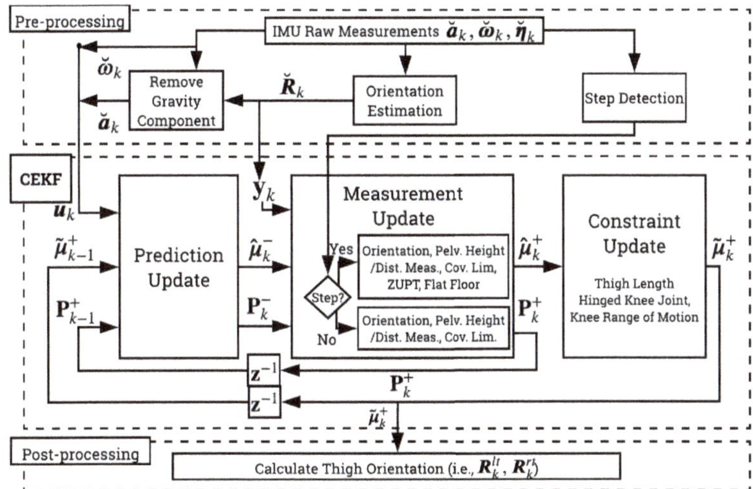

Figure 3. Algorithm overview which consists of pre-processing, CEKF, and post-processing. Pre-processing calculates the body segment orientation, inertial body acceleration (calculated by expressing IMU acceleration with respect world frame using its current orientation estimate then subtracting gravity), and step detection from raw acceleration, $\breve{a}_k$, angular velocity, $\breve{\omega}_k$, and magnetic north heading, $\breve{\eta}_k$, measured by the IMU. The CEKF-based state estimation consists of a prediction (kinematic equation), measurement (orientation, pelvis height/inter-IMU distance measurement, covariance limiter, intermittent zero-velocity update, and flat-floor assumption), and constraint update (thigh length, hinge knee joint, and knee range of motion). Post-processing calculates the left and right thigh orientations, $\boldsymbol{R}_k^{lt}$ and $\boldsymbol{R}_k^{rt}$.

3.1. System, Measurement, and Constraint Models

The system, measurement, and constraint models are presented below

$$\boldsymbol{X}_k = f(\boldsymbol{X}_{k-1}, \boldsymbol{u}_k, \boldsymbol{n}_k) = \boldsymbol{X}_{k-1} \exp([\Omega(\boldsymbol{X}_{k-1}, \boldsymbol{u}_k) + \boldsymbol{n}_k]_G^\wedge) \tag{16}$$

$$\boldsymbol{Z}_k = h(\boldsymbol{X}_k) \exp([\boldsymbol{m}_k]_G^\wedge), \quad \boldsymbol{D}_k = c(\boldsymbol{X}_k) \tag{17}$$

where k is the time step. $\boldsymbol{X}_k \in G$ is the system state, an element of state Lie group G. $\Omega\left(\boldsymbol{X}_{k-1}, \boldsymbol{u}_k\right) : G \to \mathbb{R}^p$ is a non-linear function which describes how the model acts on the state and input, $\boldsymbol{u}_{k-1}$, where p is the number of dimensions of the compact vector representation for Lie algebra $\mathfrak{g}$. $\boldsymbol{n}_k$ is a zero-mean process noise vector with covariance matrix $\mathcal{Q}$ (i.e., $\boldsymbol{n}_k \sim \mathcal{N}_{\mathbb{R}^p}(\boldsymbol{0}_{p \times 1}, \mathcal{Q})$). $\boldsymbol{Z}_k \in G_m$ is the system measurement, an element of the measurement Lie group G_m. $h\left(\boldsymbol{X}_k\right) : G \to G_m$ is the measurement function. $\boldsymbol{m}_k$ is a zero-mean measurement noise vector with covariance matrix $\mathcal{R}_k$ (i.e., $\boldsymbol{m}_k \sim \mathcal{N}_{\mathbb{R}^q}(\boldsymbol{0}_{q \times 1}, \mathcal{R}_k)$ where q is the number of dimensions of available measurements). $\boldsymbol{D}_k \in G_c$ is the constraint state, an element of constraint Lie group G_c. $c\left(\boldsymbol{X}_k\right) : G \to G_c$ is the equality constraint function the state $\boldsymbol{X}_k$ must satisfy (i.e., $c\left(\boldsymbol{X}_k\right) = \boldsymbol{D}_k$). Similar to [23,36], the state distribution of $\boldsymbol{X}_k$ is assumed to be a concentrated Gaussian distribution on Lie groups (i.e., $\boldsymbol{X}_k = \mu_k \exp_G [\epsilon]_G^\wedge$, where μ_k is the mean of $\boldsymbol{X}_k$ and Lie algebra error $\epsilon \sim \mathcal{N}_{\mathbb{R}^p}(\boldsymbol{0}_{p \times 1}, \mathbf{P})$) [19].

The Lie group state variables $\boldsymbol{X}_k$ model the position, orientation, and velocity of the three instrumented body segments (i.e., pelvis and shanks) as $\boldsymbol{X}_k = \mathrm{diag}(\boldsymbol{T}^p, \boldsymbol{T}^{ls}, \boldsymbol{T}^{rs}, \exp([[(\boldsymbol{v}^p)^T (\boldsymbol{v}^{ls})^T (\boldsymbol{v}^{rs})^T]^T]_{\mathbb{R}^9}^\wedge)) \in G = SE(3)^3 \times \mathbb{R}^9$ where $\boldsymbol{T}^b \in SE(3)$ represents the pose (i.e., orientation and position) of body segment b relative to world frame W, and $^A\boldsymbol{v}^b$ is the velocity of body segment b relative to frame A. If frame A is not specified, assume reference to the world frame, W. The Lie algebra error is denoted as $\epsilon = [(\epsilon_{\mathbf{T}}^p)^T (\epsilon_{\mathbf{T}}^{ls})^T (\epsilon_{\mathbf{T}}^{rs})^T (\epsilon_v^{mp})^T (\epsilon_v^{la})^T (\epsilon_v^{ra})^T]^T$ where the first three variables correspond to the Lie group in $SE(3)$ while the latter three are for

$\mathbb{R}^9$. $[\,]_G^{\vee}$, $\exp([\,]_G^{\wedge})$, $[\log(\,)]_G^{\vee}$, $\mathrm{Ad}_G(X_k)$, and $\Phi_G(\,)$ are constructed similarly as X_k (e.g., $\mathrm{Ad}_G(X_k) = \mathrm{diag}(\mathrm{Ad}_{SE(3)}(T^p), \mathrm{Ad}_{SE(3)}(T^{ls}), \mathrm{Ad}_{SE(3)}(T^{rs}), \mathrm{Ad}_{\mathbb{R}^9}(\exp([[(v^p)^T \ (v^{ls})^T \ (v^{rs})^T]^T]_{\mathbb{R}^9}^{\wedge})))$. Refer to Sections 2.2 and 2.3 for definition of $SE(3)$ and $\mathbb{R}^n$ operators.

3.2. Lie Group Constrained EKF (LG-CEKF)

The a priori, a posteriori, and constrained state estimate (i.e., estimated mean of X_k) for time step k are denoted by $\hat{\mu}_k^-$, $\hat{\mu}_k^+$, and $\tilde{\mu}_k^+$, respectively. Note that the true state X_k can be expressed as $\mu_k \exp([\epsilon]_G^{\wedge})$ where μ_k is one of the state means just mentioned with error, $[\epsilon]_G^{\wedge}$. The a priori and a posteriori error covariance matrices are denoted as $\mathbf{P}_k^-$ and $\mathbf{P}_k^+$, respectively. Note the error covariance is not updated at the constrain update step. The KF is based on the Lie group EKF, as defined in [36], where the state means ($\hat{\mu}_k^-$, $\hat{\mu}_k^+$, and $\tilde{\mu}_k^+$) and state error covariance matrices ($\mathbf{P}_k^-$ and $\mathbf{P}_k^+$) are propagated by the KF at each time step.

3.2.1. Prediction Step

Prediction step estimates the a priori state $\hat{\mu}_k^-$ at the next time step. The mean propagation of the three instrumented body segments is governed by Equation (18) where $\Omega(\tilde{\mu}_{k-1}^+, u_k)$ (Equation (19)) is the motion model for the three instrumented body segments. Note that the state propagation may not necessarily respect the biomechanical constraints, so joints may become dislocated after this step. The input u_k is defined in Equation (20), where the orientation and acceleration as obtained by the IMU attached to segment b with respect world frame W are denoted as $\check{R}_k^b$ and $\check{a}_k^b$ for $b \in \{p, ls, rs\}$, while the angular velocity as obtained by the IMU attached to segment b expressed in frame b is denoted as ${}^b\check{\omega}_k$.

$$\hat{\mu}_k^- = \tilde{\mu}_{k-1}^+ \exp([\Omega(\tilde{\mu}_{k-1}^+, u_k)]_G^{\wedge}) \tag{18}$$

$$\Omega(\tilde{\mu}_{k-1}^+, u_k) = \Big[\underbrace{(\Delta t\,\tilde{v}_{k-1}^{mp+} + \tfrac{\Delta t^2}{2}\check{a}_k^p)^T \check{R}_k^p}_{1 \times 3} \quad \underbrace{\Delta t\,{}^p\check{\omega}_k^T}_{1 \times 3} \quad \underbrace{(\Delta t\,\tilde{v}_{k-1}^{la+} + \tfrac{\Delta t^2}{2}\check{a}_k^{ls})^T \check{R}_k^{ls}}_{1 \times 3} \quad \underbrace{\Delta t\,{}^{ls}\check{\omega}_k^T}_{1 \times 3}$$
$$\underbrace{(\Delta t\,\tilde{v}_{k-1}^{ra+} + \tfrac{\Delta t^2}{2}\check{a}_k^{rs})^T \check{R}_k^{rs}}_{1 \times 3} \quad \underbrace{\Delta t\,{}^{rs}\check{\omega}_k^T}_{1 \times 3} \quad \underbrace{\Delta t(\check{a}_k^{mp})^T \quad \Delta t(\check{a}_k^{la})^T \quad \Delta t(\check{a}_k^{ra})^T}_{1 \times 9}\Big]^T \tag{19}$$

$$u_k = \begin{bmatrix} \check{R}_k^p & \check{R}_k^{ls} & \check{R}_k^{rs} & \check{a}_k^p & \check{a}_k^{ls} & \check{a}_k^{rs} & {}^p\check{\omega}_k & {}^{ls}\check{\omega}_k & {}^{rs}\check{\omega}_k \end{bmatrix} \tag{20}$$

The state error covariance matrix propagation is governed by Equation (21), where $\mathcal{F}_k$ represents the matrix Lie group equivalent to the Jacobian of $f(X_{k-1}, n_{k-1})$, $\mathcal{Q}$ is the covariance matrix of the process noise, and $\mathscr{C}_k = \frac{\partial}{\partial \epsilon}\Omega(\tilde{\mu}_{k-1}^+ \exp([\epsilon]_G^{\wedge}), u_k)|_{\epsilon=0}$ represents the linearization of the motion model with an infinitesimal perturbation ϵ. The process noise covariance matrix, $\mathcal{Q}$, is calculated from the input-to-state matrix $\mathcal{G}$ and the noise variances of the measured acceleration and angular velocity, σ_a^2 and σ_ω^2, respectively.

$$\mathbf{P}_k^- = \mathcal{F}_k \mathbf{P}_{k-1}^+ \mathcal{F}_k^T + \Phi_G(\Omega(\tilde{\mu}_{k-1}^+, u_k)) \mathcal{Q} \Phi_G(\Omega(\tilde{\mu}_{k-1}^+, u_k))^T \tag{21}$$

$$\mathcal{F}_k = \mathrm{Ad}_G(\exp_G(-[\Omega(\tilde{\mu}_{k-1}^+, u_k)]_G^{\wedge})) + \Phi_G(\Omega(\tilde{\mu}_{k-1}^+, u_k))\mathscr{C}_k \tag{22}$$

$$\mathcal{Q} = \mathcal{G}\,\mathrm{diag}(\sigma_a^2, \sigma_\omega^2)\,\mathcal{G}^T \tag{23}$$

$$\mathscr{C}_k = \frac{\partial}{\partial \epsilon}\Omega(\tilde{\mu}_{k-1}^{+}\exp([\epsilon]_G^{\wedge}), u_k)|_{\epsilon=0}$$

$$= \frac{\partial}{\partial \epsilon}\Big[\ (\Delta t(\tilde{v}_{k-1}^{mp+} + \epsilon_v^{mp}) + \tfrac{\Delta t^2}{2}\breve{a}_k^{p})^T\ \breve{R}_k^{p}\ \Delta t\ ^{p}\breve{\omega}_k^{T}\ (\Delta t(\tilde{v}_{k-1}^{la+} + \epsilon_v^{la}) + \tfrac{\Delta t^2}{2}\breve{a}_k^{ls})^T\ \breve{R}_k^{ls}\ \Delta t\ ^{ls}\breve{\omega}_k^{T}$$

$$(\Delta t(\tilde{v}_{k-1}^{ra+} + \epsilon_v^{ra}) + \tfrac{\Delta t^2}{2}\breve{a}_k^{rs})^T\ \breve{R}_k^{rs}\ \Delta t\ ^{rs}\breve{\omega}_k^{T}\ \Delta t(\breve{a}_k^{mp})^T\ \Delta t(\breve{a}_k^{la})^T\ \Delta t(\breve{a}_k^{ra})^T\ \Big]^T\Big|_{\epsilon=0}$$

$$\mathscr{C}_k = \begin{bmatrix} \mathbf{0}_{18\times 18} & \begin{matrix} \Delta t(\breve{R}_k^{p})^T & \mathbf{0}_{3\times3} & \mathbf{0}_{3\times3} \\ \mathbf{0}_{3\times3} & \mathbf{0}_{3\times3} & \mathbf{0}_{3\times3} \\ \mathbf{0}_{3\times3} & \Delta t(\breve{R}_k^{ls})^T & \mathbf{0}_{3\times3} \\ \mathbf{0}_{3\times3} & \mathbf{0}_{3\times3} & \mathbf{0}_{3\times3} \\ \mathbf{0}_{3\times3} & \mathbf{0}_{3\times3} & \Delta t(\breve{R}_k^{rs})^T \\ \mathbf{0}_{3\times3} & \mathbf{0}_{3\times3} & \mathbf{0}_{3\times3} \end{matrix} \\ \hline \multicolumn{2}{c}{\mathbf{0}_{9\times27}} \end{bmatrix} \tag{24}$$

$$\mathcal{G} = \begin{bmatrix} \begin{matrix} \Delta t^2/2\,\mathbf{I}_{3\times3} & \mathbf{0}_{3\times3} & \mathbf{0}_{3\times3} \\ \mathbf{0}_{3\times3} & \mathbf{0}_{3\times3} & \mathbf{0}_{3\times3} \\ \mathbf{0}_{3\times3} & \Delta t^2/2\,\mathbf{I}_{3\times3} & \mathbf{0}_{3\times3} \\ \mathbf{0}_{3\times3} & \mathbf{0}_{3\times3} & \mathbf{0}_{3\times3} \\ \mathbf{0}_{3\times3} & \mathbf{0}_{3\times3} & \Delta t^2/2\,\mathbf{I}_{3\times3} \\ \mathbf{0}_{3\times3} & \mathbf{0}_{3\times3} & \mathbf{0}_{3\times3} \end{matrix} & \begin{matrix} \mathbf{0}_{3\times3} & \mathbf{0}_{3\times3} & \mathbf{0}_{3\times3} \\ \Delta t\mathbf{I}_{3\times3} & \mathbf{0}_{3\times3} & \mathbf{0}_{3\times3} \\ \mathbf{0}_{3\times3} & \mathbf{0}_{3\times3} & \mathbf{0}_{3\times3} \\ \mathbf{0}_{3\times3} & \Delta t\mathbf{I}_{3\times3} & \mathbf{0}_{3\times3} \\ \mathbf{0}_{3\times3} & \mathbf{0}_{3\times3} & \mathbf{0}_{3\times3} \\ \mathbf{0}_{3\times3} & \mathbf{0}_{3\times3} & \Delta t\mathbf{I}_{3\times3} \end{matrix} \\ \hline \multicolumn{1}{c}{\Delta t\mathbf{I}_{9\times9}} & \multicolumn{1}{c}{\mathbf{0}_{9\times9}} \end{bmatrix} \tag{25}$$

3.2.2. Measurement Update

Measurement update estimates the state at the next time step by: (i) updating the orientation state using new orientation measurements of body segments from IMUs; by (ii) encouraging pelvis position to be above the feet, as informed by either some pseudo-measurement or inter-IMU distance measurements; and by (iii) enforcing ankle velocity to reach zero, and the ankle z position to be near the floor level, z_f when step is detected. When only IMU measurements are available, (iia) pelvis z position is encouraged to be close to initial standing height, z_p. When inter-IMU distance measurements are available, (iia) is not used. Instead, (iib) ankle distance is directly incorporated while pelvis position is inferred from inter-IMU distance measurements assuming hinged knee joints and constant body segment lengths. The a posteriori state mean $\hat{\mu}_k^{+}$ is calculated following the Lie EKF equations below. Note that $[\log(h(\hat{\mu}_k^{-})^{-1}\mathbf{Z}_k)]_{G_m}^{\vee}$ in Equation (27) is akin to the KF innovation/residual, where $h(\hat{\mu}_k^{-})^{-1}\mathbf{Z}_k$ (derived from Equation (17) assuming $m_k = 0$ and $\mathbf{X}_k = \hat{\mu}_k^{-}$, i.e., $\mathbf{Z}_k = h(\hat{\mu}_k^{-})$) is the innovation/residual in Lie group G_m brought to the vector representation of the Lie algebra space using the inverse exponential (i.e., logarithm) mapping.

$$\hat{\mu}_k^{+} = \hat{\mu}_k^{-}\exp_G([v_k]_G^{\wedge}) \tag{26}$$

$$v_k = \mathbf{K}_k([\log(h(\hat{\mu}_k^{-})^{-1}\mathbf{Z}_k)]_{G_m}^{\vee}) \tag{27}$$

$$\mathbf{K}_k = \mathbf{P}_k^{-}\mathcal{H}_k^{T}(\mathcal{H}_k\mathbf{P}_k^{-}\mathcal{H}_k^{T} + \mathcal{R}_k)^{-1} \tag{28}$$

$\mathcal{H}_k$ can be seen as the matrix Lie group equivalent to the Jacobian of $h(\mathbf{X}_k)$, and is defined as the concatenation of $\mathcal{H}_{ori}$ and $\mathcal{H}_{mp,k}$ when inter-IMU distance measurement is not available. When inter-IMU distance measurement is available, $\mathcal{H}_{mp,k}$ is replaced by $\mathcal{H}_{dist,k} = [\mathcal{H}_{pla,k}^{T}\ \ \mathcal{H}_{pra,k}^{T}\ \ \mathcal{H}_{lra,k}^{T}]^T$. $\mathcal{H}_{ls,k}$ and/or $\mathcal{H}_{rs,k}$ are also concatenated to $\mathcal{H}_k$ when the left and/or right foot contact (FC) is detected (See Equation (9) of [15]). Each component matrix will be described later. The measurement matrix

$\mathbf{Z}_k \in G_m$, measurement function $h(\mathbf{X}_k) \in G_m$, and measurement covariance noise $\mathcal{R}_k$ are constructed similarly to $\mathcal{H}_k$, but combined using diag instead of concatenation (e.g., $\mathcal{R}_k = \mathrm{diag}(\sigma^2_{ori}, \sigma^2_{mp})$).

$$\mathcal{H}_k = \frac{\partial}{\partial \epsilon} \left[\log \left(h(\hat{\boldsymbol{\mu}}_k^-)^{-1} h(\hat{\boldsymbol{\mu}}_k^- \exp([\epsilon]_G^\wedge)) \right) \right]_{G_m}^{\vee} \Big|_{\epsilon=0}$$

$$= \begin{cases} [\mathcal{H}_{ori}^T \quad \mathcal{H}_{mp/dist}^T]^T & \text{no FC} \\ [\mathcal{H}_{ori}^T \quad \mathcal{H}_{mp/dist}^T \quad \mathcal{H}_{ls,k}^T]^T & \text{left FC} \\ [\mathcal{H}_{ori}^T \quad \mathcal{H}_{mp/dist}^T \quad \mathcal{H}_{rs,k}^T]^T & \text{right FC} \\ [\mathcal{H}_{ori}^T \quad \mathcal{H}_{mp/dist}^T \quad \mathcal{H}_{ls,k}^T \quad \mathcal{H}_{rs,k}^T]^T & \text{both FC} \end{cases} \tag{29}$$

Orientation Update

The orientation update utilizes the orientation measurement to update the state estimate as defined by Equation (30), with measurement noise variance σ^2_{ori} (9×1 vector).

$$h_{ori}(\mathbf{X}_k) = \mathrm{diag}(\mathbf{R}_k^p, \mathbf{R}_k^{ls}, \mathbf{R}_k^{rs}) \in SO(3)^3, \quad \mathbf{Z}_{ori} = \mathrm{diag}(\check{\mathbf{R}}_k^p, \check{\mathbf{R}}_k^{ls}, \check{\mathbf{R}}_k^{rs}) \tag{30}$$

$\mathcal{H}_{ori}$ along with other components of $\mathcal{H}_k$ are calculated by applying Equation (29) to their corresponding measurement functions, followed by tedious algebraic manipulation and first order linearization (i.e., $\exp([\epsilon]^\wedge) \approx \mathbf{I} + [\epsilon]^\wedge$). The derivation for $\mathcal{H}_{ori}$ (Equation (31)) can be solved trivially as $[\log(h_{ori}(\hat{\boldsymbol{\mu}}_k^-)^{-1} h_{ori}(\hat{\boldsymbol{\mu}}_k^- \exp([\epsilon]_G^\wedge)))]^{\vee} = [(\epsilon_\phi^p)^T \ (\epsilon_\phi^{ls})^T \ (\epsilon_\phi^{rs})^T]^T$, where $\epsilon_{\mathbf{T}}^b = [(\epsilon_\rho^b)^T \ (\epsilon_\phi^b)^T]^T$ for body segment $b \in \{p, ls, rs\}$.

$$\mathcal{H}_{ori} = \begin{bmatrix} \mathbf{0}_{3\times3} \ \mathbf{I}_{3\times3} & & & \\ & \mathbf{0}_{3\times3} \ \mathbf{I}_{3\times3} & & \mathbf{0}_{9\times9} \\ & & \mathbf{0}_{3\times3} \ \mathbf{I}_{3\times3} & \end{bmatrix} \tag{31}$$

Pelvis Height Assumption

The pelvis height assumption softly constrains the pelvis z position to be close to initial standing height z_p as defined by Equation (32) (represented in vector space of its Lie algebra) and Equation (33), with measurement noise variance σ^2_{mp} (1×1 vector). This assumption is used only when inter-IMU distance measurement is not available. i_x, i_y, i_z, and i_0 denote 4×1 vectors whose 1st to 4th rows, respectively, are 1, while the rest are 0; they are used below to select rows, columns, or elements from matrices.

$$[\log(h_{mp}(\mathbf{X}_k))]^{\vee} = i_z^T \, \mathbf{T}_k^p i_0 = \begin{bmatrix} 0 & 0 & 1 & 0 \end{bmatrix} \begin{bmatrix} \mathbf{R}_k^p & \mathbf{p}_k^p \\ \mathbf{0}_{1\times3} & 1 \end{bmatrix} \begin{bmatrix} 0 \\ 0 \\ 0 \\ 1 \end{bmatrix} = \begin{bmatrix} 0 & 0 & 1 & 0 \end{bmatrix} \begin{bmatrix} \mathbf{p}_k^p \\ 1 \end{bmatrix} = p_{z,k}^p \in \mathbb{R} \tag{32}$$

$$[\log(\mathbf{Z}_{mp})]^{\vee} = z_p \in \mathbb{R} \tag{33}$$

The derivation of $\mathcal{H}_{mp,k} = \frac{\partial}{\partial \epsilon}[\log(h_{mp}(\hat{\boldsymbol{\mu}}_k^-)^{-1} h_{mp}(\hat{\boldsymbol{\mu}}_k^- \exp([\epsilon]_G^\wedge)))]^{\vee}|_{\epsilon=0}$ is shown in Equations (34)–(36). Taking best estimate $\mathbf{X}_k = \hat{\boldsymbol{\mu}}_k^-$ gives us Equation (34).

$$[\log(h_{mp}(\hat{\boldsymbol{\mu}}_k^-))]^{\vee} = i_z^T \, \hat{\mathbf{T}}_k^{p-} i_0 \tag{34}$$

$$\begin{aligned} [\log(h_{mp}(\hat{\boldsymbol{\mu}}_k^- \exp([\epsilon]_G^\wedge)))]^{\vee} &= i_z^T \, \hat{\mathbf{T}}_k^{p-} \exp([\epsilon_{\mathbf{T}}^p]^\wedge) i_0 \\ &\approx i_z^T \, \hat{\mathbf{T}}_k^{p-} i_0 + i_z^T \, \hat{\mathbf{T}}_k^{p-} [\epsilon_{\mathbf{T}}^p]^\wedge i_0, \quad \text{1st order linearization} \\ & \text{Use Equation (11), } [a]^\wedge b = [b]^\odot a, \text{ to bring } \epsilon_{\mathbf{T}}^p \text{ to right of } i_0 \\ &= [\log(h_{mp}(\hat{\boldsymbol{\mu}}_k^-))]^{\vee} + i_z^T \, \hat{\mathbf{T}}_k^{p-} [i_0]^\odot \epsilon_{\mathbf{T}}^p \end{aligned} \tag{35}$$

Remember ϵ_T^p is a subvector of ϵ as defined in Section 3.1 and is the Lie algebra error of the state in its compact vector representation. Note that if measurement function $h_a(\mathbf{X}_k) \in$ Lie group $\mathbb{R}^b$, then $[\log(h_a(\hat{\boldsymbol{\mu}}_k^-)^{-1}h_a(\mathbf{X}_k))]^\vee = [\log(h_a(\mathbf{X}_k))]^\vee - [\log(h_a(\hat{\boldsymbol{\mu}}_k^-))]^\vee = [\log(h_a(\hat{\boldsymbol{\mu}}_k^- \exp([\epsilon]_G^\wedge)))]^\vee - [\log(h_a(\hat{\boldsymbol{\mu}}_k^-))]^\vee$ by applying Equations (15) and (13) (inverse of Lie group $\mathbb{R}^n$). Finally, $\mathcal{H}_{mp,k}$ is calculated as shown in Equation (36).

$$
\begin{aligned}
\mathcal{H}_{mp,k} &= \tfrac{\partial}{\partial \epsilon}[\log(h_{mp}(\hat{\boldsymbol{\mu}}_k^-)^{-1}h_{mp}(\hat{\boldsymbol{\mu}}_k^- \exp([\epsilon]_G^\wedge)))]^\vee |_{\epsilon=0} \\
&= \tfrac{\partial}{\partial \epsilon}\left([\log(h_{mp}(\hat{\boldsymbol{\mu}}_k^- \exp([\epsilon]_G^\wedge)))]^\vee - [\log(h_{mp}(\hat{\boldsymbol{\mu}}_k^-))]^\vee\right)|_{\epsilon=0} \\
&= \left[\underbrace{i_z^T \, \hat{T}_k^{p-}[i_0]^\odot \quad \mathbf{0}_{1\times6} \quad \mathbf{0}_{1\times6} \mid \mathbf{0}_{1\times9}}_{1\times6} \right]
\end{aligned}
\tag{36}
$$

Zero Velocity Update and Flat Floor Assumption

When step is detected, the ankle velocity is enforced to be zero and the ankle z position is brought to near the floor level, z_f (i.e., flat floor assumptions). The corresponding measurement function is defined by Equation (37), with measurement noise variance σ_{ls}^2 (4×1 vector).

$$
[\log(h_{ls}(\mathbf{X}_k))]^\vee = \begin{bmatrix} v^{ls} \\ i_z^T \, T_k^{ls} i_0 \end{bmatrix} = \begin{bmatrix} v^{ls} \\ p_{z,k}^{ls} \end{bmatrix} \in \mathbb{R}^4, \quad [\log(\mathbf{Z}_{ls})]^\vee = \begin{bmatrix} \mathbf{0}_{3\times1} \\ z_f \end{bmatrix}
\tag{37}
$$

The zero velocity part of $\mathcal{H}_{ls,k}$ (Equation (38)) and $\mathcal{H}_{rs,k}$ can also be calculated trivially, while the flat floor assumption can be calculated similarly as $\mathcal{H}_{mp,k}$ but the z position set to floor height, z_f, instead of the pelvis standing height, z_p.

$$
\begin{aligned}
\mathcal{H}_{ls,k} &= \tfrac{\partial}{\partial \epsilon}[\log(h_{ls}(\hat{\boldsymbol{\mu}}_k^-)^{-1}h_{ls}(\hat{\boldsymbol{\mu}}_k^- \exp([\epsilon]_G^\wedge)))]^\vee |_{\epsilon=0} \\
&= \tfrac{\partial}{\partial \epsilon}\left([\log(h_{ls}(\hat{\boldsymbol{\mu}}_k^- \exp([\epsilon]_G^\wedge)))]^\vee - [\log(h_{ls}(\hat{\boldsymbol{\mu}}_k^-))]^\vee\right)|_{\epsilon=0} \\
&= \begin{bmatrix} \underbrace{\mathbf{0}_{3\times6} \quad \mathbf{0}_{3\times6} \quad \mathbf{0}_{3\times6} \mid \mathbf{0}_{3\times3} \quad \mathbf{I}_{3\times3} \quad \mathbf{0}_{3\times3}}_{} \\ \underbrace{\mathbf{0}_{1\times6} \quad i_z^T \, \hat{T}_k^{ls-}[i_0]^\odot \quad \mathbf{0}_{1\times6} \mid \mathbf{0}_{1\times3} \quad \mathbf{0}_{1\times3} \quad \mathbf{0}_{1\times3}}_{} \end{bmatrix}
\end{aligned}
\tag{38}
$$

$$
\underbrace{\phantom{\mathbf{0}_{3\times6} \quad \mathbf{0}_{3\times6} \quad \mathbf{0}_{3\times6}}}_{\text{pose states in } SE(3)} \qquad \underbrace{\phantom{\mathbf{0}_{3\times3} \quad \mathbf{I}_{3\times3} \quad \mathbf{0}_{3\times3}}}_{\text{velocity states}}
$$

Left and Right Ankle Distance Measurement

When the inter-IMU distance between the ankles, d_k^{lra}, is available, ankle distance measurement is incorporated as a soft distance constraint. The measurement function is defined by Equation (40), with measurement noise variance σ_{lra}^2 (1×1 vector). $\tau_{lra}(\mathbf{X}_k)$ (Equation (39)) is the vector that points from the right ankle to the left ankle, where $^{ls}p^{la}$ is the position of the left ankle expressed in left shank frame, and $^{rs}p^{ra}$ is the position of the right ankle expressed in right shank frame. We have chosen that the ankles are at the origin of their respective shank frames. Note that matrix $\mathbf{E}$ converts homogeneous 4×1 coordinates to standard 3×1 coordinates (i.e., drops the 1 from the end of the 4×1 vector).

$$
\tau_{lra}(\mathbf{X}_k) = \underbrace{\begin{bmatrix} \mathbf{I}_{3\times3} & \mathbf{0}_{3\times1} \end{bmatrix}}_{\mathbf{E}}(\underbrace{T_k^{ls} \, {}^{ls}p^{la}}_{\text{left ankle in } W} - \underbrace{T_k^{rs} \, {}^{rs}p^{ra}}_{\text{right ankle in } W}), \quad {}^{ls}p^{la} = {}^{rs}p^{ra} = \underbrace{\begin{bmatrix} 0 & 0 & 0 & 1 \end{bmatrix}^T}_{\text{origin of frame}}
\tag{39}
$$

By taking the squared Euclidean distance of $\tau_{lra}(\mathbf{X}_k)$ (i.e., $\|\tau_{lra}(\mathbf{X}_k)\|^2$), we can get the ankle distance measurement model.

$$
[\log(h_{lra}(\mathbf{X}_k))]^\vee = (\tau_{lra}(\mathbf{X}_k))^T \tau_{lra}(\mathbf{X}_k) \in \mathbb{R}, \quad [\log(\mathbf{Z}_{lra})]^\vee = (d_k^{lra})^2
\tag{40}
$$

To solve for $\mathcal{H}_{lra,k}$ (Equation (44)), we first solved for $[\log(h_{lra}(\mathbf{X}_k))]^\vee$ at $\mathbf{X}_k = \hat{\boldsymbol{\mu}}_k^-$ (Equation (41)).

$$
\tau_{lra}(\hat{\boldsymbol{\mu}}_k^-) = \mathbf{E}(\hat{T}_k^{ls-} \, {}^{ls}p^{la} - \hat{T}_k^{rs-} \, {}^{rs}p^{ra}), \quad [\log(h_{lra}(\hat{\boldsymbol{\mu}}_k^-))]^\vee = (\tau_{lra}(\hat{\boldsymbol{\mu}}_k^-))^T \tau_{lra}(\hat{\boldsymbol{\mu}}_k^-)
\tag{41}
$$

Then solve for $\tau_{lra}(\hat{\mu}_k^- \exp([\epsilon]_G^\wedge))$ and $[\log(h_{lra}(\hat{\mu}_k^- \exp([\epsilon]_G^\wedge)))]^\vee$ as shown in Equations (42) and (43).

$$\tau_{lra}(\hat{\mu}_k^- \exp([\epsilon]_G^\wedge)) = \mathbf{E}(\hat{T}_k^{ls-} \exp([\epsilon_T^{ls}]^\wedge)\,{}^{ls}p^{la} - \hat{T}_k^{rs-} \exp([\epsilon_T^{rs}]^\wedge)\,{}^{rs}p^{ra})$$

$$\text{Take the 1st order approximation}$$

$$\approx \mathbf{E}(\hat{T}_k^{ls-}\,{}^{ls}p^{la} - \hat{T}_k^{rs-}\,{}^{rs}p^{ra} + \hat{T}_k^{ls-}[\epsilon_T^{ls}]^\wedge\,{}^{ls}p^{la} - \hat{T}_k^{rs-}[\epsilon_T^{rs}]^\wedge\,{}^{rs}p^{ra}) \tag{42}$$

$$= \tau_{lra}(\hat{\mu}_k^-) + \overbrace{\mathbf{E}(\hat{T}_k^{ls-}[{}^{ls}p^{la}]^\odot \epsilon_T^{ls} - \hat{T}_k^{rs-}[{}^{rs}p^{ra}]^\odot \epsilon_T^{rs})}^{\Gamma_{lra}}, \quad \text{Using Equation (11)}$$

$$[\log(h_{lra}(\hat{\mu}_k^- \exp([\epsilon]_G^\wedge)))]^\vee = \left(\tau_{lra}(\hat{\mu}_k^-) + \Gamma_{lra}\right)^T \left(\tau_{lra}(\hat{\mu}_k^-) + \Gamma_{lra}\right)$$

$$\text{Assume 2nd order error} \approx 0$$

$$= \tau_{lra}(\hat{\mu}_k^-)^T \tau_{lra}(\hat{\mu}_k^-) + 2\,\tau_{lra}(\hat{\mu}_k^-)^T \Gamma_{lra} + \cancelto{\approx 0}{\Gamma_{lra}^T \Gamma_{lra}} \tag{43}$$

$$= [\log(h_{lra}(\hat{\mu}_k^-))]^\vee$$

$$+ 2\,\tau_{lra}(\hat{\mu}_k^-)^T \mathbf{E}(\hat{T}_k^{ls-}[{}^{ls}p^{la}]^\odot \epsilon_T^{ls} - \hat{T}_k^{rs-}[{}^{rs}p^{ra}]^\odot \epsilon_T^{rs})$$

$$\mathcal{H}_{lra,k} = \frac{\partial}{\partial \epsilon}[\log(h_{lra}(\hat{\mu}_k^-)^{-1} h_{lra}(\hat{\mu}_k^- \exp([\epsilon]_G^\wedge)))]^\vee|_{\epsilon=0}$$

$$= \frac{\partial}{\partial \epsilon}\left([\log(h_{lra}(\hat{\mu}_k^- \exp([\epsilon]_G^\wedge)))]^\vee - [\log(h_{lra}(\hat{\mu}_k^-))]^\vee\right)|_{\epsilon=0}$$

$$= \frac{\partial}{\partial \epsilon}\left(2\,\tau^{lra}(\hat{\mu}_k^-)^T \mathbf{E}(\hat{T}_k^{ls-}[{}^{ls}p^{la}]^\odot \epsilon_T^{ls} - \hat{T}_k^{rs-}[{}^{rs}p_0^{rs}]^\odot \epsilon_T^{rs})\right)|_{\epsilon=0} \tag{44}$$

$$= \left[\; \mathbf{0}_{1\times 6} \;\middle|\; \underbrace{2\,\tau_{lra}(\hat{\mu}_k^-)^T \mathbf{E}\,\hat{T}_k^{ls-}[{}^{ls}p^{la}]^\odot}_{1\times 6} \;\middle|\; \underbrace{-2\,\tau_{lra}(\hat{\mu}_k^-)^T \mathbf{E}\,\hat{T}_k^{rs-}[{}^{rs}p^{ra}]^\odot}_{1\times 6} \;\middle|\; \mathbf{0}_{1\times 9} \;\right]$$

Pelvis-to-Ankle Distance Measurement

In addition to the soft ankle distance constraint, the ankle to pelvis vector is inferred from the ankle to pelvis distance measurements while assuming hinged knee joints and constant body segment lengths. The measurement function is defined by Equation (45), with measurement noise variance σ_{pla}^2 (3×1 vector), where ${}^p p^{mp}$ is the position of the mid-pelvis expressed in pelvis frame, and ${}^{ls}p^{la}$ is the position of the left ankle expressed in left shank frame. We have chosen that the mid-pelvis and ankle are at the origin of their corresponding reference frames.

$$[\log(h_{pla}(\mathbf{X}_k))]^\vee = \mathbf{E}(\ \overbrace{T_k^p\,{}^p p^{mp}}^{\text{mid-pelvis in } W} - \overbrace{T_k^{ls}\,{}^{ls}p^{la}}^{\text{left ankle in } W}\ ) \in \mathbb{R}^3, \quad {}^p p^{mp} = {}^{ls}p^{la} = \begin{bmatrix} 0 & 0 & 0 & 1 \end{bmatrix}^T \tag{45}$$

The measurement pelvis to left ankle vector can be calculated from the measured pelvis to left ankle distance, d_k^{pla} as shown in Equation (46) which is the Lie Group reformulation of [31] (Equation (4)). In essence, Equation (47) calculates the most probably knee angle assuming hinged knee joint and constant body segment lengths, then Equation (46) adds the thigh (expressed in shank coordinate system with knee angle $\hat{\theta}_k^{lk}$) and shank long axis to the hips to obtain the pelvis-to-ankle vector. See Appendix A for derivation. There are two solutions for $\hat{\theta}_k^{lk}$ due to the inverse cosine in Equation (47). We chose the $\hat{\theta}_k^{lk}$ value as that closer to the current left knee angle estimate from the prediction step. Note that this measurement function could also be formulated as a linearized Euclidean distance between the pelvis and ankle (i.e., similar to Equation (44)); however, a preliminary exploration of this approach showed poorer performance.

$$[\log(\mathbf{Z}_{pla,k})]^\vee = \overbrace{\frac{d^p}{2}\hat{T}_k^{p-} i_y - d^{ls}\hat{T}_k^{ls-} i_z}^{\psi_{pla} = \text{half pelvis } y\text{-axis} + \text{shank } z\text{-axis}} + \overbrace{d^{lt}\hat{T}_k^{ls-}(i_x \sin(\hat{\theta}_k^{lk}) - i_z \cos(\hat{\theta}_k^{lk}))}^{\text{thigh } z\text{-axis in shank frame}} \in \mathbb{R}^3 \tag{46}$$

$$\hat{\theta}_k^{lk} = \cos^{-1}\left(\frac{\alpha\gamma \pm \beta\sqrt{\alpha^2+\beta^2-\gamma^2}}{\alpha^2+\beta^2}\right) \text{ where } \begin{array}{l} \alpha = -2d^{lt}\psi_{pla}^T \hat{T}_k^{ls-} i_z, \quad \beta = 2d^{lt}\psi_{pla}^T \hat{T}_k^{ls-} i_x, \\[4pt] \gamma = (d_k^{pla})^2 - \psi_{pla}^T \psi_{pla} - (d^{lt})^2 \end{array} \tag{47}$$

To calculate for $\mathcal{H}_{pla,k}$, we first solved for $[\log(h_{pla}(\mathbf{X}_k))]^\vee$ at $\mathbf{X}_k = \hat{\boldsymbol{\mu}}_k^-$ similar to Equation (41).

$$[\log(h_{pla}(\hat{\boldsymbol{\mu}}_k^-))]^\vee = \tau_{pla}(\hat{\boldsymbol{\mu}}_k^-) = \mathbf{E}(\hat{T}_k^{p-}\,{}^p p^{mp} - \hat{T}_k^{ls-}\,{}^{ls}p^{la}) \tag{48}$$

Then solve for $[\log(h_{pla}(\hat{\boldsymbol{\mu}}_k^- \exp([\epsilon]_G^\wedge)))]^\vee$ similar to Equation (42) (i.e., distance between mid-pelvis and left ankle) giving us $[\log(h_{pla}(\hat{\boldsymbol{\mu}}_k^- \exp([\epsilon]_G^\wedge)))]^\vee = \tau_{pla}(\hat{\boldsymbol{\mu}}_k^-) + \boldsymbol{\Gamma}_{pla}$. $\mathcal{H}_{pla,k}$ is then calculated as shown in Equation (49). The right side of the pelvis-to-ankle distance measurement (i.e., $h_{pra}(\hat{\boldsymbol{\mu}}_k^-)$, $\mathbf{Z}_{pra}$, $\mathcal{H}_{pra,k}$) can be solved similarly to the left side.

$$
\begin{aligned}
\mathcal{H}_{pla,k} &= \tfrac{\partial}{\partial \epsilon}[\log(h_{pla}(\hat{\boldsymbol{\mu}}_k^-)^{-1}h_{pla}(\hat{\boldsymbol{\mu}}_k^- \exp([\epsilon]_G^\wedge)))]^\vee|_{\epsilon=0} \\
&= \tfrac{\partial}{\partial \epsilon}\left([\log(h_{pla}(\hat{\boldsymbol{\mu}}_k^- \exp([\epsilon]_G^\wedge)))]^\vee - [\log(h_{pla}(\hat{\boldsymbol{\mu}}_k^-))]^\vee\right)|_{\epsilon=0} \\
&= \tfrac{\partial}{\partial \epsilon}\left(\tau_{pla}(\hat{\boldsymbol{\mu}}_k^-) + \boldsymbol{\Gamma}_{pla} - \tau_{pla}(\hat{\boldsymbol{\mu}}_k^-)\right)|_{\epsilon=0} \\
&= \tfrac{\partial}{\partial \epsilon}\left(\boldsymbol{\Gamma}_{pla}\right)|_{\epsilon=0} = \tfrac{\partial}{\partial \epsilon}\left(\mathbf{E}(\hat{T}_k^{p-}[^p p^{mp}]^\odot \epsilon_{\mathbf{T}}^p - \hat{T}_k^{ls-}[^{ls}p^{la}]^\odot \epsilon_{\mathbf{T}}^{ls})\right)|_{\epsilon=0} \\
&= \left[\ \underbrace{\mathbf{E}\,\hat{T}_k^{p-}[^p p^{mp}]^\odot}_{1\times 6}\ \bigm|\ \underbrace{-\mathbf{E}\,\hat{T}_k^{ls-}[^{ls}p^{la}]^\odot}_{1\times 6}\ \bigm|\ \mathbf{0}_{1\times 6}\ \bigm|\ \mathbf{0}_{1\times 9}\ \right]
\end{aligned}
\tag{49}
$$

Covariance Limiter

Lastly, the error covariance of the position estimates of the three instrumented body segments must be prevented from growing unbounded and/or becoming badly conditioned, as will occur naturally when tracking global position of objects without any global position reference. At this step, a pseudo-measurement equal to the current state $\hat{\boldsymbol{\mu}}_k^+$ is used (implemented by Equation (50)) with some measurement noise of variance σ_{lim} (9×1 vector). The covariance $\mathbf{P}_k^+$ is then calculated through Equations (51)–(53).

$$
\mathcal{H}_{lim} =
\begin{bmatrix}
\overbrace{\mathbf{I}_{3\times3}\ \mathbf{0}_{3\times3}}^{\text{mp pos.}} & \overbrace{\phantom{\mathbf{I}_{3\times3}\ \mathbf{0}_{3\times3}}}^{\text{la pos.}} & \overbrace{\phantom{\mathbf{I}_{3\times3}\ \mathbf{0}_{3\times3}}}^{\text{ra pos.}} & \\
 & \mathbf{I}_{3\times3}\ \mathbf{0}_{3\times3} & & \mathbf{0}_{9\times9} \\
 & & \mathbf{I}_{3\times3}\ \mathbf{0}_{3\times3} &
\end{bmatrix}
\tag{50}
$$

$$\mathcal{H}_k' = [\mathcal{H}_k^T\ \ \mathcal{H}_{lim}^T]^T, \quad \mathcal{R}_k' = \mathrm{diag}(\sigma_k^2, \sigma_{lim}^2) \tag{51}$$

$$\mathbf{K}_k' = \mathbf{P}_k^- \mathcal{H}_k'^T(\mathcal{H}_k'\mathbf{P}_k^-\mathcal{H}_k'^T + \mathcal{R}')^{-1} \tag{52}$$

$$\mathbf{P}_k^+ = \Phi_G(\nu_k)(\mathbf{I} - \mathbf{K}_k'\mathcal{H}_k')\mathbf{P}_k^- \Phi_G(\nu_k)^T \tag{53}$$

3.2.3. Satisfying Biomechanical Constraints

After the preceding updates, the joint positions or angles may be beyond their allowed range (i.e., knee hyperflexion). The constraint update corrects the kinematic state estimates to satisfy the biomechanical constraints of the human body by projecting the current a posteriori state estimate $\hat{\boldsymbol{\mu}}_k^+$ onto the constraint surface, guided by our uncertainty in each state variable, which is encoded by $\mathbf{P}_k^+$. The following biomechanical constraint equations are enforced: (i) estimated thigh long axis vector lengths equal the thigh lengths; (ii) both knees act as hinge joints (formulation similar to Section 2.3, Equation (4) of [9]); and (iii) the knee joint angle is within realistic range. The constraint functions are similar to Section II-E.3 of [15] but expressed under $SE(3)$ state variables. The constrained state $\bar{\boldsymbol{\mu}}_k^+$ can be calculated using the equations below, similar to the measurement update of [36] with zero noise, where $C_k = [\mathcal{C}_{L,k}^T\ \ \mathcal{C}_{R,k}^T]^T$. $\mathcal{C}_{L,k}$ is the concatenation of $\mathcal{C}_{ltl,k}$, $\mathcal{C}_{lkh,k}$, and $\mathcal{C}_{lkr,k}$; the last matrix is not concatenated when the knee angle, α_{lk}, is within its allowed range (i.e., $\alpha_{lk,min} \leq \alpha_{lk} \leq \alpha_{lk,max}$). $\mathcal{C}_{ltl,k}$, $\mathcal{C}_{lkh,k}$, and $\mathcal{C}_{lkr,k}$ corresponds to the biomechanical constraint for the left thigh length (ltl), left knee hinged joint (*lkh*), and left knee angle ROM (*lkr*), respectively, which will be described more later. $\mathcal{C}_{R,k}$ can be derived similarly, while $\mathbf{D}_k$ and $c(\hat{\boldsymbol{\mu}}_k^+))$ are constructed similarly to $\mathbf{Z}_k$.

$$\tilde{\boldsymbol{\mu}}_k^+ = \hat{\boldsymbol{\mu}}_k^+ \exp([\boldsymbol{v}_k]_G^\wedge) \tag{54}$$

$$\boldsymbol{v}_k = \mathbf{K}_k([\log(c(\hat{\boldsymbol{\mu}}_k^+)^{-1}\mathbf{D}_k)]_{G_c}^\vee \tag{55}$$

$$\mathbf{K}_k = \mathbf{P}_k^+ \mathcal{C}_k^T (\mathcal{C}_k \mathbf{P}_k^+ \mathcal{C}_k^T)^{-1}) \tag{56}$$

$$\mathcal{C}_k = \tfrac{\partial}{\partial \epsilon}[\log(c(\hat{\boldsymbol{\mu}}_k^+)^{-1}c(\hat{\boldsymbol{\mu}}_k^+ \exp([\epsilon]_G^\wedge)))]_{G_c}^\vee|_{\epsilon=0} \tag{57}$$

Thigh Length Constraint

Firstly, the thigh length constraint is shown in Equation (59), where $\tau_z^{lt}(\mathbf{X}_k)$ (Equation (58)) denotes the thigh long axis vector and d^{lt} denotes the measured thigh length during calibration. $^p p^{lh}$ is the position of the left hip expressed in pelvis frame, and $^{ls} p^{lk}$ is the position of the left knee expressed in left shank frame. We have chosen that the left hip to be $\frac{d^p}{2}$ to the left of the mid-pelvis origin, and the left knee to be d^{ls} from the left shank origin (i.e., from the left ankle).

$$\tau_z^{lt}(\mathbf{X}_k) = \mathbf{E}(\ \overbrace{\mathbf{T}^p\,^p p^{lh}}^{\text{hip jt. pos. in } W} - \overbrace{\mathbf{T}^{ls}\,^{ls} p^{lk}}^{\text{knee jt. pos. in } W}\ ), \quad ^p p^{lh} = \begin{bmatrix} 0 & \tfrac{d^p}{2} & 0 & 1 \end{bmatrix}^T, \quad ^{ls} p^{lk} = \begin{bmatrix} 0 & 0 & d^{ls} & 1 \end{bmatrix}^T \tag{58}$$

$$[\log(c_{ltl}(\mathbf{X}_k))]^\vee = (\tau_z^{lt}(\mathbf{X}_k))^T \tau_z^{lt}(\mathbf{X}_k) \in \mathbb{R}, \quad [\log(\mathbf{D}_{ltl})]^\vee = (d^{lt})^2 \tag{59}$$

$\mathcal{C}_{ltl,k}$ is calculated using Equation (60).

$$\begin{aligned}\mathcal{C}_{ltl,k} &= \tfrac{\partial}{\partial \epsilon}[\log(c_{ltl}(\hat{\boldsymbol{\mu}}_k^+)^{-1}c_{ltl}(\hat{\boldsymbol{\mu}}_k^+ \exp([\epsilon]_G^\wedge)))]^\vee|_{\epsilon=0} \\ &= \tfrac{\partial}{\partial \epsilon}\left([\log(c_{ltl}(\hat{\boldsymbol{\mu}}_k^+ \exp([\epsilon]_G^\wedge)))]^\vee - [\log(c_{ltl}(\hat{\boldsymbol{\mu}}_k^+))]^\vee\right)|_{\epsilon=0}\end{aligned} \tag{60}$$

Following similar procedure to $\mathcal{H}_{lra,k}$, we obtain $\tau_z^{lt}(\hat{\boldsymbol{\mu}}_k^+ \exp([\epsilon]_G^\wedge)) = \tau_z^{lt}(\hat{\boldsymbol{\mu}}_k^+) + \Gamma_{ltz}$ (similar to Equation (42)), and $[\log(c_{ltl}(\hat{\boldsymbol{\mu}}_k^+ \exp([\epsilon]_G^\wedge)))]^\vee = [\log(c_{ltl}(\hat{\boldsymbol{\mu}}_k^+))]^\vee + 2(\tau_z^{lt}(\hat{\boldsymbol{\mu}}_k^+))^T \mathbf{E}(\ \hat{\mathbf{T}}_k^{p+}[^p p^{lh}]^\odot \epsilon_{\mathbf{T}}^p - \hat{\mathbf{T}}_k^{ls+}[^{ls} p^{lk}]^\odot \epsilon_{\mathbf{T}}^{ls})$ (similar to Equation (43)), which if we substitute in Equation (60) gives us Equation (61)

$$\begin{aligned}\mathcal{C}_{ltl,k} &= \tfrac{\partial}{\partial \epsilon}\left(2\,\tau_z^{lt}(\hat{\boldsymbol{\mu}}_k^+)^T \mathbf{E}(\hat{\mathbf{T}}_k^{p+}[^p p^{lh}]^\odot \epsilon_{\mathbf{T}}^p - \hat{\mathbf{T}}_k^{ls+}[^{ls} p^{lk}]^\odot \epsilon_{\mathbf{T}}^{ls})\right)|_{\epsilon=0} \\ &= \Big[\underbrace{2(\tau_z^{lt}(\hat{\boldsymbol{\mu}}_k^+))^T \mathbf{E}\,\hat{\mathbf{T}}_k^{p+}[^p p^{lh}]^\odot}_{1\times 6}\ \underbrace{-2(\tau_z^{lt}(\hat{\boldsymbol{\mu}}_k^+))^T \mathbf{E}\,\hat{\mathbf{T}}_k^{ls+}[^{ls} p^{lk}]^\odot}_{1\times 6}\ \mathbf{0}_{1\times 6}\ \mathbf{0}_{1\times 9}\Big]\end{aligned} \tag{61}$$

Hinge Knee Joint Constraint

Secondly, the hinge knee joint constraint as defined by Equation (62) is enforced by having the long (z) axis of the thigh to be perpendicular to the mediolateral axis (y) of the shank. For example, on the left leg, we would want r_y^{ls} be perpendicular to the thigh long axis vector $\tau_z^{lt}(\hat{\boldsymbol{\mu}}_k^+)$ (i.e., the dot product of r_y^{ls} and $\tau_z^{lt}(\hat{\boldsymbol{\mu}}_k^+)$ should be 0). Refer to Figure 2 for visualization. This formulation is similar to Section 2.3, Equation (4) of [9].

$$[\log(c_{lkh}(\mathbf{X}_k))]^\vee = (\mathbf{E}\,\mathbf{T}^{ls}i_y)^T \tau_z^{lt}(\mathbf{X}_k) = (r_y^{ls})^T \tau_z^{lt}(\mathbf{X}_k) \in \mathbb{R}, \quad [\log(\mathbf{D}_{lkh})]^\vee = 0 \tag{62}$$

Following similar procedure to $\mathcal{C}_{ltl,k}$ and taking $\mathbf{X}_k = \hat{\boldsymbol{\mu}}_k^+$, $[\log(c_{lkh}(\hat{\boldsymbol{\mu}}_k^+))]^\vee$ and $[\log(c_{lkh}(\hat{\boldsymbol{\mu}}_k^+ \exp([\epsilon]_G^\wedge)))]^\vee$ can be calculated as shown in Equations (63) and (64), respectively.

$$[\log(c_{lkh}(\hat{\boldsymbol{\mu}}_k^+))]^\vee = (\mathbf{E}\,\hat{\mathbf{T}}^{ls+}i_y)^T \tau_z^{lt}(\hat{\boldsymbol{\mu}}_k^+) \tag{63}$$

$$[\log(c_{lkh}(\hat{\pmb{\mu}}_k^+ \exp([\epsilon]_G^\wedge)))]^\vee = (\mathbf{E}\,\hat{\pmb{T}}^{ls+}\exp([\epsilon_{\mathbf{T}}^{ls}]^\wedge)\pmb{i}_y)^T(\tau_z^{lt}(\hat{\pmb{\mu}}_k^+) + \pmb{\Gamma}_{ltz})$$

Taking 1st order approximation of exp

$$\approx (\mathbf{E}(\hat{\pmb{T}}^{ls+} + \hat{\pmb{T}}^{ls+}[\epsilon_{\mathbf{T}}^{ls}]^\wedge)\pmb{i}_y)^T(\tau_z^{lt}(\hat{\pmb{\mu}}_k^+) + \pmb{\Gamma}_{ltz})$$

Assume 2nd order error ≈ 0

$$
\begin{aligned}
&= (\mathbf{E}\,\hat{\pmb{T}}^{ls+}\pmb{i}_y)^T\tau_z^{lt}(\hat{\pmb{\mu}}_k^+) + (\mathbf{E}\,\hat{\pmb{T}}^{ls+}\pmb{i}_y)^T\pmb{\Gamma}_{ltz}\\
&\quad + (\tau_z^{lt}(\hat{\pmb{\mu}}_k^+))^T\mathbf{E}\,\hat{\pmb{T}}^{ls+}[\epsilon_{\mathbf{T}}^{ls}]^\wedge\pmb{i}_y + \underbrace{(\mathbf{E}\,\hat{\pmb{T}}^{ls+}[\epsilon_{\mathbf{T}}^{ls}]^\wedge\pmb{i}_y)^T\pmb{\Gamma}_{ltz}}_{\approx\,0}\\
&= [\log(c_{lkh}(\hat{\pmb{\mu}}_k^+))]^\vee + (\mathbf{E}\,\hat{\pmb{T}}^{ls+}\pmb{i}_y)^T\mathbf{E}(\,\hat{\pmb{T}}_k^{p+}[^p\pmb{p}^{lh}]^\odot\epsilon_{\mathbf{T}}^p - \hat{\pmb{T}}_k^{ls+}[^{ls}\pmb{p}^{lk}]^\odot\epsilon_{\mathbf{T}}^{ls})\\
&\quad + (\tau_z^{lt}(\hat{\pmb{\mu}}_k^+))^T\mathbf{E}\,\hat{\pmb{T}}^{ls+}[\pmb{i}_y]^\odot\epsilon_{\mathbf{T}}^{ls}, \text{ by expanding }\pmb{\Gamma}_{ltz}\text{ and using Equation (11)}
\end{aligned}
$$

(64)

$C_{lkh,k}$ can be calculated using Equation (65).

$$
\begin{aligned}
C_{lkh,k} &= \frac{\partial}{\partial\epsilon}[\log(c_{lkh}(\hat{\pmb{\mu}}_k^+)^{-1}c_{lkh}(\hat{\pmb{\mu}}_k^+\exp([\epsilon]_G^\wedge)))]^\vee|_{\epsilon=0}\\
&= \frac{\partial}{\partial\epsilon}\left([\log(c_{lkh}(\hat{\pmb{\mu}}_k^+\exp([\epsilon]_G^\wedge)))]^\vee - [\log(c_{lkh}(\hat{\pmb{\mu}}_k^+))]^\vee\right)|_{\epsilon=0}
\end{aligned}
$$

(65)

Substituting Equations (63) and (64) into Equation (65) gives us Equation (66).

$$C_{lkh,k} = \left[\,\underbrace{(\mathbf{E}\,\hat{\pmb{T}}^{ls+}\pmb{i}_y)^T\mathbf{E}\,\hat{\pmb{T}}^{p+}[^p\pmb{p}^{lh}]^\odot}_{1\times 6}\,\Big|\,\underbrace{-(\mathbf{E}\,\hat{\pmb{T}}^{ls+}\pmb{i}_y)^T\mathbf{E}\,\hat{\pmb{T}}^{ls+}[^{ls}\pmb{p}^{lk}]^\odot + (\tau_z^{lt}(\hat{\pmb{\mu}}_k^+))^T\mathbf{E}\,\hat{\pmb{T}}^{ls+}[\pmb{i}_y]^\odot}_{1\times 6}\,\Big|\,\pmb{0}_{1\times 15}\,\right]$$

(66)

Knee Range of Motion Constraint

Thirdly, the knee ROM constraint is defined by Equation (69) and is only enforced if the knee angle, α_{lk}, is outside the allowed ROM. The bounded knee angle, α'_{lk}, is calculated by Equation (67). Equation (69) is obtained by expanding Equation (67) to Equation (68) which when rearranged gives us $[\log(c_{lkr}(\mathbf{X}_k))]^\vee$ (i.e., Lie group representation of Equation (26) in [15]). Note that $^{ls}\pmb{r}_z^{lt}$ is the normalized thigh long axis expressed in the left shank frame.

$$\alpha_{lk} = \tan^{-1}\left(\frac{-(\pmb{r}_z^{ls})^T\pmb{r}_z^{lt}}{-(\pmb{r}_x^{ls})^T\pmb{r}_z^{lt}}\right) + \frac{\pi}{2},\quad \alpha'_{lk} = \min(\alpha_{lk,max},\max(\alpha_{lk,min},\alpha_{lk}))$$

(67)

$$\frac{-\pmb{r}_z^{lt}\cdot\pmb{r}_z^{ls}}{-\pmb{r}_z^{lt}\cdot\pmb{r}_x^{ls}} = \frac{\sin(\alpha'_{lk}-\frac{\pi}{2})}{\cos(\alpha'_{lk}-\frac{\pi}{2})}$$

(68)

$$\overbrace{}^{^{ls}\pmb{r}_z^{lt}=\text{long axis of left thigh in shank frame}}$$

$$[\log(c_{lkr}(\mathbf{X}_k))]^\vee = (\mathbf{E}\,\pmb{T}^{ls}(\pmb{i}_z\cos(\alpha'_{lk}-\tfrac{\pi}{2}) - \pmb{i}_x\sin(\alpha'_{lk}-\tfrac{\pi}{2})))^T\tau_z^{lt}(\mathbf{X}_k) \in \mathbb{R},\quad [\log(\mathbf{D}_{lkr})]^\vee = 0$$

(69)

Following a similar procedure to $C_{lkh,k}$ (i.e., replace $\pmb{i}_y$ in Equation (64) with $^{ls}\pmb{r}_z^{lt}$) and taking $\mathbf{X}_k = \hat{\pmb{\mu}}_k^+$, $C_{lkr,k}$ can be calculated from $c_{lkr}(\hat{\pmb{\mu}}_k^+\exp([\epsilon]_G^\wedge)) = [\log(c_{lkr}(\hat{\pmb{\mu}}_k^+))]^\vee + (\mathbf{E}\,\hat{\pmb{T}}^{ls+\,ls}\pmb{r}_z^{lt})^T\mathbf{E}(\,\hat{\pmb{T}}_k^{p+}[^p\pmb{p}^{lh}]^\odot\epsilon_{\mathbf{T}}^p - \hat{\pmb{T}}_k^{ls+}[^{ls}\pmb{p}^{lk}]^\odot\epsilon_{\mathbf{T}}^{ls}) + (\tau_z^{lt}(\hat{\pmb{\mu}}_k^+))^T\mathbf{E}\,\hat{\pmb{T}}^{ls+}[^{ls}\pmb{r}_z^{lt}]^\odot\epsilon_{\mathbf{T}}^{ls}$, as shown in Equation (70).

$$
\begin{aligned}
C_{lkr,k} &= \frac{\partial}{\partial\epsilon}[\log(c_{lkr}(\hat{\pmb{\mu}}_k^+)^{-1}c_{lkr}(\hat{\pmb{\mu}}_k^+\exp([\epsilon]_G^\wedge)))]^\vee|_{\epsilon=0}\\
&= \frac{\partial}{\partial\epsilon}\left([\log(c_{lkr}(\hat{\pmb{\mu}}_k^+\exp([\epsilon]_G^\wedge)))]^\vee - [\log(c_{lkr}(\hat{\pmb{\mu}}_k^+))]^\vee\right)|_{\epsilon=0}\\
&= \left[\,\underbrace{(\mathbf{E}\,\hat{\pmb{T}}^{ls+\,ls}\pmb{r}_z^{lt})^T\mathbf{E}\,\hat{\pmb{T}}^{p+}[^p\pmb{p}^{lh}]^\odot}_{1\times 6}\,\Big|\,\underbrace{-(\mathbf{E}\,\hat{\pmb{T}}^{ls+\,ls}\pmb{r}_z^{lt})^T\mathbf{E}\,\hat{\pmb{T}}^{ls+}[^{ls}\pmb{p}^{lk}]^\odot + (\tau_z^{lt}(\hat{\pmb{\mu}}_k^+))^T\mathbf{E}\,\hat{\pmb{T}}^{ls+}[^{ls}\pmb{r}_z^{lt}]^\odot}_{1\times 6}\,\Big|\,\pmb{0}_{1\times 15}\,\right]
\end{aligned}
$$

(70)

3.3. Post-Processing

The orientation of the pelvis and shanks are obtained from the state $\tilde{\pmb{\mu}}_k^+$. The orientation of the left thigh, $\tilde{\pmb{R}}^{lt+}$, can be calculated using $\tilde{\pmb{R}}^{lt+} = [\,\tilde{\pmb{r}}_y^{ls+}\times\tilde{\pmb{r}}_z^{lt+}\,|\,\tilde{\pmb{r}}_y^{ls+}\,|\,\tilde{\pmb{r}}_z^{lt+}\,] = [\,[\mathbf{E}\,\tilde{\pmb{T}}_k^{ls+}\pmb{i}_y]_{SO(3)}^\wedge\tilde{\pmb{r}}_z^{lt+}\,|\,(\mathbf{E}\,\tilde{\pmb{T}}_k^{ls+}\pmb{i}_y)\,|\,\tilde{\pmb{r}}_z^{lt+}\,]$, where $\tilde{\pmb{r}}_z^{lt+} = \tau_z^{lt}(\tilde{\pmb{\mu}}_k^+)/\|\tau_z^{lt}(\tilde{\pmb{\mu}}_k^+)\|$. The orientation of the right thigh, $\tilde{\pmb{R}}^{rt+}$, is calculated similarly.

4. Experiment

An extension of the dataset from [15] was used to evaluate our L5S based algorithms. The movements involved are listed in Table 1 (note the addition of dynamic movements), and were collected from from nine healthy subjects (7 men and 2 women, weight 63.0 ± 6.8 kg, height 1.70 ± 0.06 m, age 24.6 ± 3.9 years old), with no known gait abnormalities. Raw data were captured using a commercial IMC (i.e., Xsens Awinda) at 100 Hz sampling rate with IMUs attached to the pelvis and ankles, compared against a benchmark OMC (i.e., the setup followed Vicon Plug-in Gait protocol in a $\sim 4 \times 4$ m^2 capture area). The experiment was approved by the Human Research Ethics Board of the University of New South Wales (UNSW) with approval number HC180413.

Table 1. Types of movements done in the validation experiment.

Movement	Description	Duration	Group
Walk	Walk straight and return	$\sim$30 s	F
Figure-of-eight	Walk along figure-of-eight path	$\sim$60 s	F
Zig-zag	Walk along zig-zag path	$\sim$60 s	F
5-minute walk	Unscripted walk and stand	$\sim$300 s	F
Speedskater	Speedskater on the spot	$\sim$30 s	D
TUG	Timed up-and-go test	$\sim$30 s	D
Jog	Jog straight and return	$\sim$30 s	D
Jumping jacks	Jumping jacks on the spot	$\sim$30 s	D
High-knee jog	High-knee jog on the spot	$\sim$30 s	D

F denotes free walk, D denotes dynamic movements.

Frame alignment and yaw offset calibrations are similar to Section III-B of [15]. The experiments and algorithm were implemented using Matlab 2020a, with initial state $\tilde{\mu}_0^+$ (i.e., position, orientation, and velocity) obtained from the OMC system (i.e., Vicon) and initial error covariance $\mathbf{P}_0^+$ set to $0.5\mathbf{I}_{27 \times 27}$. The variance parameters for the process and measurement error covariance matrix $\mathcal{Q}$ and $\mathcal{R}$ are shown in Table 2.

Table 2. Parameters for error covariance matrices, $\mathcal{Q}$ and $\mathcal{R}$.

$\mathcal{Q}$ Parameters				$\mathcal{R}$ Parameters			
σ_a^2 $(\mathrm{m}^2 \cdot \mathrm{s}^{-4})$	σ_ω^2 $(\mathrm{rad}^2 \cdot \mathrm{s}^{-2})$	σ_{ori}^2 (rad^2)	σ_{mp}^2 (m^2)	σ_{ls}^2 and σ_{rs}^2 $(\mathrm{m}^2 \cdot \mathrm{s}^{-2}$ and m$^2)$	σ_{dl}^2 and σ_{dr}^2 (m^2)	σ_{da}^2 (m^2)	σ_{lim}^2 (m^2)
$10^2\mathbf{1}_9$	$10^3\mathbf{1}_9$	$1\mathbf{1}_9$	0.1	$[0.01\mathbf{1}_3 \quad 10^{-4}]$	10	1	101_{18}

where $\mathbf{1}_n$ is an $1 \times n$ row vector with all elements equal to 1.

The inter-IMU distance measurements, $\breve{d}^{pla}$, $\breve{d}^{pra}$, and $\breve{d}^{lra}$, were simulated by calculating the distance from the mid-pelvis to the left and right ankles and adding normally distributed positional noise with different standard deviations (i.e., $\sigma_{dist} \in \{0, 0.01, \ldots, 0.1, 0.15, 0.2\}$ m). Each trial was simulated five times.

Lastly, the evaluation was done using the following metrics: (1) Mean position and orientation root-mean-square error (RMSE) (e.g., similar to [15,17] as shown in Equations (71) and (72)), where p_k^b and R_k^b are obtained from the benchmark OMC system, $\tilde{p}_k^{b+}$ and $\tilde{R}_k^{b+}$ are obtained from the algorithm. Note that as the global position of the estimate is still prone to drift due to the absence of an external global position reference, the root position of our system was set equal to that of the benchmark system (i.e., the mid-pelvis is placed at the origin in the world frame for all RMSE calculations). (2) Joint angles RMSE with bias removed (i.e., the mean difference between the angles over each entire trial was subtracted) and correlation coefficient (CC) of the hip in the sagittal (Y), frontal (X), and transverse (Z) planes and of the knee in the sagittal (Y) plane. Note that these joint angles are commonly used parameters in gait analysis. (3) Spatiotemporal gait parameters (e.g., total travelled distance (TTD) deviation, average stride length, and gait speed of the foot). Refer to Section III of [15] for more details.

$$e_{pos,k} = \frac{1}{N_{pos}} \sum_{b \in \mathbb{DP}} || \boldsymbol{p}_k^b - \tilde{\boldsymbol{p}}_k^{b+} ||, \quad N_{pos} = 6, \quad \mathbb{DP} = \{lh, rh, lk, rk, la, ra\} \tag{71}$$

$$e_{ori,k} = \frac{1}{N_{ori}} \sum_{b \in \mathbb{DO}} ||[\log(\boldsymbol{R}_k^b(\tilde{\boldsymbol{R}}_k^{b+})^T)]^\vee ||, \quad N_{ori} = 2, \quad \mathbb{DO} = \{lt, rt\} \tag{72}$$

5. Results

5.1. Mean Position and Orientation RMSE, Joint Angle RMSE and CC

In this experiment, multiple variations of the algorithm were tested as shown in Table 3. Firstly, L5S-3IMU is the algorithm described in this paper (Section 3) with parameters listed in Table 2. The parameter for L5S-3IMU were selected by taking the best joint CC (i.e., mean of free walk and dynamic movements) from a grid search of parameters $\sigma_\omega^2 = \{1, 10, 10^2, 10^3\}$ rad^2/s^2 and $\sigma_{ori}^2 = \{10^{-2}, 10^{-1}, 1, 10\}$ rad^2. Secondly, CKF-3IMU and CKF-3IMU+D were the algorithms described in [15,31], respectively. Thirdly, CKF-3I-KB is a modified CKF-3IMU using similar parameters, measurement, and constraint functions as L5S-3IMU. The key difference between CKF-3IMU and CKF-3I-KB is that CKF-3I-KB allows knee bending, denoted by the suffix KB, during the constraint update. Fourthly, L5S-3I-NO is a variation of L5S-3IMU with $\sigma_\omega^2 = 10^7$ rad^2/s^2, $\sigma_{ori}^2 = 10^{-1}$ rad^2, and $^b\breve{\omega}_k = 0$ rad. The parameters were chosen to have high uncertainty on the tracked orientation (i.e., effectively not using the orientation measurements at all), leading to a variation of L5S-3IMU that is similar to our prior work CKF-3IMU which assumed orientation measurements were noise-free. Lastly, the black box output (i.e., pelvis, thigh, and shank orientations) from the MVN Studio software (denoted as OSPS), which illustrates the performance of a widely-accepted commercial wearable IMC system with an OSPS configuration. For the first to fourth variations, the +D suffix means simulated inter-IMU distance measurements ($\sigma_{dist} = 0.1$ m) was used instead of the pelvis height assumption.

Table 3. The experiment was tested on the following algorithm variations.

Algorithm	Inter-IMU Distance	Summary Description
L5S-3IMU	N	Tracks position and orientation as described in Section 3 with
L5S-3IMU+D	Y	parameters listed in Table 2.
CKF-3IMU [15]	N	
CKF-3IMU+D [31]	Y	Only tracks position using a constrained KF.
CKF-3I-KB	N	Modified CKF-3IMU using similar parameters as L5S-3IMU (Table 2).
CKF-3I-KB+D	Y	This also allows knee bending during the constraint update.
L5S-3I-NO	N	L5S-3IMU with parameters that assume noise-free orientation (NO)
L5S-3I-NO+D	Y	measurements like CKF-3IMU.
OSPS	N	Output from a commercial OSPS wearable IMC system.

Figure 4 shows the mean position and orientation RMSE, mean knee Y and hip joint angle RMSE (bias removed) and CC of different variations of CKF-3IMU and L5S-3IMU for both free walking and dynamic motions. Y, X, and Z refers to the sagittal, frontal, and transverse planes, respectively. CKF-3IMU performed well with free walking ($e_{pos} = 4.27$ cm, $e_{ori} = 15.85°$, CC $= 0.66$) [15]. However, a more extensive evaluation showed that it performed poorly for certain dynamic movements (e.g., high-knee jog with $e_{pos} = 18.15$ cm, $e_{ori} = 24.87°$, CC $= 0.02$). Removing the no-knee-bending assumption during the constraint update fixed this issue, as shown by the performance of CKF-3I-KB (e.g., high-knee jog improved by $\sim$9 cm e_{pos}, $\sim$9° e_{ori}, $\sim$0.4 CC). L5S-3I-NO which is the L5S version of CKF-3IMU expectedly have similar performance with CKF-3I-KB (i.e., $\Delta e_{pos} < 0.5$ cm, $\Delta e_{ori} < 1°$, and ΔCC 0.02 differences). L5S-3IMU, which tracked both position and orientation while assuming there is noise in the orientation measurements, had a slightly better performance (e.g., improved jumping jacks and high-knee jog by $\sim$0.1 CC, $<$0.03 CC difference with other movement types). The use of simulated distance measurement with σ_{dist} = 0.1 m on CKF-3I-KB, L5S-3I-NO, and L5S-3IMU had slight effects for free walking, and a significant improvement for dynamic movements. For free walking, joint angle RMSE and CC of L5S-3IMU+D compared to L5S-3IMU improved by $\sim$1° and $<$0.01 CC, while e_{pos} and e_{ori} slightly disimproved ($<$0.5 cm and $<$1 °). The similar results suggest that inferring pelvis position from simulated distance measurement (σ_{dist} = 0.1 m) is comparable to our pelvis height assumption at least for free walking. For dynamic movements, the e_{pos}, e_{ori}, joint angle RMSE, and CC of L5S-3IMU+D improved by 2–16 cm, 0–40 °, 1–9 °, and $<$0.42, respectively—more significantly for movements TUG and high-knee jog.

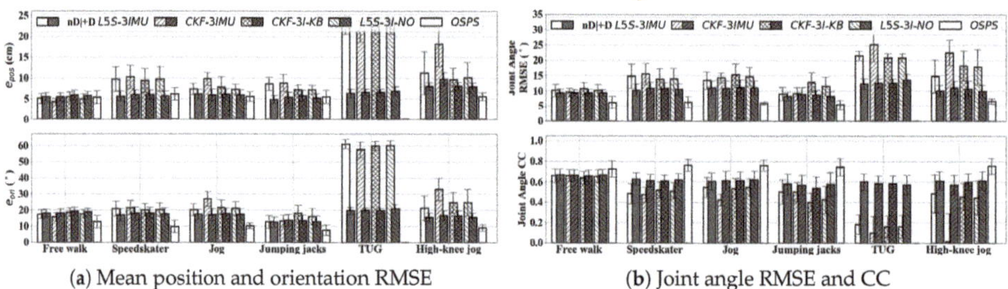

Figure 4. The performance of CKF, L5S, and OSPS with and without using inter-IMU distance measurements at each motion type.

To give insight on how the accuracy of the simulated inter-IMU distance measurements affect pose estimation performance, Figure 5 shows the mean of knee Y and hip joint angle RMSE and CC at different σ_{dist} values. At σ_{dist} = 0.1 m, the simulation showed comparable performance between L5S-3IMU, which implements pelvis height assumption, and L5S-3IMU+D, which implements inter-IMU distance measurement to supplement the pelvis position estimate, for free walking. Significant improvement for dynamic movements can be seen even for σ_{dist} = 0.2 m. These results suggest that the actual distance measurement sensor must have noise standard deviation $\sigma_{dist} \leq 0.1$ m to improve pose estimate performance. Note that the +D variation in Figure 4 and in the experiments that follow were evaluated at σ_{dist} = 0.1 m.

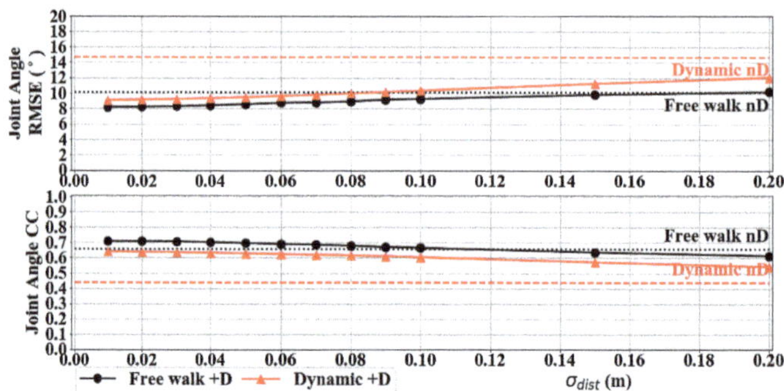

Figure 5. Joint angle RMSE (top) and CC (bottom) of free walk and dynamic movements at different noise level σ_{dist}. The broken lines represent L5S-3IMU results (denoted as nD) where inter-IMU distance measurements were not used. The solid lines represent L5S-3IMU+D results (denoted as +D) where we can observe slight and great improvements for free walk and dynamic movements, respectively.

5.2. Hip and Knee Joint Angle RMSE and CC

Figure 6 shows the knee and hip joint angle RMSE (bias removed) and CC of L5S-3IMU and L5S-3IMU+D compared against the OMC output. Y, X, and Z refers to the sagittal, frontal, and transverse planes, respectively. Turning movements and half steps were manually removed from the per-step result of Walk movement and was denoted as Straight Walk. Note that sensor-to-body calibration was only done at the beginning of trial, not for each step. Between L5S-3IMU and L5S-3IMU+D, there was minimal hip and knee joint angle RMSE and CC improvement for free walking ($\sim 1°$ RMSE and ~ 0.03 CC difference). However, there was significant improvement for most dynamic movements, specifically, speed-skater, jog, high-knee jog, and TUG (e.g., $4°$–$17°$ knee Y and hip Y joint angle RMSE improvements). Furthermore, the CC for dynamic movements started to reach similar performance with the free walk movement, indicating that inter-IMU distance measurements have indeed made the pose estimator capable of tracking more ADLs and not just walking.

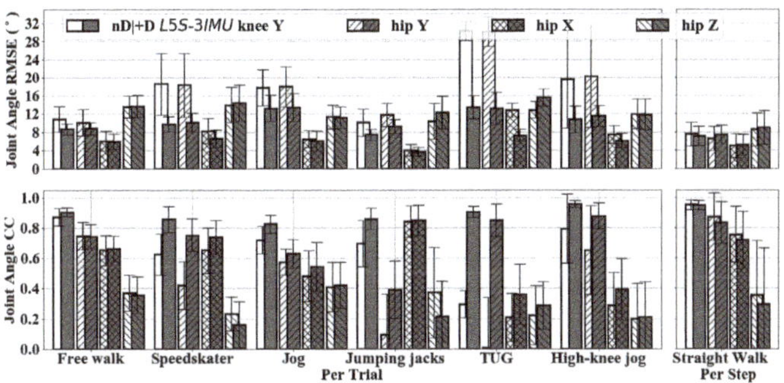

Figure 6. The CC of knee (Y) and hip (Y, X, Z) joint angles for L5S-3IMU (denoted as nD) and L5S-3IMU+D (denoted as +D) at each motion type.

Figure 7 shows a sample walk trial. At the peaks of knee Y angle, the distance between the pelvis and ankle positions of L5S-3IMU+D were a few cm shorter (i.e., pelvis position was lower than actual while ankle position was higher) than the actual distance resulting in higher knee Y angle peaks. Violations of our biomechanical constraints are also apparent at $t = 4$ to 5.5 s, where the subject makes a $180°$ turn. After the turn, L5S-3IMU and L5S-3IMU+D were able to recover during the straight walking ($t = 5.5$ to 9.74 s of Figure 7). Notice that the bias between OSPS and OMC can be observed at $t = 0$ of the hip Y joint angle.

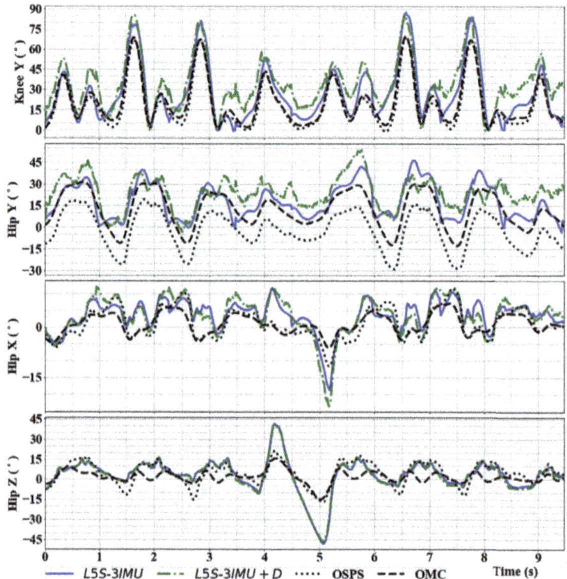

Figure 7. Knee (Y) and hip (Y, X, Z) joint angle output of L5S-3IMU in comparison with the benchmark system (Vicon) for a Walk trial. The subject walked straight from $t = 0$ to 3 s, turned 180° around from $t = 3$ to 5.5 s, and walked straight to the original starting point from 5.5 s until the end.

5.3. Spatiotemporal Gait Parameters

Table 4 shows the TTD, stride length, and gait speed accuracy computed from the global ankle position estimate of L5S-3IMU, L5S-3IMU+D, and the OMC system for free walk, jogging, and TUG. The use of inter-IMU distance measurements ($\sigma_{dist} = 0.1$ m) helped improve the TTD, stride length, and gait speed accuracy of free walk and TUG (e.g., TTD improved from ~9% to ~5%). Refer to the code repository for links to videos of sample trials.

Table 4. Total travelled distance (TTD) deviation from optical motion capture (OMC) system at the ankles.

Algo.	Side	TTD Error	Stride Length (cm) Actual		Stride Length (cm) Error		Gait Speed (cm.s⁻¹) Actual		Gait Speed (cm.s⁻¹) Error	
		% dev	μ	med	$\mu \pm \sigma$	RMS	μ	med	$\mu \pm \sigma$	RMS
Freewalk	L	8.97%	91	99	-8.1 ± 6	9.9	70	74	-6.0 ± 5	7.7
L5S-3IMU	R	9.00%	93	99	-8.3 ± 6	10.3	71	75	-6.2 ± 5	8.2
Freewalk	L	5.23%	91	99	-4.7 ± 7	8.3	70	74	-3.6 ± 6	6.6
L5S-3IMU+D	R	5.85%	93	99	-5.4 ± 8	9.4	71	75	-4.1 ± 6	7.4
Jog	L	21.35%	81	86	-17.4 ± 23	28.5	107	118	-19.2 ± 33	38.0
L5S-3IMU	R	26.79%	85	97	-22.9 ± 25	33.8	111	124	-26.4 ± 34	43.1
Jog	L	22.40%	81	86	-18.2 ± 22	28.4	107	118	-21.6 ± 30	37.0
L5S-3IMU+D	R	26.70%	85	97	-22.8 ± 24	32.8	111	124	-27.5 ± 31	41.4
TUG	L	18.20%	74	76	-13.5 ± 18	22.1	58	60	-10.0 ± 15	18.0
L5S-3IMU	R	20.98%	79	90	-16.6 ± 15	22.5	63	67	-13.1 ± 13	18.4
ine TUG	L	3.80%	74	76	-2.8 ± 6	6.7	58	60	-2.3 ± 5	5.9
L5S-3IMU+D	R	4.22%	79	90	-3.3 ± 6	6.8	63	67	-2.7 ± 5	5.6

where μ and σ denote mean and standard deviation. Error denotes estimate minus actual value, while TTD % dev denotes abs(*error*)/actual TTD.

6. Discussion

In this paper, a Lie group EKF algorithm for lower body pose estimation using only three IMUs, ergonomically placed on the ankles and sacrum to facilitate continuous recording outside the laboratory, was described and evaluated. The algorithm utilizes fewer sensors than other approaches reported in the literature, at the cost of reduced accuracy.

6.1. Mean Position and Orientation RMSE

The mean position and orientation RMSE of L5S-3IMU, L5S-3IMU+D, and related literature (sparse orientation poser (SOP) and sparse inertial poser (SIP) [17]) are listed in Table 5. SOP used orientation measured by IMUs and biomechanical constraints, while SIP used similar information but with the addition of acceleration. Both SOP and SIP were benchmarked against an OSPS system tracking the full body while our algorithm was benchmarked against an OMC system tracking only the lower body. The e_{pos} and e_{ori} (no bias) performance of L5S-3IMU and compared to SOP for free walking and jogging were comparable (Δe_{pos} 0.1–0.5 cm and Δe_{ori} 2.5°–3° differences). The e_{pos} and e_{ori} (no bias) of SIP was better than L5S-3IMU and L5S-3IMU+D for free walking ($\sim$2.1–2.5 cm and 6.5°–7° difference). Although this improvement was expected, as SIP optimizes the pose over multiple frames whereas our algorithm, like CKF-3IMU, optimizes the pose for each individual frame. For jumping jacks, the e_{ori} of L5S-3IMU and L5S-3IMU+D was significantly ($\sim$4°–8°) better than SOP's and SIP's. However, this difference is probably because both SOP and SIP were evaluated on the full body (our algorithm was only evaluated on the lower body) and errors in arm pose estimation may have increased e_{ori} for the SOP and SIP algorithms.

Table 5. Mean position and orientation RMSE of L5S-3IMU, L5S-3IMU+D, OSPS, sparse orientation power (SOP) and sparse inertial poser (SIP) [17].

	e_{pos} (cm)			e_{ori}, No Bias (cm)		
	Free Walk	Jog	Jumping Jacks	Free Walk	Jog	Jumping Jacks
L5S-3I	5.1 ± 1.2	7.3 ± 1.4	8.7 ± 1.6	$17.5 \pm 2.7°$	$20.2 \pm 3.8°$	$12.8 \pm 4.0°$
L5S-3I+D	5.5 ± 1.0	6.2 ± 1.1	4.9 ± 0.9	$18.0 \pm 2.5°$	$17.4 \pm 3.2°$	$12.6 \pm 3.2°$
OSPS	5.4 ± 1.5	5.6 ± 1.2	5.5 ± 1.6	$12.9 \pm 4.0°$	$10.3 \pm 1.8°$	$7.6 \pm 3.3°$
SOP [17]	$\sim$5.0	$\sim$8.0	$\sim$8.0	$\sim$15.0°	$\sim$22.0°	$\sim$20.0°
SIP [17]	$\sim$3.0	$\sim$5.0	$\sim$4.0	$\sim$11.0°	$\sim$16.0°	$\sim$16.0°

Comparing processing times, L5S-3IMU and L5S-3IMU+D were slower than CKF-3IMU, but can still be used in real-time; specifically, CKF-3IMU, L5S-3IMU, and L5S-3IMU+D processed a 1000-frame sequence (i.e., 10 s long) in $\sim$0.7, $\sim$2, $\sim$3.5 s, respectively, on an Intel Core i5-6500 3.2 GHz CPU [15], while SIP [17] took 7.5 min on a quad-core Intel Core i7 3.5 GHz CPU. All set-ups used single-core non-optimized Matlab code. Albeit slower than CKF-3IMU, L5S-3IMU and L5S-3IMU+D could also be used to provide real-time gait parameter measurement to inform actuation of assistive or rehabilitation robotic devices.

6.2. Hip and Knee Joint Angle RMSE and CC

The knee and hip joint angle RMSEs (no bias) and CCs of L5S-3IMU, L5S-3IMU+D, OSPS and related literature for straight walking (i.e., per step evaluation) are shown in Table 6 [7,15,37,38]. Similar to IMC based systems, L5S-3IMU and L5S-3IMU+D also follows the trend of having sagittal (Y axis) joint angles similar to that captured by OMC systems (0.95 knee Y and >0.83 hip Y CCs), but with significant difference in frontal and transverse (X and Z axis) joint angles [15,37]. CKF-3IMU performed slightly better (e.g., 0.03 knee Y, 0.09 hip Y CC), which is expected as the biomechanical constraint (i.e., no-knee-bending) assumption of CKF-3IMU was designed specifically for walking, at the cost of being less accurate for other dynamic movements. Both L5S-3IMU and L5S-3IMU+D were comparable, and at times even better (within 2.5° RMSE, 0.1 CC difference) than the results of Hu et al. and Tadano et al., indicating excellent per-step reconstruction in the sagittal plane [7,38]. Hu et al. used 4 IMUs (two at the pelvis and one on each foot) and Tadano et al. used an OSPS configuration. Both systems can only estimate the pose in the sagittal plane.

Despite the promising performance when using inter-IMU distance measurements, further validation with actual hardware implementation is needed, as the sensor noise in the real world may not necessarily follow a normal distribution and may be non-stationary.

Table 6. Knee and hip angle RMSE no bias (top) and CC (bottom) of CKF-3IMU, OSPS, and related literature for free walk.

Joint Angle RMSE (°)	Knee Sagittal	Hip Sagittal	Hip Frontal	Hip Transverse
L5S-3IMU	7.6 ± 2.6	6.6 ± 2.7	5.0 ± 2.6	8.6 ± 3.6
L5S-3IMU+D	7.1 ± 2.1	7.5 ± 2.1	5.1 ± 2.3	8.9 ± 3.7
OSPS	5.0 ± 1.8	3.6 ± 1.7	4.1 ± 2.2	11.9 ± 4.3
CKF-3IMU [15]	5.7 ± 2.2	4.4 ± 1.9	5.5 ± 2.6	9.0 ± 3.8
Cloete et al. [37]	8.5 ± 5.0	5.8 ± 3.8	7.3 ± 5.2	7.9 ± 4.9
Hu et al. [38]	4.9 ± 3.5	6.8 ± 3.0	-	-
Tadano et al. [7]	10.1 ± 1.0	7.9 ± 1.0	-	-
Joint Angle CC	**Knee Sagittal**	**Hip Sagittal**	**Hip Frontal**	**Hip Transverse**
L5S-3IMU	0.95 ± 0.03	0.87 ± 0.16	0.76 ± 0.18	0.36 ± 0.36
L5S-3IMU+D	0.95 ± 0.03	0.83 ± 0.14	0.72 ± 0.19	0.29 ± 0.37
OSPS	0.97 ± 0.04	0.95 ± 0.06	0.72 ± 0.19	0.26 ± 0.20
CKF-3IMU [15]	0.98 ± 0.03	0.96 ± 0.08	0.73 ± 0.17	0.26 ± 0.39
Cloete et al. [37]	0.89 ± 0.15	0.94 ± 0.08	0.55 ± 0.40	0.54 ± 0.20
Hu et al. [38]	0.95 ± 0.04	0.97 ± 0.04	-	-
Tadano et al. [7]	0.97 ± 0.02	0.98 ± 0.01	-	-

For reference, portable ultrasound-based distance measurement can achieve millimetre accuracy with a sampling rate of 125 Hz [30], while a commercial UWB-based distance measurement devices can achieve $\sim$10 cm accuracy with a sampling rate of 200 Hz [39,40].

Lastly, despite L5S-3IMU and L5S-3IMU+D achieving 0.95 joint angle CCs in the sagittal plane, the unbiased joint angle RMSE ($>5°$) makes its utility in clinical applications uncertain [41]. Although the algorithm is expected to work on pathological gait where our biomechanical assumptions are satisfied, overall performance still needs more improvement. To achieve clinical utility, one may either use more accurate sensors or average out cycle-to-cycle variation in estimation errors over many gait cycles; for example, use a more accurate distance measurement sensor ($\sigma_{dist} < 0.1$ m). Furthermore, the accuracy must also be validated on a larger and more diverse cohort to quantify its ultimate clinical utility. The evaluation of how these solutions can bridge the gap to clinical application for the proposed system will be part of future work.

6.3. Spatiotemporal Gait Parameters

The focus of the proposed algorithms, L5S-3IMU and L5S-3IMU+D, are to estimate joint kinematics. However, as L5S-3IMU and L5S-3IMU+D both track the global position of the ankles, it is also capable of calculating spatiotemporal gait parameters (performance listed in Table 4). The TTD deviation of our algorithms compared against the gold standard OMC were not as good as CKF-3IMU [15] (3.6–3.81% TTD deviation) or other state-of-the-art dead reckoning algorithms [42,43] (0.2–1.5% TTD deviation). Two possible sources of inaccuracy lies (1) in the dead reckoning approximation done in the prediction step, and (2) in the assumption that the velocity of the shank IMU is zero when the associated foot touches the floor, but of course this IMU continues to move with some small velocity on the lower shank during the stance phase. To illustrate the dead reckoning approximation, let us look at the predicted pelvis pose in Equation (73). In our algorithm, we assumed $\psi_p \approx \mathbf{I}_{3\times3}$ (note that $\Phi(-\Delta t\,^P\breve{\omega}_k) \approx \mathbf{I}_{3\times3}$ and $\tilde{R}^{p+}_{k-1}(\breve{R}^p_k)^T \approx \mathbf{I}_{3\times3}$ since $\Delta t\,^P\breve{\omega}_k$ is small) which did not significantly affect the joint kinematic estimate, but slightly affected the global position estimate. Nevertheless, body drift has been reduced substantially compared to Marcard et al.'s SIP [17].

$$
\hat{T}^{p-}_k = \tilde{T}^{p+}_{k-1}\exp\left(\left[\begin{matrix}(\breve{R}^p_k)^T(\Delta t\,\tilde{v}^{mp+}_{k-1} + \frac{\Delta t^2}{2}\breve{a}^p_k)\\ \Delta t\,^P\breve{\omega}_k\end{matrix}\right]^\wedge\right)
$$
$$
= \left[\begin{matrix}\tilde{R}^{p+}_{k-1}\exp([\Delta t\,^P\breve{\omega}_k]^\wedge)\ \tilde{p}^{mp+}_{k-1} + \overbrace{\tilde{R}^{p+}_{k-1}\Phi(-\Delta t\,^P\breve{\omega}_k)(\breve{R}^p_k)^T}^{\psi_p \approx \mathbf{I}_{3\times3}}(\Delta t\,\tilde{v}^{mp+}_{k-1} + \frac{\Delta t^2}{2}\breve{a}^p_k)\\ \mathbf{0}_{1\times3} \qquad\qquad 1\end{matrix}\right]
$$

$$(73)$$

6.4. Limitations and Future Work

L5S has similar pelvis drift, covariance matrix numerical issue, and flat floor limitation as CKF-3IMU, which is expected as L5S implements the same measurement and constraint update as CKF-3IMU, albeit formulated using Lie group representation instead of vectors and quaternions [15]. The pelvis height and flat floor assumption helps prevent the pelvis and the ankles from drifting towards each other (i.e., pelvis drift downward while ankles drift upward). However, it will also prevent accurate pose estimation of motions such as sitting, lying down, or standing on one leg, where the pose is maintained for a duration much longer than that of a typical gait cycle. The covariance limiter (Section 3.2.2) helps prevent the covariance becoming badly conditioned (i.e., singular), especially for longer duration trials (e.g., 5-minute walk) where the position uncertainty grows at a faster rate for the pelvis position than the ankle position. As can be observed from Figure 6, substituting the pelvis height assumption with inter-IMU distance measurements can increase the algorithm's accuracy especially for tracking dynamic movements. If the distance measurement is accurate enough (i.e., smaller σ^2_{dist}), the inter-IMU distance measurement update may be enough to limit the growth of pelvis position uncertainly and possibly making the covariance limiter not needed.

Figure 6 shows that the optimized performance of L5S-3IMU, even if it allows the tracked orientation to be corrected by inter-IMU distance measurements and the tracked position estimate, was only slightly better than CKF-3IMU/L5S-3I-NO, which effectively assumed the measurement input from the orientation estimation algorithm to be perfect (i.e., trusted the tracked orientation less). As L5S-3IMU requires more computing resources, such result suggests that CKF-3IMU may be more suitable to use when computing power is limited. To fully leverage the advantages brought by the Lie group representation, additional sensor measurements that can help correct tracked orientation will be needed (e.g., estimating angle of arrival between two sensors [44] or using fish eye cameras to improve pose estimate [45]).

Additional sensor measurements provide new opportunities for automatic calibration even under RSC configuration. IMC systems typically need anthropometric measurements (i.e., measurement of body segments such as d^{ls}) beforehand. By taking the initial distance measurement at some predetermined posture, anthropometric measurements can be automatically inferred. The formulation for a hinge joint with two IMUs on both sides has been leveraged to enable automatic sensor-to-segment calibration (i.e., align sensor frame to body frame) and even a completely magnetometer free orientation estimation [46,47]. Magnetometer free orientation estimation rids us of the yaw offset issue from an inhomogeneous magnetic field in indoor environments, typically with stronger disturbances closer to the floor [48]. An approach using a hinge joint with two IMUs may not be applicable to RSC configurations (e.g., our algorithm only has one IMU on one side of the hinge joint). However, distance measurements may be use to compensate for the missing IMU information from the uninstrumented segment, and a modified version may be developed for a RSC configuration.

Enabling longer-term tracking of ADL in the subject's natural environment may lead to novel investigations of movement disorder progression and the identification of early intervention opportunities. This work is just one of the early steps towards seamless remote gait monitoring.

Developing solutions to further increase accuracy, increase the number of body segments tracked (e.g., track full body under RSC [17]), or use even fewer IMUs (tracking lower body using two IMUs [49]) will be investigated in the future.

7. Conclusions

This paper presented a Lie group CEKF-based algorithm (L5S-3IMU) to estimate lower limb kinematics using a RSC configuration of IMUs, supplemented by inter-IMU distance measurements in one implementation. The knee and hip joint angle RMSEs in the sagittal plane for straight walking were $7.6° \pm 2.6°$ and $6.6° \pm 2.7°$, respectively, while the CCs were 0.95 ± 0.03 and 0.87 ± 0.16, respectively. We also showed that inter-IMU distance measurement is a promising new source of information to improve the pose estimation of IMC under a RSC configuration. Simulations show that performance improved dramatically for dynamic movements even at higher noise levels (e.g., $\sigma_{dist} = 0.2$ m), and that similar performance to L5S-3IMU was achieved at $\sigma_{dist} = 0.1$ m for free walk movements. However, further validation is recommended with actual distance measurement from real sensors. The source code for the L5S algorithm, and links to sample videos will be made available at https://git.io/JTRQ3.

Author Contributions: conceptualization, L.W.F.S.; methodology, L.W.F.S.; software, L.W.F.S.; validation, L.W.F.S.; formal analysis, L.W.F.S. and S.J.R.; investigation, L.W.F.S.; resources, S.J.R and N.H.L.; data curation, L.W.F.S.; writing–original draft preparation, L.W.F.S.; writing–review and editing, L.W.F.S., S.J.R., and N.H.L.; visualization, L.W.F.S.; supervision, S.J.R. and N.H.L.; project administration, S.J.R. All authors have read and agreed to the published version of the manuscript.

Funding: This research received no external funding.

Acknowledgments: This research was supported by an Australian Government Research Training Program (RTP) Scholarship.

Conflicts of Interest: The authors declare no conflict of interest.

Abbreviations

The following abbreviations are used in this manuscript:

OMC	Optical Motion Capture
IMC	Inertial Motion Capture
IMU	Inertial Measurement Unit
OSPS	One Sensor per Body Segment
RSC	Reduced-Sensor-Count
KF	Kalman Filter
EKF	Extended Kalman Filter
CEKF	Constrained Extended Kalman Filter
ADL	Activities of Daily Living
TTD	Total Travelled Distance
SOP	Sparse Orientation Poser
SIP	Sparse Inertial Poser

Appendix A. Derivation of Pelvis-to-Ankle Distance Measurement

This section explains the derivation of the measurement pelvis-to-ankle vector (Equation (46)) as obtained from pelvis-to-ankle distance measurements, d_k^{pla} and d_k^{pra}, while assuming hinged knee joints and constant body segment lengths. For the sake of brevity, only the left side formulation is shown. The right side (i.e., pelvis to right ankle vector) can be calculated similarly.

First, we solve for an estimated left knee angle, $\hat{\theta}_k^{lk}$ (Equation (47)), from the measured pelvis to left ankle distance, d_k^{pla}. The pelvis to left ankle vector, $\tau_m^{pla}(\hat{\mu}_k^-, \theta_k^{lk})$ (Equation (A6)), can be defined as the sum of the mid-pelvis to hip, thigh long axis, and shank long axis vectors.

$$\tau_{pla}(\hat{\mu}_k^-, \theta_k^{lk}) = \overbrace{\frac{d^p}{2}\,\hat{T}_k^{p-}\,\boldsymbol{i}_y - d^{ls}\,\hat{T}_k^{ls-}\,\boldsymbol{i}_z}^{\psi_{pla}\,=\,\text{half pelvis } y\text{-axis + shank } z\text{-axis}} + d^{lt}\,\hat{T}_k^{ls-}\overbrace{(\boldsymbol{i}_x \sin(\theta_k^{lk}) - \boldsymbol{i}_z \cos(\theta_k^{lk}))}^{\text{thigh } z\text{-axis in shank frame}} \tag{A1}$$

By definition of $(d_k^{pla})^2$ and expanding $\tau_m^{pla}(\hat{\mu}_k^-, \theta_k^{lk})$ with Equation (A1), we obtain

$$\begin{aligned}
(d_k^{pla})^2 &= (\tau_{pla}(\hat{\mu}_k^-, \theta_k^{lk}))^T\, \tau_{pla}(\hat{\mu}_k^-, \theta_k^{lk}) \\
&= \psi_{pla}^T \psi_{pla} - 2d^{lt}\psi_{pla}^T\,\hat{T}^{ls-}\,\boldsymbol{i}_z \cos(\theta_k^{lk}) + 2d^{lt}\psi_{pla}^T\,\hat{T}^{ls-}\,\boldsymbol{i}_x \sin(\theta_k^{lk}) + (d^{lt})^2
\end{aligned} \tag{A2}$$

Equation (A2) can be rearranged in the form of Equation (A3) with α, β, γ as shown in Equation (A4).

$$\alpha \cos(\theta_k^{lk}) + \beta \sin(\theta_k^{lk}) = \gamma \tag{A3}$$

$$\alpha = -2d^{lt}\psi_{pla}^T\,\hat{T}_k^{ls-}\,\boldsymbol{i}_z, \quad \beta = 2d^{lt}\psi_{pla}^T\,\hat{T}_k^{ls-}\,\boldsymbol{i}_x, \quad \gamma = (d_k^{pla})^2 - \psi_{pla}^T\psi_{pla} - (d^{lt})^2 \tag{A4}$$

Solving for $\hat{\theta}_k^{lk}$ from Equation (A3) gives us a quadratic equation with two solutions as shown in Equations (A5) and (47). Between the two solutions, $\hat{\theta}_k^{lk}$ is set as the $\hat{\theta}_k^{lk}$ whose value is closer to the current left knee angle estimate from the prediction step. This solution serves as a pseudomeasurement of the knee angle.

$$\hat{\theta}_k^{lk} = \cos^{-1}\left(\frac{\alpha\gamma \pm \beta\sqrt{\alpha^2 + \beta^2 - \gamma^2}}{\alpha^2 + \beta^2}\right) \tag{A5}$$

Finally, $\boldsymbol{Z}_{pla,k}$, the KF measurement shown in Eqs. (A6) and (46), is the inter-IMU vector between the pelvis and left ankle, calculated using Equation (A1) with input $\hat{\theta}_k^{lk}$.

$$\boldsymbol{Z}_{pla,k} = \tau_m^{pla}(\hat{\mu}_k^-, \hat{\theta}_k^{lk}) \tag{A6}$$

References

1. Merriaux, P.; Dupuis, Y.; Boutteau, R.; Vasseur, P.; Savatier, X. A Study of Vicon System Positioning Performance. *Sensors* **2017**, *17*, 1591. [CrossRef]
2. Glowinski, S.; Łosiński, K.; Kowiański, P.; Waśkow, M.; Bryndal, A.; Grochulska, A. Inertial sensors as a tool for diagnosing discopathy lumbosacral pathologic gait: A preliminary research. *Diagnostics* **2020**, *10*, 342. [CrossRef] [PubMed]
3. Rovini, E.; Maremmani, C.; Cavallo, F. A wearable system to objectify assessment of motor tasks for supporting parkinson's disease diagnosis. *Sensors* **2020**, *20*, 2630. [CrossRef] [PubMed]
4. Lloréns, R.; Gil-Gómez, J.A.; Alcañiz, M.; Colomer, C.; Noé, E. Improvement in balance using a virtual reality-based stepping exercise: A randomized controlled trial involving individuals with chronic stroke. *Clin. Rehabil.* **2015**, *29*, 261–268. [CrossRef] [PubMed]
5. Shull, P.; Lurie, K.; Shin, M.; Besier, T.; Cutkosky, M. Haptic gait retraining for knee osteoarthritis treatment. In Proceedings of the 2010 IEEE Haptics Symposium, Waltham, MA, USA, 25–26 March 2010; pp. 409–416.
6. Roetenberg, D.; Luinge, H.; Slycke, P. Xsens MVN: Full 6DOF human motion tracking using miniature inertial sensors. *Xsens Motion Technologies BV Tech. Rep.* **2009**, *3*, 1–9.
7. Tadano, S.; Takeda, R.; Miyagawa, H. Three dimensional gait analysis using wearable acceleration and gyro sensors based on quaternion calculations. *Sensors* **2013**, *13*, 9321–9343. [CrossRef] [PubMed]
8. Yoon, P.K.; Zihajehzadeh, S.; Kang, B.S.; Park, E.J. Robust Biomechanical Model-Based 3-D Indoor Localization and Tracking Method Using UWB and IMU. *IEEE Sens. J.* **2017**, *17*, 1084–1096. [CrossRef]
9. Meng, X.L.; Zhang, Z.Q.; Sun, S.Y.; Wu, J.K.; Wong, W.C. Biomechanical model-based displacement estimation in micro-sensor motion capture. *Meas. Sci. Technol.* **2012**, *23*, 055101. [CrossRef]

10. Del Rosario, M.B.; Lovell, N.H.; Redmond, S.J. Quaternion-based complementary filter for attitude determination of a smartphone. *IEEE Sens. J.* **2016**, *16*, 6008–6017. [CrossRef]

11. Del Rosario, M.B.; Khamis, H.; Ngo, P.; Lovell, N.H.; Redmond, S.J. Computationally efficient adaptive error-state Kalman filter for attitude estimation. *IEEE Sens. J.* **2018**, *18*, 9332–9342. [CrossRef]

12. Tautges, J.; Zinke, A.; Krüger, B.; Baumann, J.; Weber, A.; Helten, T.; Müller, M.; Seidel, H.; Eberhardt, B. Motion reconstruction using sparse accelerometer data. *ACM Trans. Graph. (TOG)* **2011**, *30*, 18. [CrossRef]

13. Huang, Y.; Kaufmann, M.; Aksan, E.; Black, M.J.; Hilliges, O.; Pons-Moll, G. Deep inertial poser: Learning to reconstruct human pose from sparse inertial measurements in real time. In *SIGGRAPH Asia 2018 Technical Papers, Tokyo, Japan*; Association for Computing Machinery, Inc.: New York, NY, USA, 2018.

14. Salarian, A.; Burkhard, P.R.; Vingerhoets, F.J.G.; Jolles, B.M.; Aminian, K. A novel approach to reducing number of sensing units for wearable gait analysis systems. *IEEE Trans. Biomed. Eng.* **2013**, *60*, 72–77. [CrossRef] [PubMed]

15. Sy, L.W.; Raitor, M.; Del Rosario, M.B.; Khamis, H.; Kark, L.; Lovell, N.H.; Redmond, S.J.; Estimating lower limb kinematics using a reduced wearable sensor count. *IEEE Trans. Biomed. Eng.* **2020** In Press. [CrossRef] [PubMed]

16. Lin, J.F.; Kulić, D. Human pose recovery using wireless inertial measurement units. *Physiol. Meas.* **2012**, *33*, 2099–2115. [CrossRef] [PubMed]

17. von Marcard, T.; Rosenhahn, B.; Black, M.J.; Pons-Moll, G. Sparse inertial poser: Automatic 3D human pose estimation from sparse IMUs. In *Computer Graphics Forum*; Wiley Online Library: Hoboken, NJ, USA, 2017; Volume 36, pp. 349–360.

18. Barfoot, T.D. *State Estimation for Robotics*; Cambridge University Press: Cambridge, UK, 2017.

19. Wang, Y.; Chirikjian, G.S. Error propagation on the Euclidean group with applications to manipulator kinematics. *IEEE Trans. Robot.* **2006**, *22*, 591–602. [CrossRef]

20. Barfoot, T.D.; Furgale, P.T. Associating uncertainty with three-dimensional poses for use in estimation problems. *IEEE Trans. Robot.* **2014**, *30*, 679–693. [CrossRef]

21. Bourmaud, G.; Mégret, R.; Arnaudon, M.; Giremus, A. Continuous-discrete extended Kalman filter on matrix Lie groups using concentrated Gaussian distributions. *J. Math. Imaging Vis.* **2014**, *51*, 209–228. [CrossRef]

22. Brossard, M.; Bonnabel, S.; Condomines, J.P. Unscented Kalman filtering on Lie groups. In Proceedings of the IEEE International Conference on Intelligent Robots and Systems, Vancouver, BC, Canada, 24–28 September 2017; pp. 2485–2491.

23. Ćesić, J.; Joukov, V.; Petrović, I.; Kulić, D. Full body human motion estimation on lie groups using 3D marker position measurements. In Proceedings of the IEEE-RAS International Conference on Humanoid Robots, Cancun, Mexico, 15–17 November 2016; pp. 826–833.

24. Joukov, V.; Cesic, J.; Westermann, K.; Markovic, I.; Petrovic, I.; Kulic, D. Estimation and observability analysis of human motion on Lie groups. *IEEE Trans. Cybern.* **2019**, *50*, 1–12. [CrossRef]

25. Joukov, V.; Cesic, J.; Westermann, K.; Markovic, I.; Kulic, D.; Petrovic, I. Human motion estimation on Lie groups using IMU measurements. In Proceedings of the IEEE International Conference on Intelligent Robots and Systems, Vancouver, BC, Canada, 24–28 Septmber 2017; pp. 1965–1972.

26. Hol, J.D.; Dijkstra, F.; Luinge, H.; Schon, T.B. Tightly coupled UWB/IMU pose estimation. In Proceedings of the 2009 IEEE International Conference on Ultra-Wideband, Vancouver, BC, Canada, 9–11 September 2009; pp. 688–692.

27. Malleson, C.; Gilbert, A.; Trumble, M.; Collomosse, J.; Hilton, A. Real-time full-body motion capture from video and IMUs. In Proceedings of the International Conference on 3D Vision (3DV), Qingdao, China, 10–12 October 2017.

28. Gilbert, A.; Trumble, M.; Malleson, C.; Hilton, A.; Collomosse, J. Fusing visual and inertial sensors with semantics for 3D human pose estimation. *Int. J. Comput. Vis.* **2019**, *127*, 381–397. [CrossRef]

29. Helten, T.; Muller, M.; Seidel, H.P.; Theobalt, C. Real-time body tracking with one depth camera and inertial sensors. In Proceedings of the IEEE International Conference on Computer Vision, Sydney, Australia, 8–12 April 2013; pp. 1105–1112.

30. Vlasic, D.; Adelsberger, R.; Vannucci, G.; Barnwell, J.; Gross, M.; Matusik, W.; Popović, J. Practical motion capture in everyday surroundings. *ACM Trans. Graph. (TOG)* **2007**, *26*, 35. [CrossRef]

31. Sy, L.; Lovell, N.H.; Redmond, S.J. Estimating lower limb kinematics using distance measurements with a reduced wearable inertial sensor count. In Proceedings of the 2020 42nd Annual International Conference of the IEEE Engineering in Medicine and Biology Society (EMBC), Montreal, QC, Canada, 20–24 July 2020.
32. Sy, L.; Lovell, N.H.; Redmond, S.J. Estimating lower limb kinematics using a Lie group constrained EKF and a reduced wearable IMU count. In Proceedings of the 2020 8th IEEE International Conference on Biomedical Robotics and Biomechatronics (Biorob), New York, NY, USA, 29 November–1 December 2020.
33. Selig, J.M. Lie groups and Lie algebras in robotics. In *Computational Noncommutative Algebra and Applications*; Springer: Berlin/Heidelberg, Germany, 2004; pp. 101–125.
34. Stillwell, J. *Naive Lie Theory*; Springer Science & Business Media: Berlin/Heidelberg, Germany, 2008.
35. Chirikjian, G. *Stochastic Models, Information Theory, and Lie Groups. II: Analytic Methods and Modern Applications*; Springer Science & Business Media: Berlin/Heidelberg, Germany, 2012.
36. Bourmaud, G.; Megret, R.; Giremus, A.; Berthoumieu, Y. Discrete extended Kalman filter on Lie groups. In Proceedings of the European Signal Processing Conference, Marrakech, Morocco, 9–13 September 2013; pp. 1–5.
37. Cloete, T.; Scheffer, C. Benchmarking of a full-body inertial motion capture system for clinical gait analysis. In Proceedings of the 2008 30th Annual International Conference of the IEEE Engineering in Medicine and Biology Society, Vancouver, BC, Canada, 20–25 August 2008; pp. 4579–4582.
38. Hu, X.; Yao, C.; Soh, G.S. Performance evaluation of lower limb ambulatory measurement using reduced inertial measurement units and 3R gait model. In Proceedings of the IEEE International Conference on Rehabilitation Robotics, Singapore, 11–14 August 2015; pp. 549–554.
39. Malajner, M.; Planinsic, P.; Gleich, D. UWB ranging accuracy. In Proceedings of the 2015 22nd International Conference on Systems, Signals and Image Processing, London, UK, 10–12 September 2015; pp. 61–64.
40. Ledergerber, A.; D'Andrea, R. Ultra-wideband range measurement model with Gaussian processes. In Proceedings of the 1st Annual IEEE Conference on Control Technology and Applications, Hawaii, HI, USA, 27–30 August 2017; pp. 1929–1934.
41. McGinley, J.L.; Baker, R.; Wolfe, R.; Morris, M.E. The reliability of three-dimensional kinematic gait measurements: A systematic review. *Gait Posture* **2009**, *29*, 360–369. [CrossRef] [PubMed]
42. Jimenez, A.R.; Seco, F.; Prieto, J.C.; Guevara, J. Indoor Pedestrian navigation using an INS/EKF framework for yaw drift reduction and a foot-mounted IMU. In Proceedings of the 2010 7th Workshop on Positioning, Navigation and Communication, Dresden, Germany, 11–12 March 2010; pp. 135–143.
43. Zhang, W.; Li, X.; Wei, D.; Ji, X.; Yuan, H. A foot-mounted PDR System Based on IMU/EKF+HMM+ ZUPT+ZARU+HDR+compass algorithm. In Proceedings of the 2017 International Conference on Indoor Positioning and Indoor Navigation, Sapporo, Japan, 18–21 September 2017; pp. 1–5.
44. Dotlic, I.; Connell, A.; Ma, H.; Clancy, J.; McLaughlin, M. Angle of arrival estimation using decawave DW1000 integrated circuits. In Proceedings of the 2017 14th Workshop on Positioning, Navigation and Communications, Bremen, Germany, 25–26 October 2017; pp. 1–6.
45. Xu, W.; Chatterjee, A.; Zollh, M.; Rhodin, H.; Fua, P.; Seidel, H.p.; Theobalt, C. Mo2Cap2: Real-time mobile 3D motion capture with a cap-mounted fisheye camera. *IEEE Trans. Vis. Comput. Graph.* **2019**, *25*, 2093–2101. [CrossRef] [PubMed]
46. Laidig, D.; Schauer, T.; Seel, T. Exploiting kinematic constraints to compensate magnetic disturbances when calculating joint angles of approximate hinge joints from orientation estimates of inertial sensors. In Proceedings of the IEEE International Conference on Rehabilitation Robotics, London, UK, 17–20 July 2017; pp. 971–976.
47. Eckhoff, K.; Kok, M.; Lucia, S.; Seel, T. Sparse magnetometer-free inertial motion tracking—A condition for observability in double hinge joint systems. *arXiv* **2020**, arXiv:2002.00902.
48. de Vries, W.H.; Veeger, H.E.; Baten, C.T.; van der Helm, F.C. Magnetic distortion in motion labs, implications for validating inertial magnetic sensors. *Gait Posture* **2009**, *29*, 535–541. [CrossRef] [PubMed]

49. Li, T.; Wang, L.; Li, Q.; Liu, T. Lower-body walking motion estimation using only two shank-mounted inertial measurement units (IMUs). In Proceedings of the IEEE/ASME International Conference on Advanced Intelligent Mechatronics, Boston, MA, USA, 6–9 July 2020; pp. 1143–1148.

Publisher's Note: MDPI stays neutral with regard to jurisdictional claims in published maps and institutional affiliations.

 *sensors*

 MDPI

Letter

Use of Functional Linear Models to Detect Associations between Characteristics of Walking and Continuous Responses Using Accelerometry Data

William F. Fadel [1,*], Jacek K. Urbanek [2], Nancy W. Glynn [3] and Jaroslaw Harezlak [4,*]

[1] Department of Biostatistics, Fairbanks School of Public Health, Indiana University, Indianapolis, IN 46202, USA

[2] Division of Geriatric Medicine and Gerontology, Department of Medicine, School of Medicine, Johns Hopkins University, Baltimore, MD 21205, USA; jurbane2@jhu.edu

[3] Department of Epidemiology, Graduate School of Public Health, University of Pittsburgh, Pittsburgh, PA 15261, USA; epidnwg@pitt.edu

[4] Department of Epidemiology and Biostatistics, Indiana University, Bloomington, IN 47405, USA

* Correspondence: wffadel@iu.edu (W.F.F.); harezlak@iu.edu (J.H.)

Received: 1 October 2020; Accepted: 6 November 2020; Published: 9 November 2020

Abstract: Various methods exist to measure physical activity. Subjective methods, such as diaries and surveys, are relatively inexpensive ways of measuring one's physical activity; however, they are prone to measurement error and bias due to self-reporting. Wearable accelerometers offer a non-invasive and objective measure of one's physical activity and are now widely used in observational studies. Accelerometers record high frequency data and each produce an unlabeled time series at the sub-second level. An important activity to identify from the data collected is walking, since it is often the only form of activity for certain populations. Currently, most methods use an activity summary which ignores the nuances of walking data. We propose methodology to model specific continuous responses with a functional linear model utilizing spectra obtained from the local fast Fourier transform (FFT) of walking as a predictor. Utilizing prior knowledge of the mechanics of walking, we incorporate this as additional information for the structure of our transformed walking spectra. The methods were applied to the in-the-laboratory data obtained from the Developmental Epidemiologic Cohort Study (DECOS).

Keywords: accelerometry; physical activity; Fourier transform; functional linear model

---

1. Introduction

Use of wearable accelerometers has become increasingly common in studies of physical activity, aging, and obesity [1–8]. Self-reported measures, such as questionnaires, have been widely used to assess physical activity (PA) previously [6,9]. One reason why we care about using accelerometry over self-reporting is because there are some populations whose self-reported measures can be inaccurate [10]. One such population is older adults. Questionnaires require individuals to recall their daily activities, which can be extremely difficult, particularly for older individuals [11]. Schrack et al. [12] showed there are changes in the daily patterns and amount of PA as people age. Accelerometry is also an important tool that can be applied to the general population, because walking is the most popular form of aerobic physical activity [13].

Accelerometers offer a non-invasive and objective alternative to self-reporting methods. Advancements in data processing allow for the analysis of specific gait characteristics, such as cadence and asymmetry [14]. Likewise, advancements in statistical methodology for analysis of high dimensional data have opened up new paths for analyzing more complex and potentially more informative summaries of accelerometry data. Accelerometers are electro-mechanical devices that measure acceleration along three orthogonal axes. They are often worn on a person's waist or wrist, and they provide high frequency, high-throughput data represented by three time series of acceleration measurements [15,16]. The data are typically collected at the sub-second level (usually between 10 to 100 observations per second); however, most studies aggregate the data over one minute epochs, or windows, and often, using a user specified threshold, the data are characterized into activity counts per minute. While thresholding methods are useful in describing the timing and duration for certain levels of PA, many nuances of the data are lost. For example, it is not possible to evaluate how a person is walking from activity count summaries. Especially in certain populations, such as older or obese populations, this raises the question as to whether a more detailed quantification of the walking signal can provide additional information. For example, if an older person's legs are bothering them, we may detect signs of limping that could help explain the low levels of PA. Our strategy is to extract detailed information from the raw accelerometry signal during periods of walking. We transform this information into useful quantities, and then we build regression models to associate characteristics of walking with continuous responses.

To illustrate the complex nature of the data collected, Figure 1 shows the raw data collected from a triaxial accelerometer (Actigraph GT3X+) for a single individual during an in-laboratory 400 m walk. The top left panel of Figure 1 presents the entire 400 m walk, where each axis is shown in a different color. With just over 5 min worth of data, the characteristics of the signal are nearly impossible to visualize. The top right panel of Figure 1 shows a 10 s window of the same data. At this scale, we can discern a fairly periodic signal. In the bottom row of Figure 1, we present the vector magnitude of the same data presented in the top row. We can see that the periodic nature of the data are preserved while information about the three dimensional direction is lost. However, in free-living data collection, it is difficult to control the orientation of the device when participants are able to remove the device. For this reason, the magnitude of the signal is sufficient and relatively stable to capture the gait characteristics described in this manuscript. The periodic characteristics of walking naturally lend themselves to a frequency analysis approach for quantifying the features of walking. By utilizing the methods described in Urbanek et al. [15], we can extract estimates of cadence (steps per second) and average magnitude from windows of raw data. In addition to these more common features, we also utilize the spectra obtained from the local fast Fourier Transform (FFT) as a functional predictor for modeling the association of walking with continuous responses such as age and body mass index (BMI). By incorporating the walking spectra as a predictor, we gain additional information about the characteristics of a person's gait which may be associated with the response of interest. For example, an individual with a very smooth walking stride would result in most of the energy from walking concentrated around the frequencies near the cadence. However, an individual with an interrupted stride (e.g., a limp) would result in energy being more dispersed through higher frequencies.

Several methods exist for fitting a scalar-on-regression function, such as $y = \int_{\mathcal{I}} W(s)\beta(s)ds + \epsilon$, where $W(\cdot)$ is a functional predictor and y is a scalar response variable. Several methods for estimating $\beta(\cdot)$ are based on the eigenfunctions associated with some covariance operator defined by the predictors [17]. Due to the periodic nature of walking, we have strong reason to believe the vast majority of information contained in the walking spectra will be located around the harmonics centered at multiples of the dominant frequency. The PEER method developed by Randolph et al. [17] allows for the incorporation of presumed structure directly into the estimation process and is preferable to a purely empirical estimator. This particular method has been widely used in other areas of research, for example, in heritability and evolution studies [18,19] and in microbiome analysis [20]. However, to the best of our knowledge, this is a novel use of the PEER method in the analysis of the raw accelerometry data.

In this manuscript, we propose a novel application of recently developed statistical methods for the analysis of accelerometry data by associating continuous responses, such as age and BMI, with the Fourier spectrum of walking. The purpose of this manuscript is to serve as a proof of concept for researchers seeking to utilize more information from the accelerometry data in modeling associations between characteristics of walking and health related outcomes. We show how this additional information can be combined with scalar predictors in a linear regression model framework. The remainder of this manuscript is structured as follows. In Section 2, we describe the data collection and pre-processing procedures. In Section 3, we describe the functional linear model used to fit the data and how the estimation is performed. In Section 4, we apply the proposed model to data collected in the laboratory from a study of an aging adult population. In Section 5, we conclude with a discussion.

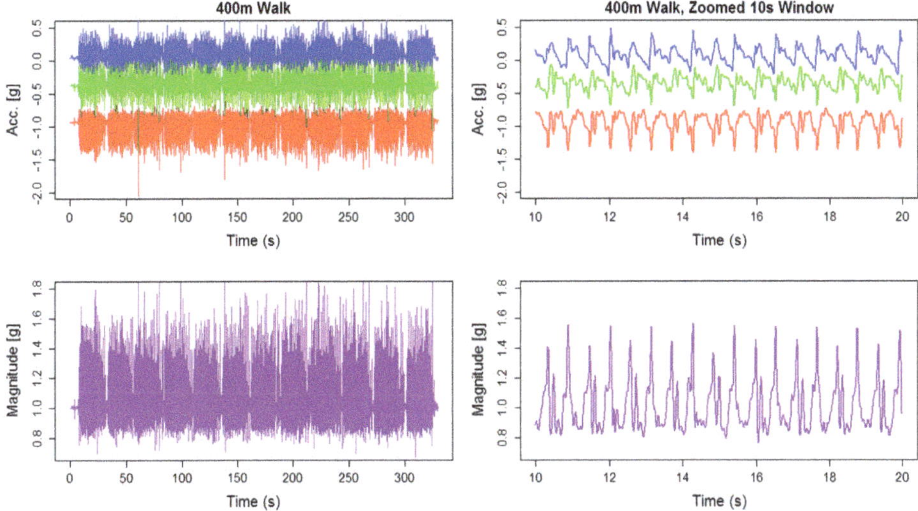

Figure 1. Triaxial accelerometer data from the 400 m walk for a single individual (**top left**) and a zoomed 10 s window (**top right**). Vector magnitude from the 400 m walk for same individual (**bottom left**) and zoomed 10 s window (**bottom right**).

2. Data Collection and Pre-Processing

Eighty-nine community-dwelling older adults were recruited from the Pittsburgh, Pennsylvania area for the National Institute on Aging, Aging Research Evaluating Accelerometry (AREA) project, part of the Developmental Epidemiologic Cohort Study (DECOS) [3]. AREA was a cross-sectional methodological initiative designed to examine the impact of accelerometry wear location on assessment of physical activity and sedentary behavior among 89 older adults enrolled between March and May of 2010 [21]. The report included data from 51 healthy participants (25 men and 26 women) who had complete the "in-the-lab" ($N = 46$) or "in-the-wild" ($N = 48$) accelerometry data. Individuals were excluded from the DECOS study for the following reasons: hip fracture, stroke in the past 12 months, cerebral hemorrhage in the past 6 months, heart attack, angioplasty, heart surgery in the past 3 months, chest pain during walking the past 30 days, current treatment for shortness of breath or a lung condition, usual aching, stiffness, or pain in their lower limbs and joints, and bilateral difficulty in bending or straightening the knees fully [22]. This paper focuses on the $N = 46$ participants that completed the "in-the-lab" fast-paced 400 m walk. Data were collected with an Actigraph GT3X+ worn at the right hip. The devices collected raw accelerometry data along three orthogonal axes at a sampling frequency of 80 Hz. A summary of the demographic data for all $N = 46$ participants is provided in Table 1.

The first step in pre-processing the data is to split the observed triaxial signal from the 400 m walk into 10 s non-overlapping windows. For each window, we transform the raw triaxial signal into vector magnitude (VM), where VM is defined as the root sum of squares of the three axes, i.e.,

$vm(t) = \sqrt{x_1(t)^2 + x_2(t)^2 + x_3(t)^2}$. The vector magnitude count (VMC) is then calculated as the mean absolute deviation of the VM:

$$v_\tau(t) = \frac{1}{\tau} \sum_{u=t-\tau/2}^{t+\tau/2} |vm(u) - \overline{vm}|, \tag{1}$$

where τ is the window size expressed as number of sampling points [15]. We then transform the VM from the time domain into the frequency domain using the local FFT, or short time Fourier transform (STFT). Similarly to Urbanek et al. [15], we define the STFT at time t of the $vm(t)$ as

$$X(t, f; \tau) = \sum_{u=[t-\tau/2]}^{[t+\tau/2]} vm(u)h(u)e^{-i2\pi f u/\tau}, \tag{2}$$

where f is the frequency index and τ is a tuning parameter specifying the number of observations in the interval centered at t. The Hanning weights, defined as, $h(u; \tau) = 0.5[1 - cos\{2\pi u/(\tau - 1)\}]$ are used to avoid a blurring of the obtained spectra which can happen as a result of the windowing process. The spectrum is then defined as the absolute value of the STFT, $|X(t, f; \tau)|$.

Table 1. DECOS participant characteristics ($N = 46$).

Characteristic	Summary Statistics
Male (n (%))	22 (47.8%)
Age (Mean (SD))	78.24 (5.74)
BMI (Mean (SD))	26.73 (3.94)
Cadence (Mean (SD))	2.06 (0.17)
VMC (Mean (SD))	0.25 (0.09)

For each spectrum obtained, we then identify the fundamental frequency (cadence) as the location of the largest peak in the spectrum. Since the reported frequency of walking is between 1.4 and 2.5 Hz [23], which corresponds to 1.4 to 2.5 steps per second, we look for the cadence in a conservative range of 1.2–4.0 Hz to be consistent with Urbanek et al. [15]. The frequency axis used is from 0 to 39.9 Hz sampled every 0.1 Hz which ensures every individual's spectra will contain at least 10 multiples of their dominant frequency, or cadence. In the top left panel of Figure 2, we display all spectra for a single participant. Although these spectra appear similar, it is evident that there is variability between spectra obtained in different windows. Therefore, once the spectra are all obtained, and we have identified their fundamental frequencies, it is imperative that we align each spectrum for aggregation.

In order to align all spectra at their fundamental frequency, we further transform each spectrum from the frequency domain into the order domain by scaling the frequency axis by the fundamental frequency for each spectrum. Linear interpolation is then used to place each spectrum back on the same sampling grid. This ensures that all spectra are aligned and sampled at equally spaced points. The top right panel of Figure 2 illustrates how the realigned spectra for a single participant. As can be seen, all spectra are perfectly aligned at the dominant frequency.

However, each spectrum is sampled discretely, therefore, further harmonics may be slightly misaligned in the order domain. To compensate, we average the spectra across all windows for each participant in order to obtain a global estimate of walking features for each individual. Each spectrum is restricted to 546 points between 0.3 and 5.75 in the order domain to avoid modeling signal noise at the beginning and end of the spectra. The bottom left panel of Figure 2 shows the averaged spectra for all $N = 46$ participants. The peaks of the average spectra for each individual are now nearly perfectly aligned in the order domain at multiples of the fundamental frequency.

Finally, we scale each individual's average spectrum by the magnitude of the spectrum at the cadence. By scaling the spectra in this way, the magnitude at each harmonic can be interpreted as a ratio to the magnitude at the cadence. This process is illustrated in bottom right panel of Figure 2, and this entire process is fully detailed in Algorithm 1 . These steps for pre-processing the raw accelerometry data are essential in order to properly fit the statistical model described in Section 3.

Algorithm 1: Steps for pre-processing the raw accelerometry data.

Input : $x(t)$—accelerometry signal, f_s—sampling frequency, $s_{min} = 1.2$ Hz, $s_{max} = 4.0$ Hz
Output: FFT—scaled average FFT spectrum, VMC—average VMC, *Cadence*—average cadence

1 Divide accelerometry signal into 10 s non-overlapping windows.
2 Transform accelerometry signal into vector magnitude $vm(t)$.
3 Compute vector magnitude count, $v_{10}(t)$, for each window.
4 Compute Fourier spectrum for each window.
5 Estimate cadence as frequency centered under the largest peak in spectrum.
6 Transform spectra from frequency domain to order domain by scaling frequency axis by the frequency of the cadence.
7 Average vector magnitude count, cadence, and order domain spectra across all windows.
8 Restrict spectra to points between 0.3 and 5.75 multiples of the cadence frequency.
9 Scale spectra by magnitude of the average spectra at the cadence.

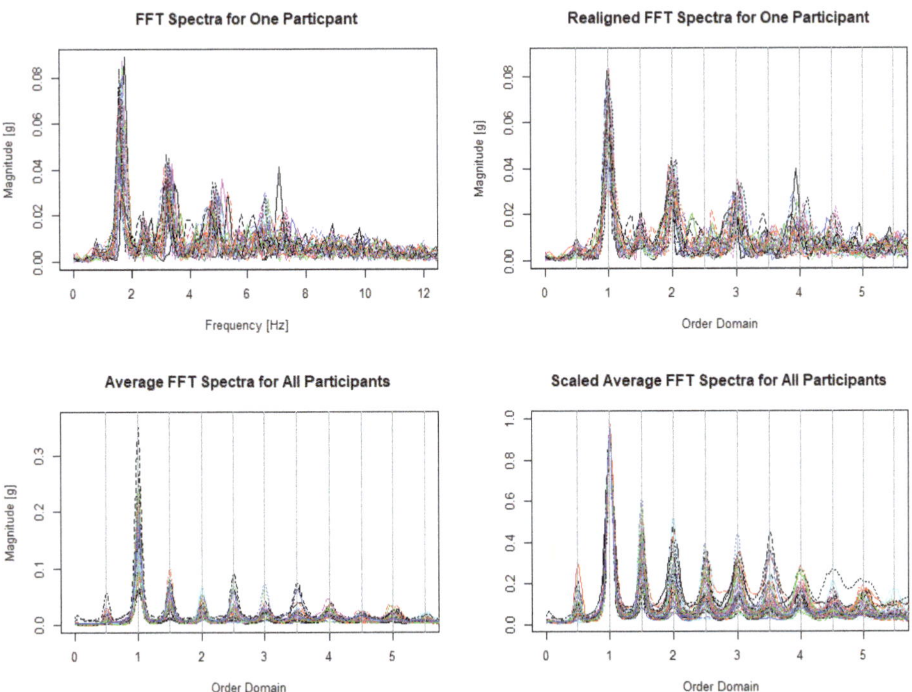

Figure 2. Pre-processing data. Observed FFT spectra for one participant as described in step 4 of Algorithm 1 (**top left**). Observed spectra realigned into order domain for the same participant as described in step 6 of Algorithm 1 (**top right**). Average realigned spectra for all participants as described in step 7 of Algorithm 1 (**bottom left**). Scaled average spectra for all participants as described in step 9 of Algorithm 1 (**bottom right**).

3. Statistical Methods

Frequently, the methods applied to analyze the data arising from the raw accelerometry signal rely on the discrete features extracted from such data, leading to possible loss of information. In contrast, we take the full time series signal into account and work with its continuous properties, namely, the spectrum obtained from the walking portion of the accelerometry signal. We summarize the approach taken below.

Let $W_i(\cdot)$ denote a functional predictor (e.g., a scaled average FFT spectrum) from the ith study participant where $(i = 1, \ldots, N)$. We will assume that each observed predictor is obtained as a discretized version of an idealized function at p equally-spaced points, $s_1, \ldots, s_p$, as can be seen in the walking spectra in Figure 2. We let $w_i := [w_i(s_1), \ldots, w_i(s_p)]^T$ be the $p \times 1$ vector of values sampled from $W_i(\cdot)$. Then our observed data take the form $\{y_i; x_i; w_i\}$ where y_i is a scalar response, x_i is a $K \times 1$ vector of measurements from K scalar predictors (e.g., sex or average cadence), and w_i is the functional predictor from the ith participant. We denote the true coefficient function by $\beta(\cdot)$, and then, the functional regression model of interest is given by

$$y_i = x_i^T \gamma + \int_{\mathcal{I}} W_i(s)\beta(s)ds + \epsilon_i \tag{3}$$

where $\epsilon_i \sim N(0, \sigma_\epsilon^2)$. Here $x_i^T \gamma$ is the linear effect from K scalar predictors and $\int_{\mathcal{I}} W_i(s)\beta(s)ds$ is the functional effect.

3.1. Estimation of Parameters

Several approaches can be used to estimate the association between the scalar x_i and functional $w_i(\cdot)$ predictors with the outcome y_i. In our work, we utilize the approach proposed in Randolph et al. [17], which incorporates functional structure into the estimation of $\beta(\cdot)$. Specifically, the properties of the estimated spectra, i.e., their continuity, smoothness, and common behavior, are taken into the estimation procedure explicitly. To represent our model in a compact form, we combine the data from all N participants and express Equation (3) as

$$y = X\gamma + W\beta + \epsilon \tag{4}$$

where $y = [y_1, \ldots, y_N]^T$ is an $N \times 1$ vector of responses; $X = [x_1^T, \ldots, x_N^T]^T$ is an $N \times K$ design matrix corresponding to the scalar predictors with coefficient vector γ; $W = [w_1^T, \ldots, w_N^T]^T$ is an $N \times p$ design matrix corresponding to the functional predictors with functional coefficient vector β.

Given the periodic nature of the walking behavior, if the walking spectral properties are associated with the outcomes, the relevant information contained in the walking spectra is localized around the harmonics at multiples of the dominant frequency. We thus want to estimate β while imposing this prior information on its functional structure. We achieve this by using the penalty operator, L [17], which is created from the basis functions in the right panel of Figure 3. The penalized estimates of γ and β are obtained as the solution to the following criterion

$$[\tilde{\gamma}, \tilde{\beta}_{\lambda,L}]^T = \underset{\gamma,\beta}{\arg\min}\{||y - X\gamma - W\beta||^2 + \lambda||L\beta||_{L^2}^2\}, \tag{5}$$

where we only penalize the functional coefficient vector β.

Given some prior knowledge about the structure of our functional predictor, the penalty is defined utilizing a subspace containing this information [17]. We define this informative space, Q, to be a span of basis functions (right panel of Figure 3) emphasizing the relevant features of $\beta(\cdot)$ and consider the orthogonal projection $P_Q = QQ^+$. As described in Randolph et al. [17], we define the decomposition-based penalty as

$$L \equiv L_Q = a(I - P_Q) + P_Q \tag{6}$$

for some $a > 0$. When $a > 1$ the estimate is penalized more in the non-informative space orthogonal to Q. When $a = 1$, the estimate is simply an ordinary ridge regression estimate. Therefore, a generalized ridge estimate of γ and β can be obtained as

$$[\tilde{\gamma}, \tilde{\beta}]^T = (X_o^T X_o + \lambda L_o^T L_o)^{-1} X_o^T y, \tag{7}$$

where $X_o = [X\ W]$ and $L_o = \text{blockdiag}\{0, L_Q^T L_Q\}$. The tuning parameter, λ, is estimated in a principled way via a linear mixed model equivalence, as described in Ruppert et al. [24]. Specifically, the optimization criterion (5) is written in an equivalent linear mixed model form with the coefficients γ being fixed and the coefficients β being random with a distribution $\beta \sim N(0, \sigma_\beta^2)$. The estimate of the tuning parameter, λ, is then simply the ratio of the variances σ_ϵ^2 and σ_β^2.

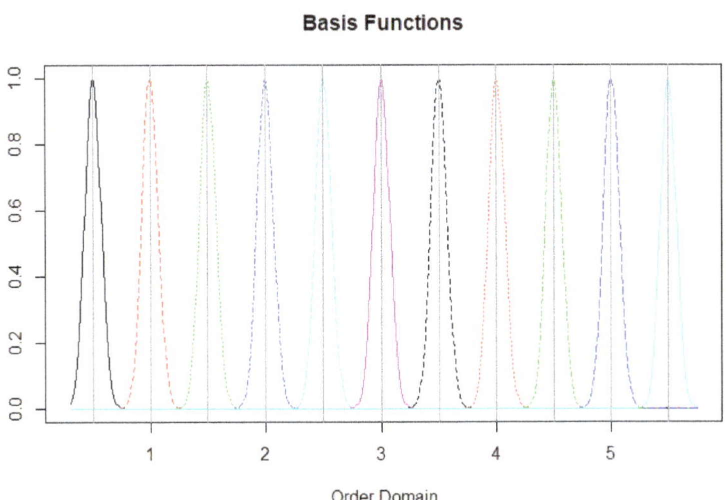

Figure 3. Pre-processed walking spectra (**top**) and basis functions used for modeling (**bottom**). The *x*-axis represents multiples of the frequency of the cadence.

4. DECOS Example

We applied the methods discussed in Section 3 to the data described in Section 2 to study the associations of walking spectra obtained from the fast-paced 400 m walk with age and BMI [3]. The fast-paced 400 m walk is often used in epidemiological studies of older adults to assess aerobic fitness [25]. The most common protocol implemented for the fast-paced 400 m walk is the long distance corridor walk (LDCW) [26]. The pre-processed walking spectra described in Section 2 were each sampled at $k = 546$ distinct sampling points within 0.3 and 5.75 of the order domain. This range was chosen because there is little energy contained in the spectra beyond 13.5 Hz. Assuming an average cadence of 2.0 Hz, this range sufficiently covers the relevant features of walking. There were $N = 46$ participants that completed the LDCW. In addition to each participant's scaled average walking spectra, an estimate of their average cadence was used as a predictor in the proposed models to control for participant-specific walking speeds. In addition, VMC was used to control for the energy magnitude each participant produced. For example, an individual with a very controlled and smooth walking style would have shown lower magnitude than an individual with a heavy stomp in their walk. We also adjusted each model for gender differences.

In order to use our prior knowledge about the structure of the walking spectra, we define a penalty L_Q as given in Equation (6) (with $a = 2$). We define our basis functions as normal density functions centered at multiples of the cadence from $0.5 \times cadence$ to $5.5 \times cadence$ using steps of $0.5 \times cadence$. We chose a standard deviation such that the distributions were nearly orthogonal. Scaled average walking spectra and basis functions are displayed in Figure 3.

Following the general formulation of the functional regression model (3), we fit the following model to these data:

$$y_i = \gamma_0 + Male_i * \gamma_1 + Cadence_i * \gamma_2 + VMC_i * \gamma_3 + \int_{\mathcal{I}} Spectrum_i(s)\beta(s)ds + \epsilon_i \qquad (8)$$

where y_i is either age or BMI, $Male_i$ is a binary variable, and $Cadence_i$ and VMC_i are the cadence and vector magnitude count for participant i, respectively. $Spectrum_i(\cdot)$ is the scaled average walking spectrum for participant i as described in Section 2. We assume that $\epsilon_i \sim N(0, \sigma_\epsilon^2)$. Regression coefficients γ and the regression function $\beta(s)$ are estimated via the procedure described in Section 3.1 using the peer() function from the refund package in R [27,28]. Thus, the scalar outcome y_i is predicted via a weighted sum of the products of the collected data (indicator variable of male sex, cadence, VMC, and a spectrum) and their respective estimated regression coefficients $\tilde{\gamma}$'s and $\tilde{\beta}(\cdot)$.

Figure 4 displays the estimates of $\beta(\cdot)$ along with the pointwise 95% confidence bands for the two models described in Equation (8). These figures show that the estimated regression function is different from zero at different multiples of the cadence. The regression function for age (top) shows that the coefficient function, $\tilde{\beta}$, is negative at the multiples 1.5 and 3.5 and positive at the multiples 4, 4.5, and 5, whereas for the other multiples the estimated coefficients are not significantly different from zero. These results indicate that younger individuals have larger magnitude in the lower harmonics relative to the magnitude at their cadence which indicates a heavier stomp component and controlled walking motion. Older individuals have larger magnitudes in the higher harmonics relative to the magnitudes of their cadence, which indicates a less controlled compensatory walking motion (e.g., a limp). These results seem to be consistent with findings of prior research [29–35]. However additional research is needed to validate our conclusions.

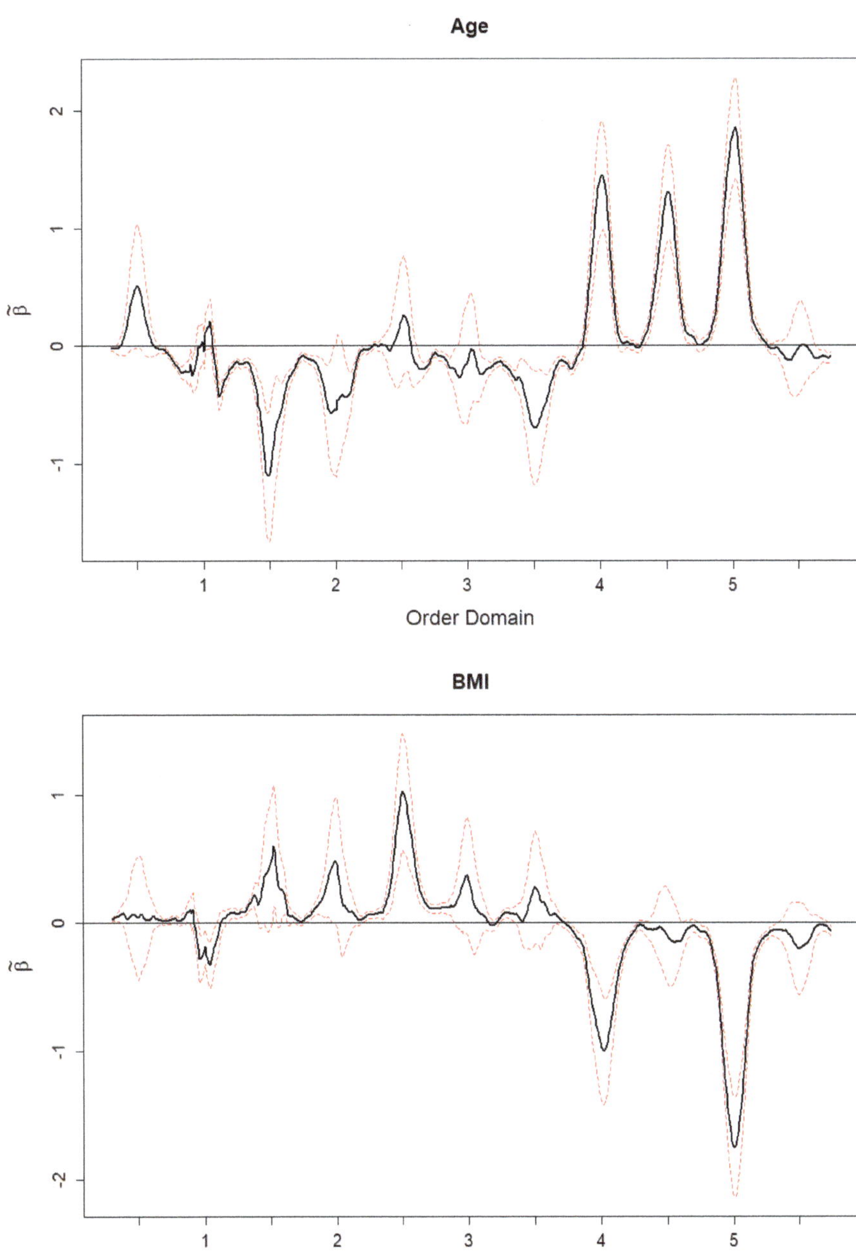

Figure 4. Estimates of the coefficient function, $\tilde{\beta}$, (with 95% point-wise confidence band) for the association of walking with age and BMI, as described in Section 4. The x-axis represents multiples of the frequency of the cadence.

The regression results for BMI (bottom) show that the coefficient function, $\tilde{\beta}$, is positive at the multiple 2.5 and negative at the multiples 4 and 5, whereas for the other multiples the estimated coefficients are not significantly different from zero. These results could indicate that overweight or obese individuals tend to walk with a heavier stomp component, resulting in higher magnitudes of the lower harmonics than the normal weight individuals. Leaner individuals walk with a lighter stomp component, resulting in walking characteristics with lower magnitudes in the lower harmonics and higher magnitudes in the higher harmonics. Further discussion and possible limitations of this interpretation are provided in Section 5.

Both results described above show that the information provided by the penalty reduces the number of spurious findings and at the same time emphasizes the signal content of the scaled Fourier spectra.

5. Discussion

In this paper we proposed a novel application of existing functional linear model methods to the study of physical activity data collected by accelerometers. We proposed an algorithm for pre-processing the raw data collected from accelerometers to quantify the characteristics of walking in a more detailed manner than is typically used with activity count summaries. By utilizing the periodic characteristics of walking, we were able to reduce the dimensionality of the raw data into a form that retained some details of the original signal while allowing us to use existing statistical methods for analyses. We applied these methods to the in-the-laboratory data collected from a study of an older adult population.

While FFT has been widely used for pre-processing accelerometry data, the features extracted from such methods have been applied to the problem of classification of activity types as opposed to associating characteristics of walking to continuous response variables [15,36–38]. To our knowledge, this is the first proposed application of functional linear regression techniques to model the association of walking spectra with continuous responses. Due to the periodic characteristics of walking, the proposed method naturally lends itself to this application, wherein we can inform the penalty operator of where the relevant information is contained in the spectra. This method is not limited to the cross-sectional setting, as demonstrated in this paper, and it is easily extended to responses collected longitudinally [39]. In addition to walking speed, this more detailed quantification of walking may provide additional information as to how certain degenerative diseases (e.g., Parkinson's disease and multiple sclerosis) affect a person's ability to walk over the progression of disease. Reuter et al. [40] showed that certain walking programs can actually improve gait characteristics of individuals with Parkinson's disease over the course of a 6-month study. Gait characteristics were measured on a specialized treadmill outfitted with specialized sensors to accurately measure foot-ground contact. The application of these proposed methods could alleviate any financial restrictions of such studies to allow for much larger randomized prospective studies to determine whether these exercise therapies actually slow down the progression of such diseases. Utility of these methods can only be assessed with the inclusion of accelerometers in such studies, and they are being increasingly used.

We acknowledge that there are limitations in our analyses. For example, we did not collect data from the 3D motion capture or ground force reaction (GFR) measurements to validate the findings of our analyses. This manuscript is meant to demonstrate how researchers can utilize existing statistical methodology to analyze finer features of the raw accelerometry data retaining additional information about characteristics and changes in gait that are usually lost due to over-aggregation of the raw data collected. It was beyond the scope of this study to obtain information from specialized laboratory equipment, but we do feel that further research is warranted to validate our findings and draw associations between the additional characteristics of gait observed in our analyses with gait measurements obtained in a laboratory. Ko et al. [33] found that older adults with obesity modify their gait patterns compared to normal weight counterparts while walking at normal and fast speeds. For example, they found that obese participants had lower mechanical work expenditure (MWE) in the ankle and significantly higher MWE in the knee and hip compared to normal weight participants. Paired with the raw accelerometry data, these findings could identify the specific mechanisms validating the additional associations we found in our analyses. Menz et al. [31] observed that older participants exhibited a more conservative gait pattern characterized by reduced velocity, shorter step length, and increased step timing variability which could be contributing to the portion of the signals observed at the higher frequencies of the walking spectra in our study.

In conclusion, we acknowledge that age and BMI are easy to measure, but the strength of this study is that it serves as a proof of concept for how researchers can utilize the extracted walking characteristics in the presence of more relevant health related outcomes, e.g., fatigability or neurocognitive function. In addition, we studied only relatively healthy elderly individuals. Thus, generalizability of the findings to either healthy younger individuals or unhealthy older individuals needs to be studied. Given the small sample size of our study and utilization of the data from the laboratory experiment only, additional research is needed to establish similar associations between the health outcomes and the free-living walking data characteristics.

Author Contributions: W.F.F. conceptualization, formal analysis, investigation, writing, and visualization; J.K.U. writing and resources; N.W.G. writing and resources; J.H. conceptualization, writing, resources, and supervision. All authors have read and agreed to the published version of the manuscript.

Funding: This research was supported by Pittsburgh Claude D. Pepper Older Americans Independence Center Research Registry and Developmental Pilot Grant (PI: Glynn)—NIH P30 AG024827 and a National Institute on Aging Professional Services Contract HHSN271201100605P to Dr. Glynn. The project was also supported, in part, by the Intramural Research Program of the National Institute on Aging. Doctors Harezlak and Fadel were supported in part by the NIMH grant R01 MH108467 and with support from the Indiana Clinical and Translational Sciences Institute Design and Biostatistics Pilot Grant, funded, in part, by grant UL1TR001108, from the National Institutes of Health, National Center for Advancing Translational Sciences, Clinical and Translational Sciences Award.

Acknowledgments: We would like to thank Tamara Harris from NIA for her support on the DECOS study.

Conflicts of Interest: The authors declare no conflict of interest.

References

1. Cooper, R.; Huang, L.; Hardy, R.; Crainiceanu, A.; Harris, T.; Schrack, J.A.; Crainiceanu, C.; Kuh, D. Obesity history and daily patterns of physical activity at age 60–64 years: Findings from the MRC National Survey of Health and Development. *J. Gerontol. Ser. Biomed. Sci. Med. Sci.* **2017**, *72*, 1424–1430. [CrossRef] [PubMed]

2. Keane, E.; Li, X.; Harrington, J.M.; Fitzgerald, A.P.; Perry, I.J.; Kearney, P.M. Physical activity, sedentary behavior and the risk of overweight and obesity in school-aged children. *Pediatr. Exerc. Sci.* **2017**, *29*, 408–418. [CrossRef] [PubMed]

3. Lange-Maia, B.S.; Newman, A.B.; Strotmeyer, E.S.; Harris, T.B.; Caserotti, P.; Glynn, N.W. Performance on fast-and usual-paced 400-m walk tests in older adults: are they comparable? *Aging Clin. Exp. Res.* **2015**, *27*, 309–314. [CrossRef] [PubMed]

4. Healy, G.N.; Winkler, E.A.; Brakenridge, C.L.; Reeves, M.M.; Eakin, E.G. Accelerometer-derived sedentary and physical activity time in overweight/obese adults with type 2 diabetes: Cross-sectional associations with cardiometabolic biomarkers. *PLoS ONE* **2015**, *10*, e0119140. [CrossRef] [PubMed]

5. Copeland, J.L.; Esliger, D.W. Accelerometer assessment of physical activity in active, healthy older adults. *J. Aging Phys. Act.* **2009**, *17*, 17–30. [CrossRef]

6. Pruitt, L.A.; Glynn, N.W.; King, A.C.; Guralnik, J.M.; Aiken, E.K.; Miller, G.; Haskell, W.L. Use of accelerometry to measure physical activity in older adults at risk for mobility disability. *J. Aging Phys. Act.* **2008**, *16*, 416. [CrossRef]

7. Gardner, A.W.; Poehlman, E.T. Assessment of free-living daily physical activity in older claudicants: Validation against the doubly labeled water technique. *J. Gerontol. Ser. Biol. Sci. Med. Sci.* **1998**, *53*, M275–M280. [CrossRef] [PubMed]

8. Richardson, C.A.; Glynn, N.W.; Ferrucci, L.G.; Mackey, D.C. Walking energetics, fatigability, and fatigue in older adults: The study of energy and aging pilot. *J. Gerontol. Ser. Biol. Sci. Med. Sci.* **2015**, *70*, 487–494. [CrossRef]

9. Losina, E.; Yang, H.; Stanley, E.; Katz, J.; Collins, J. Physical activity as a novel outcome of total knee replacement: Comparing self-report and objective PA assessments. *Osteoarthr. Cartil.* **2019**, *27*, S218–S219. [CrossRef]

10. Sabia, S.; Cogranne, P.; van Hees, V.T.; Bell, J.A.; Elbaz, A.; Kivimaki, M.; Singh-Manoux, A. Physical activity and adiposity markers at older ages: Accelerometer vs questionnaire data. *J. Am. Med. Dir. Assoc.* **2015**, *16*, 438.e7–438.e13. [CrossRef]

11. Baranowski, T. Validity and reliability of self report measures of physical activity: An information-processing perspective. *Res. Q. Exerc. Sport* **1988**, *59*, 314–327. [CrossRef]

12. Schrack, J.A.; Zipunnikov, V.; Goldsmith, J.; Bai, J.; Simonsick, E.M.; Crainiceanu, C.; Ferrucci, L. Assessing the "physical cliff": Detailed quantification of age-related differences in daily patterns of physical activity. *J. Gerontol. Ser. Biol. Sci. Med. Sci.* **2014**, *69*, 973–979. [CrossRef]

13. Centers for Disease Control and Prevention. More People Walk to Better Health. 2013. Available online: https://www.cdc.gov/vitalsigns/Walking/ (accessed on 8 November 2020).

14. Moe-Nilssen, R.; Helbostad, J.L. Estimation of gait cycle characteristics by trunk accelerometry. *J. Biomech.* **2004**, *37*, 121–126. [CrossRef]

15. Urbanek, J.K.; Zipunnikov, V.; Harris, T.; Fadel, W.; Glynn, N.; Koster, A.; Caserotti, P.; Crainiceanu, C.; Harezlak, J. Prediction of sustained harmonic walking in the free-living environment using raw accelerometry data. *Physiol. Meas.* **2018**, *39*, 02NT02. [CrossRef]

16. Sandroff, B.M.; Motl, R.W.; Suh, Y. Accelerometer output and its association with energy expenditure in persons with multiple sclerosis. *J. Rehabil. Res. Dev.* **2012**, *49*, 467. [CrossRef]

17. Randolph, T.W.; Harezlak, J.; Feng, Z. Structured penalties for functional linear models–partially empirical eigenvectors for regression. *Electron. J. Stat.* **2012**, *6*, 323–353. [CrossRef]

18. Kulbaba, M.W.; Clocher, I.C.; Harder, L.D. Inflorescence characteristics as function-valued traits: Analysis of heritability and selection on architectural effects. *J. Syst. Evol.* **2017**, *55*, 559–565. [CrossRef]

19. Gomulkiewicz, R.; Kingsolver, J.G.; Carter, P.A.; Heckman, N. Variation and evolution of function-valued traits. *Annu. Rev. Ecol. Evol. Syst.* **2018**, *49*, 139–164. [CrossRef]

20. Randolph, T.W.; Zhao, S.; Copeland, W.; Hullar, M.; Shojaie, A. Kernel-penalized regression for analysis of microbiome data. *Ann. Appl. Stat.* **2018**, *12*, 540. [CrossRef]

21. Urbanek, J.K.; Zipunnikov, V.; Harris, T.; Crainiceanu, C.; Harezlak, J.; Glynn, N.W. Validation of gait characteristics extracted from raw accelerometry during walking against measures of physical function, mobility, fatigability, and fitness. *J. Gerontol. Ser. A* **2017**, *73*, 676–681. [CrossRef]

22. Lange-Maia, B.S.; Strotmeyer, E.S.; Harris, T.B.; Glynn, N.W.; Simonsick, E.M.; Brach, J.S.; Cauley, J.A.; Richey, P.A.; Schwartz, A.V.; Newman, A.B.; et al. Physical activity and change in long distance corridor walk performance in the health, aging, and body composition study. *J. Am. Geriatr. Soc.* **2015**, *63*, 1348–1354. [CrossRef]

23. Pachi, A.; Ji, T. Frequency and velocity of people walking. *Struct. Eng.* **2005**, *84*, 36–40.

24. Ruppert, D.; Wand, M.P.; Carroll, R.J. *Semiparametric Regression*; Number 12; Cambridge University Press: Cambridge, UK, 2003.

25. Simonsick, E.M.; Fan, E.; Fleg, J.L. Estimating cardiorespiratory fitness in well-functioning older adults: Treadmill validation of the long distance corridor walk. *J. Am. Geriatr. Soc.* **2006**, *54*, 127–132. [CrossRef] [PubMed]

26. Simonsick, E.M.; Montgomery, P.S.; Newman, A.B.; Bauer, D.C.; Harris, T. Measuring fitness in healthy older adults: The Health ABC Long Distance Corridor Walk. *J. Am. Geriatr. Soc.* **2001**, *49*, 1544–1548. [CrossRef]

27. Goldsmith, J.; Scheipl, F.; Huang, L.; Wrobel, J.; Gellar, J.; Harezlak, J.; McLean, M.W.; Swihart, B.; Xiao, L.; Crainiceanu, C.; Reiss, P.T. *Refund: Regression with Functional Data*; R Package Version 0.1-22. 2020. Available online: https://CRAN.R-project.org/package=refund (accessed on 8 November 2020).

28. R Core Team. *R: A Language and Environment for Statistical Computing*; R Foundation for Statistical Computing: Vienna, Austria, 2019.

29. Gabell, A.; Nayak, U. The effect of age on variability in gait. *J. Gerontol.* **1984**, *39*, 662–666. [CrossRef]

30. Öberg, T.; Karsznia, A.; Öberg, K. Basic gait parameters: Reference data for normal subjects, 10–79 years of age. *J. Rehabil. Res. Dev.* **1993**, *30*, 210–210.

31. Menz, H.B.; Lord, S.R.; Fitzpatrick, R.C. Age-related differences in walking stability. *Age Ageing* **2003**, *32*, 137–142. [CrossRef]

32. Kavanagh, J.J.; Menz, H.B. Accelerometry: A technique for quantifying movement patterns during walking. *Gait Posture* **2008**, *28*, 1–15. [CrossRef] [PubMed]
33. Ko, S.u.; Stenholm, S.; Ferrucci, L. Characteristic gait patterns in older adults with obesity—Results from the Baltimore Longitudinal Study of Aging. *J. Biomech.* **2010**, *43*, 1104–1110. [CrossRef]
34. Harding, G.T.; Hubley-Kozey, C.L.; Dunbar, M.J.; Stanish, W.D.; Wilson, J.L.A. Body mass index affects knee joint mechanics during gait differently with and without moderate knee osteoarthritis. *Osteoarthr. Cartil.* **2012**, *20*, 1234–1242. [CrossRef]
35. Moissenet, F.; Leboeuf, F.; Armand, S. Lower limb sagittal gait kinematics can be predicted based on walking speed, gender, age and BMI. *Sci. Rep.* **2019**, *9*, 9510. [CrossRef] [PubMed]
36. Zhang, S.; Rowlands, A.V.; Murray, P.; Hurst, T.L. Physical activity classification using the GENEA wrist-worn accelerometer. *Med. Sci. Sports Exerc.* **2012**, *44*, 742–748. [CrossRef] [PubMed]
37. Preece, S.J.; Goulermas, J.Y.; Kenney, L.P.; Howard, D. A comparison of feature extraction methods for the classification of dynamic activities from accelerometer data. *IEEE Trans. Biomed. Eng.* **2009**, *56*, 871–879. [CrossRef]
38. Mannini, A.; Intille, S.S.; Rosenberger, M.; Sabatini, A.M.; Haskell, W. Activity recognition using a single accelerometer placed at the wrist or ankle. *Med. Sci. Sports Exerc.* **2013**, *45*, 2193–2203. [CrossRef]
39. Kundu, M.G.; Harezlak, J.; Randolph, T.W. Longitudinal functional models with structured penalties. *Stat. Model.* **2016**, *16*, 114–139. [CrossRef] [PubMed]
40. Reuter, I.; Mehnert, S.; Leone, P.; Kaps, M.; Oechsner, M.; Engelhardt, M. Effects of a flexibility and relaxation programme, walking, and nordic walking on Parkinson's disease. *J. Aging Res.* **2011**, *2011*, 232473. [CrossRef] [PubMed]

Publisher's Note: MDPI stays neutral with regard to jurisdictional claims in published maps and institutional affiliations.

sensors

MDPI

Article

Automatically Determining Lumbar Load during Physically Demanding Work: A Validation Study

Charlotte Christina Roossien [1,*], **Christian Theodoor Maria Baten** [2,3], **Mitchel Willem Pieter van der Waard** [2,3], **Michiel Felix Reneman** [1] and **Gijsbertus Jacob Verkerke** [1,3]

[1] Department of Rehabilitation Medicine, University Medical Center Groningen, University of Groningen, Hanzeplein 1, 9713 GZ Groningen, The Netherlands; m.f.reneman@umcg.nl (M.F.R.); g.j.verkerke@gmail.com (G.J.V.)

[2] Department Rehabilitation Technology, Roessingh Research and Development, Roessinghsbleekweg 33b, 7522 AH Enschede, The Netherlands; c.baten@rrd.nl (C.T.M.B.); M.vanderWaard@rrd.nl (M.W.P.v.d.W.)

[3] Department of Biomechanical Engineering, University of Twente, Drienerlolaan 5, 7522 NB Enschede, The Netherlands

* Correspondence: c.c.roossien@umcg.nl

Abstract: A sensor-based system using inertial magnetic measurement units and surface electromyography is suitable for objectively and automatically monitoring the lumbar load during physically demanding work. The validity and usability of this system in the uncontrolled real-life working environment of physically active workers are still unknown. The objective of this study was to test the discriminant validity of an artificial neural network-based method for load assessment during actual work. Nine physically active workers performed work-related tasks while wearing the sensor system. The main measure representing lumbar load was the net moment around the L5/S1 intervertebral body, estimated using a method that was based on artificial neural network and perceived workload. The mean differences (MDs) were tested using a paired t-test. During heavy tasks, the net moment (MD = 64.3 ± 13.5%, $p = 0.028$) and the perceived workload (MD = 5.1 ± 2.1, $p < 0.001$) observed were significantly higher than during the light tasks. The lumbar load had significantly higher variances during the dynamic tasks (MD = 33.5 ± 36.8%, $p = 0.026$) and the perceived workload was significantly higher (MD = 2.2 ± 1.5, $p = 0.002$) than during static tasks. It was concluded that the validity of this sensor-based system was supported because the differences in the lumbar load were consistent with the perceived intensity levels and character of the work tasks.

Keywords: physically active workers; low back pain; inertial motion units

Citation: Roossien, C.C.; Baten, C.T.M.; van der Waard, M.W.P.; Reneman, M.F.; Verkerke, G.J. Automatically Determining Lumbar Load during Physically Demanding Work: A Validation Study. *Sensors* **2021**, *21*, 2476. https://doi.org/10.3390/s21072476

Academic Editor: Angelo Maria Sabatini

Received: 29 January 2021
Accepted: 30 March 2021
Published: 2 April 2021

Publisher's Note: MDPI stays neutral with regard to jurisdictional claims in published maps and institutional affiliations.

1. Introduction

Physically active workers sometimes can experience muscle and spinal overload while performing their physically demanding jobs [1]. Such an overload is hypothesized to be due to a misbalance between the physical workload and the individual capacity of each worker [2]. This misbalance may cause health problems among these workers, such as musculoskeletal disorders like lower back pain [3–7]. These problems usually result in the loss of productivity [8–10], loss of quality and safety [1,11], and absenteeism [1,12]. Hence, to help prevent these health problems, to improve rehabilitation, and promote return to work and sustainable employability, it is important to investigate and optimize the musculoskeletal load while performing physically demanding jobs [13–15].

There is a need for a device that can measure the individual work-related lumbar load exposure objectively while performing a physically demanding job [16,17]. Typically, this lumbar load is represented by the net moment around the center of the intervertebral body at the spinal level L5/S1. Various methods have been developed to estimate the net moment in the lower back under known load-handling conditions [18–22]. All of these methods use 3D body segment kinematics data acquired using a marker-based motion analysis system, and load conditions were "known" through the direct measurement of the reaction forces exerted to the feet or hands or from detailed information on the loads handled.

Inertial motion capture systems seem useful for assessing work-related back load exposure in workers in industrial environments through monitoring the working postures of individuals and through driving biomechanical-linked segment models to estimate 3D net moments and forces in the lower spine [18,19,23–30]. Using inertial magnetic measurement units (IMMUs) allows for more freedom in 3D kinematics assessments in comparison to marker-based motion analysis systems [31]. However, the need to measure all the forces exerted on the human's body, e.g., by force plates embedded in the lab floor, severely limits their practical applications. A more mobile alternative is to use instrumented shoes that measure ground reaction forces while walking around [19,20,32,33]. This system provides more freedom of movement but requires that no other external forces be exerted to the lower body (e.g., by leaning against a table or supporting the load being handled). Additionally, the relatively large weight of these shoes and their current design make them less usable in practice. Therefore, alternative methods for estimating lumbar load assessment were developed, that do not require force assessment. In this study, an artificial neural network (ANN)-based method was applied to estimate 3D net moments (L5/S1) which was driven by electromyographic (EMG) and trunk kinematics data. It was trained in supervised mode during the initial part of each session with a limited set of calibration trials [18,19,21,31]. Target net moments for training were generated by direct estimation using a linked segment model (LSM-based method [24,29]) scaled by the height and weight of the subject. The LSM-based method was just driven by the kinematics data from IMMUs on the trunk and arms. The weight and inertial properties of the load handled during the calibration trials have to be known and their kinematics are derived from hand kinematics. In the actual trials, a trained ANN-based method was used to estimate the net moments from the EMG magnitude data and the IMMU kinematics data and of a subject during actual work. This ANN-based method has been developed over several iterations from 1993 until the current date and has been evaluated in several studies, e.g., in direct comparison against a state-of-the-art laboratory-based method driven by marker-based kinematics and force plates, the results of which were found to be promising [24]. The same group also successfully applied and validated the same approach for estimating shoulder joint load exposure estimation, confirming feasibility [34]. This ANN-based method, driven by trunk muscles EMG amplitude and IMMU kinematics, is therefore considered potentially useful for monitoring mechanical workload in the context of the worker's postures and movements, while performing physically demanding jobs [19,21,22]. Ultimately, this may represent a tool that provides workers and ergonomists with instant feedback, which may contribute to preventing excess load exposure for a worker without back complaints. It might also be used for workers returning to work after an injury, and/or as part of their rehabilitation. However, the validity and usability of this system in the uncontrolled real-life working environment of physically active workers are still unknown.

The objective of this study was to test the discriminant validity of an ANN-based method for load assessment during actual work. The research questions were: (1) What is the base quality of this ANN-based method in estimating lumbar load when applied to the trial data of this study with known load handling? (2) Can this ANN-based method detect differences in load intensity and perceived workload during light and heavy tasks? (3) Can the system detect differences in load-variability during the static and dynamics tasks? (4) Can the system detect (a) symmetrical lumbar load difference around the anterior–posterior, mediolateral and longitudinal axes?

2. Materials and Methods

2.1. Subjects

A total of 23 subjects participated in this study, all of whom were physically active workers recruited through flyers distributed within selected companies. These selected companies were active in medical disinfection care, industrial chemical cleaning, and technical services. All subjects were informed about the study through an information letter and received a verbal explanation before the start of the study. The inclusion criterion was being a physically active worker aged between 18 and 67. The exclusion criteria were having any cardiovascular diseases; using pacemakers or other vital electronic devices; having high levels of pain or injuries in the back, shoulders, or upper extremities; or being at an advanced stage (around 20 weeks) of pregnancy. The Medical Ethics Committee of the University Medical Center Groningen, the Netherlands, issued a waiver for this study, stating that it did not involve medical research according to the Dutch law (M17.208063), and all subjects signed an informed consent form.

2.2. Study Design and Procedures

Each session with every subject comprised three phases: (1) trials for upper body segment calibration; (2) trials with known loads for supervised training and training quality validation; and (3) trials illustrating performance during a set of work-related tasks.

In phase 1, the subjects were asked to perform a set of movements to calibrate the orientation of the IMMUs relative to the body segments, while wearing the complete sensor set-up with IMMUs (and bipolar surface electromyography (sEMG) electrodes). The resulting segment calibration parameters were used to translate all sensor casing kinematics within a session to body segment and joint kinematics. The set contained $90°$ trunk bending, $45°$ trunk lateroflexion, $45°$ trunk rotation, $45°$ shoulder flexion, and $90°$ shoulder abduction. This was repeated five times and followed by three seconds of standing in a neutral anatomic position with the arms hanging next to the body with thumbs pointing forward.

In phase 2, the ANN-based method performance was tested against the LSM-based method by comparing the estimated moments by both methods in bending, flexion, and rotation movements while hand holding a 6 kg load (question 1). For this, the ANN was first trained with net moments directly estimated with the LSM-based method from similar tasks handling no load ("0 kg task") and holding a load of a 10 kg trunk. Based on previous experience (unpublished sensitivity study) a simple feed-forward ANN was used with 1 hidden layer of 31 elements and sigmoid transfer functions with a fixed training criterium: "sum of RMSE Mx, RMSE My and RMSE Mz < 10 Nm". This required about 10 s of computer time on a regular desktop PC for a typical input vector with about 6000 elements. The ANN was driven by EMG amplitudes from all 4 locations plus 3D kinematics only from the 2 IMMU sensors positioned on the sternum and sacrum: per segment, the 3D angular velocity and 3D linear acceleration signals were used. The rationale for not using the available 3D orientation data was that: (1) information of inclination is already present in the 3D accelerometer signal; and (2) in an industrial environment, a substantial risk exists of having large errors in IMMU orientation estimates by the disturbance of the natural earth magnetic field or permanent magnetization of the magnetometer in the IMMUs [35]. In preparation for phase 3, the ANN-based estimator was trained for each subject separately using all their respective bending, flexion and rotation tasks for all three weights (0, 6 and 10 kg).

In phase 3, questions 2, 3, and 4 about the discriminant validity of the ANN-based method were explored. The subjects performed job-specific work-related tasks for 5–10 min of different load intensity and variability. For all these tasks, net moment curves were estimated using the ANN-based network trained at the end of phase 2. Additionally, these tasks were ranked according to the checklist of physical workload [36]. Before the start of the measurements, the subjects received a questionnaire to identify the daily tasks and the frequency, duration and perceived the workload per task. This physical workload questionnaire of Peereboom and Lange [36] contained the following questions: What are tasks you perform on a daily basis at work? How often are these tasks performed on an average day? How heavy do you find these tasks? All subjects ranked tasks according to the load, starting with the heaviest task [36]. From the list of work-related tasks, four tasks were chosen, which may vary from one individual to another, which represented: (1) a light task with a low workload on the lower spine; (2) a heavy task with a high workload on the lower spine; (3) a static task (working with the lumbar region in the same posture); and (4) a dynamic (lifting) task in different (spinal) working postures [37,38]. To explore questions 2 and 4, the task with the highest workload was selected as a heavy task, and the task with the lowest workload was selected as a light task. Net moment data curve appearances were discussed with respect to the trial perceived workload. To explore questions 3 and 4, the criterion for the static task was that the lumbar region was held in the same posture or joint position for at least 4 s throughout the task with low variances in the lumbar posture when changing the posture (ISO standard 11226:2000) [37]. The criteria for the dynamic task were as follows: the task must be a lifting one and the lumbar spine should vary in posture. After every task, the subjects were asked to rate the perceived workload of the three tasks using Borg CR-10 rating scale, ranging from 0 to 10 (0 = not burdensome; 10 = extremely heavy) [39].

2.3. Materials

2.3.1. Surface Electromyography Acquisition

All sEMG recordings were performed using a wearable sEMG instrument (Polybench Dipha; Inbiolab, Roden, the Netherlands). Bipolar electrodes (Covidien Kendall™ H124SG Ag/AgCl electrodes; Medtronic, Minneapolis, MN, USA) with an interelectrode distance of 2 cm (heart to heart) were placed bilaterally on the longissimus thoracis muscles at L1 (±3 cm horizontal from L1) and the iliocostalis lumborum muscles at L2–L3 (±6.5 cm horizontal from L2–L3), along with a reference electrode placed at the processus spinosus of C7 (Roy et al., 1995) (see Figure 1).

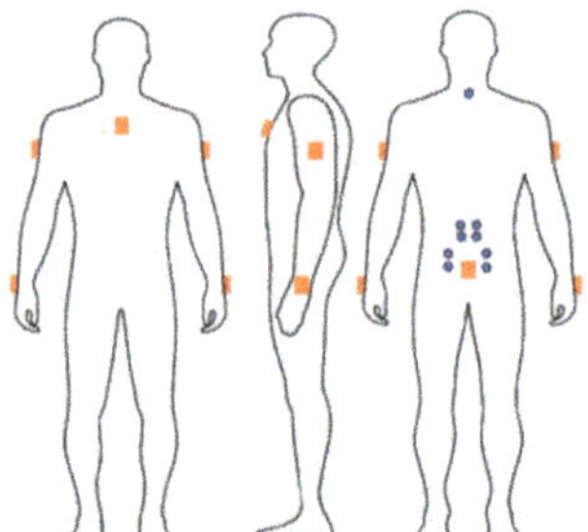

Figure 1. The positioning of the sEMG electrodes and IMMUs on the body. The surface electromyography (sEMG electrodes, blue circles) positioned on the longissimus thoracis muscles at L1 and the iliocostalis lumborum muscles at L2–L3 with a reference electrode placed at the processus spinosus of C7. The inertial magnetic measuring units (IMMUs, orange blocks) positioned on the sternum, upper, and lower (left and right) arms, and pelvis (sacrum) with a front (left), side (middle), and back view (right).

2.3.2. Kinematics Acquisition and Net Moment Estimation

Six wired IMMUs (MVN Awinda; Xsens, Enschede, the Netherlands) were used to record 3D body segment kinematics. The IMMUs were placed on the sternum, upper and lower arms, and pelvis (sacrum) [19,22,40], as shown in Figure 1. The sample rate was 50 Hz. All IMMU data-acquisition procedures, as well as the translation of IMMU casing kinematics data to body segment and joint kinematics data, were performed with the FusionTools/XCM software suite (Roessingh Research and Development, Enschede, the Netherlands) [19,23] using the Xsens application programming interface (API 4.7). Using the same software suite, EMG data preparation (amplitude estimation by smoothed rectification, intrapolated resampling to 50 Hz) and synchronization with IMMU data were performed. The IMMU and sEMG data were synchronized by tapping on two sEMG electrodes and one IMMU electrode before the start of the measurement. This created a marker (peak) in the data which was used to synchronize both systems in a semi-automated procedure indicating both markers, manually using a cursor. Estimated synchronization errors < 0.05 s.

The LSM was driven by kinematics derived from all 6 IMMUs: 3D orientation (represented by quaternions); 3D angular velocity; and 3D linear acceleration (only used in controlled environment). The ANN was driven by EMG amplitudes plus 3D kinematics from only the 2 IMMU sensors positioned on sternum and sacrum: per segment the 3D angular velocity and 3D linear acceleration were signals used. All load exposure estimations were also performed using this software suite, with both LSM-based and ANN-based methods, as well as the calculation of all the descriptive statistics of the net moment curves and root-mean-square error (RMSE) values and correlation coefficients, comparing the target and estimated net moments.

2.4. Data Analysis

To test the correlation between the ANN-based method and the LSM-based method (question 1) in phase 2, the main evaluation comprised a comparison of ANN-based estimated and target net moment trajectories in 3D. The primary outcomes were the 3D curves of the estimated and target moment, as well as of the moment norm, and their RMSE curves of the moment magnitude ($||M||$, calculated through the net moment vector norm). The evaluation was performed for every rotation axis separately and for the net moment magnitude by means of the visual inspection of data plots and also by evaluating RMSE values between the estimated and target moments and the Pearson's correlation coefficient (r with $0.1 < r < 0.5$ indicating a weak correlation, $0.5 < r < 0.7$ indicating a moderate correlation, $0.7 < r < 0.9$ indicating a good correlation and > 0.9 indicating a very good correlation [41]) and its squared value (r^2) representing the amount of explained variance in estimation. Movement in the mediolateral transverse (y) axis was named "trunk bending", and movement in the anterior–posterior (x) axis was named "trunk lateroflexion" and the combined trunk twisting movement mainly in the longitudinal (z) axis was named "trunk rotation". If a strong correlation was found between the ANN-based method and the LSM-based method, the ANN-based method was judged to be of an acceptable level.

In phase 3, the results of the questionnaire were categorized on the basis of the perceived workload, with 1 meaning a light task and 5 meaning a very heavy task [36]. According to these scores, the tasks for test questions 2, 3 and 4 were selected. To test whether this sensor system could distinguish differences between the intensity and variability of the estimated lumbar load, a discriminant validity analysis was performed. The primary parameters that we compared were the mean, peak (max) and variance (standard deviation within a subject) of the net moment in the lumbar region and results. They were also discussed in relation to the perceived workload (Borg CR-10) [39].

To explore if this ANN-based method can distinguish the estimated lumbar load differences in intensity levels (question 2), differences between the light and heavy tasks were analyzed. The hypothesis was that during heavy tasks, the mean net moments in the lumbar region would be significantly higher. Additionally, the hypothesis that the perceived workload (Borg CR-10 score) of the heavy tasks was significantly higher than the light tasks was tested. The hypothesis of question 3 (variability level) was that the variance in the net moments in the lumbar region was higher during dynamic tasks than during the static tasks.

To explore the (a) symmetrical character of working postures, the movement direction of the moment around the anterior–posterior, mediolateral and longitudinal axes (question 4) was assessed. The net moment data were divided into two segments based on the direction of movement (positive or negative movement direction); anterior–posterior with lateroflexion to the left (positive) and right (negative), mediolateral with flexion (positive) and extension (negative), longitudinal with rotation to the left (positive) and right (negative) axes. This was done based on the hypothesis that the tasks had an asymmetrical character.

These positive and negative moment segments were tested separately with the same hypothesis described for questions 2 and 3.

Questions 2, 3 and 4 were tested by firstly exploring the distribution of the data using a Shapiro–Wilk test of normality and was considered to be normally distributed if $p \leq 0.05$. Normally distributed data between the tasks were assessed using a paired t-test and non-normally distributed data were also tested using the Wilcoxon signed-rank test. A difference of the net moment was significant when $p \leq 0.05$. The results are presented as the absolute and relative mean or mean difference (MD) $\pm$ the standard deviation (SDD). All statistical analyses were performed using IBM SPSS Statistics (version 25; IBM Corp., Armonk, NY, USA). In Table 1 can be found a framework to describe per research question the activities.

Table 1. Framework study design. A framework to describe per research question the activities in terms of inertial magnetic measurement units (IMMUs) and surface electromyography sensors (sEMG)), discriminant validity and related criteria with artificial neural network (ANN) and linked segment model (LSM).

Step	Research Question	Activity	Discriminant Validity	Criteria
1	1	Calibration measurement with IMMUs and sEMG	ANN vs. LSM method	Mean r^2 of subject ≥ 0.5
2	2, 4	Physical workload checklist	Light vs. heavy task	Light: lowest workload per job Heavy: highest workload per job
	3, 4		Static vs. dynamic task	Static: same posture at least 4 s Dynamic: different working postures
3	2, 3, 4	Performing work-related tasks with IMMUs and sEMG	Test research questions	$r \geq 0.5$ $p \leq 0.005$
4	2, 3, 4	Perceived workload questionnaire	Test research questions	$p \leq 0.005$
5	1, 2, 3, 4	Calibration measurement with IMMUs and sEMG	Check measurement quality	N/A

3. Results

Out of the 23 workers who participated in this study, the data of 12 subjects were not useable because of data-acquisition errors in either the IMMUs or sEMG hardware during essential trials for ANN training. In addition, the data of another two subjects performing dynamic and/or heavy tasks contained data-acquisition errors. Therefore, these 14 subjects were excluded from the analysis, leaving a set of data of nine subjects (eight males, one female): four medical disinfectant care workers, three maintenance engineers, and two industrial chemical cleaners. Their mean age was 33.7 ± 10.3 years, height 185 ± 9 cm, and weight 93 ± 12 kg. Eight subjects were right-handed, and one subject was left-handed.

3.1. Question 1: Base Quality of ANN-Based Method

The base quality was studied by examining the correlation between the corresponding moment curves from the ANN-based method and the LSM-based method. Table 2 (and Appendix A.1) shows the correlations. For one subject (subject 6), the data of the calibration movement with 6 kg were not usable and the correlation with the LSM-based model could not be calculated. Therefore, the results of subject 6 were not included in the mean results in Table 2. Hence, this subject was included in the study. The trained ANN method estimation base quality differs from one subject to another.

Table 2. ANN-based method performance in handling known loads. Shown are the correlations between the net moment at L5/S1 estimated with the ANN-based method and with the LSM-based method. Correlation is represented by the Pearson correlation coefficient (r) and determination coefficient (r^2) are shown only for the axis of movement in each task (i.e., in the mediolateral axis (y) for the trunk bending tasks, in the anterior–posterior axis (x) for the lateroflexion tasks, and the longitudinal axis (z) for the rotation tasks, respectively. Shown are individual values for each subject plus the mean and standard deviation (SDD) over all subjects. No valid data were obtained for subject number 6 for reasons of partially missing data in the 6 kg trial.

Movement	Axis		Subject 1	2	3	4	5	6	7	8	9	Mean	SDD
Bending	y	r	0.97	0.98	0.96	0.97	0.98	-	0.96	0.35	0.98	0.89	0.22
		r^2	0.94	0.95	0.93	0.93	0.96	-	0.92	0.12	0.96	0.84	0.29
Latero-flexion	x	r	0.94	0.93	0.95	0.95	0.94	-	0.85	0.32	0.92	0.85	0.22
		r^2	0.89	0.87	0.91	0.90	0.88	-	0.72	0.10	0.84	0.76	0.27
Rotation	z	r	0.63	0.23	0.84	0.84	0.81	-	0.76	0.38	0.92	0.68	0.24
		r^2	0.40	0.06	0.70	0.70	0.65	-	0.58	0.15	0.84	0.51	0.28

Good correlations were observed in the trunk bending (mean $r = 0.89 \pm 0.22$) and lateroflexion (mean $r = 0.85 \pm 0.22$). All subjects but one (subject 8) showed good correlations ($r \geq 0.85$). Subject 8 showed a weak correlation ($r \leq 0.38$) for both movements. A moderate correlation was observed during the trunk rotation in the longitudinal axis ($r = 0.68 \pm 0.24$) and the anterior–posterior axis ($r = 0.65 \pm 0.24$). However, subject 2 showed a weak correlation ($r = 0.23$). Overall, the results were within the acceptable range ($r > 0.5$).

Table 3 compares the RMSE values between the ANN-based method and the LSM-based method. These results indicate a mean estimation error of 9.25 ± 6.01 Nm, relative to the typical peak net moment range from 150 to 220 Nm (see Table 4).

Table 3. RMSE between the ANN-based method and the biomechanical model. The root means square error (RMSE) between the net moment curves at L5/S1 was estimated by the ANN-based method and the LSM-based method. Values are shown for the three movements trunk bending, lateroflexion and rotation with the anterior–posterior axis (x), mediolateral axis (y) and longitudinal axis (z). Shown are individual values for each subject plus the mean and standard deviation (SDD) over all subjects. No valid data were obtained for subject number 6 for reasons of partially missing data in the 6 kg trial.

Movement	Axis	Subject 1	2	3	4	5	6	7	8	9	Mean	SDD
Bending	x	12.5	5.00	10.7	4.49	7.09	-	5.24	17.2	4.84	8.38	4.63
	y	8.95	8.65	9.85	8.56	10.4	-	21.1	9.31	6.17	10.4	4.51
	z	6.49	3.91	5.66	3.15	5.87	-	6.70	9.64	2.61	5.50	2.27
Lateroflexion	x	17.2	10.9	12.5	7.12	13.4	-	19.5	31.3	11.9	15.5	7.44
	y	23.7	11.2	11.7	6.83	16.6	-	13.2	19.5	11.7	14.3	5.37
	z	11.3	8.88	10.4	3.48	9.83	-	8.73	24.0	6.92	10.4	5.98
Rotation	x	2.98	2.55	3.27	3.89	6.07	-	5.15	18.9	2.73	5.68	5.47
	y	4.28	4.10	3.73	4.06	6.26	-	6.05	21.5	4.33	6.79	6.02
	z	5.01	5.73	4.41	4.12	6.32	-	6.90	12.8	4.87	6.27	2.80

Table 4. Tasks per job type. Based on the results of the checklist for the physical workload.

Task Perception	Medical Disinfect Care Worker	Maintenance Engineer	Industrial Chemical Cleaner
Light	Changing personal protective working clothing before or after working in contaminated space	Administration of technical maintenance service	Disassemble parts of a gas mask as preparation for cleaning
Static	Assembly or lamination of surgical instruments	Tinkering under a machine to fix or loosen components	Cleaning chemical hazard suit in sink
Heavy and dynamic	Carrying bins of 3 up to 10 kg over a distance of about 1 m	Moving (pushing and/or pulling) bin with wastewater of 1000 kg or carrying a toolbox of 35 kg over a distance of about 50 m	Carrying bins of 5–10 kg over a distance of 50 m

3.2. Question 2: Capability to Distinguish Task Intensity

Table 4 shows the tasks per job according to the results of the checklist of physical workload. All subjects, except for one industrial chemical cleaner, perceived the dynamic task as the heaviest task of their job. Small differences in the checklist for physical workload scores were observed, which were related to the diversity in the individual job description.

Table 5 presents the estimated net moments per task together with the experienced workload of the tasks according to the subjects. During all the tasks, the mean net lumbar moment was 25.2 ± 16.8 Nm, the mean peak moment was 179.5 ± 152.9 Nm, and the mean variance was 15.5 ± 11.5 Nm.

Table 5. Net moments, load ranking, and perceived workload for each task. The questionnaire workload factor with 1 = light work and 5 = very heavy task. The experienced workload (Borg CR-10) according to the subjects with 0 = not burdensome and 10 = extremely heavy.

Task Perception	Net Moment (Nm)			Questionnaire Load Factor (1–5)	Perceived Workload (Borg 0–10)
	Mean	Peak	Variance		
Light task	18.7 ± 8.1	166.4 ± 195.5	13.0 ± 10.6	1.0 ± 0.0	0.9 ± 0.8
Static task	26.3 ± 19.2	153.5 ± 106.1	13.7 ± 9.9	3.6 ± 1.3	3.8 ± 1.6
Heavy and dynamic task	30.7 ± 20.0	218.5 ± 154.4	19.8 ± 13.8	4.8 ± 0.4	6.0 ± 2.0

Table 6 summarizes the differences between the light and heavy tasks presented for all subjects, with a typical example in Figure 2. The graphs of all subjects can be found in Appendix A.2. It can be seen that the net moments estimated using the ANN-based method exhibit overall higher moments during heavy tasks with more variances than during light tasks. The differences in the mean net moment between light and heavy tasks were significant (MD = $64.3 \pm 72.1\%$, $p = 0.028$), whereas other differences were not. The perceived workload was significantly higher during the heavy tasks than during light tasks (MD = 5.1 ± 2.1, $p < 0.001$).

Table 6. Light vs. heavy tasks. Shown are the absolute (Nm) and relative (%) differences between the light tasks and the heavy tasks through mean difference (MD: heavy–light task) and its standard deviation (SDD) plus 95% confidence interval (CI) of these differences for the moment magnitude ($||M||$) with p-value of paired t-test.

	Absolute (Nm)		Relative (%)		
Parameter	MD ± SDD	(95% CI)	MD ± SDD	(95% CI)	p
Mean	12.0 ± 13.5	[1.7;22.4]	64.3 ± 13.5	[8.9;119.8]	0.028
Peak	52.1 ± 256.9	[−145.4;249.6]	23.9 ± 117.6	[−66.5;114.2]	0.560
Variance	6.8 ± 9.4	[−0.4;14.0]	52.1 ± 71.8	[−3.1;107.3]	0.061

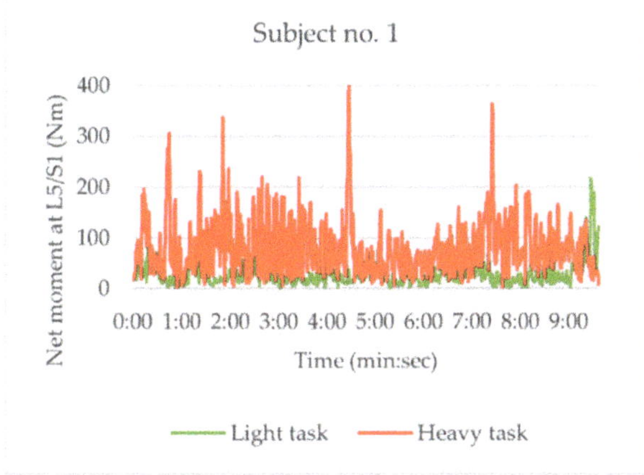

Figure 2. Typical example of net moment curves during light (green) and heavy (red) tasks (Subject 1).

3.3. Question 3: Capability to Distinguish Static/Dynamic Task Variance

Table 7 summarizes the differences between the static and dynamic tasks presented for all subjects, with a typical example in Figure 3. The graphs of all subjects can be found in Appendix A.3. It can be seen that the mean net moments of the magnitude estimated using the ANN-based method exhibit overall higher values during dynamic tasks with more variances than during static tasks. The difference in the variance between static and dynamic tasks was significant (MD = 44.8 ± 48.9%, p = 0.025). The perceived workload was significantly higher during the dynamic tasks than during static tasks (MD = 2.2 ± 1.5, p = 0.002).

Table 7. Static vs. dynamic tasks. Shown are the absolute (Nm) and relative (%) differences between the static tasks and the dynamic tasks through mean difference (MD: dynamic-static task) and its standard deviation (SDD) plus 95% confidence interval (CI) of these differences for the moment magnitude ($||M||$) with p-value of paired t-test.

	Absolute (Nm)		Relative (%)		
Parameter	MD ± SDD	(95% CI)	MD ± SDD	(95% CI)	p
Mean	4.42 ± 8.03	[−1.8;10.6]	16.8 ± 30.6	[−6.7;40.4]	0.137
Peak	65.0 ± 138.6	[−41.5;171.5]	42.3 ± 90.3	[−27.0;111.7]	0.197
Variance	6.13 ± 6.69	[1.0;11.3]	44.8 ± 48.9	[7.2;82.4]	0.025

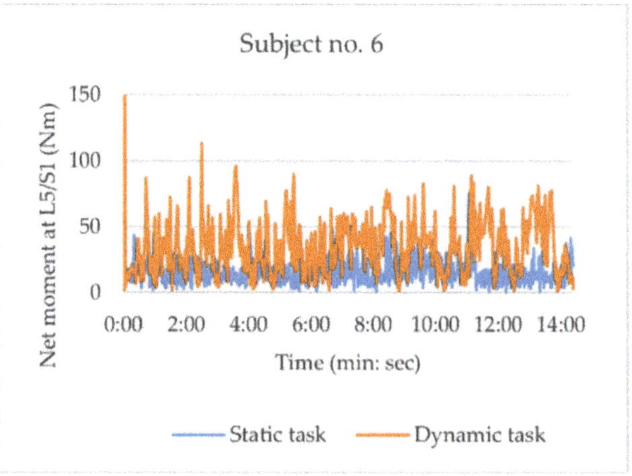

Figure 3. Typical example of net moment curves during static (blue) and dynamic (orange) tasks (Subject 6).

3.4. Question 4: (a) Symmetrical Lumbar Load

Table 8 summarizes the differences between light and heavy tasks presented while taking into account the direction of the movement around the axis (e.g., flexion vs. extension). Similar results of the intensity levels were observed as shown in Table 6. It was observed that the mean net moment of the magnitude was significantly higher during heavy tasks than during light tasks (MD $\geq$ 92.3 $\pm$ 105.4%, $p \leq 0.030$) as well as in the singular anterior–posterior axis (MD $\geq$ 56.8 $\pm$ 44.9%, $p \leq 0.016$). All the other differences were not significant.

Table 8. Direction of the net moment of the magnitude and around the axis of the light vs. heavy tasks. Shown are the absolute (Nm) and relative (%) differences between the light tasks and the heavy tasks through mean (MD: heavy–light task), standard deviation (SDD), and 95% confidence interval (CI) of these differences for the anterior–posterior (Mx) with lateroflexion to the left (positive) and right (negative), mediolateral (My) with flexion (positive) and extension (negative), longitudinal (Mz) with rotation to the left (positive) and right (negative) axes separately and for the moment magnitude (| |M| |) with a p-value of paired t-test.

Parameter	Direction	Axis	Absolute (Nm) MD ± SDD	Absolute (Nm) (95% CI)	Relative (%) MD ± SDD	Relative (%) (95% CI)	p
Mean	Positive	| |M| |	152.0 ± 92.8	[80.6;223.4]	1324.0 ±808.6	[702.4;1945.5]	0.001
		Mx	5.6 ± 4.4	[2.2;8.9]	56.8 ± 44.9	[22.3;91.3]	0.005
		My	−0.6 ± 8.6	[−7.2;6.0]	−2.5 ± 33.6	[−28.4;23.3]	0.828
		Mz	1.3 ± 3.6	[−1.5;4.1]	15.6 ± 43.7	[−18.0;49.1]	0.317
	Negative	| |M| |	9.5 ± 10.9	[1.8;17.9]	92.3 ± 105.4	[11.3;173.3]	0.030
		Mx	8.4 ± 8.3	[2.1;14.8]	95.0 ± 93.6	[23.1;166.91]	0.016
		My	3.4 ± 13.5	[−5.7;12.5]	24.4 ± 97.2	[−41.0;90.4]	0.424
		Mz	4.8 ± 7.2	[−0.7;10.3]	53.5 ± 79.7	[−7.8;114.8]	0.079
Variance	Positive	| |M| |	6.6 ± 13.5	[−2.5;15.7]	37.3 ± 76.8	[−14.2;88.9]	0.138
		Mx	1.3 ± 7.5	[−3.8;6.4]	11.9 ± 68.0	[−33.8;57.5]	0.575
		My	1.1 ± 11.2	[−6.5;8.6]	6.8 ± 72.4	[−41.8;55.5]	0.762
		Mz	−1.1 ± 6.7	[−5.6;3.5]	−12.2 ± 76.7	[−63.7;39.3]	0.610
	Negative	| |M| |	2.2 ± 9.5	[−4.2;8.6]	17.3 ± 73.7	[−32.2;66.8]	0.453
		Mx	5.3 ± 9.6	[−1.2;11.8]	38.1 ± 69.4	[−8.6;84.7]	0.099
		My	1.1 ± 8.6	[−4.7;6.8]	9.9 ± 79.0	[−43.1;63.0]	0.686
		Mz	2.3 ± 8.4	[−3.3;8.0]	24.3 ± 88.4	[−35.0;83.7]	0.382

Table 9 summarizes the differences between the static and dynamic tasks while taking into account the direction of the movement around the axis. When the direction of the moment around the separated axis was considered, significantly less variance was observed during rotation to the left around the longitudinal axis during dynamic tasks (MD = $-90.8 \pm 10.5\%$, $p = 0.031$). In addition, significant differences in the mean moment ($||M||$, M_y, and M_z) were observed in the positive direction ($-47.1\% \leq$ MD $\leq 302.7\%$, $p \leq 0.042$), but were not significant in the negative direction ($-54.0\% \leq$ MD $\leq -276.5\%$, $p \geq 0.326$).

Table 9. Direction of the net moment of magnitude and around the axis of the static vs. dynamic tasks. Shown are the absolute (Nm) and relative (%) differences between the static tasks and the dynamic tasks through mean difference (MD: static–dynamic) and its standard deviation (SDD) plus 95% confidence interval (CI). These are shown for the net moments around each of the three axes separately (in the right-handed axes frame): for the moments around the anterior–posterior axis (Mx), with the net moment positive for direction of rotation to the left; for the moments around the mediolateral axis (My), with the net moment positive in the forward flexion direction; for the moments around the longitudinal axis (Mz), with the net moment positive in the direction of rotation to the left and for the moment magnitude ($||M||$) with a p-value of paired *t*-test.

Parameter	Direction	Axis	Absolute (Nm)		Relative (%)		p
			MD ± SDD	(95% CI)	MD ± SDD	(95% CI)	
Mean	Positive	$||M||$	100.5 ± 132.8	[11.3;189.7]	302.7 ± 399.7	[34.1;571.2]	0.031
		Mx	−6.6 ± 14.8	[−16.6;3.3]	−34.5 ± 76.9	[−86.1;17.1]	0.168
		My	−18.1 ± 25.7	[−35.4;−0.8]	−47.1 ± 66.9	[−92.0;−2.1]	0.042
		Mz	−6.5 ± 8.6	[−12.3;−0.7]	−45.2 ± 60.2	[−85.7;−4.8]	0.032
	Negative	$||M||$	−44.9 ± 156.2	[−149.9;60.0]	−276.5 ± 960.7	[−921.9;368.9]	0.362
		Mx	−7.7 ± 38.6	[−33.6;18.3]	−54.0 ± 272.6	[−237.1;129.1]	0.526
		My	−22.4 ± 78.0	[−74.8;30.0]	−161.7 ± 561.8	[−539.1;215.7]	0.362
		Mz	−6.7 ± 21.4	[−21.0;7.7]	−59.0 ± 189.5	[−186.3;68.3]	0.326
Variance	Positive	$||M||$	−6.9 ± 44.9	[−37.1;23.2]	−39.4 ± 254.7	[−210.5;131.7]	0.619
		Mx	−5.1 ± 12.5	[−13.6;3.3]	−46.2 ± 112.8	[−122.0;29.6]	0.204
		My	−4.8 ± 13.1	[−13.6;3.9]	−31.2 ± 84.3	[−87.8;25.4]	0.248
		Mz	−8.0 ± 10.5	[12.1;0.9]	−90.8 ± 10.5	[−171.2;10.4]	0.031
	Negative	$||M||$	−27.2 ± 97.5	[−92.7;38.3]	−210.5 ± 755.0	[−717.7;296.8]	0.377
		Mx	−1.5 ± 25.3	[−18.5;15.5]	−10.7 ± 182.2	[−133.1;111.7]	0.849
		My	−6.4 ± 26.1	[−23.9;11.1]	−59.0 ± 240.5	[−220.6;102.5]	0.434
		Mz	−3.3 ± 17.6	[−15.1;8.5]	−34.4 ± 183.7	[−157.8;89.1]	0.549

4. Discussion

The results showed that the ANN-based method can estimate the net moments of the 6 kg test trials with an accuracy of about 9 Nm in comparison with the LSM-based method after being trained with 0 and 10 kg test trials. These results are in line with the research of Baten et al. [19,22], Dolan [25] and Kingma et al. [24] and support the notion that the ANN-based method can be used for evaluating lumbar load exposure patterns and exposure levels in real-life work settings. The feasibility of this approach is also supported by the results of others [42,43], who also used an ANN to predict static postures and net moments driven by static posture data (3D Euler angles) and EMG data and by the results of applying the same method to predict shoulder joint load exposure [34]. The results of the base quality differ per subject due to the individual character of the results. The calibration differs from one subject to another, which can be due to differences between the three calibration sets with three different weights or due to an unidentified event. This resulted in nonperfect calibration for subject 8, with an overall medium to small correlations. This ANN-based method seemed to be capable of distinguishing differences in the intensity level between light and heavy tasks which are in line with the perceived workload. Additionally, it can distinguish the differences in variance level between static and dynamic tasks. When the direction of the moment around the (anterior–posterior, lateroflexion or longitudinal) axis was considered, similar results were observed between light and heavy tasks. However, between static and dynamic tasks, the variance differences of the net moment were not observed. This is because of the differences due to the direction of the moment around the axis. This indicates that it is important to analyze and interpret net moments with different signs separately and thereby acknowledge the (a) symmetrical character of the net moment exertion when analyzing the lumbar load. This study showed that this system can measure the lumbar load in 3D, separately for the direction of the net moment in uncontrolled real-life working conditions.

In studies under more controlled conditions (Kingma et al., 2001, Baten, 2000), a more consistent high estimation accuracy was found, which suggests higher generalizability than the data in this study. The fact that accuracy is less consistent in this study is probably caused by the less controlled experimental conditions and environment, resulting in a lower quality and consistency in the signals recorded and driving the ANN. This is already signaled by the large loss of data due to the technical failures during acquisition. Still, in the end, this study reports the accuracy achieved in spite of the technically challenged acquisition conditions during the actual working conditions at actual work sites.

Like in other (state-of-the-art) research, this study provides insights into the usability and validity of this ANN-based method. This method only requires EMG and IMMU data for monitoring the lumbar load of physically active workers, without requiring any a priori knowledge regarding the load. Tasks were selected on the basis of actual working activities performed in the natural environments of physically demanding jobs. This study also explored the effect of the direction of the moment around the axis with which asymmetrical working routines can be investigated. For example, these could be the case of a paver who uses only one hand to lift tiles or fabric workers who mainly rotate in one direction or axis. Insight into these movements and moments can provide useful information that can help to effectively prevent musculoskeletal complaints. It should also be mentioned that the diversity in the jobs, related workloads, and selected tasks was a challenging aspect, for example, in selecting uniform tasks for the workers because of the differences in their working activities. Real-life tasks are not merely light, heavy, static, or dynamic; rather, a static or dynamic task may also be light or heavy, which results in an overlap between tasks.

The main weakness of this study was the lack of a reference method during actual work. Currently, the mechanical workload of the physically active workers was assessed by observations (video), questionnaires, performance tests, or combinations of motion trackers and force sensors [4,20,44]. Both observations and questionnaires were indirect methods and do not provide information about the working posture or related lumbar load. Reference systems, such as Vicon motion-capture cameras, are not practical or allowed in the real-life working environment [4,27,40,41]. The closest option for a reference method is the method that uses IMMUs and ground reaction forces assessed using an instrumented shoe [20,32,40]. This method, however, has two major drawbacks. The first drawback is that it yields erroneous results every time an external force other than the ground reaction force is exerted on the lower body (e.g., external forces resulting from supporting loads handled with any part of the lower body, or from leaning against a table or workbench, etc.). Another drawback is that it requires wearing heavy and somewhat bulky instrumented shoes, which constitute potential hazards and noncompliance with shoe and work safety functionality and regulations.

The current physical setup is not usable in real-life physically demanding jobs. Gathering data during actual work seemed to be technically challenging and resulted in the data of 14 out of the 23 subjects not being used in the study analysis. According to a power analysis based on the lab study, at least six subjects must be included to validate this model. The remaining nine subjects were sufficient for a first validation in real-life working conditions. However, this large number of dropouts and data processing issues must be prevented. To improve the usability an improved data-acquisition, (hardware) setup is required for future studies and applications. Additionally, further validation in follow-up research with more subjects and in more real-life working situations would increase insight into the quality of the assessment and may help develop insights into how to further improve this method. Preferably, this should be done in selected work situations in which the instrumented shoe method can serve as a reference and/or situations in which all external forces are known in some way. The environment's influence on the system (e.g., disturbance in the observability of the Earth's magnetic field) also needs to be further explored [35] and dealt with. Moreover, the design (mechanical) of the system needs to be improved (e.g., by integrating the sensors in the clothes). In addition, the monitored working posture and lumbar load exposure should be linked to ergonomic guidelines to obtain feedback regarding exceeding acceptable loading levels or loading patterns in a way like already pioneered by Baten et al. (e.g., [31,45]). As suggested in these papers, this information can be provided to the user (e.g., using a traffic light model based on EU or NIOSH ergonomic norms, to indicate areas of overload risk). Preferably, the estimating software should do this as fully automated with instant feedback to the user. Such a system would provide workers with feedback regarding their working behaviors and/or work task design and/or working conditions and/or workplace design, which can help improve workers' behavior, but also work task, organization and conditions and decrease their complaints. Both work postures and their net moments as well as the link to ergonomic guidelines need to, and can, be further explored in follow-up research with the ambulatory methods and tools studied in this paper. Estimated net moment data seem to be very suitable for exposure variance analysis. Adding load exposure pattern analysis to the current analysis of only the amount of load exposure [46] can be useful by means of generating a 2D graph of the joint distribution of intensity of the net moment data dynamically. This exposure variance analysis can be used to link the results to ergonomic guidelines and provide the user with feedback.

The results of this study suggest that not only can the ANN-based method be used in monitoring lumbar back load exposure in physically demanding jobs, but also that it may have the potential to be used in other occupational rehabilitation applications, such as office workers. For office workers, the lumbar load during sedentary behavior could be monitored and investigated, aiming to prevent health problems and physical discomfort related to static working postures. Additionally, this method has the potential to be extended and made usable for full body monitoring or specific areas. Other clinical applications are in the fields of rehabilitation medicine and sports. It may be used individually to assess muscular overload causing or contributing to an (individual) problem and help patients and clinicians to tailor treatments. The ANN-based method could be used in back pain rehabilitation, rehabilitation involving improvement of knee loading behavior, e.g., anterior cruciate ligament (ACL) surgery recovery. Additionally, with this method, injuries could be predicted or prevented in rehabilitation and sports applications and the performance could be improved.

5. Conclusions

Lumbar loads could be distinguished with the ANN-based method in terms of intensity and variance levels. The moments in the lumbar region are significantly higher during heavy tasks than during light tasks and the amount of variance is significantly higher during dynamic tasks than during static tasks. The estimated net moments were consistent with the perceived intensity levels and character of the work task in physically demanding occupations. It was concluded that the validity of this sensor-based system was supported, because the differences in the lumbar load were consistent with the perceived intensity levels and character of the work tasks.

Author Contributions: Conceptualization, C.C.R., C.T.M.B., M.F.R. and G.J.V.; methodology, C.C.R., C.T.M.B., M.F.R. and G.J.V.; software, C.T.M.B. and M.W.P.v.d.W.; validation, C.C.R., C.T.M.B. and M.W.P.v.d.W.; formal analysis, C.C.R., C.T.M.B., M.W.P.v.d.W., M.F.R. and G.J.V.; investigation, C.C.R., C.T.M.B., M.W.P.v.d.W., M.F.R. and G.J.V.; resources, C.C.R., C.T.M.B., M.W.P.v.d.W., M.F.R. and G.J.V.; data curation, C.C.R.; writing—original draft preparation, C.C.R. and C.T.M.B.; writing—review and editing, C.T.M.B., M.W.P.v.d.W., M.F.R. and G.J.V.; visualization, C.C.R., C.T.M.B. and M.W.P.v.d.W.; supervision, M.F.R. and G.J.V.; project administration, M.F.R. and G.J.V.; funding acquisition, M.F.R. and G.J.V. All authors have read and agreed to the published version of the manuscript.

Sensors **2021**, 21, 2476

Funding: This research was funded by the European Regional Development Fund, a co-founder of INCAS3 (Assen, The Netherlands), an independent, non-profit research institute focused on technology from science and industry, the province and municipality of Groningen, and the province of Drenthe, Grant [number T-3036, 2013].

Institutional Review Board Statement: The study was conducted according to the guidelines of the Declaration of Helsinki, and approved by the Institutional Review Board (or Ethics Committee) of the University Medical Center Groningen, the Netherlands (M17.208063, 1 March 2017).

Informed Consent Statement: Informed consent was obtained from all subjects involved in the study.

Data Availability Statement: The data presented in this study are available on request from the corresponding author. The data are not publicly available due to it possibly containing information that could compromise the privacy of some research participants.

Acknowledgments: We would like to thank Alejandro Miguel Reina Machena, Sylvia Nauta, Annemarie Boer, Alienke Noordhuis, and Wim Grootens for their support and help during this study. We also would like to thank Leo van Eykern and Jurryt Vellinga of Inbiolab for sharing their knowledge, skills, and equipment.

Conflicts of Interest: The authors declare no conflict of interest. The study was conducted according to the guidelines of the Declaration of Helsinki, and approved by the Medical Ethics Committee of the University Medical Center of Groningen, the Netherlands (M17.208063, 1 March 2017).

Appendix A

Appendix A.1. Question 1: Base Quality of ANN-Based Method

Table A1. ANN-based method performance in handling known loads. Shown are the correlations between the net moment at L5/S1 estimated with the ANN-based method and with the LSM-based method. Correlations are represented by the Pearson correlation coefficient (r) and the determination coefficient (r^2) and are shown by the axes anterior–posterior (x), mediolateral axis (y) and the longitudinal axis (z). Shown are individual values for each subject plus the mean and standard deviation (SDD) over all subjects. No valid data were obtained for subject number 6 for reasons of partially missing data in the 6 kg trial.

Movement	Axis		1	2	3	4	5	6	7	8	9	Mean	SD
Flexion	x	r	0.08	0.86	−0.17	0.25	0.41	-	0.25	0.37	0.52	0.32	0.30
		r^2	0.01	0.75	0.03	0.06	0.17	-	0.06	0.13	0.27	0.18	0.24
	y	r	0.97	0.98	0.96	0.97	0.98	-	0.96	0.35	0.98	0.89	0.22
		r^2	0.94	0.95	0.93	0.93	0.96	-	0.92	0.12	0.96	0.84	0.29
	z	r	−0.20	0.83	−0.13	0.13	0.03	-	0.42	0.46	0.48	0.25	0.35
		r^2	0.04	0.69	0.02	0.02	0.00	-	0.18	0.21	0.23	0.17	0.23
Abduction	x	r	0.94	0.93	0.95	0.95	0.94	-	0.85	0.32	0.92	0.85	0.22
		r^2	0.89	0.87	0.91	0.90	0.88	-	0.72	0.10	0.84	0.76	0.27
	y	r	0.43	0.59	0.22	−0.25	−0.22	-	−0.16	0.42	0.36	0.17	0.33
		r^2	0.19	0.35	0.05	0.06	0.05	-	0.03	0.18	0.13	0.13	0.11
	z	r	0.44	0.05	0.02	0.39	−0.02	-	0.13	−0.33	0.17	0.11	0.24
		r^2	0.20	0.00	0.00	0.15	0.00	-	0.02	0.11	0.03	0.06	0.08
Rotation	x	r	0.74	0.69	0.86	0.52	0.86	-	0.40	0.89	0.23	0.65	0.24
		r^2	0.54	0.48	0.73	0.27	0.73	-	0.16	0.80	0.05	0.47	0.28
	y	r	0.00	0.44	0.09	0.56	0.01	-	0.43	−0.06	0.24	0.21	0.24
		r^2	0.00	0.19	0.01	0.31	0.00	-	0.19	0.00	0.06	0.10	0.12
	z	r	0.63	0.23	0.84	0.84	0.81	-	0.76	0.38	0.92	0.68	0.24
		r^2	0.40	0.06	0.70	0.70	0.65	-	0.58	0.15	0.84	0.51	0.28

The column header group "Subject" spans columns 1–9.

Appendix A.2. Question 2: Capability to Distinguish Task Intensity

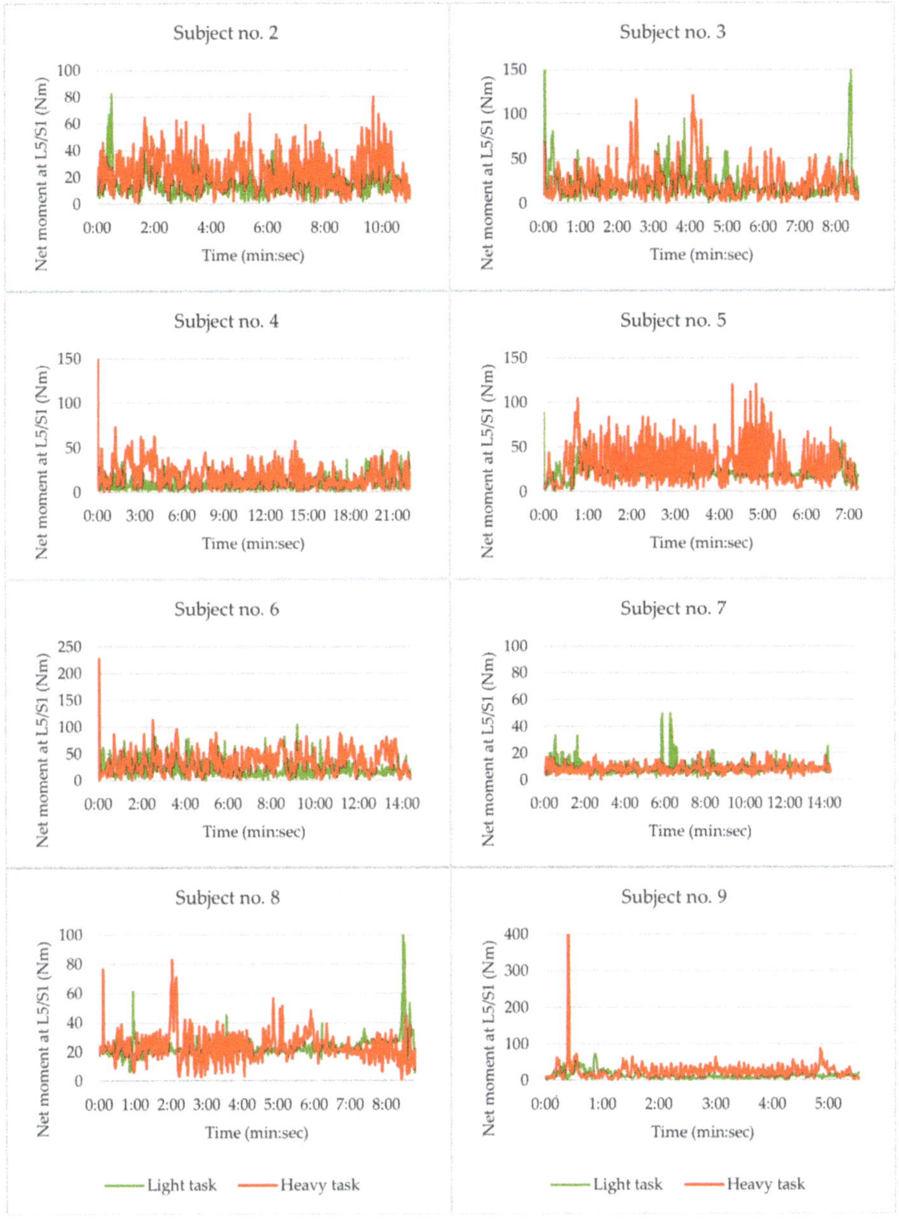

Figure A1. Net moment curves during light (green) and heavy (red) tasks.

Appendix A.3. Question 3: Capability to Distinguish Static/Dynamic Task Variance

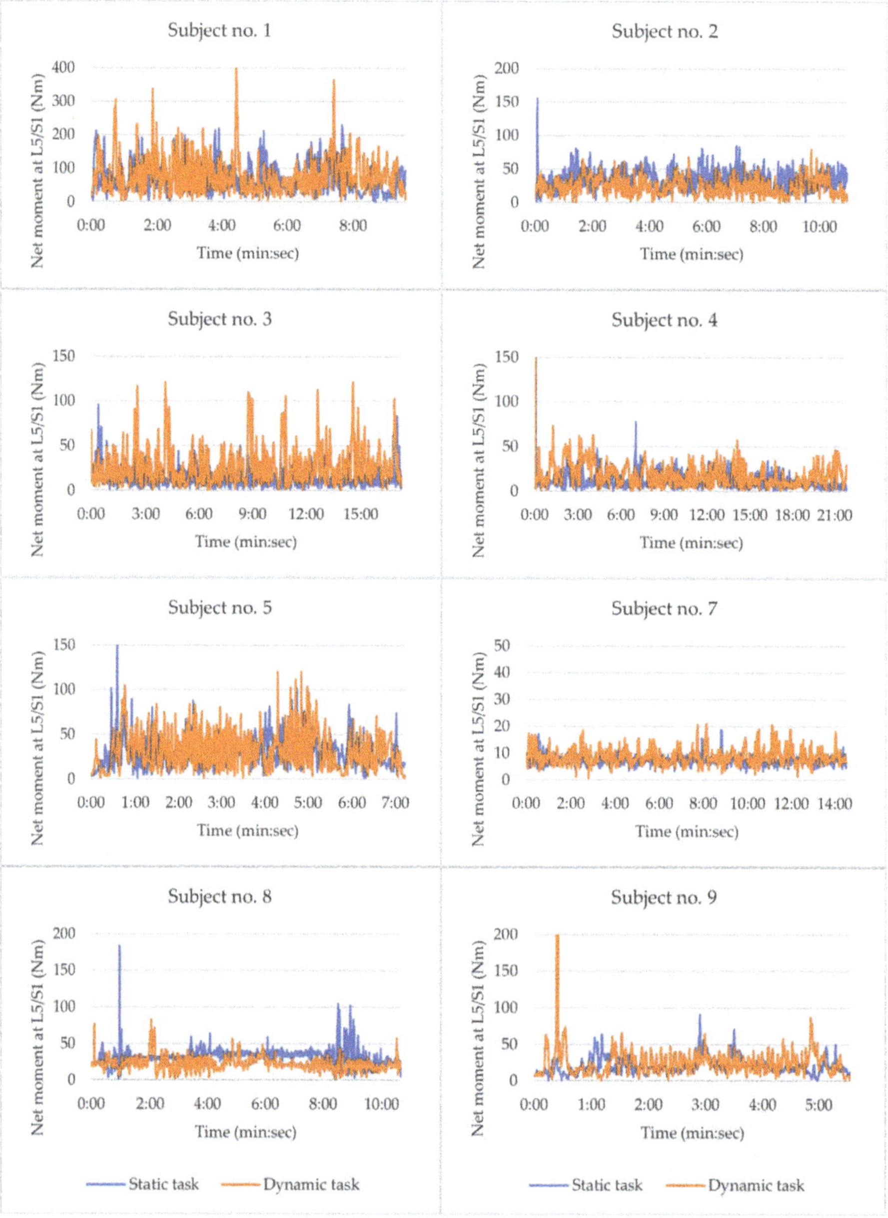

Figure A2. Net moment curves during static (blue) and dynamic (orange) tasks.

References

1. Rom, W.N.; Markowitz, S.B. (Eds.) *Occupational and Environmental Medicine*; Lippincott Williams & Wilkins: Philadelphia, PA, USA, 2007; pp. 546–552.
2. Wu, H.; Wang, M. Relationship between maximal acceptable work time and physical workload. *Ergonomics* **2002**, *45*, 280–289. [CrossRef]

3. Coenen, P.; Douwes, M.; van den Heuvel, S.; Bosch, T. Towards exposure limits for working postures and musculoskeletal symptoms–a prospective cohort study. *Ergonomics* **2016**, *59*, 1–11. [CrossRef] [PubMed]
4. Jorgensen, M.; Korshoj, M.; Lagersted-Olsen, J.; Villumsen, M.; Mortensen, O.; Skotte, J.; Sogaard, K.; Madeleine, O.; Thomsen, B.; Holtermann, A. Physical activities at work and risk of musculoskeltal pain and ites consequences: Protocol for a study with objective field measures among blue-collar workers. *MBC Musculoskelet. Disord.* **2013**, *14*, 1–9.
5. Holterman, A.; Clausen, T.; Aust, B.; Mortsen, O.; Andersen, L. Risk for low back pain from different frequencies, load mass and trunk postures of lifting and carrying among female healthcare worker. *Int. Arch. Occup. Environ. Health* **2013**, *86*, 463–470. [CrossRef]
6. Andersen, J.H.; Haahr, J.P.; Frost, P. Risk factors for more severe regional musculoskeletal symptons: A two-year prospective study of a general working population. *Arthritis Rheum.* **2007**, *56*, 1355–1364. [CrossRef] [PubMed]
7. Bakker, E.; Verhagen, A.; Trijffel, E.; van Lucas, A.; Koes, B. Spinal mechnical load as a risk factor for low back pain—A systematic review of prospective cohort studies. *Spine* **2009**, *34*, E281–E293. [CrossRef]
8. Weerding, W.; IJzelenberg, W.; Koopmans, M.; Severens, J.; Burdorf, A. Health problems lead to considerable productivity loss at work among workers with high physical work loads jobs. *J. Clin. Epidemology* **2005**, *58*, 517–523. [CrossRef]
9. Pransky, G.; Benjamin, K.; Savageau, J.; Currivan, D.; Fletcher, K. Outcomes in work-related injuries: A comparison of older and younger workers. *Am. J. Ind. Med.* **2005**, *47*, 104–112. [CrossRef]
10. Karpansalo, M.; Manninen, P.; Lakka, T.; Kauhanen, J.; Rauramaa, R.; Salonen, J. Physical workload and risk of early retirement: Prospective population-based study among middle-aged men. *J. Occup. Enviromental Med.* **2002**, *44*, 930–939. [CrossRef]
11. Varianou-Mikellidou, C.; Boustras, G.; Dimopoulos, C.; Wybo, J.; Guldenmund, F.; Nicolaidou, O.; Anyfantis, I. Occupational he alth and safety management in the context of an ageing workforce. *Saf. Sci.* **2019**, *116*, 231–244. [CrossRef]
12. Kenny, G.; Yardley, J.; Martineau, L.; Jay, O. Physical work capacity in older adults: Implications for the aging worker. *Am. J. Ind. Med.* **2008**, *51*, 610–625. [CrossRef]
13. Nath, N.; Akhavian, R.; Behzadan, A. Ergonomic analysis of construction worker's body postures using wearable mobile sensors. *Appl. Ergon.* **2017**, *62*, 107–117. [CrossRef] [PubMed]
14. Heerkens, Y.; Engels, J.; Kuiper, C.; van der Gulden, J.; Oostendorp, R. The use of the ICF to describe work related factors influencing the health of employees. *Disabil. Rehabil.* **2004**, *26*, 1060–1066. [CrossRef] [PubMed]
15. Costa-Black, K.; Feuerstein, M.; Loisel, P. Work Disability Models: Past and Present. In *Handbook of Work Disability: Prevention and Management*; Springer Science+Business Media: New York, NY, USA, 2013; pp. 71–93.
16. Ranavolo, A.; Draiccio, F.; Varrecchia, T.; Silvetti, A.; Iaviacoli, S. Wearable monitoring devices for biomechanical risk assessment at work: Current status and future challenges—A systematic review. *Int. J. Environ. Res. Public Health* **2018**, *15*, 2001. [CrossRef] [PubMed]
17. Coenen, P.; Gouttebarge, V.; van der Burgth, A.; van Dieen, J.; Frings-Dresen, M.; van der Beek, A.; Burdorf, A. The effect of lifting during work on low back pain: A health impact assessment based on a meta-analysis. *Occup. Enviromental Med.* **2014**, *12*, 871–877. [CrossRef] [PubMed]
18. Baten, C.; Hamberg, H.; Veltink, P.; Hermens, H. Calibration of low back load exposure estimation through surface EMG signals with the use of artificial, neural network technology. In Proceedings of the IEEE 17th International Conference of the Engineering in Medicine and biology Society, Montreal, QC, Canada, 20–23 September 1995.
19. Baten, C.; van der Aa, R.; Verkuyl, A. Effect of wearable trunk support for working in sustained stooped posture on low back net extension moments. In Proceedings of the ISB 2015—25th Congress of the International Society of Biomechanics 2015, Glasgow, UK, 12–16 July 2015.
20. Faber, F.; Chang, C.; Kingma, I.; Dennerelein, J.; van Dieen, J. Estimating 3D L5/S1 moments and ground reaction forces during trunk bending using a full-body ambulatory inertial capture system. *J. Biomech.* **2015**, *49*, 904–912. [CrossRef]
21. Baten, C.; Oosterhoff, P.; de Looze, M.; Dolan, P.; Veltink, P.; Hermens, H. Ambulatory back load estimation—validation in lifting. In Proceedings of the 11th Congress of the International Society of Electrophysiology and Kinesiology (ISEK), Enschede, The Netherlands, 27–30 October 1996.
22. Baten, C. Ambulatory low back load exposure estimation. In Proceedings of the Human Factors and Ergonomics Society Annual Meeting 2000, San Diego, CA, USA, 30 July–4 August 2000.
23. Baten, C.; de Vries, W.; Schaake, L.; Witteveen, J.; Scherly, D.; Stadler, K.; Hidalgo Sanchez, A.; Rocon, E.; Plass-Oude Bos, D.; Linssen, J. XoSoft Connected Monitor (XCM) Unsupervised Monitoring and Feedback in Soft Exoskeletons of 3D Kinematics, Kinetics, Behavioral Context and Control System Status. In *International Symposium on Wearable Robotics*; Springer: Cham, Switzerland, 2018.
24. Kingma, I.; Baten, C.; Dolan, P.; Toussaint, H.; van Dieen, J.; de Looze, M.; Adams, M. Lumbar loading during lifting: A comparative study of three measurement techniques. *J. Electromyogr. Kinesiol.* **2001**, *11*, 337–345. [CrossRef]
25. Dolan, P. Measuring inertial forces acting on the lumbar spine during lifting. *J. Biomech.* **1998**, *31*, 1101. [CrossRef]
26. Valevicus, A.; Jun, P.; Hebert, J.; Vette, A. Use of optical motion capture for the analysis of normative upper body kinematics during functional upper limb tasks: A systematic review. *J. Electromyogr. Kinesiol.* **2018**, *40*, 1–15. [CrossRef]
27. Xu, X.; Chang, C.; Faber, G.; Kingma, I.; Dennerlein, J. Estimating 3-D L5/S1 moments during manual lifting using a video coding system; Validity and interrater reliability. *Hum. Factors* **2012**, *54*, 1053–1065. [CrossRef]

28. Coenen, P.; Kingma, I.; Boot, C.; Bongers, P.; van Dieen, J. Cumulatieve mechanical low-back load at work is a determinant of low-back pain. *Occup. Enviromental Med.* **2014**, *71*, 332–337. [CrossRef]

29. Baten, C.; Oosterhoff, P.; Kingma, I.; Veltink, P.; Hermens, H. Inertial sensing in ambulatory back load estimation. In Proceedings of the 18th Annual Interantional Conference of the IEEE Engineering in Medicine and Biology Society 2, Amsterdam, The Netherlands, 31 October–3 November 1996.

30. Monaco, M.; Fiori, L.; Marchesi, A.; Greco, A.; Ghiboudo, L.; Spada, S.; Caputo, F.; Miraglia, N.; Silvetti, A.; Draiccion, F. Biomechanical overload evaluation in manufacturing: A novel approach with sEMG and inertial motion capture system integration. *Adv. Intell. Syst. Comput.* **2019**, *818*, 719–726.

31. Baten, C. Advancements in sensor-based ambulatory 3D motion analysis. *J. Biomech.* **2007**, *40*, S422. [CrossRef]

32. Schepers, H.; Koopman, H.; Baten, C.; Veltink, P. Ambulatory measurement of ground reaction force and estimation of ankle and foot dynamics. *J. Biomech.* **2007**, *40*, S436. [CrossRef]

33. Schepers, H.; van Asseldonk, E.; Baten, C.; Veltink, P. Ambulatory estimation of foot placement during walking using inertial sensors. *J. Biomech.* **2010**, *43*, 3138–3143. [CrossRef]

34. De Vries, W.; Veeger, H.; Baten, C.; van der Helm, F. Can shoulder joint reaction forces be estimated by neural networks? *J. Biomech.* **2016**, *49*, 73–79. [CrossRef]

35. De Vries, W.; Veeger, D.; Baten, C.; van der Helm, F. Magnetic distortion in motion labs, implications for validating internal magnetic sensors. *Gait Posture* **2009**, *29*, 535–541. [CrossRef]

36. Peereboom, K.; de Langen, N. *Handboek fysiek Werkbelasting*; SDU Uitgevers: The Hague, The Netherlands, 2012.

37. Arbo. *Vormen van Fysieke Belasting*; Vakmedianet: Alpen aan den Rijn, The Netherlands, 2013.

38. Dutch Ministry of Social Affairs and Employment. *Health and Safety Catalog*; Dutch Ministry of Social Affairs and Employment: Den Haag, The Netherlands, 2017.

39. Borg, G. *Borg's Perceived Exertion and Pain Scales*; Human Kinetics: Champaign, IL, USA, 1998.

40. Koopman, A.; Kingma, I.; Faber, G.; Bornmann, J.; van Dieen, J. Estimating the L5S1 flexion/extension moment in symmetrical lifting using a simplified ambulatory measurement system. *J. Biomech.* **2018**, *70*, 242–248. [CrossRef]

41. Kristiansen, M.; Rasmussen, G.; Sloth, M.; Voigt, M. Inter- and intra-individual variability in the kinematics of the back squat. *Hum. Mov. Sci.* **2019**, *67*, 102510. [CrossRef]

42. Aghazadeh, F.; Arjmand, N.; Nasrabadi, A. Coupled artificial neural networks to estimate 3D whole-body posture, lumbosacral moments, and spinal loads during load-handling activities. *J. Biomech.* **2020**, *102*, 109332. [CrossRef]

43. Gholipour, A.; Arjmand, N. Artificial neural networks to predict 3D spinal posture in reaching and lifting activities; Applications in biomechanical models. *J. Biomech.* **2016**, *49*, 2946–2952. [CrossRef] [PubMed]

44. Vieira, E.; Kumar, S. Working posture: A literature review. *J. Occup. Rehabil.* **2004**, *14*, 143–159. [CrossRef]

45. Baten, C. Advances in 3D Analysis of human Movement. In Proceedings of the International Conference on Ambulatory Monitoring of Physical Activity and Movement, Rotterdam, The Netherlands, 21–24 May 2008.

46. Mathiassen, S.; Winkel, J. Quantifying variation in physical load using exposure-vs-time data. *Ergonomics* **1991**, *34*, 1455–1468. [CrossRef]

 *sensors*

Article

An Inertial Measurement Unit-Based Wireless System for Shoulder Motion Assessment in Patients with Cervical Spinal Cord Injury: A Validation Pilot Study in a Clinical Setting

Riccardo Bravi [1,*,†], Stefano Caputo [2,†], Sara Jayousi [2], Alessio Martinelli [2], Lorenzo Biotti [2], Ilaria Nannini [3], Erez James Cohen [1], Eros Quarta [1], Stefano Grasso [1], Giacomo Lucchesi [3], Gabriele Righi [3], Giulio Del Popolo [3], Lorenzo Mucchi [2,*] and Diego Minciacchi [1]

[1] Department of Experimental and Clinical Medicine, University of Florence, Largo Brambilla 3, 50134 Florence, Italy; erezjames.cohen@unifi.it (E.J.C.); equarta@unifi.it (E.Q.); stefano.grasso@stud.unifi.it (S.G.); diego.minciacchi@unifi.it (D.M.)

[2] Department of Information Engineering, University of Florence, via di S. Marta 3, 50139 Florence, Italy; stefano.caputo@unifi.it (S.C.); sara.jayousi@pin.unifi.it (S.J.); alessio.martinelli@unifi.it (A.M.); lorenzo.biotti@unifi.it (L.B.)

[3] Spinal Unit, Azienda Ospedaliero-Universitaria Careggi, Largo Piero Palagi 1, 50139 Florence, Italy; nanninii@aou-careggi.toscana.it (I.N.); lucchesig@aou-careggi.toscana.it (G.L.); righiga@aou-careggi.toscana.it (G.R.); delpopolog@aou-careggi.toscana.it (G.D.P.)

* Correspondence: riccardo.bravi@unifi.it (R.B.); lorenzo.mucchi@unifi.it (L.M.); Tel.: +39-055-275-8020 (R.B.); +39-055-275-8539 (L.M.)

† Co-first authors.

Citation: Bravi, R.; Caputo, S.; Jayousi, S.; Martinelli, A.; Biotti, L.; Nannini, I.; Cohen, E.J.; Quarta, E.; Grasso, S.; Lucchesi, G.; et al. An Inertial Measurement Unit-Based Wireless System for Shoulder Motion Assessment in Patients with Cervical Spinal Cord Injury: A Validation Pilot Study in a Clinical Setting. *Sensors* **2021**, *21*, 1057. https://doi.org/10.3390/s21041057

Academic Editor: G.R.H. (Ruben) Regterschot

Received: 29 December 2020
Accepted: 2 February 2021
Published: 4 February 2021

Publisher's Note: MDPI stays neutral with regard to jurisdictional claims in published maps and institutional affiliations.

Abstract: Residual motion of upper limbs in individuals who experienced cervical spinal cord injury (CSCI) is vital to achieve functional independence. Several interventions were developed to restore shoulder range of motion (ROM) in CSCI patients. However, shoulder ROM assessment in clinical practice is commonly limited to use of a simple goniometer. Conventional goniometric measurements are operator-dependent and require significant time and effort. Therefore, innovative technology for supporting medical personnel in objectively and reliably measuring the efficacy of treatments for shoulder ROM in CSCI patients would be extremely desirable. This study evaluated the validity of a customized wireless wearable sensors (Inertial Measurement Units—IMUs) system for shoulder ROM assessment in CSCI patients in clinical setting. Eight CSCI patients and eight healthy controls performed four shoulder movements (forward flexion, abduction, and internal and external rotation) with dominant arm. Every movement was evaluated with a goniometer by different testers and with the IMU system at the same time. Validity was evaluated by comparing IMUs and goniometer measurements using Intraclass Correlation Coefficient (ICC) and Limits of Agreement (LOA). inter-tester reliability of IMUs and goniometer measurements was also investigated. Preliminary results provide essential information on the accuracy of the proposed wireless wearable sensors system in acquiring objective measurements of the shoulder movements in CSCI patients.

Keywords: inertial measurement unit; wireless sensors network; motion tracking; kinematics; range of motion; shoulder; goniometer; spinal cord injury; tetraplegia; clinical setting

1. Introduction

Spinal cord injury (SCI) is a debilitating neurological condition which can result in a total or partial motor and sensory function impairment below the site of the injury, associated with a various degree of bladder, bowel and sexual dysfunctions [1]. The sensory and/or motor impairment is caused by the loss of communication between the brain areas devoted to motor/sensory processing and the axons of neural cells controlled by spinal cord levels below the injury site and innervating the body surface. The site and completeness of the injury largely determine the clinical outcomes of SCI [2]. Lesions at lower spinal segments (i.e., sacral, lumbar, or thoracic) cause the loss of motor and/or sensory function in lower limbs and trunk (paraplegia) while more rostral lesions (i.e., cervical) involve lower limbs, trunk, and upper limbs (tetraplegia).

Regarding severity, spinal cord injury can be graded on the basis of motor and sensory dysfunction as follows: no motor or sensory function is preserved; sensory function preserved but not motor function; both motor and sensory functions are—partially—preserved below the neurological level [3].

For CSCI individuals the residual motion of upper limbs is a key-element to perform activities of daily living and participate in community [4–6].

Shoulder range of motion (ROM) is involved in basic self-management skills (e.g., bathing, feeding or dressing), independent transferring skills (e.g., getting in/out of bed or moving to/from wheelchair) as well as sport and leisure activities participation [7].

In an attempt to increase functional independence of CSCI individuals, different surgical interventions and rehabilitation programs have been developed to restore deficits in motor control provoked by spinal cord injuries [8]. Therefore, tools to objectively and reliably measure the efficacy of a treatment process would be extremely desirable.

At the present moment, in clinical settings the assessment of the shoulder function in patients with cervical spinal cord injury (CSCI) is commonly limited to simple clinical observations by the examiner or based on rating scales, which are not only inefficient for detecting small differences but also inclined to produce subjective errors [9,10].

Alongside the observational examination, goniometer is a portable, inexpensive and handy tool traditionally used in clinical settings to assist doctors for the evaluation of shoulder movements [11]. However, conventional goniometric measurements can vary among testers [12,13] and require substantial time and effort in clinics.

In addition, clinical assessments using simple tools such as goniometer allow us to measure only the ROM and are not sufficient when attempting to assess complex patterns of upper limb movement in persons with CSCI. Thus, dynamic kinematic tests have been developed to integrate ROM evaluation while providing a more complete monitoring of upper limb functional movements [14,15].

Several movement analysis devices are available to assist objective and accurate kinematic measurements varying from simple video cameras to complex optical motion capture systems (e.g., [16–19]). Such evaluation systems generally demand highly expensive equipment, well-skilled personnel for operational procedure, as well as dedicated spaces, which is not realistically suitable for clinical settings in hospital or space-limited environments. This paved the way for the introduction of wearable, low-cost and user-friendly devices that can provide objective information of movement characteristics, with the possibility of monitoring these parameters in daily life, in accordance to future healthcare system [20,21].

In the last few years, a new generation of inertial measurement units (IMUs) has emerged, significantly impacting on motion tracking research due to their ease of use, relative low cost, and portability [22–24]. This technology, taking advantage of recent progress in miniature inertial sensors, consists of lightweight, non-invasive, wearable, wireless sensing units, which open a new opportunity to capture human motion in different settings [13,25].

An IMU device estimates orientation of human body segments based on data combined from multiple electromechanical sensors as accelerometers, tri-axial gyroscopes, and magnetometers through the use of sensor fusion algorithms such as a Kalman filter or complementary weighting algorithm [26–28].

The combination of information from different inertial sensor units is considered an advance in minimizing measurement errors of one component-based orientation (e.g., linear acceleration interference and inertial sensor drift), thus guaranteeing a more accurate and reliable estimation of motion [26,29–31]. Therefore, IMU systems have been increasingly adopted to efficiently measure joint ROM of several body parts such as lower limbs [32], trunk [33], as well as arm/shoulder [34–36].

In spite of the increasing application of IMU systems in shoulder joint measurements, only a few studies tested the validity of IMU in comparison with goniometer [13,37]. In [37]

IMU and goniometer were compared by passively positioning the shoulder at specific angles and good concurrent validity of IMU was found, though they showed relatively low agreement at elevated shoulder positions. Also, in [13] an excellent agreement was exhibited between goniometer- and IMU-based measurements and the two methods were shown to be interchangeable for measuring active shoulder ROM. Overall, these investigations provided preliminary evidence of the reliability of IMUs in accurately measuring shoulder ROM by comparing it with a goniometer tool. Nonetheless, the tested populations in these studies consisted of healthy participants as the most of studies evaluating IMUs systems [38] and, as suggested by Rigoni et al. [13], their results might not be replicable in pathological individuals. It should be considered that the monitoring of active shoulder ROM in patients with CSCI can occur in a very different context if compared to healthy subjects. The introduction of a wheelchair in the setup is necessary when testing CSCI patients. However, this assistive device can interfere with the arm motion to be performed. Deviations from single plane of unrestricted movements were shown in a previous study to potentially influence the accuracy of technology used to measure shoulder ROM [39]. As such the wheelchair, along with deficits in motor control of upper limbs in this special population [40], could be a source of measurement error for IMU sensors when testing active shoulder movements. Moreover, electronic controls of motorized wheelchairs can reduce the accuracy of some components of IMU sensors (e.g., magnetometer). All these context-related factors could limit the generalization of the results reported by Rigoni to a specific population as the patients with CSCI while active shoulder movements are performed. Since IMU devices should ultimately support clinicians and patients suffering from pathological motor diseases, additional research is needed to understand the reliability of IMU systems on these special populations.

The objective of this pilot study was to investigate the validity of a customized, wireless wearable IMU-based sensors system in evaluating active shoulder movements in CSCI patients, while seated in a wheelchair, in a clinical setting. To achieve this, we compared the accuracy of the IMU system to goniometer method in measuring shoulder ROM during the performance of flexion, abduction, and extra- and intra-rotation movements of the upper limb. The IMU sensors used for this study were composed of an accelerometer and a tri-axial gyroscope without the implementation of the magnetometer, so as to reduce possible interference due to electronic controls in motorized wheelchairs. Also, according to previous studies investigating concurrent validity of IMU devices in monitoring ROM on pathological populations [39,41], our experimental design included an additional group of healthy subjects. In order to obtain uniform measurements between the different populations, even the healthy subjects performed the active shoulder movements while seated in a wheelchair. Finally, as it is important to know to what extent the measurement made by an instrument is dependent on the person (i.e., operator) who carries it out, in this study it was also investigated the levels of consistency between repeated goniometer measurements and repeated IMU measurements taken by different operators during the same movement.

2. Materials and Method

2.1. Shoulder Movements to Evaluate Range of Motion

This section describes the movements to evaluate shoulder ROM in individuals with spinal cord injury and healthy subjects. The tested movements and measurement procedures to evaluate shoulder ROM have been identified in forward flexion, abduction, external and internal rotation at 90° abduction (See Figure 1), following the study of Frye et al. [42].

For shoulder forward flexion, subject was asked to raise the arm straight up in front of him/her with the forearm in neutral position, i.e., the palm of the hand toward the mid-line (See Figure 1A). For shoulder abduction, the starting hand position was the same of flexion maneuver and the subject was required to raise the arm at the side (See Figure 1B).

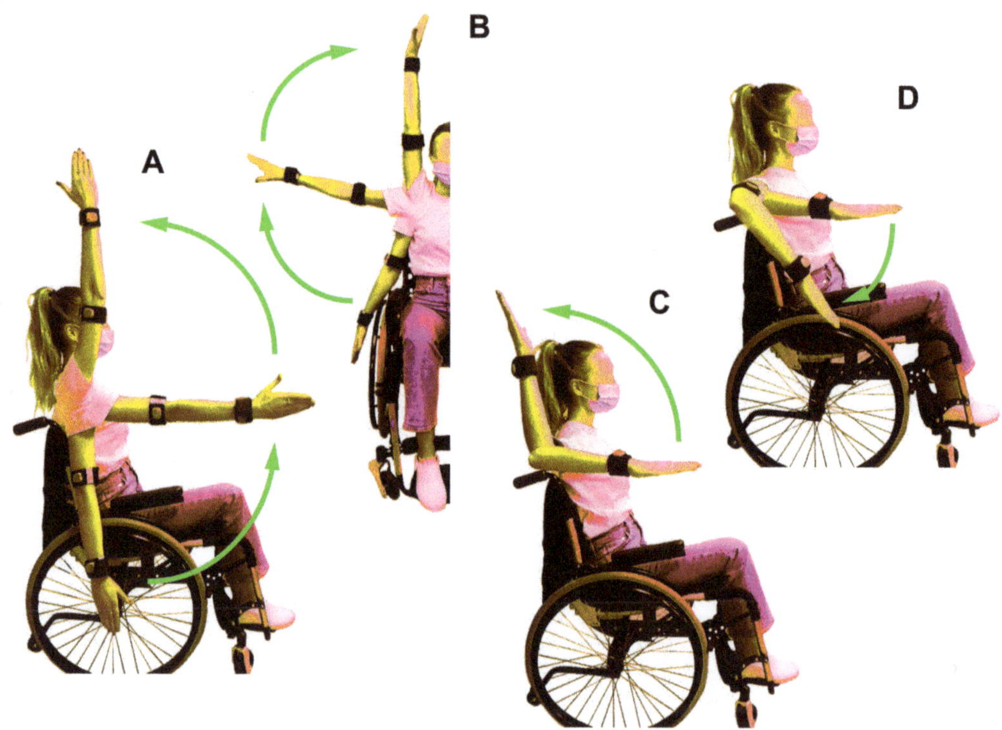

Figure 1. The four recorded active shoulder maneuvers: forward flexion (**A**), abduction (**B**), external rotation (**C**) and internal rotation (**D**).

For external and internal rotation at the 90° shoulder abduction, the shoulder and elbow were positioned in approximately 90° of abduction and flexion, respectively, with the forearm parallel to the floor.

For external rotation, subject was asked to rotate the forearm upwards, whereas in the internal rotation subject was required to rotate the forearm downwards (See Figure 1C,D). Scapular rotation was allowed during shoulder forward flexion and abduction. In addition, subjects were assisted by a clinician to keep the shoulder and the elbow at the initial position during external and internal rotation at the 90° shoulder abduction.

2.2. Goniometer Measurement Method

To measure the ROM for forward flexion, abduction, external and internal rotation at 90° of shoulder abduction (see Figure 1), a standard plastic goniometer (Gima Co., Gessate, Italy), characterized by two arms that align with the angle of the joint, was used to provide the degree of movement in that joint. The active movements of shoulder were calculated by identified landmarks (see Table 1). The methods used to measure ROM for each tested shoulder maneuver [39,43–46] are described below.

The flexion angle was calculated by lateral aspect of the glenohumeral joint and aligning its stationary arm parallel to the midline of the trunk and its moving arm with the lateral epicondyle of the humerus.

The abduction angle was calculated by placing the center fulcrum of goniometer on posterior aspect of the glenohumeral joint and aligning its stationary arm along the trunk (parallel to the spine) and its moving arm with the lateral epicondyle of the humerus.

The external and internal rotation angles at 90° of shoulder abduction were calculated by placing the center fulcrum of goniometer on olecranon process ulna, and aligning stationary arm parallel to the floor and its moving arm with the ulna styloid process.

Table 1. Movement and Goniometer Landmarks.

Shoulder Movement	Goniometer Landmarks		
	Center Fulcrum	Stationary Arm	Moving Arm
Flexion	Lateral aspect of the glenohumeral joint	Parallel to the midline of the trunk	Lateral epicondyle of the humerus
Abduction	Posterior aspect of the glenohumeral joint	Laterally along the trunk, parallel to the spine.	Lateral epicondyle of the humerus
External Rotation at 90° abduction	Olecranon process ulna	Parallel to the floor	Ulna styloid process
Internal Rotation at 90° abduction	Olecranon process ulna	Parallel to the floor	Ulna styloid process

2.3. Wearable Sensors System for Motion Assessment

A wearable sensors system was developed for motion assessment during the experimental tests. The system aims to create an innovative protocol for patient monitoring to be used both in hospital ward and/or in patient's house. The system is composed of 3 main modules, as shown in Figure 2:

- **Hardware** module is the part of the system (for complete details see Appendix A.1) composed of IMU sensors and a gateway (Raspberry Pi);
- **Software** module is the part of the system (for complete details see Appendix A.3) composed of software components, which run on the gateway and provide the following functionalities: IMU sensors synchronization, data collection, and data processing to obtain the kinematics parameters used for medical evaluation of the movement (for complete details see Appendix A.2);
- **Data Visualization** module is the display part, showing data in real time to clinicians. Since this part is not necessary for the experimental campaign, it will be deployed as future development.

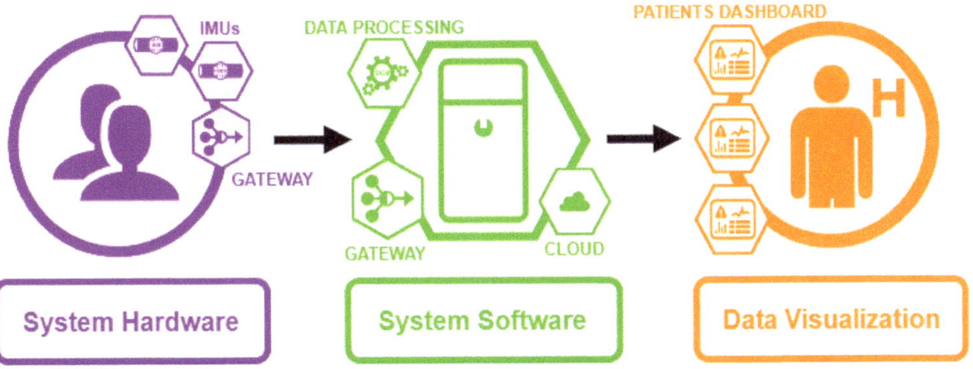

Figure 2. General scheme of the system consisting of 3 main modules (represented with different colors). The round element corresponds to an end user (human), while hexagonal elements (both hardware and software) corresponds to the system. In purple, the components of system hardware module; in green, the components of system software module and in orange, the components of data aggregation/visualization module.

In Figure 2, the gateway symbol appears twice, in the hardware part and in the software part. In the hardware part it includes a communication protocol to synchronize the connection with the IMU sensors and collect the data. In the software part it includes a software program to process the data and store them into a cloud database for offline analysis.

Two sensors were used to measure shoulder ROM. The sensor on the arm was used to measure the angle during the flexion and abduction movements (Figure 1A,B), while the sensor on the wrist was used during the rotation movement at the 90° shoulder abduction (Figure 1C,D).

2.4. Experimental Campaign: Setup and Protocols

Experimental data were gathered in a room of the Careggi Hospital Spinal Unit, Florence. The four recorded active shoulder maneuvers (Figure 1) during the session consisted of: forward flexion, abduction, external and internal rotation at 90° of shoulder abduction [42]. The shoulder maneuvers were assessed on each participant's dominant shoulder [47].

Each movement was performed by CSCI subjects sitting in their own wheelchair, as well as healthy controls who were provided with a wheelchair by the hospital [48]. This avoided possible differences in range of motion between the two groups due to potential interference with the wheelchair [48,49]. The wheelchair wheels were blocked during the tests execution [50].

At the beginning of the session, all participants were informed about the aim of the study and that they would perform 4 different active shoulder movements. After having signed the informed consent to participate in the experiment (See Section 2.4.2), they were fitted with the IMU-based system on the dominant arm.

Two IMU sensors were attached to subject's arm using velcro straps. One of them was placed on the posterior surface of wrist while the other one was located on the arm, approximately at 10 cm distance to the lateral epicondyle (Figure 3). Each IMU sensor was securely attached to the participant's body with a self-adhering strap.

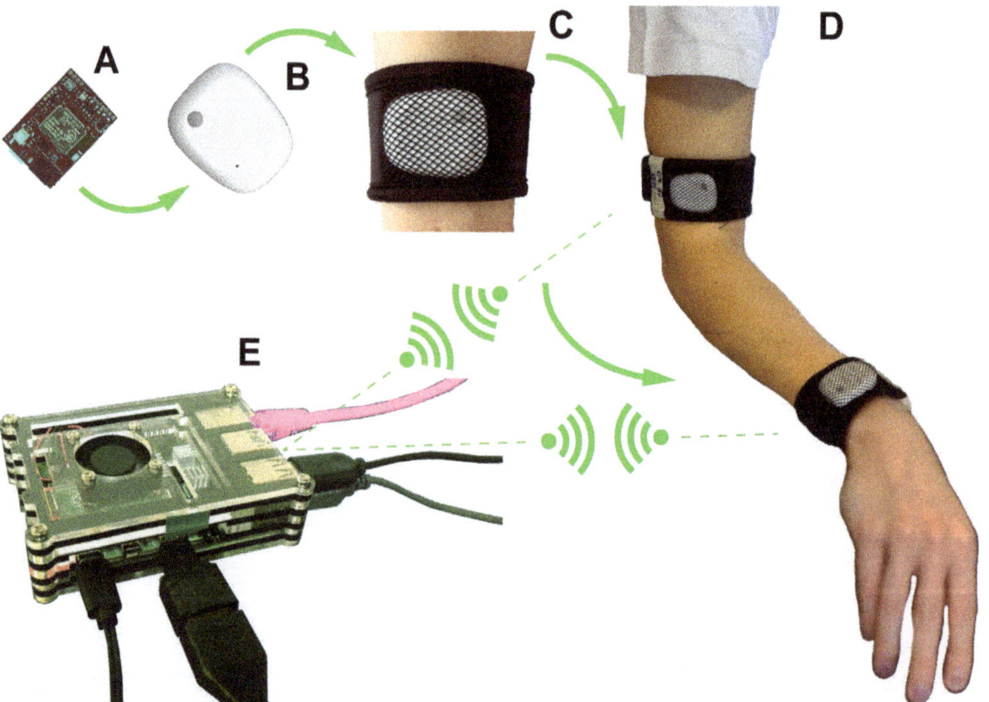

Figure 3. Wearable sensors system for motion assessment installation procedure: MetaMotionR (MMR) sensors boards (**A**) were put inside the cases (**A**→**B**); the cases were put in the velcro bands (**B**→**C**); the bands were placed as shown in (**D**); than the software on the Raspberry Gateway was run to collect data from sensors (**E**).

The procedure of application of the IMU-based system on the subject was performed by the same operator for all participants to the study. Then, subjects were prepared to execute the measurements of forward flexion, abduction, external and internal rotation at 90° of shoulder abduction. For each motion direction, participants were instructed to move the arm as far as they comfortably could, maintaining unchanged the initial arm position for the entire arc of movement. In addition, participants were required to perform shoulder movements at their self-selected speed [51,52]. Prior to measurements the correct execution of each maneuver was verbally explained and demonstrated by one of the investigators (I.N.). The subject was given some time to familiarize with the movement [53] until the researcher judged it to be correct.

Subsequently, the calibration procedure of the IMU-based system for a specific maneuver (e.g., forward flexion) was performed and data were gathered. Active shoulder movements were randomized across participants. Every participant was required to perform three repetitions of each maneuver. During each repetition two consecutive pairs of goniometer and IMU-based measurements were collected one after the other. For each pair of measurements, goniometer- and IMU-based assessments were performed simultaneously. In total, 6 goniometer and 6 IMU measurements were taken for each shoulder maneuver. Goniometer measurements were acquired by two raters (Medical Doctors). During each repetition one rater (E.J.C., RATER 1) made the first goniometer measurement and the second one (G.L., RATER 2), who was blind to the first measurement, repeated the same procedure.

IMU recordings were performed by two researchers (S.C. and L.B., IMU 1 and IMU 2, respectively) who were blind to goniometer measurements. Data were transmitted wirelessly to the gateway and subsequently processed by a specific software to calculate ROM. The RATER 1 and IMU 1 measurements, and similarly the RATER 2 and IMU 2 measurements, were acquired simultaneously. An operator (I.N.) assisted the subject to maintain the position of his/her arm to allow for consistency of measurements as well as effort relief for participant.

A rest interval was also given to participants between shoulder maneuvers to avoid fatigue effects [50,54,55]. The order of raters and researchers in collecting data was randomized among shoulder maneuvers and subjects.

2.4.1. Participant Recruitment

Eight healthy control (mean age 44 ± 18 years, 1 female) and 8 CSCI patients (mean age 50 ± 12 years, 1 female) who had suffered traumatic spinal cord injuries were recruited in the study. Patients with CSCI were enrolled from the Spinal Unit of the Florence University Hospital between July and October 2020. The following inclusion criteria were adopted to select the sample:

- subjects over 18 years of age;
- C4–C7 cervical lesion level;
- at least one month post-injury;
- subjects with intact cognitive abilities;
- no joint contracture or severe spasticity in the affected upper limb (modified Ashworth scale greater than 3);
- sufficient Italian language skills.

The exclusion criteria were neuropsychiatric comorbidities and orthopedic impairments/or symptoms such as pain when moving their arm. The healthy controls were hospital staff or students from the University of Florence. Healthy individuals were all volunteers and they did not experience any type of shoulder disease. The demographics and clinical features of CSCI and healthy control groups are reported in Table 2.

Table 2. Characteristics of the participants to the experimental tests. Abbreviations used: P stands for patient; HC stands for healthy control; R stands for Right; L stands for Left; AIS stands for American Spinal Injury Association Impairment Scale; T stands for traumatic; SURG stands for Surgery Intervention; BTOX stands for Botulin Toxin Intervention.

Patients	Sex	Age (Years)	Dominant Arm	Lesion Level	AIS Grade	Etiology	Severity of Lesion	Time since Injury (Years)	Shoulder Intervention
P1	Male	56	R	C4	D	T	Incomplete	1	/
P2	Male	63	R	C6–C7	C	T	Incomplete	2	/
P3	Male	63	R	C5–C6	B	T	Incomplete	5	SURG
P4	Male	32	R	C4	D	T	Incomplete	1	BTOX
P5	Male	52	L	C6	D	T	Incomplete	2	/
P6	Male	43	R	C6–C7	A	T	Complete	21	BTOX
P7	Male	59	R	C4	D	T	Incomplete	5	SURG
P8	Female	34	R	C4–C5	D	T	Incomplete	3	/
HC1	Male	65	R						
HC2	Female	27	R						
HC3	Male	27	R						
HC4	Male	34	L						
HC5	Male	49	R						
HC6	Male	36	R						
HC7	Male	37	R						
HC8	Male	76	R						

2.4.2. Ethical Consideration

All procedures were conducted according to the Declaration of Helsinki [56] and were approved by the Institutional Ethics Committee (Area Vasta Centro AOU Careggi, Florence, Italy—ref:17768_oss) following streamlined approval process for low-risk observational studies [57]. All Participants provided informed written consent (see the template in Supplementary Materials).

The wireless sensors (IMU) system does not need a direct interaction with the patient. The small size of the sensor and the stretch band where it is housed make it easily wearable and not bulky. These sensors are not equipped with additional actuators, such as buzzers, so they do not disturb the patient during the movement. Battery can last for more than two working days before a re-charge. The wireless connection assures that cables do not obstruct the patient and the doctor. Due to the restrictions imposed by Covid-19, the entire software development phase, including the first tests, was done entirely remotely. With regard to the experimental tests carried out at the Spinal Unit of Careggi Hospital in Florence, all protocols and provisions for the safety of patients and medical and university staff have been followed. It is important to stress that the experimental tests were performed with a smaller number of subjects than planned due to the anti-Covid protocol. Only 8 patients and 8 healthy individuals have participated to the experimental tests.

2.5. Metrics for Statistical Analysis

Concurrent validation of IMU-based sensor system for measuring the ROM active shoulder movements in CSCI patients and matched healthy control was tested by comparing the IMU system to the goniometer method. To test for concurrent validity, we evaluated the agreement between IMU-based ROM measurements and goniometer-based ROM measurements using Intraclass Correlation Coefficient (ICC) and Bland-Altman analysis, two methods that provide a measure of relative and absolute reliability of an instrument, respectively [58]. In addition, the degree of agreement between the measurements (Rater 1 versus Rater 2 and IMU 1 versus IMU 2 measurements, respectively) was assessed for every tested movement using ICC and Bland–Altman analysis.

A two-way random effect model, absolute agreement, multiple measurements ICCs (model 2, m) was used to test the inter-instrument and inter-tester reliability [59]. The guidelines for interpretation of ICC inter-rater agreement measurement proposed by Cicchetti [60] are exclusive for individual ICC, i.e., in this case ICC(2, 1). The Spearman-Brown formula [61,62] permits to evaluate the m-average ICC thresholds based on the individual ICC:

$$ICC(2, m) = \frac{m \cdot ICC(2, 1)}{1 + (m - 1) \cdot ICC(2, 1)} \tag{1}$$

In this work, ICC(2, 2) and ICC(2, 4) has been used. The thresholds calculated by Equation (1) are shown in Table 3 for equal to 1, 2 and 4.

Table 3. Thresholds for interpretation of Intraclass Correlation Coefficient (ICC) between measurements.

Reliability	ICC(2, 1)	ICC(2, m)	
		m = 2	m = 4
Poor	<0.4	<0.57	<0.73
Fair	0.4–0.6	0.57–0.75	0.73–0.86
Good	0.6–0.75	0.75–0.85	0.86–0.92
Excellent	>0.75	>0.85	>0.92

Bland-Altman analysis quantifies the amount of agreement between two methods of measurement by constructing 95% limits of agreement (LOA), which provides an estimate of the interval where 95% of the differences between both methods fall [63,64] The 95% LOA is computed by using the mean difference (δ_0) and the standard deviation (σ) of the differences between two measurements methods, and is defined by Carmona-Perez et al. in [41] as:

$$LOA = \delta_0 \pm 1.96 \cdot \sigma \tag{2}$$

The LOA in (2) defines two values: upper bound (UB) and low bound (LB). Criteria are required to assume acceptable agreement of two instruments [65]. As suggested by literature, acceptable agreement between measurements requires LOA to be within 10° of no difference between measurements [13]. However, as suggested in Mullaney's study [66], when shoulder ROM measurements are taken by different raters using goniometer, LOA can be expected to be within 15° to consider acceptable agreement between raters. The sample size required to test the concurrent validity between IMU and goniometer methods was based on ICC.

Statistical analysis was performed in MATLAB© and IBM SPSS Statistics software package (version 26).

3. Results

3.1. Accuracy of the IMU-Based System: Laboratory Tests

The evaluation of the accuracy of the proposed IMU-based system has been carried out by testing it in our research laboratory. Static test was performed by mounting the IMU sensor on the arm of a goniometer. The accuracy test was executed by measuring the angle provided by the IMU sensor compared to the reference angle set with the goniometer. In particular, the goniometer moving arm was set and kept at a specific reference angle and data from IMU sensor was recorded for 3 min (see Figure 4). Ten different reference angles were tested, ranging from 0° to 180°. The reference angles and the IMU measurements are shown in Table 4, together with the average error. The mean difference of the IMU system and goniometer measurements was consistently below 3°, which indicated a quite good accuracy of the system in measuring the angle.

Both ICC and Bland-Altman analysis for the static test were performed and results are shown in Table 4. ICC mean value indicated an excellent agreement between IMU system and goniometer measurements. The electronic noise of IMU sensor can be evaluated in Table 4, where LOA between −3° and 5° indicates the noise influence in IMU measurements with the 95% of confidence interval.

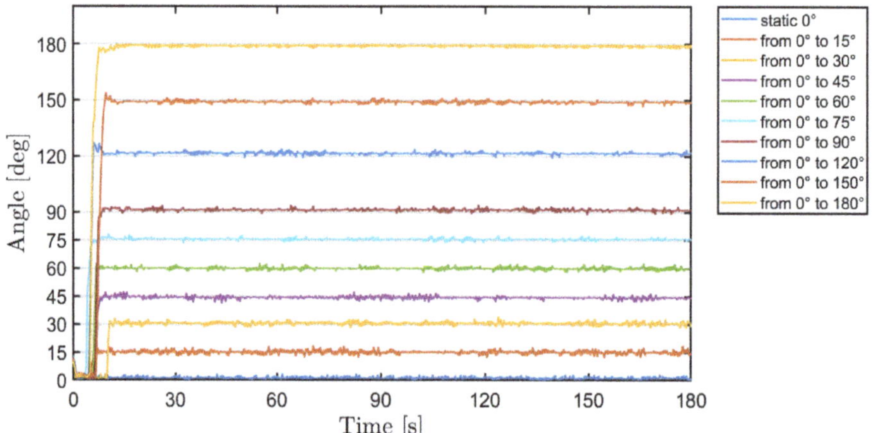

Figure 4. Results of experimental measurement campaign in laboratory.

Table 4. Main statistical results of the laboratory measurement campaign.

Goniometer	IMUs Average (σ)	Difference	ICC (95% CI)	LOA
0°	1.45° (0.77°)	1.45°		
15°	15.19° (0.96°)	0.19°		
30°	30.90° (0.94°)	0.90°		
45°	45.23° (1.19°)	0.23°		
60°	61.51° (1.64°)	1.51°	0.9996	−3.19°; 4.92°
75°	76.11° (0.82°)	1.11°	(0.9994; 0.9997)	
90°	92.06° (1.16°)	2.06°		
120°	122.74° (1.42°)	2.74°		
150°	151.40° (2.56°)	1.40°		
180°	177.02° (2.00°)	2.98°		

Abbreviations: σ stands for standard deviation; ICC stands for Intraclass Correlation Coefficient; CI stands for Confidence Interval (95% for this work); LOA stands for Limits of Agreement.

3.2. Accuracy of IMU-Based and Goniometer Systems: Clinical Tests

Overall, ICCs for whole group of persons ranged from 0.94 to 0.97, indicating an excellent inter-rater reliability between the two raters and the two IMU measurements, as defined in Section 2.5. These four measurements were considered as four different proofs, for each of the arm movements (Table 5). When the groups were evaluated separately results showed different levels of reliability. In the CSCI group, an excellent agreement (ICCs ranging from 0.94 to 0.97) was confirmed for all tested movements. In healthy group, the agreement for flexion movement results fair (ICC of 0.84), whereas for the others movements the ICCs ranging from 0.94 to 0.97 was confirmed.

Table 5. Inter-rater reliability indicators for two raters and two inertial measurement units (IMUs) measurements as four different judges, it calculated as ICC(2, m) with m = 4 and the confidence interval is 95%.

	Whole Group n = 48			CSCI Group n = 24			Healthy Group n = 24		
	ICC(2, m)	LB	UB	ICC(2, m)	LB	UB	ICC(2, m)	LB	UB
Flexion	0.94	0.86	0.97	0.94	0.82	0.98	0.84	0.65	0.93
Abduction	0.97	0.96	0.98	0.97	0.95	0.99	0.94	0.88	0.97
External Rotation	0.97	0.95	0.98	0.95	0.91	0.98	0.97	0.94	0.98
Internal Rotation	0.95	0.91	0.97	0.96	0.90	0.98	0.94	0.86	0.97

Abbreviations: LB: Lower Bound, UB: Upper Bound; n, measurements per method.

It is required to separate the measurements in different cluster to deeply investigate the reason of these assessment agreements identified by the analysis of the data in Table 5. The measurements have been divided in following clusters:

- Inter-instrument reliability and accuracy (See Section 3.2.1)
    - The two raters and the two IMUs measurements are considered as two different judges (Table 6),
    - RATER 1 and IMU 1 measurements are considered as two different judges (Table 7),
    - RATER 2 and IMU 2 measurements are considered as two different judges (Table 8).
- Inter-tester reliability and accuracy (See Section 3.2.2)
    - The two raters' measurements are considered as two different judges (Table 9),
    - The two IMUs measurements are considered as two different judges (Table 10).

3.2.1. IMU versus Goniometric Measurements

Table 6 displays the inter-instrument reliability through ICCs, mean differences and LOAs considering the two raters and the two IMUs measurements as two different judges. IMU-system and goniometer showed excellent agreement (ICCs ranging from 0.86 to 0.97, higher than threshold 0.85) for all tested movements in both groups with exception for flexion movements in healthy group, where the agreements between the two methods were fair (ICCs of 0.61).

In CSCI group mean differences between goniometer and IMU system were small for abduction and external rotation (bias of −1° and 1°, respectively), whereas larger values were observed for flexion and internal rotation (bias of 7° and −7°, respectively). In healthy group large mean differences were evidenced for flexion and abduction (bias of 4° and −6°, respectively). Lastly, LOAs were wide for both groups, consistently greater than 10° of the mean difference, indicating a non-homogeneus behaviour between goniometer and IMU method in measuring shoulder ROM for all tested movements, as explained in Section 2.5.

Additionally, in Tables 7 and 8 is shown the inter-instrument reliability through ICCs, mean differences and LOAs when comparing each rater's performance to the respective IMU measurement. Agreement between Rater 1 and IMU 1 measurements (Table 7), as Rater 2 and IMU 2 measurements (Table 8), were good to excellent for almost all movements in both groups (ICCs ranging from 0.77 to 0.98). Also, in healthy group a fair agreement was shown between Rater 1 and IMU 1 and Rater 2 and IMU 2 for flexion movement (ICC of 0.62), confirming the results shown in Table 6.

However, mean differences between Rater 2 and IMU 2 measurements were larger than those between Rater 1 and IMU 1 measurements in most of tested movements (Tables 7 and 8). This trend was specifically evidenced for flexion and internal rotation in CSCI group and for flexion in healthy group. For flexion movement in CSCI group the mean difference between Rater 1 and IMU 1 measurements was found to be 1°, whereas the mean difference between the Rater 2 and IMU 2 measurements was 14°; this was also observed for internal rotation in which mean differences between rater 1 and IMU

1 measurements and rater 2 and IMU 2 measurements were −5° and −9°, respectively. Similarly, in healthy group for flexion movement rater 1 and IMU 1 measurements showed a mean difference of −2° and rater 2 and IMU 2 measurements showed a mean difference of 9°.

Table 6. Inter-instrument reliability indicators for goniometer and IMU measurements in the whole group, cervical spinal cord injury (CSCI) patients and healthy group.

	Goniometer Average (σ)	IMUs Average (σ)	Difference (σ)	ICC (95% CI)	LOA
Whole group (n = 96)					
Flexion	140° (18°)	134° (20°)	6° (12°)	0.86 (0.75; 0.92)	−19°; 30°
Abduction	146° (20°)	149° (21°)	−3° (8°)	0.95 (0.91; 0.97)	−19°; 13°
External Rotation	79° (16°)	78° (18°)	1° (8°)	0.94 (0.91; 0.96)	−15°; 16°
Internal Rotation	53° (14°)	56° (13°)	−3° (8°)	0.90 (0.81; 0.94)	−19°; 12°
CSCI group (n = 48)					
Flexion	131° (20°)	124° (20°)	7° (13°)	0.86 (0.65; 0.93)	−17°; 32°
Abduction	135° (20°)	136° (19°)	−1° (9°)	0.95 (0.91; 0.97)	−18°; 15°
External Rotation	71° (13°)	70° (18°)	1° (10°)	0.88 (0.78; 0.93)	−19°; 20°
Internal Rotation	51° (16°)	58° (16°)	−7° (8°)	0.90 (0.55; 0.96)	−22°; 8°
Healthy group (n = 48)					
Flexion	149° (9°)	145° (14°)	4° (12°)	0.61 (0.31; 0.78)	−20°; 28°
Abduction	156° (13°)	162° (13°)	−6° (7°)	0.87 (0.53; 0.95)	−20°; 9°
External Rotation	87° (14°)	86° (14°)	1° (5°)	0.97 (0.94; 0.98)	−9°; 11°
Internal Rotation	55° (12°)	55° (10°)	0° (7°)	0.89 (0.81; 0.94)	−14°; 13°

Abbreviations: IMU, Inertial Measurement Unit; ICC, Intraclass Correlation Coefficient; CI, confidence interval; LOA, Limit of Agreement; n, measurements per method; σ, Standard deviation.

Table 7. Inter-instrument reliability indicators for goniometer and IMU measurements in CSCI patients and healthy group for RATER 1 and IMU 1.

	Goniometer Average (σ) RATER 1	IMUs Average (σ) IMU 1	Difference (σ)	ICC (95% CI)	LOA
Whole group (n = 48)					
Flexion	134° (16°)	135° (21°)	0° (11°)	0.90 (0.83; 0.95)	−22°; 21°
Abduction	144° (18°)	148° (21°)	−5° (8°)	0.94 (0.86; 0.97)	−20°; 12°
External Rotation	77° (13°)	78° (17°)	−2° (10°)	0.89 (0.80; 0.94)	−21°; 17°
Internal Rotation	56° (13°)	57° (13°)	−1° (9°)	0.89 (0.80; 0.94)	−18°; 15°

Table 7. *Cont.*

	Goniometer Average (σ) RATER 1	IMUs Average (σ) IMU 1	Difference (σ)	ICC (95% CI)	LOA
	CSCI group (n = 24)				
Flexion	124° (18°)	124° (20°)	1° (11°)	0.92 (0.81; 0.96)	−20°; 22°
Abduction	133° (17°)	136° (19°)	−3° (8°)	0.94 (0.85; 0.97)	−20°; 14°
External Rotation	68° (11°)	71° (18°)	−2° (13°)	0.77 (0.49; 0.90)	−27°; 22°
Internal Rotation	53° (15°)	58° (16°)	−5° (8°)	0.91 (0.72; 0.96)	−21°; 11°
	Healthy group (n = 24)				
Flexion	144° (7°)	146° (14°)	−2° (12°)	0.62 (0.10; 0.83)	−24°; 21°
Abduction	156° (11°)	161° (13°)	−6° (8°)	0.81 (0.33; 0.93)	3°; 22°
External Rotation	85° (9°)	86° (13°)	−1° (5°)	0.94 (0.86; 0.97)	−12°; 9°
Internal Rotation	58° (11°)	56° (10°)	2° (7°)	0.85 (0.66; 0.94)	−12°; 17°

Abbreviations: IMU, Inertial Measurement Unit; ICC, Intraclass Correlation Coefficient; CI, confidence interval; LOA, Limit of Agreement; n, measurements per method; σ, Standard deviation.

Table 8. *Inter-instrument reliability indicators for goniometer and IMU measurements in CSCI patients and healthy group for RATER 2 and IMU 2.*

	Goniometer Average (σ) RATER 2	IMUs Average (σ) IMU 2	Difference (σ)	ICC (95% CI)	LOA
	Whole group (n = 48)				
Flexion	146° (18°)	134° (20°)	12° (11°)	0.83 (0.09; 0.94)	−10°; 33°
Abduction	147° (21°)	149° (21°)	−2° (8°)	0.96 (0.93; 0.98)	−17°; 14°
External Rotation	81° (18°)	78° (19°)	3° (5°)	0.97 (0.92; 0.99)	−6°; 13°
Internal Rotation	50° (15°)	56° (14°)	−6° (6°)	0.91 (0.56; 0.97)	−18°; 7°
	CSCI group (n = 24)				
Flexion	137° (21°)	123° (20°)	14° (11°)	0.82 (−0.15; 0.95)	−7°; 35°
Abduction	137° (22°)	136° (20°)	1° (8°)	0.96 (0.91; 0.98)	−15°; 17°
External Rotation	73° (14°)	69° (18°)	4° (6°)	0.96 (0.85; 0.98)	−8°; 15°
Internal Rotation	49° (18°)	58° (16°)	−9° (6°)	0.91 (0.08; 0.98)	−21°; 4°
	Healthy group (n = 24)				
Flexion	154° (8°)	145° (14°)	9° (11°)	0.62 (−0.12; 0.85)	−12°; 30°
Abduction	157° (15°)	162° (13°)	−5° (7°)	0.91 (0.64; 0.97)	−18°; 8°
External Rotation	89° (17°)	87° (15°)	3° (4°)	0.98 (0.92; 0.99)	−5°; 10°
Internal Rotation	51° (11°)	54° (10°)	−3° (5°)	0.93 (0.79; 0.97)	−13°; 7°

Abbreviations: IMU, Inertial Measurement Unit; ICC, Intraclass Correlation Coefficient; CI, confidence interval; LOA, Limit of Agreement; n, measurements per method; σ, Standard deviation.

3.2.2. Goniometer vs. Goniometer and IMU vs. IMU Measurements

Focusing on the raters measurements, ICCs for whole group ranged from 0.85 to 0.93, showing good to excellent inter-tester reliability between Rater 1 and Rater 2 measurements for all movements, as shown in Table 9. A differentiated analysis for each group indicated in CSCI group excellent agreement between raters for abduction and internal rotation (ICCs of 0.90 and 0.91, respectively) and good agreement for flexion and external rotation (ICCs of 0.85 and 0.83, respectively). In healthy group, excellent to good agreement was displayed for abduction, external and internal rotation (ICCs of 0.90, 0.84, and 0.84, respectively). A poor reliability (ICC of 0.47) was observed for the flexion movement, as depicted in Table 9.

Table 9. Inter-tester/rater reliability indicators for goniometer measurements in CSCI patients and healthy group for RATER 1 and RATER 2.

	Goniometer Average (σ) RATER 1	Goniometer Average (σ) RATER 2	Difference (σ)	ICC (95% CI)	LOA
Whole group (n = 48)					
Flexion	134° (16°)	146° (18°)	−12° (8°)	0.85 (−0.15; 0.96)	−26°; 3°
Abduction	144° (18°)	147° (21°)	−3° (10°)	0.93 (0.88; 0.96)	−22°; 16°
External Rotation	77° (13°)	81° (18°)	−5° (9°)	0.89 (0.75; 0.94)	−23°; 13°
Internal Rotation	56° (13°)	50° (15°)	5° (8°)	0.89 (0.62; 0.95)	−9°; 20°
CSCI group (n = 24)					
Flexion	124° (18°)	137° (21°)	−13° (8°)	0.85 (−0.16; 0.96)	−29°; 3°
Abduction	133° (17°)	137° (22°)	−4° (11°)	0.90 (0.77; 0.96)	−27°; 18°
External Rotation	68° (11°)	73° (14°)	−5° (9°)	0.83 (0.55; 0.93)	−22°; 12°
Internal Rotation	53° (15°)	49° (18°)	4° (9°)	0.91 (0.78; 0.97)	−13°; 21°
Healthy group (n = 24)					
Flexion	144° (7°)	154° (8°)	−10° (6°)	0.47 (−0.30; 0.81)	−23°; 2°
Abduction	156° (11°)	157° (15°)	−1° (8°)	0.90 (0.77; 0.96)	−16°; 14°
External Rotation	85° (9°)	89° (17°)	−4° (10°)	0.84 (0.61; 0.93)	−24°; 15°
Internal Rotation	58° (11°)	51° (11°)	7° (6°)	0.84 (0.04; 0.95)	−5°; 19°

Abbreviations: ICC, Intraclass Correlation Coefficient; CI, confidence interval; LOA, Limit of Agreement; n, measurements per method; σ, Standard deviation.

However, mean differences between each raters' measurements were quite large for almost all movements in both groups, with the highest value shown in CSCI group for the flexion movement (bias of −13°). In the healthy group the largest mean differences was shown for flexion and internal rotation (bias of −10° and 7°, respectively). Lastly, LOAs were wide for both groups, consistently greater than 15° of the mean difference, indicating a non-homogeneus behaviour between Rater 1 and Rater 2 in measuring shoulder ROM for all tested movements.

Focusing on the IMUs measurements, ICCs for whole group ranged from 0.988 to 0.999, showing very excellent inter-tester reliability between IMU 1 and IMU 2 measurements for all movements, as depicted in Table 10. Very excellent inter-tester reliability were confirmed also when CSCI and healthy groups were separately analyzed (ICCs ranging from 0.979 to 0.998).

Table 10. Inter-tester/rater reliability indicators for IMU measurements in CSCI patients and healthy group for IMU 1 and IMU 2.

	IMUs Average (σ) IMU 1	IMUs Average (σ) IMU 2	Difference (σ)	ICC (95% CI)	LOA
	Whole group (n = 48)				
Flexion	135° (21°)	134° (20°)	0° (2°)	0.997 (0.995; 0.998)	−4°; 5°
Abduction	149° (21°)	149° (21°)	0° (2°)	0.999 (0.998; 0.999)	−3°; 3°
External Rotation	78° (17°)	78° (19°)	0° (4°)	0.988 (0.978; 0.993)	−7°; 8°
Internal Rotation	57° (13°)	56° (14°)	1° (2°)	0.993 (0.983; 0.996)	−3°; 5°
	CSCI group (n = 24)				
Flexion	124° (20°)	123° (20°)	1° (2°)	0.997 (0.993; 0.999)	−4°; 5°
Abduction	136° (19°)	136° (19°)	−1° (2°)	0.998 (0.994; 0.999)	−4°; 3°
External Rotation	71° (18°)	69° (18°)	1° (4°)	0.988 (0.972; 0.995)	−6°; 9°
Internal Rotation	58° (16°)	58° (16°)	0° (2°)	0.996 (0.990; 0.998)	−4°; 5°
	Healthy group (n = 24)				
Flexion	146° (14°)	145° (14°)	0° (2°)	0.995 (0.988; 0.998)	−4°; 5°
Abduction	162° (13°)	162° (13°)	0° (1°)	0.998 (0.995; 0.999)	−2°; 2°
External Rotation	86° (13°)	87° (15°)	−1° (4°)	0.979 (0.952; 0.991)	−8°; 7°
Internal Rotation	56° (10°)	54° (10°)	2° (2°)	0.985 (0.852; 0.996)	−2°; 5°

Abbreviations: IMU, Inertial Measurement Unit; ICC, Intraclass Correlation Coefficient; CI, confidence interval; LOA, Limit of Agreement; n, measurements per method; σ, Standard deviation.

In addition, mean differences were found very small for all tested movements in CSCI group, ranging from −1° to 1°, as well as in healthy group, ranging from −1° to 2°. Lastly, LOAs were narrow, very similar to those found in the preliminary laboratory tests, and consistently showed that the difference in IMU 1 and IMU 2 measurements was within 10° of the mean difference for approximately 95% of participants in both group.

4. Discussion

This is the first study to evaluate the validity of wireless wearable IMU-based sensors system for the assessment of shoulder ROM in patients with CSCI in clinical environment. A custom IMU-based system was developed and compared to the method used currently in clinic, i.e., a sanitary operator measuring the ROM with a goniometer. The good accuracy of the IMU-based system was proven in laboratory tests with absolute error consistently below 3° and 95% LOA within 10°. Together with the accuracy of the proposed system, we aimed to provide a measure of the accordance between the two measuring methods: the IMU-based and the rater-based method. According to the results, the concurrent validity of IMU system was partially confirmed. ICCs values were found to be very high for all tested shoulder maneuvers except for flexion in healthy group, showing an excellent relative validity of the IMU system. However, given that the mean differences between IMU and goniometer measures were shown to be relatively high for the majority of movements alongside a high distance between LOAs, the absolute validity of the IMU system is not strongly supported. The two methods might not seem interchangeable. In addition, while two raters' measurements showed a considerably low inter-rater reliability, IMU system was proven to have excellent agreement when compared one measurement to the other.

Within limited literature regarding the concurrent validation of IMU sensors against goniometer in assessing shoulder ROM, we found little similarities with other studies though the differences in research methodology may interfere with comparisons. Yoon et al. [37] assessed the validity of IMU sensors in evaluating upper extremity movements while passively positioning subjects' shoulder at specific angles using a goniometer (shoulder flexion 0°–170°, abduction 0°–170°, external rotation 0°–90°, and internal rotation 0°–60° angles). The authors showed large mean bias and a high distance between LOA for most of shoulder position except for flexion (0–135°), abduction (0°), external (0°) and internal rotation (0°–45°), though the relative reliability between goniometer and IMU by means of ICC analysis was not evaluated. In a more recent study, Rigoni and coworkers [13] validated the IMU system by testing active shoulder ROM on healthy subjects and, similar to our study, showed high ICC values for all tested movements (flexion, abduction, external and internal rotation at 90 shoulder abduction). However, they showed absolutely reliability of IMU system through very small mean bias (approximately 0°) and narrow LOA (−4.5° to 3.2°).

Our study showed excellent relative reliability of IMU system in comparison with goniometer method in most movements performed by CSCI and healthy groups. However, some high ICC values in our results should be considered with caution as it is known that the ICC values are influenced by the range of measured values (sample heterogeneity) [66]. Higher ranges (greater heterogeneity) are associated with higher ICCs, independent of actual measurement error [57,58]. In fact, in some movements where we obtained very high ICC values, there was also found a high variability of the data. For instance, in CSCI group for internal rotation the standard deviation was about 30% of the mean value of ROM. This can mask poor trial-to-trial consistency and not accurately represent the high agreement of measurements between the two methods. These interpretations of ICCs are in conformity with previous studies using IMU sensors to evaluate motion in disease of motor control [41,67].

On the contrary, our study revealed the existence of measurement differences between IMU system and goniometer in CSCI patients with 95% LOA for the two instruments ranged from −22° to 32°, showing a certain change in measuring between goniometer and IMU for all tested movements. These wider LOA were confirmed also for control group of healthy subjects (LOA ranged from −20° to 28°).

Several factors regarding the setup in our study might have influenced the results by IMU system measurements. The exclusion of magnetometer usage in our IMU sensors could be the first possible reason. This choice allowed to avoid local magnetic disturbances which could affect IMUs accuracy due to magnetometers sensitivity to the magnetic fields [68–71]. In our clinical tests the hospital environment and electronic controls of motorized wheelchair could influence the efficiency of magnetometer. The downside of this decision, on the other hand, is that an IMU system without a magnetometer could have provided lower accuracy in orientation estimation, specifically while testing a mobile joint like shoulder with several degrees of freedom due to bony constraint scarcity and soft tissue function [72]. Another possible source of measurement error for our IMU system could be the presence of wheelchair in the setup used to test CSCI patients, which seems to interfere in performing motion in one single plane specifically for flexion and abduction movements. Indeed, a previous study by Lee and colleagues [39], in which the validity of Kinetec against goniometer was tested in measuring active and passive shoulder ROM movements, showed that the deviation from planes of unrestricted motion was an important source in increasing 95% limit of agreement between two measurement methods. Moreover, for maneuvers like external and internal rotation with forearm abduction at 90°

an additional potential cause for discrepancies of measurements between goniometer and IMU system includes differences in starting positions of the forearm among trials. In fact, even with the assistance of an operator, holding a perfect position of forearm parallel to the floor by the subject was not completely guaranteed and it could provoke differences in measurement starting point between IMU system and raters who were instructed to maintain the stationary arm of goniometer fixed consistently parallel to the floor (see Table 1 for goniometer landmarks).

Additionally, when a deeper analysis comparing each rater's performance to the respective IMU measurement was executed, larger differences of measurements (mean bias) were found between Rater 2 and IMU 2 than between Rater 1 and IMU 1, especially for flexion and internal rotation movements (see Tables 7 and 8). Moreover, inter-tester reliability for IMU measurements showed perfect accordance between IMU 1 and IMU 2 (ICCs from 0.979 to 0.998; mean bias from $-1°$ to $2°$; and LOA within $10°$) for both groups (see Table 10), whereas inter-tester reliability for goniometer measurements evidenced larger differences (mean bias from $-13°$ to $7°$; LOA wider than $15°$) between Rater 1 and Rater 2, even though ICC values appeared to be good to excellent except for flexion in healthy group (see Table 9). Thus, it is possible to assume that inconsistency between raters partially contributes to the differences observed in measuring shoulder ROM between IMU and goniometer. This also confirmed similar findings from previous literature which evaluated inter-rater reliability of goniometer measurements for shoulder ROM and showed that reliability of goniometer is lower when measurements are taken by different raters [43].

Furthermore, concerning the inter-tester reliability between two IMU measurements, Rigoni and coworkers [13] found low agreement between measurements when one goniometer and one IMU measurement were taken by each assessor for each movement. They suggested that IMU measurements recorded by one assessor could not be exchanged for another one's measurements. Differently, as we conducted three repetition for each tested movement and recorded pairs of two goniometer- and two IMU-based measurements simultaneously during every repetition, a very high inter-tester reliability between two IMU measurements was found. This implicates that differences of IMU measurements between assessors found in Rigoni's study might be ascribed to the inconsistency of movements performed over time rather than the assessor's performance. This could suggest that, despite its level of accuracy, our IMU system is a stable and operator-independent method to measure shoulder ROM in patients with CSCI.

Altogether, our custom IMU-based system was shown to be a promising tool to assess shoulder ROM in CSCI patients according to several benefits. First, IMU could be introduced in clinical setting as a stable tool which is independent from assessor's availability and manual skills. Second, the simplicity of setup could allow the easy self-applying procedure of IMU system and increase the variety of body contexts in which it is possible to collect measurements. Third, IMU-based system could allow clinicians to remotely monitor patients' movements while these latter staying out of routine clinical system (e.g., house), and thus make the caring more convenient and economical in such patients with motor difficulties by easy accessibility [73–75]. Finally, IMU system could also provide patients with feedback on their performance and progress during rehabilitation [13].

Despite the promising findings, the current study showed some limitations, which could suggest future research. The setup of IMU system without magnetometer might have affected the accuracy in measuring active unrestricted movements of shoulder. The magnetometer integration on IMU system could be introduced in future studies due to the fact that it is essential to evaluate the possible implementation of IMU system for distant monitoring of patients. In addition, the number of participants was relatively small which did not allow us to detect statistically significant differences between CSCI and healthy groups in each shoulder ROM via IMU system. Furthermore, the current study focused only on the evaluation of ROM of simple movements, which limits the applicability of the results to more complex and functional tasks which are more identical to daily-life activities (e.g., reach-to-grasp upper limb movements) [40]. Finally, our study did not analyze dynamic movement characteristics of upper limbs, including angular velocity and acceleration, that could provide a more comprehensive clinical assessment of patients [40]. All these factors should be considered in future studies on assessing the validity of IMU system [64].

5. Conclusions

This work aimed to provide a methodological study on the validity of a customized wireless wearable IMU-based sensors system to measure the shoulder ROM in patients with cervical spinal cord injury. Laboratory and clinical trials have been performed to evaluate the accuracy of the IMU-based system, and to compare it to the goniometer-based measurements taken by sanitary operators. The results showed an excellent relative reliability between the IMU-based system and goniometer-operator method, even though the two methods might not seem interchangeable. Nonetheless, unlike the goniometer method, the accuracy of IMU system is not influenced by the operator who carries on the measurement. Therefore, the proposed system can be a potential tool to be integrated in clinical settings for monitoring shoulder ROM in patients with cervical spinal cord injury.

It is worth mentioning that, in addition to static angles, the IMU sensors also capture dynamic kinematic parameters (e.g., angular velocity and acceleration) that could be used with these patients to quantify muscle stress and effort during movements by estimating fatigue-related tremor. Real-time measurements from the sensors could be further used to provide feedback to the patient during the execution of motions in physical therapy.

Furthermore, our validation work poses the base for a possible use of the IMU system for a remote monitoring of patients at home, and this feature could be crucial, in particular during emergency situations such as COVID-19 pandemic. In fact, a wireless wearable IMU system automatically could collect data on the movements of patient's arms and allow us to create a database with all the performance.

Possible future developments of the proposed system span from the inclusion of the magnetometer measure to improve the accuracy, to the fusion of the kinematics data to provide a deeper interpretation in quantifying motion recovery status of patient during the rehabilitation process, and, finally, to the use of artificial intelligence (AI) to automatically manage the everyday activities of the patient.

Author Contributions: Conceptualization, R.B., E.J.C., E.Q., S.G. and D.M.; methodology, R.B., E.J.C., G.L., I.N., G.R. and D.M.; software, S.C., L.B., S.J., A.M. and L.M.; validation, R.B., E.J.C. and D.M.; formal analysis, S.C., L.B. and L.M.; investigation, R.B., S.C., E.J.C., L.B., G.L. and I.N.; resources, G.L., G.D.P., L.M. and D.M.; data curation, S.C. and L.B.; writing—original draft preparation, R.B.; writing—review and editing, S.C., S.J., A.M., E.Q., S.G. and G.R.; visualization, R.B., S.C. and L.B.; supervision, G.L., G.D.P., L.M. and D.M.; project administration, G.L., G.D.P., L.M. and D.M.; funding acquisition, G.L., G.D.P., L.M. and D.M. All authors have read and agreed to the published version of the manuscript.

Funding: This work was supported in part by the ECRF foundation (Florence, Italy) under the ARTEMIS project grant.

Institutional Review Board Statement: The study was conducted according to the guidelines of the Declaration of Helsinki and approved by the Ethics Committee of the Florence University Hospital (protocol code: 17768 OSS, date of approval: 4 June 2020).

Informed Consent Statement: Informed consent was obtained from all subjects involved in the study.

Data Availability Statement: The data presented in this study are available on request from the corresponding author. The data are not publicly available due to privacy issue.

Conflicts of Interest: The authors declare no conflict of interest.

Appendix A. Technical Information about Wearable Sensors System for Motion Assessment

Appendix A.1. System Hardware Architecture

The Hardware system is composed of commercial IMU sensors, which are connected to a gateway (Raspberry Pi). The two sensors are positioned one on the arm and one on the wrist of the individual (Figure A1). The Bluetooth Low Energy (BLE) is the communication protocol used to transmit data from sensor to the gateway, in compliance with standard IEEE 802.15.6, typically used in Wireless Body Area Networks (WBANs). Once data are received by the gateway, they are processed by a custom application and stored on a cloud database.

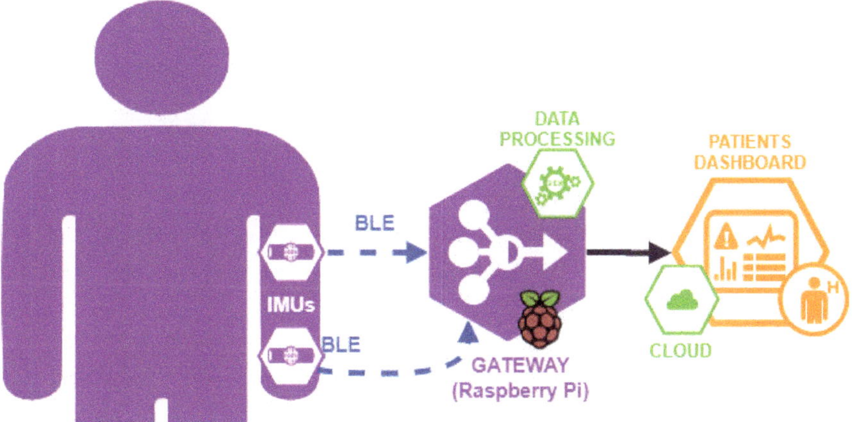

Figure A1. Scheme of the Hardware Module of the system. In purple, the components of system hardware module; in green, the components of system software module and in orange, the components of data aggregation/visualization module.

Among the many IMU sensors on the market, the MbientLab MetaMotion R [76] was selected, since it shows sufficient accuracy for this application (from datasheet [76], the accuracy is less than 1 degree RMS), and it is characterized by small dimensions and low cost, compared to other similar sensors. The accuracy of IMU-based system has been verified in Section 3.1.

The manufacturer provides a mobile application, which does not include synchronization functionality between multiple sensors, thus we had to implement a customized gateway for the synchronization of sensors, the collection and the analysis of the data.

Appendix A.2. Data Processing and ROM Calculation with IMU-Based System

Notation: Scalar variables are displayed in sans serif, lowercase letters; their realizations in serif, italic fonts, x. Vectors and matrices are denoted by bold lowercase and uppercase letters, respectively, e.g., $\mathbf{x}$ indicates a vector and $\mathbf{X}$ indicates a matrix. Quaternions are denoted by bold lowercase letters with tilde sign as a hat $\tilde{\mathbf{x}}$. Coordinate frame are displayed in sans serif, italic fonts, uppercase letters X. A pre-subscript denotes the source coordinate frame and a pre-superscript denotes the destination coordinate frame, $_{A}^{B}\tilde{\mathbf{x}}$ denoted a quaternion from coordinate frame A to B. In the case that only a pre-superscript is present in the quaternion, it means that the quaternion is measured and represented in the same coordinate frame $^{B}\tilde{\mathbf{x}}$.

The range of motion (ROM) can be calculated by using accelerometer or gyroscope measurements, both present in the IMU sensors. A fusion of those data can enhance the accuracy of the ROM calculation. This fusion is provided by the Madgwick filtering [77]. In this study, the monitoring of the arm movement has been done in the vertical plane.

For this reason, the fusion of the data coming from the accelerometer and the gyroscope by using the Madgwick filter algorithm [77] is assumed to be enough for ROM assessment.

Typically, to represent orientations and rotations of generic object in three dimensions, two different mathematical notations are used: Euler angles or quaternions. Usually, data fusion filter algorithms use quaternions due to lower computational cost than Euler angles. A quaternion is a mathematical iper-complex entity ($\widetilde{\mathbf{q}} \in \mathbb{C}^4$), composed of a scalar part and a vector part. A quaternion is generally represented as

$$\widetilde{\mathbf{q}} = a + b\mathbf{i} + c\mathbf{j} + d\mathbf{k} \tag{A1}$$

where a, b, c and d are real numbers and $\mathbf{i}$, $\mathbf{j}$, $\mathbf{k}$ are orthogonal versors.

Denoting with $\mathbf{u}$ and α the rotation axis and angle, respectively, Equation (A1) can be rewritten as

$$\widetilde{\mathbf{q}} = \cos\frac{\alpha}{2} + \mathbf{u}\sin\frac{\alpha}{2} \tag{A2}$$

where $\mathbf{u}$ is described by the versors $\mathbf{i}$, $\mathbf{j}$, $\mathbf{k}$.

The Madgwick filter algorithm is deployed in the Raspberry Pi module to reduce the centrifugal acceleration component and to reduce the drift error due to the angular velocity (gyroscope) integration. The Madgwick filtering is based on the estimated quaternion $^I_W\widetilde{\mathbf{q}}_{est,t+1}$ from fixed global coordinate (W) to inertial coordinate (I), given the accelerometer measure, the gyroscope measure and the estimated quaternion $^I_W\widetilde{\mathbf{q}}_{est,t}$ at previous step time t

$$^I_W\widetilde{\mathbf{q}}_{est,t+1} = \int_t^{t+1} \left(\underbrace{\frac{1}{2} {}^I_W\widetilde{\mathbf{q}}_{est,t} \otimes [0, {}^I\boldsymbol{\omega}_{t+1}]}_{\text{Gyroscope component}} - \beta \underbrace{\frac{\nabla f({}^I_W\widetilde{\mathbf{q}}_{est,t}, {}^W\widetilde{\mathbf{g}}, {}^I\widetilde{\mathbf{a}}_{t+1})}{\| \nabla f({}^I_W\widetilde{\mathbf{q}}_{est,t}, {}^W\widetilde{\mathbf{g}}, {}^I\widetilde{\mathbf{a}}_{t+1}) \|}}_{\text{Accelerometer component}} \right) \cdot dt \tag{A3}$$

where

- β is a parameter which weights the accelerometer contribution on the quaternion estimate;
- $\otimes$ indicates quaternions product;
- $\nabla f(\cdot)$ indicates an error direction on the solution surface defined by the objective function, $f({}^I_W\widetilde{\mathbf{q}}_{est,t}, {}^W\widetilde{\mathbf{g}}, {}^I\widetilde{\mathbf{a}}_{t+1})$ (function defined as in (A4)) and its Jacobian;
- $\| \nabla f(\cdot) \|$ indicates a norm of function $\nabla f({}^I_W\widetilde{\mathbf{q}}_{est,t}, {}^W\widetilde{\mathbf{g}}, {}^I\widetilde{\mathbf{a}}_{t+1})$;
- ω is conversion from gyroscope 3D measurements into a quaternion;
- $\widetilde{\mathbf{g}}$ is conversion from gravity 3D vector into a quaternion;
- $\widetilde{\mathbf{a}}$ is conversion from accelerometer 3D measurements into a quaternion;

The gyroscope component in Equation (A3) represents the quaternion that describes the rate of change of the earth frame (inertial frame of reference with the gravity vector as the vertical) relative to sensor frame based on gyroscope measurements, $^I\boldsymbol{\omega}_{t+1}$. The acceleration component in Equation (A3) represents the error correction function based on orientation increment given by the accelerometer measurements, $^I\widetilde{\mathbf{a}}_{t+1} = [0, \mathbf{a}_{t+1}]$, compared to gravity component, $^W\widetilde{\mathbf{g}} = [0, 0, 0, 1]$. In fact, the term $f({}^I_W\widetilde{\mathbf{q}}_{est,t}, {}^W\widetilde{\mathbf{g}}, {}^I\widetilde{\mathbf{a}}_{t+1})$ is defined as

$$f({}^I_W\widetilde{\mathbf{q}}_{est,t}, {}^W\widetilde{\mathbf{g}}, {}^I\widetilde{\mathbf{a}}_{t+1}) = {}^I_W\widetilde{\mathbf{q}}^*_{est,t} \otimes {}^W\widetilde{\mathbf{g}} \otimes {}^I_W\widetilde{\mathbf{q}}_{est,t} - {}^I\widetilde{\mathbf{a}}_{t+1} \tag{A4}$$

where $^I_W\widetilde{\mathbf{q}}^*_{est,t}$ indicates a conjugate quaternion of $^I_W\widetilde{\mathbf{q}}_{est,t}$

For further details about these formulas, their full proofs and explanations, see [77].

The procedure to evaluate the angle of motion with the IMU-based system begins with a calibration procedure, in which the sensor has been positioned with its y-axis along the arm, i.e., the y-axis in sensor coordinate frame corresponds to one principal axis of inertia in arm coordinate frame. The initial position of the arm defines a versor ($\mathbf{v}_0$), which has been saved before the begin of every movements.

In the initial position, the arm has been positioning in vertical position to reduce the influence of gyroscope drift. During the test at every display refresh time (50 frames per seconds, so 0.02 s), the arm position defines a versor (**v**) and the developed gateway shows in a monitor the angle (θ) between $\mathbf{v_0}$ and **v**, calculated by:

$$\theta = \arccos(\langle \mathbf{v}, \mathbf{v_0} \rangle) \tag{A5}$$

where $\langle \mathbf{v}, \mathbf{v_0} \rangle$ indicates scalar product between versors **v** and $\mathbf{v_0}$.

Appendix A.3. System Software Architecture

A custom Python software was deployed to connect multiple sensors simultaneously, to synchronize the sensor readings, and to send data to the gateway for storing and processing. The general scheme of the software is represented by using the Unified Modeling Language (UML) in Figure A2 and it is composed of:

- **MadgwickFilter** represents the mathematical tool to update quaternion value from accelerometer and gyroscope measurements. It has two main attributes:
    - *madgwickQuaternion* is an instance of the Quaternion class. It is modified every times is called the function *update* using Equation (A4) to evaluate the angle measured by each sensor.
    - *beta* is the editable parameter to change the accelerometer contribution weight, as shown in Equation (A3).
- **Quaternion** is the implemented mathematical library to help with operations with quaternions, e.g., product between quaternions, conjugate, apply rotation to vector, etc. This mathematical library is been used by MadgwickFilter class to update its quaternion.
- **MetaMotion** is the main class of the schema in Figure A2, it represents an abstraction of the sensors used for this work. As these devices use the Bluetooth protocol to communicate, they are uniquely identified by their MAC address. The sensor unit has different sensors, in particular accelerometer and gyroscope, used for this work [76].
- **SmartGateway (*Raspberry*)** is the back end unit of data (post) processing. This component is in charge of synchronizing the connection with the various devices, which may be of different types. It has to connect with these devices, configure them by setting the date rate parameters, and allow us to broadcast the data stream.
  The angle of the arm movement has been calculated every display refresh, because the data acquisition rate of these devices is not time-constant. When the movement is finished, this unit has to collect and store all data into his internal storage: this operation has been implemented to analyze and compare data of past movements with the actual.

In Figure A2 both the data visualization part (the orange module of Figure 2) and the cloud database part (the green hexagonal icon with the cloud symbol inside, both in Figures 2 and A1) are missing: these two parts are currently under development. The priority has been given to the implementation of the sensor synchronization and data processing part of the system, due to only these parts are necessary for this work.

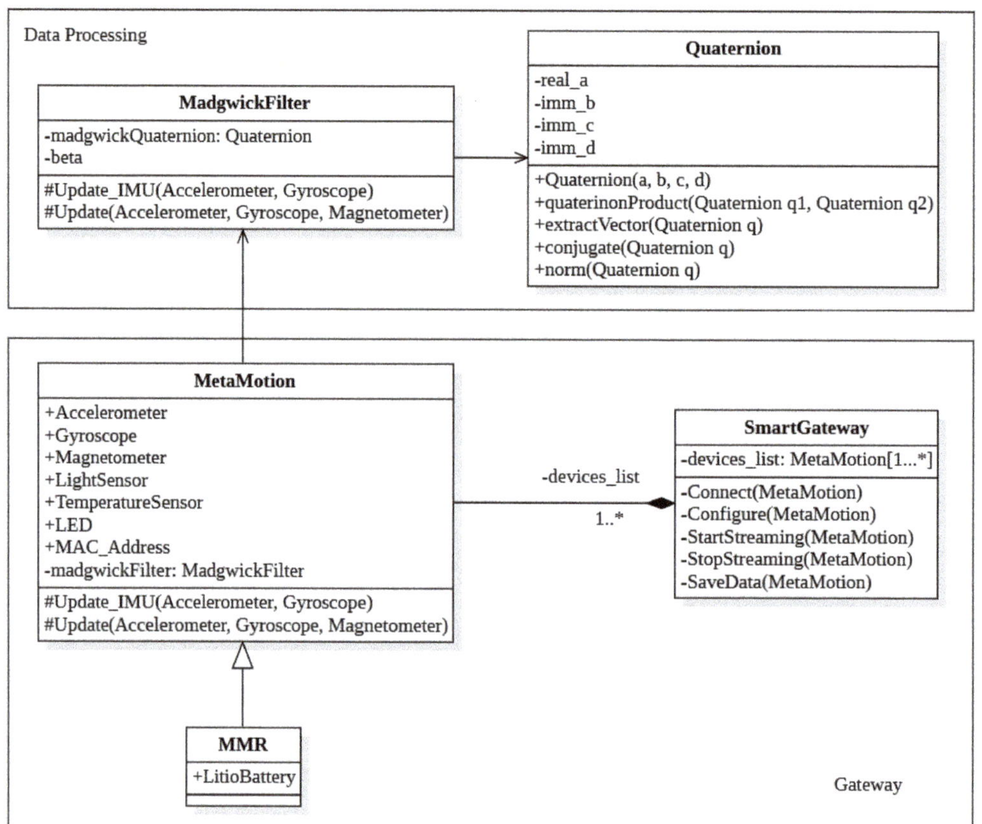

Figure A2. UML Class Diagram of the back-end *software module* of IMU sensors network system used in this work. The writing (1...*) stands for "from 1 to infinity".

References

1. Alizadeh, A.; Dyck, S.M.; Karimi-Abdolrezaee, S. Traumatic Spinal Cord Injury: An Overview of Pathophysiology, Models and Acute Injury Mechanisms. *Front. Neurol.* **2019**, *10*, 282. [CrossRef]
2. Kirshblum, S.C.; Burns, S.P.; Biering-Sorensen, F.; Donovan, W.; Graves, D.E.; Jha, A.; Johansen, M.; Jones, L.; Krassioukov, A.; Mulcahey, M.; et al. International standards for neurological classification of spinal cord injury (Revised 2011). *J. Spinal Cord Med.* **2011**, *34*, 535–546. [CrossRef] [PubMed]
3. Maynard, F.M.; Bracken, M.B.; Creasey, G.; Ditunno, J.F., Jr.; Donovan, W.H.; Ducker, T.B.; Garber, S.L.; Marino, R.J.; Stover, S.L.; Tator, C.H.; et al. International Standards for Neurological and Functional Classification of Spinal Cord Injury. *Spinal Cord* **1997**, *35*, 266–274. [CrossRef] [PubMed]
4. McDonald, J.W.; Sadowsky, C. Spinal-cord injury. *Lancet* **2002**, *359*, 417–425. [CrossRef]
5. Wagner, J.P.; Curtin, C.M.; Gater, D.R.; Chung, K.C. Perceptions of People with Tetraplegia Regarding Surgery to Improve Upper-Extremity Function. *J. Hand Surg.* **2007**, *32*, 483–490. [CrossRef] [PubMed]
6. Mortenson, W.B.; Miller, W.C.; Backman, C.L.; Oliffe, J.L. Association Between Mobility, Participation, and Wheelchair-Related Factors in Long-Term Care Residents Who Use Wheelchairs as Their Primary Means of Mobility. *J. Am. Geriatr. Soc.* **2012**, *60*, 1310–1315. [CrossRef]
7. Eriks-Hoogland, I.; de Groot, S.; Post, M.; van der Woude, L. Correlation of shoulder range of motion limitations at discharge with limitations in activities and participation one year later in persons with spinal cord injury. *J. Rehabil. Med.* **2011**, *43*, 210–215. [CrossRef] [PubMed]
8. Martin, R.; Silvestri, J. Current Trends in the Management of the Upper Limb in Spinal Cord Injury. *Curr. Phys. Med. Rehabil. Rep.* **2013**, *1*, 178–186. [CrossRef]

9. Gil-Agudo, Á.; De los Reyes-Guzmán, A.; Dimbwadyo Terrer, I.; Peñasco-Martín, B.; Bernal, A.; López-Monteagudo, P.; Del-Ama, A.; Pons, J. A novel motion tracking system for evaluation of functional rehabilitation of the upper limbs. *Neural Regen. Res.* **2013**, *8*, 1773–1782. [CrossRef]

10. Struyf, F.; Nijs, J.; Mottram, S.; Roussel, N.A.; Cools, A.M.J.; Meeusen, R. Clinical assessment of the scapula: A review of the literature. *Br. J. Sports Med.* **2014**, *48*, 883–890. [CrossRef]

11. Ahn, S.Y.; Ko, H.; Yoon, J.O.; Cho, S.U.; Park, J.H.; Cho, K.H. Determining the Reliability of a New Method for Measuring Joint Range of Motion Through a Randomized Controlled Trial. *Ann. Rehabil. Med.* **2019**, *43*, 707–719. [CrossRef] [PubMed]

12. Riddle, D.L.; Rothstein, J.M.; Lamb, R.L. Goniometric Reliability in a Clinical Setting. *Phys. Ther.* **1987**, *67*, 668–673. [CrossRef] [PubMed]

13. Rigoni, M.; Gill, S.; Babazadeh, S.; Elsewaisy, O.; Gillies, H.; Nguyen, N.; Pathirana, P.N.; Page, R. Assessment of Shoulder Range of Motion Using a Wireless Inertial Motion Capture Device—A Validation Study. *Sensors* **2019**, *19*, 1781. [CrossRef] [PubMed]

14. Grangeon, M.; Guillot, A.; Sancho, P.O.; Picot, M.; Revol, P.; Rode, G.; Collet, C. Rehabilitation of the Elbow Extension with Motor Imagery in a Patient with Quadriplegia After Tendon Transfer. *Arch. Phys. Med. Rehabil.* **2010**, *91*, 1143–1146. [CrossRef]

15. Grangeon, M.; Revol, P.; Guillot, A.; Rode, G.; Collet, C. Could motor imagery be effective in upper limb rehabilitation of individuals with spinal cord injury? A case study. *Spinal Cord* **2012**, *50*, 766–771. [CrossRef]

16. Chèze, L. The Different Movement Analysis Devices Available on the Market. In *Kinematic Analysis of Human Movement*; John Wiley & Sons, Inc.: New York, NY, USA, 2014; pp. 17–33. [CrossRef]

17. Cohen, E.J.; Bravi, R.; Minciacchi, D. 3D reconstruction of human movement in a single projection by dynamic marker scaling. *PLoS ONE* **2017**, *12*, e0186443. [CrossRef]

18. Baskwill, A.J.; Belli, P.; Kelleher, L. Evaluation of a Gait Assessment Module Using 3D Motion Capture Technology. *Int. J. Ther. Massage Bodyw.* **2017**, *10*, 3–9. [CrossRef]

19. Mustapa, A.; Justine, M.; Mustafah, N.M.; Manaf, H. The Effect of Diabetic Peripheral Neuropathy on Ground Reaction Forces during Straight Walking in Stroke Survivors. *Rehabil. Res. Pract.* **2017**, *2017*, 5280146. [CrossRef] [PubMed]

20. Mucchi, L.; Jayousi, S.; Martinelli, A.; Caputo, S.; Marcocci, P. An Overview of Security Threats, Solutions and Challenges in WBANs for Healthcare. In Proceedings of the 2019 13th International Symposium on Medical Information and Communication Technology (ISMICT), Oslo, Norway, 8–10 May 2019. [CrossRef]

21. Mucchi, L.; Jayousi, S.; Caputo, S.; Paoletti, E.; Zoppi, P.; Geli, S.; Dioniso, P. How 6G Technology Can Change the Future Wireless Healthcare. In Proceedings of the 2020 2nd 6G Wireless Summit (6G SUMMIT), Levi, Finland, 17–20 March 2020; pp. 1–6. [CrossRef]

22. Tian, Y.; Meng, X.; Tao, D.; Liu, D.; Feng, C. Upper limb motion tracking with the integration of IMU and Kinect. *Neurocomputing* **2015**, *159*, 207–218. [CrossRef]

23. Filippeschi, A.; Schmitz, N.; Miezal, M.; Bleser, G.; Ruffaldi, E.; Stricker, D. Survey of Motion Tracking Methods Based on Inertial Sensors: A Focus on Upper Limb Human Motion. *Sensors* **2017**, *17*, 1257. [CrossRef]

24. Poitras, I.; Bielmann, M.; Campeau-Lecours, A.; Mercier, C.; Bouyer, L.J.; Roy, J.S. Validity of Wearable Sensors at the Shoulder Joint: Combining Wireless Electromyography Sensors and Inertial Measurement Units to Perform Physical Workplace Assessments. *Sensors* **2019**, *19*, 1885. [CrossRef]

25. Najafi, B.; Aminian, K.; Paraschiv-Ionescu, A.; Loew, F.; Bula, C.; Robert, P. Ambulatory system for human motion analysis using a kinematic sensor: Monitoring of daily physical activity in the elderly. *IEEE Trans. Biomed. Eng.* **2003**, *50*, 711–723. [CrossRef]

26. Luinge, H.J.; Veltink, P.H. Measuring orientation of human body segments using miniature gyroscopes and accelerometers. *Med. Biol. Eng. Comput.* **2005**, *43*, 273–282. [CrossRef]

27. Sabatini, A.M. Variable-State-Dimension Kalman-Based Filter for Orientation Determination Using Inertial and Magnetic Sensors. *Sensors* **2012**, *12*, 8491–8506. [CrossRef]

28. Ligorio, G.; Sabatini, A. Extended Kalman Filter-Based Methods for Pose Estimation Using Visual, Inertial and Magnetic Sensors: Comparative Analysis and Performance Evaluation. *Sensors* **2013**, *13*, 1919–1941. [CrossRef] [PubMed]

29. Saber-Sheikh, K.; Bryant, E.C.; Glazzard, C.; Hamel, A.; Lee, R.Y. Feasibility of using inertial sensors to assess human movement. *Man. Ther.* **2010**, *15*, 122–125. [CrossRef] [PubMed]

30. Zhang, Z.Q.; Ji, L.Y.; Huang, Z.P.; Wu, J.K. Adaptive Information Fusion for Human Upper Limb Movement Estimation. *IEEE Trans. Syst. Man, Cybern. Part A Syst. Hum.* **2012**, *42*, 1100–1108. [CrossRef]

31. Fasel, B.; Sporri, J.; Chardonnens, J.; Kroll, J.; Muller, E.; Aminian, K. Joint Inertial Sensor Orientation Drift Reduction for Highly Dynamic Movements. *IEEE J. Biomed. Health Inform.* **2018**, *22*, 77–86. [CrossRef] [PubMed]

32. Fong, D.; Chan, Y.Y. The Use of Wearable Inertial Motion Sensors in Human Lower Limb Biomechanics Studies: A Systematic Review. *Sensors* **2010**, *10*, 11556–11565. [CrossRef]

33. Giansanti, D.; Maccioni, G.; Benvenuti, F.; Macellari, V. Inertial measurement units furnish accurate trunk trajectory reconstruction of the sit-to-stand manoeuvre in healthy subjects. *Med. Biol. Eng. Comput.* **2007**, *45*, 969–976. [CrossRef] [PubMed]

34. Zhou, H.; Hu, H.; Tao, Y. Inertial measurements of upper limb motion. *Med. Biol. Eng. Comput.* **2006**, *44*, 479–487. [CrossRef]

35. Zhou, H.; Hu, H. Upper limb motion estimation from inertial measurements. *Int. J. Inf. Technol.* **2007**, *13*, 1–14.

36. Zhou, H.; Stone, T.; Hu, H.; Harris, N. Use of multiple wearable inertial sensors in upper limb motion tracking. *Med. Eng. Phys.* **2008**, *30*, 123–133. [CrossRef]

37. Yoon, T.L. Validity and Reliability of an Inertial Measurement Unit-Based 3D Angular Measurement of Shoulder Joint Motion. *J. Korean Phys. Ther.* **2017**, *29*, 145–151. [CrossRef]
38. Garimella, R.; Peeters, T.; Beyers, K.; Truijen, S.; Huysmans, T.; Verwulgen, S. Capturing Joint Angles of the Off-Site Human Body. In Proceedings of the 2018 IEEE SENSORS, New Delhi, India, 28–31 October 2018. [CrossRef]
39. Lee, S.H.; Yoon, C.; Chung, S.G.; Kim, H.C.; Kwak, Y.; won Park, H.; Kim, K. Measurement of Shoulder Range of Motion in Patients with Adhesive Capsulitis Using a Kinect. *PLoS ONE* **2015**, *10*, e0129398. [CrossRef]
40. Mateo, S.; Roby-Brami, A.; Reilly, K.T.; Rossetti, Y.; Collet, C.; Rode, G. Upper limb kinematics after cervical spinal cord injury: A review. *J. Neuroeng. Rehabil.* **2015**, *12*, 9. [CrossRef] [PubMed]
41. Carmona-Pérez, C.; Garrido-Castro, J.L.; Vidal, F.T.; Alcaraz-Clariana, S.; García-Luque, L.; Alburquerque-Sendín, F.; de Souza, D.P.R. Concurrent Validity and Reliability of an Inertial Measurement Unit for the Assessment of Craniocervical Range of Motion in Subjects with Cerebral Palsy. *Diagnostics* **2020**, *10*, 80. [CrossRef] [PubMed]
42. Frye, S.K.; Geigle, P.R.; York, H.S.; Sweatman, W.M. Functional passive range of motion of individuals with chronic cervical spinal cord injury. *J. Spinal Cord Med.* **2019**, *43*, 257–263. [CrossRef]
43. Norkin, C.C.; White, D.J. *Measurement of Joint Motion: A Guide to Goniometry*, 5th ed.; F.A. Davis: Philadelphia, PA, USA, 2016; pp. 66–114.
44. Sabari, J.S.; Maltzev, I.; Lubarsky, D.; Liszkay, E.; Homel, P. Goniometric assessment of shoulder range of motion: Comparison of testing in supine and sitting positions. *Arch. Phys. Med. Rehabil.* **1998**, *79*, 647–651. [CrossRef]
45. Hayes, K.; Walton, J.R.; Szomor, Z.L.; Murrell, G.A. Reliability of five methods for assessing shoulder range of motion. *Aust. J. Physiother.* **2001**, *47*, 289–294. [CrossRef]
46. Jain, N.B.; Wilcox, R.B.; Katz, J.N.; Higgins, L.D. Clinical Examination of the Rotator Cuff. *PM&R* **2013**, *5*, 45–56. [CrossRef]
47. Maksimovic, R.; Popovic, M. Classification of tetraplegics through automatic movement evaluation. *Med. Eng. Phys.* **1999**, *21*, 313–327. [CrossRef]
48. Acosta, A.M.; Kirsch, R.F.; van der Helm, F.C.T. Three-dimensional shoulder kinematics in individuals with C5–C6 spinal cord injury. *Proc. Inst. Mech. Eng. Part H J. Eng. Med.* **2001**, *215*, 299–307. [CrossRef] [PubMed]
49. Kebaetse, M.; McClure, P.; Pratt, N.A. Thoracic position effect on shoulder range of motion, strength, and three-dimensional scapular kinematics. *Arch. Phys. Med. Rehabil.* **1999**, *80*, 945–950. [CrossRef]
50. Powers, C.; Newsam, C.; Gronley, J.; Fontaine, C.; Perry, J. Isometric shoulder torque in subjects with spinal cord injury. *Arch. Phys. Med. Rehabil.* **1994**, *75*, 761–765. [CrossRef]
51. Gronley, J.; Newsam, C.; Mulroy, S.; Rao, S.; Perry, J.; Helm, M. Electromyographic and kinematic analysis of the shoulder during four activities of daily living in men with C6 tetraplegia. *J. Rehabil. Res. Dev.* **2000**, *37*, 423–432. [PubMed]
52. Bravi, R.; Quarta, E.; Tongo, C.D.; Carbonaro, N.; Tognetti, A.; Minciacchi, D. Music, clicks, and their imaginations favor differently the event-based timing component for rhythmic movements. *Exp. Brain Res.* **2015**, *233*, 1945–1961. [CrossRef] [PubMed]
53. Bravi, R.; Ioannou, C.I.; Minciacchi, D.; Altenmüller, E. Assessment of the effects of Kinesiotaping on musical motor performance in musicians suffering from focal hand dystonia: A pilot study. *Clin. Rehabil.* **2019**, *33*, 1636–1648. [CrossRef]
54. Bravi, R.; Cohen, E.J.; Martinelli, A.; Gottard, A.; Minciacchi, D. When Non-Dominant Is Better than Dominant: Kinesiotape Modulates Asymmetries in Timed Performance during a Synchronization-Continuation Task. *Front. Integr. Neurosci.* **2017**, *11*, 21. [CrossRef]
55. Cohen, E.J.; Bravi, R.; Minciacchi, D. The effect of fidget spinners on fine motor control. *Sci. Rep.* **2018**, *8*, 3144. [CrossRef]
56. World Medical Association. World Medical Association Declaration of Helsinki: Ethical Principles for Medical Research Involving Human Subjects. *JAMA* **2013**, *310*, 2191–2194. [CrossRef] [PubMed]
57. Weir, J.P. Quantifying Test-Retest Reliability Using the Intraclass Correlation Coefficient and the SEM. *J. Strength Cond. Res.* **2005**, *19*, 231–240. [CrossRef] [PubMed]
58. Atkinson, G.; Nevill, A.M. Statistical Methods For Assessing Measurement Error (Reliability) in Variables Relevant to Sports Medicine. *Sports Med.* **1998**, *26*, 217–238. [CrossRef] [PubMed]
59. Koo, T.K.; Li, M.Y. A Guideline of Selecting and Reporting Intraclass Correlation Coefficients for Reliability Research. *J. Chiropr. Med.* **2016**, *15*, 155–163. [CrossRef] [PubMed]
60. Cicchetti, D. Guidelines, Criteria, and Rules of Thumb for Evaluating Normed and Standardized Assessment Instrument in Psychology. *Psychol. Assess.* **1994**, *6*, 284–290. [CrossRef]
61. Spearman, C. Correlation Calculated From Faulty Data. *Br. J. Psychol.* **1910**, *3*, 271–295. [CrossRef]
62. Brown, W. Some Experimental Results in the Correlation of Mental Abilities. *Br. J. Psychol.* **1910**, *3*, 296–322. [CrossRef]
63. Bland, J.M.; Altman, D. Statistical Methods For Assessing Agreement Between two methods of clinical measurement. *Lancet* **1986**, *327*, 307–310. [CrossRef]
64. Sedgwick, P. Limits of agreement (Bland-Altman method). *Br. Med. J.* **2013**, *346*, f1630. [CrossRef] [PubMed]
65. Portney, L. *Foundations of Clinical Research: Applications to Practice*; Pearson/Prentice Hall: Upper Saddle River, NJ, USA, 2015.
66. Mullaney, M.J.; McHugh, M.P.; Johnson, C.P.; Tyler, T.F. Reliability of shoulder range of motion comparing a goniometer to a digital level. *Physiother. Theory Pract.* **2010**, *26*, 327–333. [CrossRef]
67. Paulis, W.D.; Horemans, H.L.; Brouwer, B.S.; Stam, H.J. Excellent test–retest and inter-rater reliability for Tardieu Scale measurements with inertial sensors in elbow flexors of stroke patients. *Gait Posture* **2011**, *33*, 185–189. [CrossRef]

68. Roetenberg, D.; Baten, C.; Veltink, P. Estimating Body Segment Orientation by Applying Inertial and Magnetic Sensing Near Ferromagnetic Materials. *IEEE Trans. Neural Syst. Rehabil.* **2007**, *15*, 469–471. [CrossRef]

69. de Vries, W.; Veeger, H.; Baten, C.; van der Helm, F. Magnetic distortion in motion labs, implications for validating inertial magnetic sensors. *Gait Posture* **2009**, *29*, 535–541. [CrossRef]

70. Kendell, C.; Lemaire, E.D. Effect of mobility devices on orientation sensors that contain magnetometers. *J. Rehabil. Res. Dev.* **2009**, *46*, 957. [CrossRef] [PubMed]

71. Palermo, E.; Rossi, S.; Patanè, F.; Cappa, P. Experimental evaluation of indoor magnetic distortion effects on gait analysis performed with wearable inertial sensors. *Physiol. Meas.* **2014**, *35*, 399–415. [CrossRef] [PubMed]

72. Haering, D.; Raison, M.; Begon, M. Measurement and Description of Three-Dimensional Shoulder Range of Motion with Degrees of Freedom Interactions. *J. Biomech. Eng.* **2014**, *136*. [CrossRef] [PubMed]

73. Tenforde, A.S.; Hefner, J.E.; Kodish-Wachs, J.E.; Iaccarino, M.A.; Paganoni, S. Telehealth in Physical Medicine and Rehabilitation: A Narrative Review. *PM&R* **2017**, *9*, S51–S58. [CrossRef]

74. Martinez, R.N.; Hogan, T.P.; Balbale, S.; Lones, K.; Goldstein, B.; Woo, C.; Smith, B.M. Sociotechnical Perspective on Implementing Clinical Video Telehealth for Veterans with Spinal Cord Injuries and Disorders. *Telemed. e-Health* **2017**, *23*, 567–576. [CrossRef]

75. Cardinale, A.M. The Opportunity for Telehealth to Support Neurological Health Care. *Telemed. e-Health* **2018**, *24*, 969–978. [CrossRef]

76. MbientLab. Metamotion R Datasheet. Available online: https://mbientlab.com/metamotionr (accessed on 3 February 2021).

77. Madgwick, S.O.H.; Harrison, A.J.L.; Vaidyanathan, R. Estimation of IMU and MARG orientation using a gradient descent algorithm. In Proceedings of the 2011 IEEE International Conference on Rehabilitation Robotics, Zurich, Switzerland, 29 June–1 July 2011. [CrossRef]

Letter

Validity of a New 3-D Motion Analysis Tool for the Assessment of Knee, Hip and Spine Joint Angles during the Single Leg Squat

Igor Tak [1,2,*], Willem-Paul Wiertz [3,4], Maarten Barendrecht [2,5] and Rob Langhout [4,6]

1   Physiotherapy Utrecht Oost, Bloemstraat 65D, 3581 WD Utrecht, The Netherlands
2   Amsterdam Collaboration on Health and Safety in Sports, Amsterdam UMC, Vrije Universiteit Amsterdam, Department of Public and Occupational Health, Amsterdam Public Health Research Institute, Van der Boechorststraat 7, 1081 BT Amsterdam, The Netherlands; maartenbarendrecht@hotmail.com
3   Knowledge Centre of Sports and Physical Activity, 6717 LZ Ede(GLD) Horapark 4, The Netherlands; willem-paul.wiertz@kenniscentrumsportenbewegen.nl
4   Dutch Center for Allied Health Care (NPi), Berkenweg 7, 3818 LA Amersfoort, The Netherlands; rob.langhout@fysiotherapiedukenburg.nl
5   Avans+, Heerbaan 14-40, 4817 NL Breda, The Netherlands
6   Physiotherapy Dukenburg Nijmegen, Aldenhof 70-03, 6537 DZ Nijmegen, The Netherlands
*   Correspondence: igor.tak@gmail.com; Tel.: +31-6-26-58-22-64

Received: 8 July 2020; Accepted: 8 August 2020; Published: 13 August 2020

Abstract: Aim: Study concurrent validity of a new sensor-based 3D motion capture (MoCap) tool to register knee, hip and spine joint angles during the single leg squat. Design: Cross-sectional. Setting: University laboratory. Participants: Forty-four physically active (Tegner ≥ 5) subjects (age 22.8 (±3.3)) Main outcome measures: Sagittal and frontal plane trunk, hip and knee angles at peak knee flexion. The sensor-based system consisted of 4 active (triaxial accelerometric, gyroscopic and geomagnetic) sensors wirelessly connected with an iPad. A conventional passive tracking 3D MoCap (OptiTrack) system served as gold standard. Results: All sagittal plane measurement correlations observed were very strong for the knee and hip ($r = 0.929$–0.988, $p < 0.001$). For sagittal plane spine assessment, the correlations were moderate ($r = 0.708$–0.728, $p < 0.001$). Frontal plane measurement correlations were moderate in size for the hip ($\rho = 0.646$–0.818, $p < 0.001$) and spine ($\rho = 0.613$–0.827, $p < 0.001$). Conclusions: The 3-D MoCap tool has good to excellent criterion validity for sagittal and frontal plane angles occurring in the knee, hip and spine during the single leg squat. This allows bringing this type of easily accessible MoCap technology outside laboratory settings.

Keywords: validity; 3-D motion analysis; single leg squat; motion capture; clinical

---

1. Introduction

The evaluation of athletes' kinematics in functional, sports-specific situations continues to receive increasing attention [1,2]. Kinematic parameters like the range of motion (ROM), velocity and acceleration are used to quantify so-called quality of movement [1]. Quality of movement is associated with injury risk in athletes, and is evaluated in clinical practice to determine exercise progression and assist in return to play decision making after injury [1,3,4]. In sports involving running, cutting and jumping, decreased spine, hip and knee flexion have been linked to the development of patellofemoral pain and increased strain of the anterior cruciate ligament (ACL) [5–7].

Three-dimensional (3D) motion capture (MoCap) systems using reflective markers are considered the gold standard in measuring kinematics during functional performance tests [8,9]. However, feasibility and financial considerations have forced clinicians to adopt two-dimensional (2D) rather than 3D analysis for commonly used functional performance tests, such as the single leg squat [10].

Even though it is considered a useful screening tool [9,11–14], most kinematic 2D analyses of the single leg squat still require the use of multiple markers and devices like cameras and computers and its use is still time consuming [3,9,14]. Like the single leg squat, most performance tests are executed in a fixed position like single leg stance, and are thus easily reproducible. Movement quality is then often judged visually, thereby lacking quantitative data. A huge step forward in movement quality assessment would then be to track 3D movement outside the laboratory and in clinical settings with easy-to-use, low cost and time efficient systems [4]. Obtaining larger data sets is then made possible by the multi-site use of this technology [2].

Registration of 3D kinematics allows for a more comprehensive assessment of compensatory movement patterns, incorporating all anatomical planes, thus resulting in a smaller loss of relevant data [1,8,9,12,13,15]. Smart devices (tablets and phones) are now commonly equipped with cameras, Bluetooth connectivity and data processing capacity. Wearable active sensors can communicate with these devices, making 3D MoCap easily accessible for clinicians. A previous review on the reliability of motion capture systems reported the highest reliability for hip and knee sagittal and frontal plane measurements while it is lowest for the transverse plane measurements [16]. Reliability of wearable active sensor systems is moderate to excellent when compared to optical tracking systems with sagittal and frontal hip and knee measurement error between sagittal 8° and 1° [4,17,18]. Active sensors recently proved to be useful and reliable in the assessment of squatting, jumping and walking in patients after ACL reconstruction [19]. Currently, active sensor monitoring of movement is the subject of many studies in the field of orthopedic and neurologic rehabilitation and sports injury populations [17,20–22].

This new field of rapidly emerging technologic possibilities however needs further evaluation before it can be applied in the field by researchers and clinicians [23]. The aim of this study was to determine the concurrent validity of a new 3D MoCap tool for sagittal and frontal plane angles of the knee, hip and spine during a single leg squat.

2. Materials and Methods

A cross-sectional study design was used. The single leg squat task was analysed using both conventional 3D motion analysis and a newly developed 3D MoCap tool. At the time of this study this product was not yet available on the market. All procedures were carried out at the same time of the day.

2.1. Subjects

Subjects were recruited from the local student population at the campus of De Haagse Hogeschool (The Hague, The Netherlands). Subjects were included if they were aged between 18 and 45 years and were physically active (activity level ≥5/10 on the Tegner activity scale [24] for at least of 60 min per week).

Exclusion criteria were: (1) a history of knee surgery; (2) knee injuries or back, hip, knee or ankle joint pain within 3 months prior to this study; (3) lower extremity symptoms at rest or during sports participation; and (4) any neurological disorders that could affect gait. This study was conducted according the Helsinki declaration of ethical principles for medical research involving human subjects [25]. The Dutch Central Committee on Research Involving Human Subjects (CCMO) [26] confirmed exemption from ethical approval as stated in the Dutch Medical Research Involving Human Subjects Act [27]. All subjects signed informed consent before participating in this study.

2.2. Motion Capture

The primary outcome measures of this study were derived from conventional 3D motion analysis and from a new 3D MoCap tool during a repeated single leg squat test.

An inert 3D motion analysis system consisting of 12 infrared cameras (OptiTrack Flex 13, NaturalPoint Inc., Corvallis, OR, USA) was used as a reference tool. The positions of reflective markers in 3D space (x, y and z axes), were recorded by the infrared cameras at a frequency of 120 Hz.

Although radiostereometric analysis (RSA) is considered the gold standard in motion analysis, a major disadvantage is its invasive character. RSA has highest accuracy, yet comparison with optical tracking systems shows clinical acceptable validity and reliability [28,29]. Systems like Vicon (Vicon, Oxford, UK) and OptiTrack are therefore commonly used in motion analysis [3,8,9]. OptiTrack systems were found to be accurate when compared to Vicon in gait assessment with the deviations reported being up to 2.2% maximum [30,31]. Hip and knee ROM data in the sagittal and frontal plane were reported to differ 1° maximum [30].

The new 3D MoCap tool (SportsLapp, Factic BV, Enschede, The Netherlands [32]) used 4 active (triaxial accelerometric, gyroscopic and geomagnetic) sensors (Hocoma AG, Volketswil, Switzerland), attached with elastic therapeutic tape (Pinotape Pro Sport, Pino Pharmaceutical Products GmbH, Hamburg, Germany). The sensors connected wirelessly by Bluetooth, to an iPad Air 2 (Apple Inc., Cupertino, CA, USA) with the SportsLapp software running on the iOS (Apple Inc., Cupertino, CA, USA) platform. Joint angles were registered using the sensors (recording at 50 Hz) and video (recording at 120 Hz). The lower sensor data capturing frequency was graphically aligned in the movement curves with the video recording frequency by spherical linear interpolation. Data collected during one single set of 3 trials was used for analyses. This protocol is similar to other protocols reported in literature [13,15,33,34].

2.3. Testing Protocol

For the conventional motion analysis and the 3D MoCap tool, cluster markers and active sensors respectively, were placed on the following anatomical landmarks: the sternal manubrium, the sacrum, halfway on the lateral aspect of the thigh and halfway, anteromedial on the shank (bony aspect of the tibia), see Figure 1.

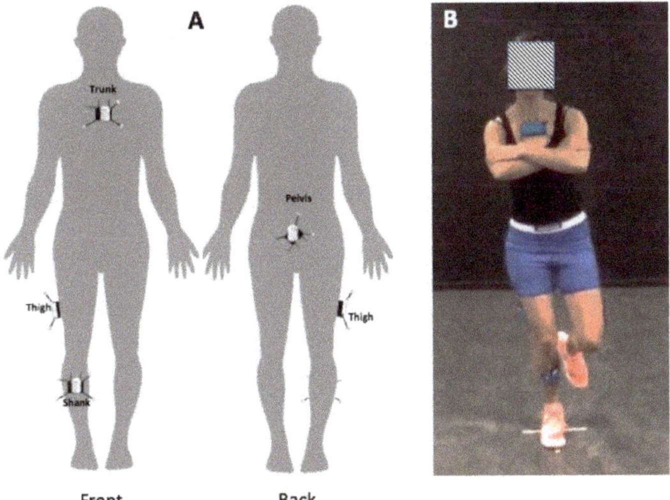

Figure 1. (**A**): The ventral view (left) and dorsal view (right) show the cluster marker and sensor positions on the sternal manubrium, the sacrum, halfway on the iliotibial tract and halfway on the tibia. (**B**): The single leg testing position with markers and sensors attached.

Prior to testing, subjects performed a standardised warming up consisting of 8 bipedal squats and 3 single leg squats on each leg. Then both systems were calibrated with the subjects standing in the upright anatomical position, while all joint angle values on both systems were set to be zero.

For the single leg squat test, the protocol as described by Dingenen et al. [3] was used. Subjects had to stand on their dominant leg (defined as the preferred leg to kick a ball) and fold their arms in front of their chest. They then squatted down in 2 s and returned to the upright position in 2 s, while maintaining balance. This low movement speed was selected to minimize the chance for trajectory gaps [31]. The non-supporting knee had to be kept parallel to the supporting knee, without touching it. During the test, squat depth was not controlled, as this better reflects a clinical setting. It was reported previously that subjects are able to produce consistent sagittal plane range of motion without monitoring [35,36]. In order to reduce the effect of movement velocity on joint angles and lower limb kinematics, a metronome was used to provide an audio cue for speed of movement [3,13,34].

2.4. Data Processing

All data from the conventional 3D motion analysis system were imported into MatLab R2015b (The MathWorks Inc., Natick, MA, USA), and marker trajectories were filtered using a 4th order low-pass Butterworth filter at 3 Hz, eliminating all signal noise with a frequency higher than 3 Hz. A custom MatLab program—including a Euler rotation sequence resolving the sagittal, the frontal and the transverse plane motion respectively—was used to calculate angles of the spine, hip and knee. Euler angles of rotation describe complex 3D kinematics of a rigid body (i.e., a limb segment) by decomposing movements into rotations about the 3 axes (x, y and z) of a fixed coordinate system that serves as a reference [37]. Spine flexion and homolateral spine lateral flexion, and hip and knee flexion, adduction and internal rotation were considered positive values.

For the new 3D MoCap tool, custom-made software (SportsLapp, Factic BV, Enschede, The Netherlands) was used, analysing the collected kinematic data and converting it into real-time joint angle curves. Both the curves and the video footage are available simultaneously in the app (see Figure 2).

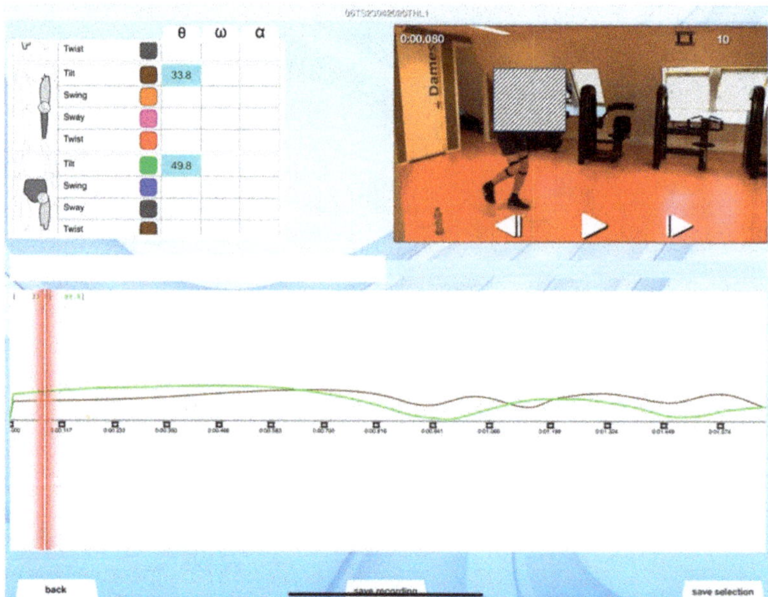

Figure 2. Screenshot of the SportsLapp app, with on-screen representation of video and kinematic data. Here left hip and knee data are visible.

Each of the sensors was assigned to a specific limb segment. As a joint consists of 2 segments, joint angles were defined as the angular difference between 2 segments. From each sensor, a quaternion was derived, expressing the orientation and rotation of its specific segment in the local coordinate frame of a primary segment (generally the proximal adjacent segment). Quaternions are a way of mathematically encoding the 3D orientation of a limb segment using 4 scalar numbers; 3 representing vectors on the axes of rotation, and 1 providing the angle of rotation [38]. In this way, they are a means of overcoming gimbal lock—a singular joint position in which 2 of the 3 axes of rotation align, and thus joint angles and kinematics cannot be described accurately [39]. The additional dimension in a quaternion provides more information on the orientation of a segment so that gimbal lock can be avoided.

For the new 3D MoCap tool, active sensors register joint angles. There is no calibration against fixed x, y, and z-axes in 3D space. Thus all joint positions are based on data obtained from 2 adjacent segments, relatively positioned against one another. To describe these 3D joint angles and movements, the 3D MoCap tool employs a new kinematic system (3D Angles, Factic BV, Enschede, The Netherlands) [40] consisting of 4 units of measurement: tilt, swing, sway and twist presented in the SportsLapp software. The amount of tilt effectively quantifies the amount of movement of a (secondary) segment with respect to its primary segment. Thus when performed in the sagittal plane, tilt describes flexion and extension of the spine, hip and knee. By analogy, sway shows movement in lateral and in medial direction. So when performed in the frontal plane, sway describes spine lateral flexion and hip abduction and adduction. Twist quantifies rotation along the longitudinal axis of the segment (transversal plane). For this study we applied sagittal plane (tilt) and frontal plane (sway) analyses (see Figure 3).

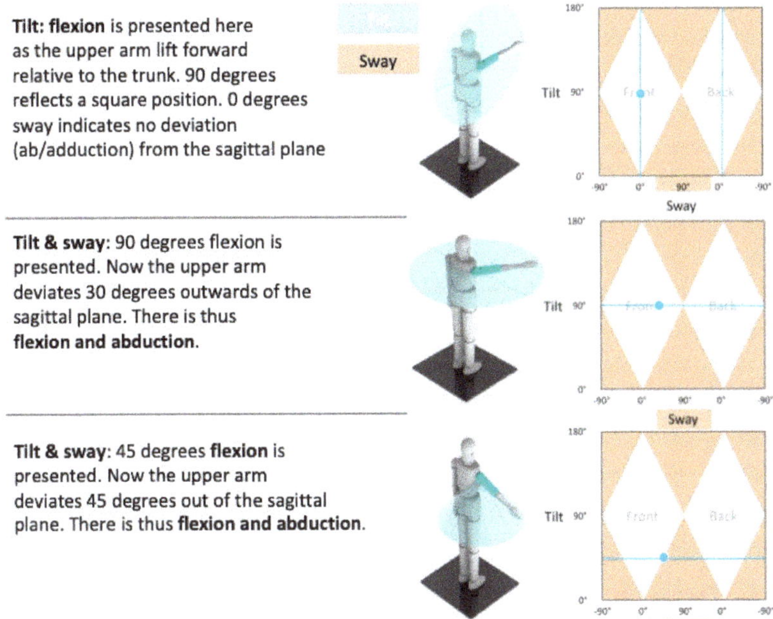

Figure 3. A graphical representation of sagittal and frontal plane movement of the SportsLapp app.

2.5. Outcome Measures

The kinematic variables collected were: knee flexion (tilt), hip flexion (tilt) and abduction/adduction (sway), spine forward flexion (tilt) and lateral flexion angles (sway). These were all presented in degrees, rounded off to 1 decimal. Both 3D motion analysis systems registered the entire movement trajectory during the single leg squat performed. In order to reduce eventual effects of velocity on joint angles registered, kinematic data at the time point of peak knee flexion were collected for analysis by the following procedure: The SportsLapp software creates a graphical curve of the joint angles. Moreover, the software allows selecting a certain part of the curve (depicting the moment where peak knee flexion position occurred) and then provides the start time and end time of the interval. By searching the raw IMU data corresponding to this interval, the magnitude of peak knee flexion and the time point at which this occurred was determined.

2.6. Statistical Analysis

In order to determine characteristics of our sample, mean, standard deviation and range were calculated for the subjects' age, height, weight, hours of sport participation per week and activity level. To investigate the concurrent validity of the new MoCap tool, correlations between outcomes on both 3D motion analysis systems were calculated. To test for normality of data distribution, Shapiro-Wilk tests were conducted. In case of normally distributed data, Pearson's r was calculated—in case of non-normally distributed data, Spearman's Rho (ρ) was used. Strength of correlations were expresses as perfect (r or $\rho = -1$ or 1), very strong ($0.8 \leq r$ to $\rho < 1$), moderate ($0.6 \leq r$ to $\rho < 0.8$), fair ($0.3 \leq r$ to $\rho < 0.6$) and poor ($0 < r$ to $\rho < 0.3$) or inversely the negative values in case of negative correlations [41]. The alfa level for statistical significance was set at $p < 0.05$ for all analyses. Statistical analyses were performed using IBM SPSS Statistics (v. 23.0, IBM Corp., Armonk, NY, USA).

3. Results

3.1. Subjects

Forty-four subjects (24 males and 20 females) took part in the experiment. Subject characteristics are presented in Table 1.

Table 1. Subject Characteristics.

	Total Group (*n* = 44)	Female (*n* = 20)	Male (*n* = 24)
Age (yrs)	22.8 (3.3) (18–30)	22.9 (3.6) (18–30)	22.7 (3.2) (18–30)
Height (cm)	175.3 (10.1) (155.0–197.0)	167 (5.7) (155–177)	182 (7.4) (170–197)
Weight (kg)	72.1 (11.6) (53.6–103.2)	64.1 (7.1) (53.6–82.2)	78.9 (10.6) (64.1–103.2)
Sport participation (hrs/wk)	5.4 (4.5) (1–25)	6.9 (5.2) (1–25)	3.1 (2.4) (1–8)
Activity level (Tegner score)	6.5 (1.3) (5–10)	5.8 (1.1) (3–7)	7.0 (1.2) (5–10)

Data are presented as mean (standard deviation) (range). Abbreviations: M = male; F = female; yrs = years; cm = centimetre; kg = kilogram; hrs = hours; wk = week.

3.2. Concurrent Validity

All single leg squats were performed 3 times resulting in a total of 132 trials being available for analyses. The data recorded per trial for both systems is presented in Table 2 as mean, standard deviation (SD) and 95% confidence interval (CI). Pearson's R (normally distributed data) or Spearman's Rho (non-normally distributed data) was calculated between data obtained with both systems including the mean differences with accompanying 95% CI.

All sagittal plane measurement correlations observed were very strong for the knee and hip ($r = 0.929$–0.988, $p < 0.001$). For sagittal plane spine assessment, the correlations were moderate ($r = 0.708$–0.728, $p < 0.001$). Frontal plane measurement correlations were moderate in size for the hip ($\rho = 0.646$–0.818, $p < 0.001$) and spine ($\rho = 0.613$–0.827, $p < 0.001$). Correlation plots of all measurement data combined (3 single leg squat trials) are presented in Figure 4.

Table 2. Correlations between measures with OptiTrack and SportsLapp motion capture systems for sagittal and frontal plane angles of knee, hip and spine.

Trial 1	Knee Flexion/Tilt *	Hip Flexion/Tilt *	Hip Ab-Adduction/Sway	Spine Flexion/Tilt *	Spine Lateral Flexion/Sway
OptiTrack (°)	73.8 (9.1) (71.1–76.6)	60.8 (14.7) (56.3–65.3)	−8.1 (17.1) (−13.3−−2.9)	13.8 (10.9) (10.5–17.1)	3.8 (5.7) (2.1–5.6)
SportsLapp (°)	73.3 (9.1) (70.6–76.1)	62.9 (13.4) (58.9–67.1)	−8.5 (18.3) (−13.9−−3.0)	18.1 (11.4) (14.6–21.6)	4.6 (10.2) (1.5–7.8)
Mean diff (°) (95%CI)	−0.5 (−1.5–0.5)	2.1 (1.3–2.9)	−0.4 (8.2) (−2.5–0.3)	−4.3 (−6.8−−1.17)	0.8 (8.1) (0.4–3.2)
Correlation	$r = 0.936$	$r = 0.985$	$\rho = 0.818$	$r = 0.713$	$\rho = 0.827$
Trial 2	Knee Flexion/Tilt *	Hip Flexion/Tilt *	Hip Ab-Adduction/Sway	Spine Flexion/Tilt *	Spine Lateral Flexion/Sway
OptiTrack (°)	75.0 (8.6) (72.4–77.6)	63.7 (14.7) (59.2–58.2)	−13.2 (14.1) (17.5−−8.9)	16.3 (12.1) (12.6–20.0)	6.3 (5.0) (4.7–7.8)
SportsLapp (°)	73.8 (8.6) (71.2–76.4)	64.9 (14.1) (60.6–69.2)	−14.5 (14.0) (−18.8−−10.3)	20.7 (11.9) (17.1–24.4)	7.4 (11.0) (4.0–10.7)
Mean diff (°) (95%CI)	−1.2 (−2.2–0.3)	1.2 (0.5–1.9)	−1.3 (−3.6–1.0)	−4.4 (−7.2−−1.6)	1.1 (−1.6–3.8)
Correlation	$r = 0.929$	$r = 0.988$	$\rho = 0.704$	$r = 0.708$	$\rho = 0.641$
Trial 3	Knee Flexion/Tilt *	Hip Flexion/Tilt *	Hip Ab-Adduction/Sway	Spine Flexion/Tilt *	Spine Lateral Flexion/Sway
OptiTrack (°)	76.0 (9.2) (73.1–78.8)	64.7 (13.6) (60.6–68.9)	−14.9 (13.1) (−18.9−−11.0)	14.7 (11.8) (11.1–18.3)	5.5 (5.3) (3.9–7.1)
SportsLapp (°)	74.6 (8.8) (71.9–77.3)	66.4 (13.0) (62.4–70.3)	−16.4 (12.5) (−20.2−−12.6)	19.8 (12.1) (16.1–23.4)	9.1 (9.4) (6.3–12.0)
Mean diff (°) (95%CI)	−1.4 (−2.3−−0.4)	1.7 (0.9–2.5)	−1.5 (−3.8–0.74)	−5.1 (−7.8−−2.4)	3.6 (1.3–5.9)
Correlation	$r = 0.944$	$r = 0.982$	$\rho = 0.646$	$r = 0.728$	$\rho = 0.613$

Data are presented in degrees as mean (standard deviation) and (95% confidence interval). Abbreviations: diff = difference; r = Pearson's R correlation; ρ = Spearman's Rho correlation. * = Normally distributed data.

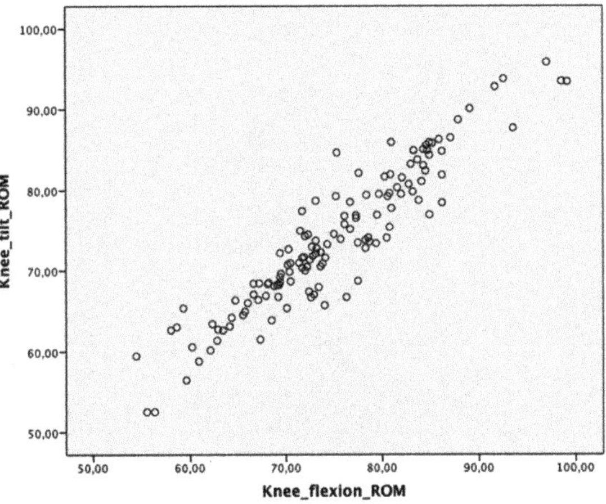

(**A**): Knee flexion and knee tilt.

Figure 4. *Cont.*

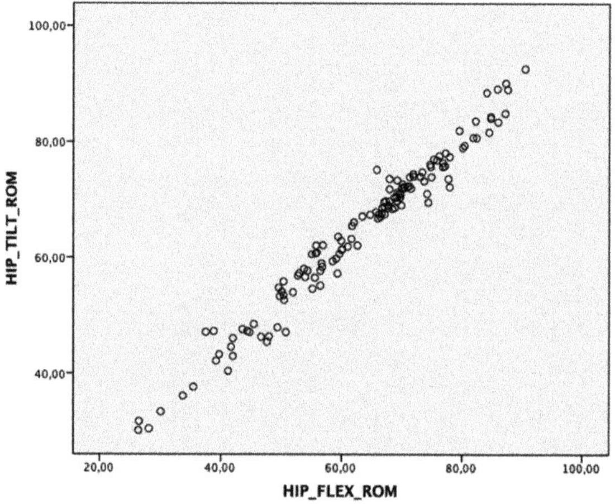

(**B**): Hip flexion and hip tilt.

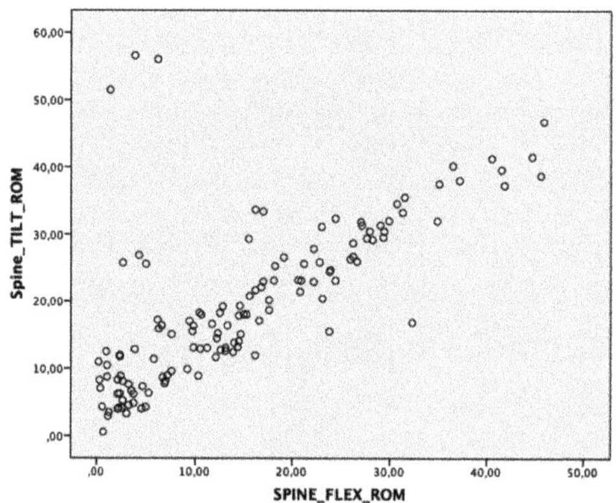

(**C**): Spine flexion and spine tilt.

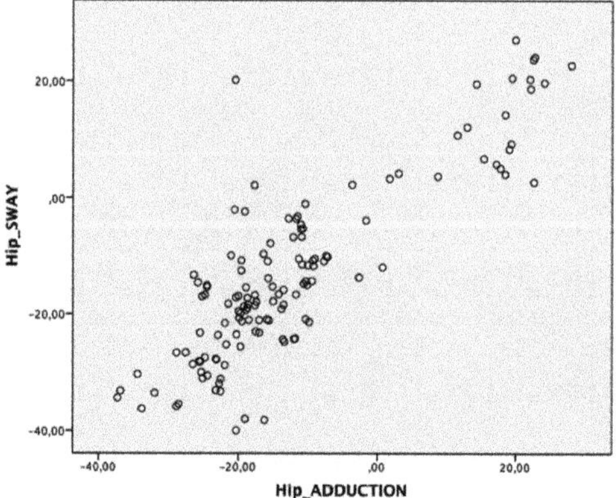

(**D**): Hip ab-adduction and hip sway.

Figure 4. *Cont.*

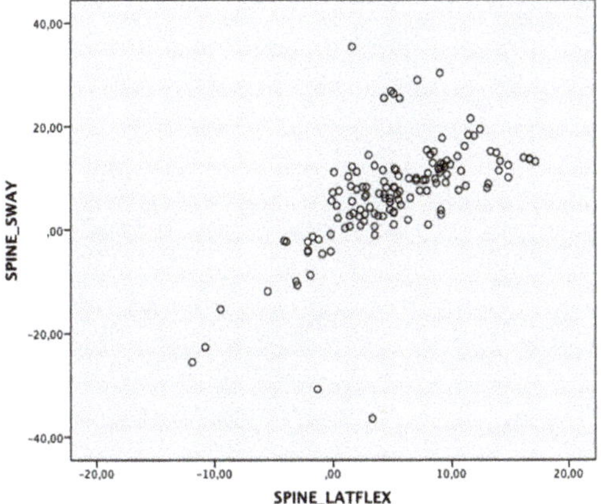

(**E**): Spine lateral flexion and spine sway.

Figure 4. Correlation plots of movement directions tested. (Please note all decimals presented to be read as "." instead of ",").

4. Discussion

4.1. Main Findings

We studied the concurrent validity of a new 3D MoCap tool during a single leg squat task performed in a standing position. This validity was found to be good to excellent for all joint angles registered in the sagittal and frontal plane during three single leg squat trials. The highest correlations between systems were observed in the sagittal plane and were found to be most consistent. Hip and knee measurements performed best with the lowest difference between both systems.

4.2. Practical Relevance

The ability of the new MoCap tool to objectively measure sagittal and frontal plane joint angles is practically relevant: decreased spine, hip and knee flexion in single leg weight bearing and landing activities have been associated with an increased risk for the development of patellofemoral pain and anterior cruciate ligament knee (re)injury [5,7]. The new 3D MoCap tool can thus be applied in injury prevention screening and in progress evaluation after exercise interventions, aimed to optimize lower extremity kinematics. Parameters obtained are likely relevant to improve clinical outcomes. An important advantage of this new tool is that it can be applied on the pitch (on field) and in rehab and sports centre settings [19]. The kinematic data are then directly available in contrast with the more time consuming conventional motion capture systems [3,9,14].

Speed to be elicited in specific actions is one of the major goals in most type of sports. The capability of motion capture is unique in quantifying magnitude, timing and symmetry of segmental velocity. This may help physicians and physiotherapists to screen for deficits in any of those parameters and subsequently identify potential risk factors for future injury [42–44].

This study shows that easily accessible technology is likely to enter the market for a broad audience of professionals from different fields. The recent research interest from orthopedic and neurologic rehabilitation specialists in this type of technology shows that there is a loud call for quantification of many parameters in their specific patient populations [2,4,9,17,19–21,31,34,45]. Every movement related health problem would thus likely have its own clinical relevant movement parameters. Future studies will show what these parameters will be. Affordable new 3D MoCap systems will accelerate the pace of these developments. Larger datasets obtained at clinics as well as the pitch side are needed to further develop new knowledge.

4.3. Validity of Motion Capture Systems

Reliability of optimal tracking systems for gait analysis has been performed previously [16]. Optimal tracking systems show high reliability for pelvis, hip and knee ROM in the sagittal and frontal plane with mean precision errors less than 6°. Errors of 2° are considered acceptable, 2–5° reasonable and over 5° misleading. Markers attached to the skin move with respect to the underlying joints they intend to measure. This error is called soft tissue artefact (STA) and therefore invasive methods using radiostereometry are considered the gold standard in investigating joint motion because of its high accuracy [29]. A systematic review compared invasive methods against optimal tracking systems and found STA arising from tissue deformation, individual physical characteristics, marker location, type of segment and the nature of the performed tasks. The magnitude of STA measured was up to 40 mm at the thigh [46]. Also MoCap systems can be expected to be subject to STA, still any comparison with invasive methods seems to be lacking.

4.4. Other Studies

Recent studies investigating systems using inertial measurement units have found that these can distinguish between abnormal and sufficient performance on lower limb exercises with moderate to excellent accuracy. However, these studies assess movement patterns rather than specific joint angles, and use visual observation as reference standard rather than conventional 3D motion analysis [47,48]. Similar to this study, other markerless 3D MoCap tools such as the Kinect V2, have demonstrated stronger validity for sagittal plane kinematics than for the frontal plane during the single leg squat [45,49]. The difference in accuracy of sagittal plane versus frontal and transverse plane measurements can be explained from a biomechanical point of view. As sagittal plane movements are the largest in terms of ROM, a difference of some degrees between both methods will affect the distribution of outcomes relatively less than in frontal and transverse plane movements, which display a smaller ROM.

Additionally, differences between both MoCap tools were expected beforehand, because of the different manner in which joint angles are obtained. The conventional gold standard utilizes reflective markers, while the new tool employs active sensors. To calculate osteokinematic joint angles from the marker trajectories in 3D space, the conventional tool uses traditional Euler rotation sequences [37]. This method, however, suffers from gimbal lock and equations may be numerically unstable [38]. To express the orientation of a segment with respect to its primary segment (i.e., the joint angle) the new 3D MoCap tool uses quaternions that are derived directly from the sensors. This method overcomes the problem of gimbal lock, and extracting angles and axes of rotation is simpler and requires less computational steps [38].

In a comparable study, Zügner et al. found a measurement error of 2.8° for hip flexion/extension with a moderate correlation (ICC 0.75) which was for knee flexion/extension an error of 0.2° with a good correlation (ICC 0.83). Although these findings for the knee are comparable to our study, differences in hip findings could be due to the different sensor positions and performance tasks [17]. Another study compared IMU with OptiTrack measuring knee motion during a dynamic task as in our study [4]. They showed an error of measurement of 8° for flexion/extension (0.5° in our study) and poor correlation (Pearson's R = 0.58 against our Pearson's R = 0.94). This being lower may be the result of the dynamic jump task that influenced angle readings by eccentric gyroscopic fluctuations. Leardini et al. found small differences between an IMU and Vicon measures reporting a 5.0° error for knee flexion during a squat performance task which was between 0.5–1.5° in our study [18]. IMU and optical tracking systems both suffer from STA, yet most studies showed accepted measurement errors below the 5° bound. In our study, all hip and knee motion errors were lower than 2° and lower than 5° for spine measurements with good to excellent correlation.

4.5. Strengths and Limitations

To better reflect a clinical setting, squat depth was not controlled for in the testing protocol. It was previously reported that subjects are able to produce consistent sagittal plane range of motion without monitoring [35,36]. Our findings are in line with those as indeed we found consistent peak hip and knee flexion angles during the three trials of single leg squat. Spine flexion and lateral flexion however were less stable on repeated single leg squatting. This is acknowledged in the clinic where aberrant spine movements distract attention of the clinician. Although these spine positional changes are present hip and knee angles remained consistent.

The application studied shows adequate validity for MoCap in a clinical and on-site test setting. This type of assessment with easily accessible and low coast technology will likely offer new opportunities for clinicians as well as researchers to capture data that where prohibited for scientists in high laboratory setting.

The frequency of 50 Hz of the SportsLapp application is lower than the OptiTrack MoCap system. The validity data obtained can thus not be generalized to higher speed movements. The velocity threshold for adequate MoCap with this new system during higher speed movements should be subject of future study.

Questions can be raised regarding the placement of markers and sensors. Units placed on muscular areas, such as the lateral thigh, may have been subject to more artefacts due to muscle contraction and skin displacement than others, which were placed on bony landmarks. Marker placement on less or non-muscular areas like the lateral femoral condyle for the upper leg marker, may be considered to further refine the current protocol. A previous study comparing inertial sensors with an optoelectronic system reported lower errors on movement tracking when assessing a prosthesis when compared to a healthy human leg [50]. On top of this, artefacts may have occurred as a result of the inertial sensors were fixated with elastic tape over the cluster markers, prohibiting a rigid connection between the two.

5. Conclusions

This study shows that a new 3D MoCap tool utilizing active sensors has good to excellent concurrent validity for sagittal and frontal plane knee, hip and spine angle measurements assessed during a single leg squat task. Future studies are needed to investigate other variables like through range movement angles and velocity parameters.

Author Contributions: I.T., W.-P.W., M.B. and R.L. were involved in conception and design of this study. W.-P.W. performed the data collection. I.T., W.-P.W., M.B. and R.L. were involved in data analysis and interpretation of the data. I.T. and W.-P.W. were responsible for drafting the manuscript. I.T., W.-P.W., M.B. and R.L. critically revised and approved the manuscript to be submitted. All authors have read and agreed to the published version of the manuscript.

Funding: This research received no external funding.

Acknowledgments: The authors would like to thank Babet van der Molen and Ingrid Diederen for their assistance during the measurements.

Conflicts of Interest: The authors declare no conflict of interest.

References

1. Kivlan, B.R. Functional Performance Testing of the Hip in Athletes. *Int. J. Sports Phys. Ther.* **2012**, *7*, 402–412. [PubMed]

2. Rojas-Valverde, D.; Gómez-Carmona, C.D.; Gutiérrez-Vargas, R.; Pino-Ortega, J. From big data mining to technical sport reports: The case of inertial measurement units. *BMJ Open Sport Exerc. Med.* **2019**, *5*, 1–3. [CrossRef] [PubMed]

3. Dingenen, B.; Malfait, B.; Vanrenterghem, J.; Verschueren, S.M.P.; Staes, F.F. The reliability and validity of the measurement of lateral trunk motion in two-dimensional video analysis during unipodal functional screening tests in elite female athletes. *Phys. Ther. Sport* **2014**, *15*, 117–123. [CrossRef]

4.	Ajdaroski, M.; Tadakala, R.; Nichols, L.; Esquivel, A. Validation of a device to measure knee joint angles for a dynamic movement. *Sensors* **2020**, *20*, 1747. [CrossRef] [PubMed]

5.	Graci, V.; Van Dillen, L.R.; Salsich, G.B. Gender differences in trunk, pelvis and lower limb kinematics during a single leg squat. *Gait Posture* **2012**, *36*, 461–466. [CrossRef]

6.	Hewett, T.E.; Torg, J.S.; Boden, B.P. Video analysis of trunk and knee motion during non-contact anterior cruciate ligament injury in female athletes: Lateral trunk and knee abduction motion are combined components of the injury mechanism. *Br. J. Sports Med.* **2009**, *43*, 417–422. [CrossRef]

7.	Pollard, C.D.; Sigward, S.M.; Powers, C.M. Limited hip and knee flexion during landing is associated with increased frontal plane knee motion and moments. *Clin. Biomech.* **2010**, *25*, 142–146. [CrossRef]

8.	McLean, S.G.; Walker, K.; Ford, K.R.; Myer, G.D.; Hewett, T.E.; Van Den Bogert, A.J. Evaluation of a two dimensional analysis method as a screening and evaluation tool for anterior cruciate ligament injury. *Br. J. Sports Med.* **2005**, *39*, 355–362. [CrossRef]

9.	Munro, A.; Herrington, L.; Carolan, M. Reliability of 2-dimensional video assessment of frontal-plane dynamic knee valgus during common athletic screening tasks. *J. Sport Rehabil.* **2012**, *21*, 7–11. [CrossRef]

10.	Munro, A.; Herrington, L.; Comfort, P. The Relationship between 2-Dimensional Knee-Valgus Angles During Single-Leg Squat, Single-Leg-Land, and Drop-Jump Screening Tests. *J. Sport Rehabil.* **2017**, *26*, 72–77. [CrossRef]

11.	Ekegren, C.L.; Miller, W.C.; Celebrin, R.G.; Eng, J.J.; MacIntyre, D.L. Reliability and validity of observational risk screening in evaluating dynamic knee valgus. *J. Orthop. Sports Phys. Ther.* **2009**, *39*, 665–674. [CrossRef] [PubMed]

12.	Stensrud, S.; Myklebust, G.; Kristianslund, E.; Bahr, R.; Krosshaug, T. Correlation between two-dimensional video analysis and subjective assessment in evaluating knee control among elite female team handball players. *Br. J. Sports Med.* **2011**, *45*, 589–595. [CrossRef]

13.	Willson, J.D.; Davis, I.S. Utility of the frontal plane projection angle in females with patellofemoral pain. *J. Orthop. Sports Phys. Ther.* **2008**, *38*, 606–615. [CrossRef] [PubMed]

14.	Willson, J.D.; Ireland, M.L.; Davis, I. Core strenght and lower extremity alignment during single leg squats. *Med. Sci. Sports Exerc.* **2006**, *38*, 945–952. [CrossRef]

15.	Stickler, L.; Finley, M.; Gulgin, H. Relationship between hip and core strength and frontal plane alignment during a single leg squat. *Phys. Ther. Sport* **2015**, *16*, 66–71. [CrossRef] [PubMed]

16.	McGinley, J.L.; Baker, R.; Wolfe, R.; Morris, M.E. The reliability of three-dimensional kinematic gait measurements: A systematic review. *Gait Posture* **2009**, *29*, 360–369. [CrossRef]

17.	Zügner, R.; Tranberg, R.; Timperley, J.; Hodgins, D.; Mohaddes, M.; Kärrholm, J. Validation of inertial measurement units with optical tracking system in patients operated with Total hip arthroplasty. *BMC Musculosketlet. Disord.* **2019**, *20*, 52. [CrossRef]

18.	Leardini, A.; Lullini, G.; Giannini, S.; Berti, L.; Ortolani, M.; Caravaggi, P. Validation of the angular measurements of a new inertial-measurement-unit based rehabilitation system: Comparison with state-of-the-art gait analysis. *J. NeuroEng. Rehabil.* **2014**, *11*, 1–7. [CrossRef]

19.	Islam, R.; Bennasar, M.; Nicholas, K.; Button, K.; Holland, S.; Mulholland, P.; Price, B.; Al-Amri, M. A Nonproprietary Movement Analysis System (MoJoXlab) Based on Wearable Inertial Measurement Units Applicable to Healthy Participants and Those With Anterior Cruciate Ligament Reconstruction Across a Range of Complex Tasks: Validation Study. *JMIR mHealth uHealth* **2020**, *8*, e17872. [CrossRef]

20.	Witchel, H.J.; Oberndorfer, C.; Needham, R.; Healy, A.; Westling, C.E.I.; Guppy, J.H.; Bush, J.; Barth, J.; Herberz, C.; Roggen, D.; et al. Thigh-derived inertial sensor metrics to assess the sit-to-stand and stand-to-sit transitions in the timed up and go (TUG) Task for quantifying mobility impairment in multiple sclerosis. *Front. Neurol.* **2018**, *9*, 9. [CrossRef]

21.	De Brabandere, A.; Emmerzaal, J.; Timmermans, A.; Jonkers, I.; Vanwanseele, B.; Davis, J. A Machine Learning Approach to Estimate Hip and Knee Joint Loading Using a Mobile Phone-Embedded IMU. *Front. Bioeng. Biotechnol.* **2020**, *8*, 1–11. [CrossRef] [PubMed]

22.	Bravi, M.; Gallotta, E.; Morrone, M.; Maselli, M.; Santacaterina, F.; Toglia, R.; Foti, C.; Sterzi, S.; Bressi, F.; Miccinilli, S. Concurrent validity and inter trial reliability of a single inertial measurement unit for spatial-temporal gait parameter analysis in patients with recent total hip or total knee arthroplasty. *Gait Posture* **2020**, *76*, 175–181. [CrossRef] [PubMed]

23. Teufl, W.; Miezal, M.; Taetz, B.; Fröhlich, M.; Bleser, G. Validity, test-retest reliability and long-term stability of magnetometer free inertial sensor based 3D joint kinematics. *Sensors* **2018**, *18*, 1980. [CrossRef] [PubMed]

24. Tegner, Y.; Lysholm, J. Rating systems in the evaluation of knee ligament injuries. *Clin. Orthop. Relat. Res.* **1985**, *198*, 43–49. [CrossRef]

25. World Medical Association. World Medical Association Declaration of Helsinki: Ethical principles for medical research involving human subjects. *JAMA* **2013**, *310*, 2191–2194. [CrossRef]

26. Central Committee on Research Involving Human Subjects. Available online: https://english.ccmo.nl/inve stigators/legal-framework-for-medical-scientific-research/laws/medical-research-involving-human-subj ects-act-wmo (accessed on 24 July 2020).

27. Medical Research Involving Human Subjects Act. In *Wet Medisch-Wetenschappelijk Onderzoek met Mensen*; Staatscourant: Cham, Switzerland, 1998. (In English)

28. Tranberg, R.; Saari, T.; Zügner, R.; Kärrholm, J. Simultaneous measurements of knee motion using an optical tracking system and radiostereometric analysis (RSA). *Acta Orthop.* **2011**, *82*, 171–176. [CrossRef]

29. Zügner, R.; Tranberg, R.; Lisovskaja, V.; Shareghi, B.; Kärrholm, J. Validation of gait analysis with dynamic radiostereometric analysis (RSA) in patients operated with total hip arthroplasty. *J. Orthop. Res.* **2017**, *35*, 1515–1522. [CrossRef]

30. Thewlis, D.; Bishop, C.; Daniell, N.; Paul, G. Next-generation low-cost motion capture systems can provide comparable spatial accuracy to high-end systems. *J. Appl. Biomech.* **2013**, *29*, 112–117. [CrossRef]

31. Carse, B.; Meadows, B.; Bowers, R.; Rowe, P. Affordable clinical gait analysis: An assessment of the marker tracking accuracy of a new low-cost optical 3D motion analysis system. *Physiotherapy* **2013**, *99*, 347–351. [CrossRef]

32. SportsLapp. Available online: http://www.sportslapp.com (accessed on 24 July 2020).

33. Crossley, K.M.; Zhang, W.J.; Schache, A.G.; Bryant, A.; Cowan, S.M. Performance on the single-leg squat task indicates hip abductor muscle function. *Am. J. Sports Med.* **2011**, *39*, 866–873. [CrossRef]

34. Nakagawa, T.H.; Moriya, E.T.U.; MacIel, C.D.; Serrão, F.V. Trunk, pelvis, hip, and knee kinematics, hip strength, and gluteal muscle activation during a single-leg squat in males and females with and without patellofemoral pain syndrome. *J. Orthop. Sports Phys. Ther.* **2012**, *42*, 491–501. [CrossRef] [PubMed]

35. Whatman, C.; Hing, W.; Hume, P. Kinematics during lower extremity functional screening tests—Are they reliable and related to jogging? *Phys. Ther. Sport* **2011**, *12*, 22–29. [CrossRef] [PubMed]

36. Zeller, B.L.; McCrory, J.L.; Ben Kibler, W.; Uhl, T.L. Differences in kinematics and electromyographic activity between men and women during the single-legged squat. *Am. J. Sports Med.* **2003**, *31*, 449–456. [CrossRef] [PubMed]

37. Slabaugh, G.G. Computing Euler Angles from a Rotation Matrix. Available online: http://gregslabaugh.nam e/publications/euler.pdf (accessed on 24 July 2020).

38. Kuipers, J.B. Quaternions and Rotation Sequences. In *Geometry, Integrability and Quantization*; Mladenov, I.M., Naber, G.L., Eds.; Coral Press: Sophia, Bulgaria, 2000; pp. 127–143.

39. Masuda, T.; Ishida, A.; Cao, L.; Morita, S. A proposal for a new definition of the axial rotation angle of the shoulder joint. *J. Electromyogr. Kinesiol.* **2008**, *18*, 154–159. [CrossRef]

40. Dijkstra, F.; den Besten, G. *3Dynamics Angles Definitions for Measuring and Presenting 3D Motions of Human Joints*; Factic, B.V., Ed.; Enschede, The Netherlands, 2018. Available online: https://www.researchgate.net/p ublication/327816713_3Dynamics_angles_definitions_for_measuring_and_presenting_3D_motions_of_hu man_joints (accessed on 12 August 2020).

41. Chan, Y.H. Biostatistics 104. Corrleational analysis. *Sing. Med. J.* **2005**, *46*, 153–160.

42. Morishige, Y.; Harato, K.; Kobayashi, S.; Niki, Y.; Matsumoto, M.; Nakamura, M.; Nagura, T. Difference in leg asymmetry between female collegiate athletes and recreational athletes during drop vertical jump. *J. Orthop. Surg. Res.* **2019**, *14*, 1–6. [CrossRef]

43. Ithurburn, M.P.; Paterno, M.V.; Thomas, S.; Pennell, M.L.; Evans, K.D.; Magnussen, R.A.; Schmitt, L.C. Change in Drop-Landing Mechanics Over 2 Years in Young Athletes After Anterior Cruciate Ligament Reconstruction. *Am. J. Sports Med.* **2019**, *47*, 2608–2616. [CrossRef]

44. King, E.; Richter, C.; Franklyn-Miller, A.; Wadey, R.; Moran, R.; Strike, S. Back to Normal Symmetry? Biomechanical Variables Remain More Asymmetrical Than Normal During Jump and Change-of-Direction Testing 9 Months After Anterior Cruciate Ligament Reconstruction. *Am. J. Sports Med.* **2019**, *47*, 1175–1185. [CrossRef]

45. Eltoukhy, M.; Kelly, A.; Kim, C.Y.; Jun, H.P.; Campbell, R.; Kuenze, C. Validation of the Microsoft Kinect® camera system for measurement of lower extremity jump landing and squatting kinematics. *Sports Biomech.* **2016**, *15*, 89–102. [CrossRef]
46. Peters, A.; Galna, B.; Sangeux, M.; Morris, M.; Baker, R. Quantification of soft tissue artifact in lower limb human motion analysis: A systematic review. *Gait Posture* **2010**, *31*, 1–8. [CrossRef]
47. Kianifar, R.; Lee, A.; Raina, S.; Kulic, D. Classification of squat quality with inertial measurement units in the single leg squat mobility test. In Proceedings of the Annual International Conference of the IEEE Engineering in Medicine and Biology Society, EMBS, Orlando, FL, USA, 16–20 August 2016; pp. 6273–6276.
48. Whelan, D.F.; O'Reilly, M.A.; Ward, T.E.; Delahunt, E.; Caulfield, B. Technology in rehabilitation: Evaluating the single leg squat exercise with wearable inertial measurement units. *Methods Inf. Med.* **2017**, *56*, 88–94. [PubMed]
49. Mentiplay, B.F.; Hasanki, K.; Perraton, L.G.; Pua, Y.H.; Charlton, P.C.; Clark, R.A. Three-dimensional assessment of squats and drop jumps using the Microsoft Xbox One Kinect: Reliability and validity. *J. Sports Sci.* **2018**, *36*, 2202–2209. [CrossRef] [PubMed]
50. Seel, T.; Raisch, J.; Schauer, T. IMU-based joint angle measurement for gait analysis. *Sensors* **2014**, *14*, 6891–6909. [CrossRef] [PubMed]

Article

Machine Learning to Quantify Physical Activity in Children with Cerebral Palsy: Comparison of Group, Group-Personalized, and Fully-Personalized Activity Classification Models

Matthew N. Ahmadi [1,2], **Margaret E. O'Neil** [3], **Emmah Baque** [1,4], **Roslyn N. Boyd** [5] and **Stewart G. Trost** [1,2,*]

[1]   Institute of Health and Biomedical Innovation at Queensland Centre for Children's Health Research, Queensland University of Technology, South Brisbane 4101, Australia; matthewnguyen.ahmadi@hdr.qut.edu.au (M.N.A.); e.baque@griffith.edu.au (E.B.)
[2]   Faculty of Health, School of Exercise and Nutrition Sciences, Queensland University of Technology, Kelvin Grove 4059, Australia
[3]   Department of Rehabilitation and Regenerative Medicine, Columbia University Irving Medical Center, New York, NY 10032, USA; mo2675@cumc.columbia.edu
[4]   School of Allied Health Sciences, Griffith University, Gold Coast 4215, Queensland, Australia
[5]   Queensland Cerebral Palsy and Rehabilitation Research Centre, UQ Child Health Research Centre, Faculty of Medicine, The University of Queensland, South Brisbane 4101, Australia; r.boyd@uq.edu.au
*   Correspondence: s.trost@qut.edu.au; Tel.: +61-7-3069-7301

Received: 18 June 2020; Accepted: 16 July 2020; Published: 17 July 2020

Abstract: Pattern recognition methodologies, such as those utilizing machine learning (ML) approaches, have the potential to improve the accuracy and versatility of accelerometer-based assessments of physical activity (PA). Children with cerebral palsy (CP) exhibit significant heterogeneity in relation to impairment and activity limitations; however, studies conducted to date have implemented "one-size fits all" group (G) models. Group-personalized (GP) models specific to the Gross Motor Function Classification (GMFCS) level and fully-personalized (FP) models trained on individual data may provide more accurate assessments of PA; however, these approaches have not been investigated in children with CP. In this study, 38 children classified at GMFCS I to III completed laboratory trials and a simulated free-living protocol while wearing an ActiGraph GT3X+ on the wrist, hip, and ankle. Activities were classified as sedentary, standing utilitarian movements, or walking. In the cross-validation, FP random forest classifiers (99.0–99.3%) exhibited a significantly higher accuracy than G (80.9–94.7%) and GP classifiers (78.7–94.1%), with the largest differential observed in children at GMFCS III. When evaluated under free-living conditions, all model types exhibited significant declines in accuracy, with FP models outperforming G and GP models in GMFCS levels I and II, but not III. Future studies should evaluate the comparative accuracy of personalized models trained on free-living accelerometer data.

Keywords: accelerometers; wearable sensors; exercise; measurement; GMFCS level

---

1. Introduction

Cerebral palsy (CP) is the most common physical disability in childhood, with a prevalence of 2.1 cases per 1000 live births [1,2]. Specifically, among children with CP, physical activity levels decrease by 34% to 47% as children progress from early childhood through adolescence and such children accumulate less physical activity than their typically developing peers [3–5]. Low physical activity levels, in addition to associated neuromotor and functional limitations, impact the long-term health and well-being of children with CP [6–8]. In light of this evidence, researchers and clinicians have focused on delivering interventions to increase habitual physical activity and decrease sedentary behavior [9–11]. Historically, the effectiveness of these interventions has been evaluated using self-reports of physical activity. Although self-reports are low-cost and easy for participants to complete, they are subject to considerable social desirability and recall bias, and therefore may not be sufficiently valid or reliable for an assessment of clinically important changes in physical activity [12]. As a result, an increasing number of studies involving children with CP are employing device-based measures of physical activity and sedentary behavior [13,14].

Accelerometer-based motion sensors have become the method of choice for assessing physical activity in children [15,16]. However, the atypical gait patterns and lower mechanical efficiency of children with CP mandate the development of bespoke algorithms for reducing the accelerometer output for physical activity metrics [12]. A number of studies have derived intensity cut-points to categorize accelerometer data as sedentary (SED), light (LPA), or moderate-to-vigorous physical activity (MVPA) among children with CP [12,17–20]. This research has enabled researchers and clinicians to quantify the physical activity levels of children with CP and examine compliance with physical activity recommendations. However, validation studies involving independent samples of children with CP have demonstrated that cut-point approaches misclassify MVPA as LPA or SED activity 30% of the time and dramatically underestimate the physical activity levels of children with more severe motor impairments [21]. As such, there is a critical need to investigate new accelerometer data processing methods that potentially provide more accurate assessments of physical activity and sedentary behavior in children with CP.

Pattern recognition methodologies, such as those utilizing machine learning approaches, have the potential to significantly improve the accuracy of accelerometer-based assessments of physical activity among children with CP. In a previous study [22], we developed machine learning activity classification models for ambulatory children with CP for accelerometers worn on the wrist, hip, and a combination of the wrist and hip [22]. The resultant Random Forest and Support Vector Machine physical activity classification models classified sedentary activities with 96% to 98% accuracy, standing utilitarian movements (e.g., folding laundry) with 83% to 97% accuracy, and walking with 90% to 97% accuracy. However, the study sample was relatively small and predominantly comprised of children functioning at Gross Motor Function Classification System (GMFCS) levels I and II. Consequently, the study could not examine the classification accuracy in children with more severe movement impairments who ambulate with the assistance of crutches and walkers. Furthermore, prior work in typically developing children has demonstrated that other accelerometer placements, such as the ankle or a combination of the ankle and wrist, provide a higher classification accuracy [23,24]. However, the utility of activity classification models trained on accelerometer data from the ankle has not been examined in children with CP.

Typically, machine learning classification models are trained using accelerometer data from groups of participants and applied to new samples assuming that the new participants have the same movement patterns as those in the training sample. This presupposition may not be valid in the CP population, given the significant heterogeneity in relation to movement impairment and functional capacity. An alternative approach, which has not been investigated in children with CP, is to train and deploy group-personalized models based on the GMFCS level. In this approach, classification models are trained on accelerometer data from a specific GMFCS group and applied to children of the same GMFCS classification. It has previously been demonstrated that GMFCS-specific decision tree models for classifying the physical activity intensity in children with CP outperform "one-size-fits-all" models by accounting for differences in the energy cost of locomotion [21].

Although group-personalized models may account for broadly defined differences in functional ability, fully-personalized or individually-calibrated models may provide even more accurate physical activity predictions by accounting for finer grained differences in functional ability related to the child's age, height, motor distribution, and movement disorder. In a study of healthy adults, personalized activity classification models, trained on less than five minutes of accelerometer data, exhibited a significantly greater classification accuracy than group models [25]. On average, the overall classification accuracy increased by 26 percentage points from 71% to 97%, while walking recognition improved from 65% to 99%. More recently, Carcreff et al. [26] reported significantly improved walking bout detection and walking speed estimation in children with CP after applying personalized thresholds for the left and right mid-swing shank angular velocity. To the best of our knowledge, the relative accuracy and utility of fully-personalized machine learning activity classification models have not been examined in children with CP.

To date, most activity classification algorithms have been developed using data collected in controlled laboratory environments, which may not be generalizable to free-living settings [27–29]. Developing and validating classification models in the laboratory may be unrealistic because only a limited number of activities are included and fixed-duration activity trials without natural transitions between activities fail to replicate the episodic nature of movement behaviors displayed in true free-living environments [28]. If accelerometer-based physical activity classification models for children with CP are to be used in field-based studies, it is important to evaluate their accuracy under scenarios that replicate free-living environments.

To address these gaps in the research literature, the purpose of the current study was to evaluate and compare the accuracy of group, group-personalized, and fully-personalized machine learning physical activity classification models in children with CP. To examine the effects of accelerometer placement, models were trained and tested using accelerometer data from the hip, wrist, and ankle, and all two- and three-placement combinations. To assess the validity under conditions that more closely replicated intervention studies conducted in real-world settings, the classification accuracy was evaluated and compared in a simulated free-living trial.

2. Materials and Methods

2.1. Participants

A total of 38 ambulatory youth with CP participated in the study. Participants were recruited from two sites—St Christopher's Hospital for Children, Philadelphia, USA and the Queensland Centre for Children's Health Research, Brisbane, Australia. The inclusion criteria were as follows: Diagnosis of CP at Gross Motor Function Classification System (GMFCS) level I, II, or III; between the ages of 6 and 18 years; and able to follow verbal instructions. Parents and/or health care providers (doctors or therapists) verified that participants were able to complete the study protocols. Participants were excluded from the study if they had undergone orthopedic surgery within the last 6 months, received lower extremity Botulinum toxin A injections within the last 3 months, or experienced a recent musculoskeletal injury or had a medical condition limiting their ability to complete the physical activity protocol. The descriptive characteristics are displayed in Table 1. The study was approved by each university's Institutional Review Board. Prior to participation, parents provided written consent and children written assent.

2.2. Individual Activity Trials

Participants were randomized to complete one of two activity protocols. The protocol was completed in a single two-hour session and consisted of five activity trials. Briefly, the activity trials included in protocol one were as follows: (1) Supine rest (lying down and resting, but not sleeping);

(2) sitting while continuously writing or coloring in a picture; (3) active video game (playing an interactive video game designed specifically for children with CP); (4) comfortable-paced walk in a 25 m course (walking at a pace with the instructions "walk like when you are at the mall or walking in your neighborhood or at school, but you are not in a hurry."); and (5) brisk-paced walk in a 25 m course (walking at a pace with the instructions "walk at a fast pace like when you are hurrying to get to class after the bell has rung or you are hurrying to cross the street"). In protocol two, sitting while continuously writing or coloring in a picture was replaced with wiping down a table (standing at a waist-height counter and spraying the counter with water and then wiping the water off). All of the activity trials were 6 min in duration. Activity trials were categorized as one of three activity classes: Sedentary (SED = supine rest and sitting); standing utilitarian movements (SUM = wiping a table and active video game); and walking (WALK = comfortable-paced walk and brisk-paced walk).

Table 1. Participant characteristics.

	GMFCS I (N = 10)	GMFCS II (N = 20)	GMFCS III (N = 8)
Sex			
Male N (%)	4 (40%)	7 (35%)	2 (25%)
Female N (%)	6 (60%)	13 (65%)	6 (75%)
Motor Distribution			
Hemiplegia N (%)	8 (80%)	13 (65%)	0 (0%)
Diplegia N (%)	2 (20%)	7 (35%)	5 (62.5%)
Quadriplegia N (%)	0 (0%)	0 (0%)	3 (37.5%)
Age (y)	10.8 ± 1.8	12.0 ± 3.1	11.8 ± 4.5
Height (cm)	145.1 ± 9.9	147.2 ± 16.7	129.7 ± 19.2
Weight (kg)	40.8 ± 12.6	42.3 ± 15.6	34.9 ± 17.8

GMFCS = Gross Motor Function Classification System.

2.3. Instrumentation

Participants wore an ActiGraph GT3X+ tri-axial accelerometer (ActiGraph Corporation, Pensacola, FL, USA) on the least impaired wrist, hip, and ankle, with the Y-axis pointing vertically. The ActiGraph GT3X+ is a small (4.6 cm × 3.3 cm × 1.5 cm) and lightweight (19 g) monitor that measures acceleration along three orthogonal axes and the sampling frequency for the current study was set to 30 Hz.

2.4. Machine Learning Activity Classification Models

Random Forest (RF) classifiers were trained to categorize activities as one of three activity classes using accelerometer data collected at each single placement (wrist, hip, and ankle), and a combination of two placements (wrist and hip = W + H; wrist and ankle = W + A; hip and ankle = H + A) and three placements (wrist and hip and ankle = W + H + A). RF is an ensemble of decision tree models. Each tree is developed based on a bootstrap sample of training data and each node in the tree is split using the best value among a randomly selected sample of features. The decisions from each tree are aggregated and a final model prediction is based on a majority vote [30].

2.4.1. Data Pre-Processing and Feature Extraction

Tri-axial accelerometer data corresponding to the start and end of each activity trial were parsed and annotated with the corresponding activity class label. Features were extracted from 10 s non-overlapping windows. In total, 15 features were extracted from each axis and included the minimum, maximum, mean, variance, standard deviation, skewness, kurtosis, percentiles (25th, 50th, and 75th), zero-crossings, energy, dominant frequency, dominant magnitude, and entropy. Cross-axis correlations and the mean vector magnitude were also extracted. The extracted features have previously been shown to be informative for activity classification in children with CP and typically developing children [22,31].

2.4.2. Model Training and Cross-Validation

Models were trained and cross-validated using the "randomForest" and "caret" packages within R (Version 3.5.3). Prior to model training, minimum Redundancy Maximum Relevance (mRMR) feature selection was used to identify features with a high discriminative ability [32]. Minimum redundancy favors features that have a low dependency on other features, without considering how important they are to the outcome variable, whereas maximum relevance selects features that are the most predictive of the outcome variable. The mRMR selection process is based on a balance between these two algorithms, selecting features that derive a high relevance and low redundancy. Feature selection was constrained to the 10, 15, and 20 best features. The number of trees for each model was kept constant at 500 and the number of features sampled at each tree node in the forest ranged from 3 to 11 features. Group models were trained on data from all 38 participants, while the group-personalized models based on the GMFCS level were only trained on data from participants at the same GMFCS level and fully-personalized models were trained on data from each individual.

For the group and group-personalized models, the out-of-sample classification accuracy was evaluated using leave-one-subject-out cross-validation (LOSO-CV). In LOSO-CV, the model is trained on data from all of the participants except one, which is "held out" and used as the test data set. The process is repeated until all participants have served as the test data, and the performance metrics are aggregated. For the fully-personalized models, the classification accuracy was evaluated using 10-fold cross-validation. With 10-fold cross-validation, the participant's data are randomly partitioned into 10 subsets and the model is trained on all subsets except one, which is used as the test data set. The process is repeated until all subsets have served as the test data, and the performance metrics are aggregated. The cross-validation classification accuracy was evaluated by computing the overall and class-level accuracy for each GMFCS level.

2.5. Simulated Free-Living Evaluation

To evaluate the classification accuracy under conditions that more closely reflected a free-living environment, participants completed a simulated free-living trial in which they performed the following sequence of activities: (1) Sitting on a bench; (2) walking 10 m whilst weaving around cones to a table; (3) standing at the table and completing a puzzle and/or playing with a toy; and (4) walking around the table and back to the bench and sitting. The entire sequence was repeated for 6 min. To obtain ground truth activity class labels, all trials were video recorded with a Go-Pro camera (GoPro, Inc., San Mateo, CA, USA). These video files were subsequently imported into the Noldus Observer XT software (Noldus Information Technology, Wageningen, The Netherlands) for coding of the participant's activity type as either "SED", "SUM", or "WALK". The Observer software generated a vector of date-time stamps corresponding to the start and finish of each activity, which were used to label the corresponding time segment of the accelerometer data. For each accelerometer placement and model type, the overall and class-level accuracy was calculated for each GMFCS level.

2.6. Statistical Evaluation

A $3 \times 3 \times 7$ repeated measures ANOVA was used to examine the effects of the model type, GMFCS level, and placement on the overall accuracy. Significant main effects and interactions were evaluated using tests of simple effects and pre-planned contrasts. Significance was set at an alpha level of 0.05.

3. Results

For the group wrist, hip, and ankle classification models, respectively, 15, 15, and 10 features were selected. For the multiple placement W + H, W + A, H + A, and W + H + A classification models, respectively, 15, 10, 10, and 20 features were selected. A complete list of the selected features for each model are reported in Supplemental Document 1.

3.1. Leave-One-Subject-Out Cross-Validation

There was a significant main effect for placement on the accuracy ($F_{6,210} = 11.26$, $p < 0.01$), with the wrist and hip exhibiting a significantly lower overall accuracy than all other placements (see Figure 1). Compared to models trained on ankle accelerometer data, there were no significant improvements in the overall accuracy for the two-placement and three-placement models.

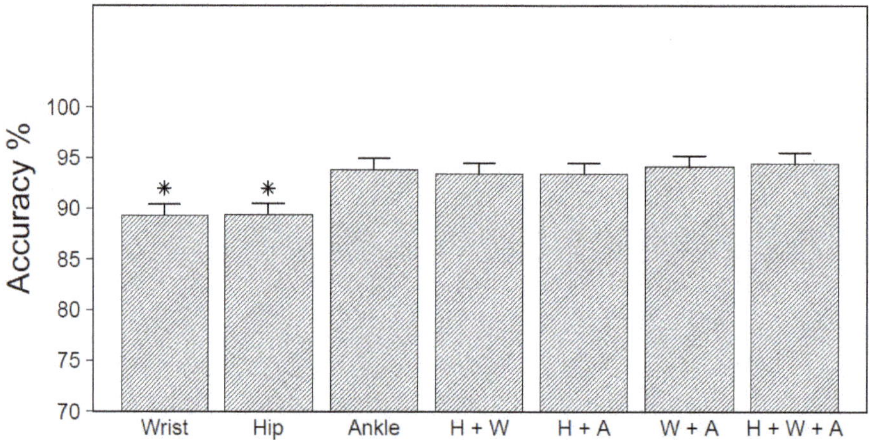

Figure 1. Overall accuracy obtained by placement for leave-one-subject-out cross-validation (LOSO-CV). * Significantly different from ankle and multi-placement models at $p < 0.05$.

Figure 2 displays the overall accuracy for group, group-personalized and fully-personalized models by GMFCS level, averaged over all placements. There was a significant model type by GMFCS level interaction ($F_{4,70} = 33.38$, $p < 0.01$), indicating that differences in the classification accuracy varied by GMFCS level. Averaged over all placements, fully-personalized models exhibited a significantly higher accuracy than group and group-personalized models, with the largest differential observed in children at GMFCS III.

Table 2 reports activity class recognition for the group, group-personalized, and fully-personalized models for each accelerometer placement combination by GMFCS level. For all placements and GMFCS levels, the fully-personalized models exhibited > 96.0% recognition accuracy for SED, SUM, and WALK. For the group and group-personalized models, the class level recognition accuracy varied by GMFCS level. Among children at GMFCS I and II, the group and group-personalized models demonstrated excellent classification accuracy for SED (89.1–98.4%) and WALK (88.3–99.6%), and good to excellent classification accuracy for SUM (76.8–93.8%). Among children at GMFCS III, the classification accuracy for the group and group-personalized models for SED (57.7–99.6%), SUM (68.6–92.1%), and WALK (53.5–96.3%) was inconsistent and ranged from poor to excellent, depending on the accelerometer placement configuration. Detailed confusion matrices for each placement, model type, and GMFCS level can be found in Supplemental Document 2.

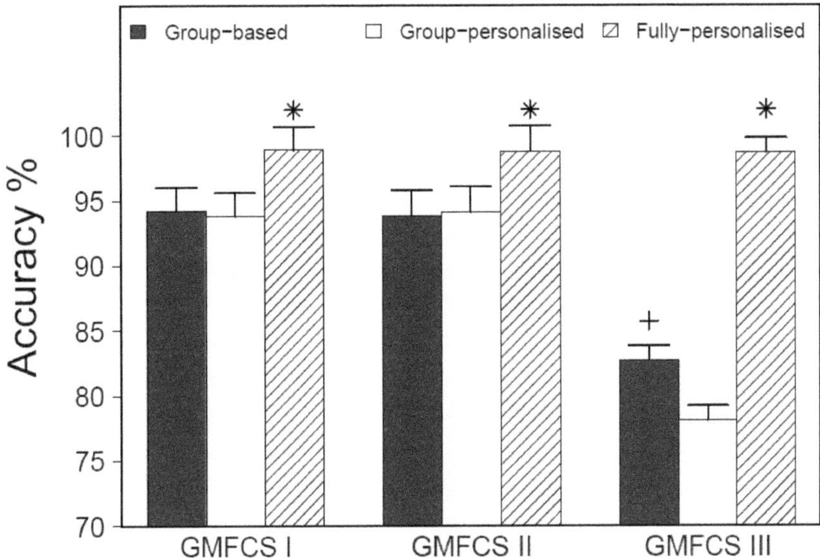

Figure 2. Overall accuracy for model type by GMFCS level for LOSO-CV. * Significantly different from group and group-personalized models; + significantly different from the group-personalized model.

Table 2. LOSO-CV overall and activity class accuracy for group, GMFCS-specific, and personalized classification models.

| | | GMFCS I | | | GMFCS II | | | GMFCS III | | |
		G Mean (SD)	GP Mean (SD)	FP Mean (SD)	G Mean (SD)	GP Mean (SD)	FP Mean (SD)	G Mean (SD)	GP Mean (SD)	FP Mean (SD)
Wrist	SED	92.7 (13.5)	94.1 (15.0)	99.3 (0.4)	94.7 (9.5)	93.2 (12.2)	99.0 (0.3)	85.0 (27.6)	57.7 (38.4)	99.1 (1.2)
	SUM	89.3 (10.0)	87.0 (9.5)	97.0 (2.3)	82.6 (14.6)	79.6 (20.1)	96.2 (1.2)	76.7 (15.0)	71.4 (25.5)	96.3 (1.1)
	WALK	96.2 (6.6)	91.9 (12.4)	98.4 (0.9)	90.4 (16.4)	92.6 (13.7)	97.2 (0.6)	84.2 (14.1)	75.5 (32.4)	96.2 (1.2)
Hip	SED	94.4 (12.5)	89.1 (16.0)	100.0 (0.1)	92.1 (13.2)	94.4 (10.2)	100.0 (0.3)	89.3 (22.7)	76.9 (34.2)	100.0 (0.4)
	SUM	83.8 (22.2)	76.8 (28.7)	98.5 (0.8)	87.6 (18.6)	85.3 (21.8)	98.4 (0.4)	75.1 (21.3)	73.3 (17.6)	98.2 (0.2)
	WALK	99.6 (1.1)	88.3 (11.2)	99.2 (0.3)	97.3 (3.9)	94.4 (10.1)	98.8 (0.9)	53.5 (34.7)	65.5 (38.5)	98.9 (0.2)
Ankle	SED	95.8 (5.3)	97.6 (2.9)	100.0 (0.1)	94.7 (12.9)	96.6 (11.6)	100.0 (0.3)	89.8 (26.9)	78.5 (37.8)	100.0 (0.1)
	SUM	91.9 (16.2)	91.1 (10.0)	98.8 (1.3)	90.4 (15.3)	93.2 (7.9)	99.2 (0.6)	71.6 (23.1)	74.5 (23.6)	99.1 (0.8)
	WALK	99.6 (1.1)	99.0 (2.3)	98.7 (1.5)	98.0 (2.4)	96.8 (5.5)	99.4 (0.5)	84.0 (16.8)	96.3 (4.4)	99.0 (1.1)
W + H	SED	89.2 (22.6)	94.0 (14.0)	100.0 (0.2)	94.4 (11.9)	94.9 (11.2)	100.0 (0.2)	99.6 (1.1)	75.0 (30.2)	100.0 (0.4)
	SUM	90.8 (16.9)	87.0 (25.4)	99.1 (1.3)	85.2 (17.6)	88.1 (20.2)	99.3 (1.2)	92.1 (10.4)	80.4 (15.2)	99.8 (0.7)
	WALK	90.9 (14.9)	98.3 (2.5)	99.3 (1.1)	94.9 (11.9)	96.2 (5.3)	99.1 (1.3)	95.8 (4.4)	94.1 (8.1)	99.7 (0.4)
W + A	SED	94.2 (12.2)	98.4 (3.2)	100.0 (0.2)	93.9 (12.1)	97.3 (10.0)	100.0 (0.1)	92.0 (18.6)	80.1 (33.4)	100.0 (0.2)
	SUM	91.6 (11.4)	91.9 (12.3)	98.1 (2.3)	93.8 (10.4)	93.7 (6.5)	99.2 (1.7)	68.6 (23.2)	81.1 (14.9)	99.5 (1.0)
	WALK	99.6 (1.1)	99.0 (2.3)	98.4 (1.4)	97.6 (3.7)	97.9 (2.8)	99.4 (1.9)	90.5 (9.9)	96.8 (3.7)	98.2 (1.7)
H + A	SED	94.6 (10.5)	96.9 (7.0)	100.0 (0.1)	96.3 (7.9)	96.5 (11.7)	100.0 (0.2)	86.9 (27.1)	76.3 (35.2)	100.0 (0.2)
	SUM	87.3 (21.7)	89.4 (14.1)	98.6 (1.6)	93.8 (12.9)	92.1 (11.1)	98.2 (1.3)	70.5 (20.6)	73.6 (24.1)	99.4 (0.7)
	WALK	99.6 (1.1)	98.6 (3.3)	99.3 (0.8)	98.3 (2.3)	97.4 (2.4)	98.9 (1.4)	87.8 (13.6)	96.3 (4.5)	99.2 (0.4)
W + H + A	SED	95.3 (10.4)	96.4 (7.9)	100.0 (0.2)	96.2 (9.6)	97.0 (7.2)	100.0 (0.2)	87.5 (27.6)	81.3 (32.7)	100.0 (0.2)
	SUM	91.9 (12.9)	91.0 (12.4)	99.3 (1.2)	93.8 (8.5)	91.4 (10.7)	99.1 (0.7)	77.2 (19.8)	82.1 (14.7)	98.8 (1.4)
	WALK	99.6 (1.1)	98.8 (2.8)	99.5 (0.4)	98.1 (2.8)	98.2 (2.6)	99.2 (0.9)	89.4 (11.1)	96.8 (3.7)	99.0 (1.1)

G = group; GP = group-personalized; GF = fully-personalized; W + H = wrist and hip; W + A = wrist and ankle; H + A = hip and ankle; W + H + A = wrist, hip, and ankle; SED = sedentary; SUM = standing utilitarian movement; WALK = walking.

3.2. Simulated Free-Living Trial

The overall classification accuracy for the different accelerometer placement configurations under simulated free-living conditions, averaged over all model types and GMFCS levels, is displayed in Figure 3. The ANOVA results indicated a significant main effect for placement ($F_{6,210} = 8.90$, $p < 0.01$). Tests of simple effects revealed that models trained on accelerometer data from the ankle, hip, or hip and ankle combined had a significantly higher accuracy than all other placement combinations. For all accelerometer placements, however, the overall accuracy statistics during the simulated free-living evaluation were substantially lower compared to the LOSO-CV evaluation and ranged from 50.9% for the H + W + A model to 61.5% for the ankle model.

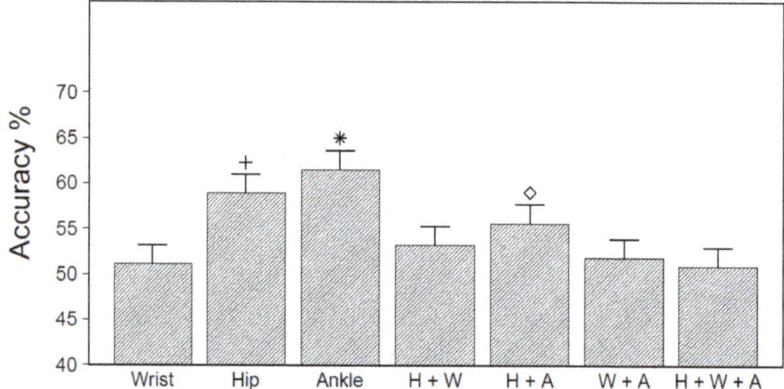

Figure 3. Overall accuracy statistics by accelerometer placement when evaluated under simulated free-living conditions. * Significantly different from all models at $p < 0.05$; + significantly different from all models except H+A; ◇ significantly different from all models except hip.

The simulated free-living accuracy results for the group, group-personalized and fully-personalized models by GMFCS level are displayed in Figure 4. There was a significant model type by GMFCS level interaction ($F_{4,70} = 17.94$, $p < 0.01$). Among children at GMFCS I and II, fully-personalized models exhibited a significantly higher overall accuracy than the group and group-personalized models. Among children at GMFCS III, however, group-personalized models exhibited a significantly greater overall accuracy than the group and fully-personalized models.

Heat map confusion matrices by model type and GMFCS level are reported in Figures 5–7. To reduce the complexity, only the results for the three best performing accelerometer placement configurations are reported—hip, ankle, and the hip and ankle combined. Detailed confusion matrices for all seven placements by GMFCS level are reported in Supplemental Document 3.

For the hip placement, the recognition of SED was poor to modest for group models (56–60%) at each GMFCS level, modest for group-personalized models (60–65%), and poor for the fully-personalized models (31–49%). For each model type, between 31% to 58% of all SED instances were misclassified as SUM. The recognition of SUM was poor for group models (29–55%) at each GMFCS level, with misclassification frequently occurring as SED (43–69%). Group-personalized models displayed very good recognition of SUM (83.2%) among children at GMFCS III; however, among children classified at GMFCS I and II, recognition was poor (33–40%) and frequently misclassified as SED (59–66%). Fully-personalized models had good to very good recognition (76–84%) and recognition increased as GMFCS function decreased from level I to II. Walking recognition for the group model ranged from modest to good for children at GMFCS levels I and II (65–77%); however, among GMFCS III children, walking recognition was very poor (28%), with 72% of walking instances being misclassified as SUM. Among children at GMFCS II and III, walking recognition for the group-personalized and fully-personalized models ranged from good to excellent (79–97%), with the accuracy increasing as GMFCS function declined from level II to III. Among children at GMFCS I, walking recognition ranged from poor to modest (57–65%) for group-personalized and fully-personalized models.

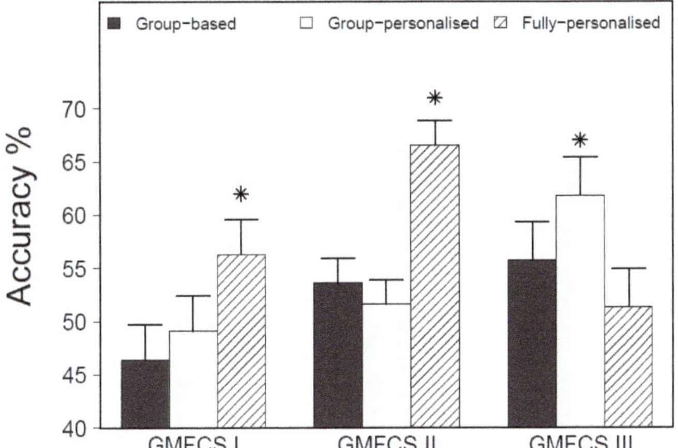

Figure 4. Overall accuracy for group, group-personalized, and fully-personalized Random Forest (RF) classifiers, by GMFCS level, during the simulated free-living evaluation. * Significantly different from other models within GMFCS level at $p < 0.05$.

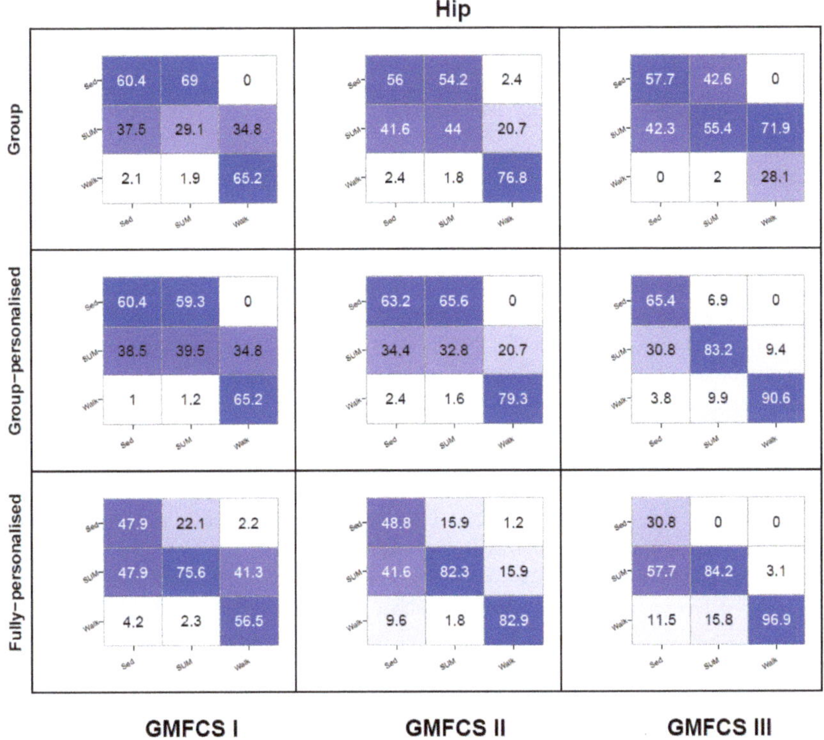

Figure 5. Hip placement activity class recognition for group, group-personalized, and fully-personalized classification models during the simulated free-living trial. Columns represent observed (%); rows represent predictions (%); values on the diagonal represent correct predictions (%).

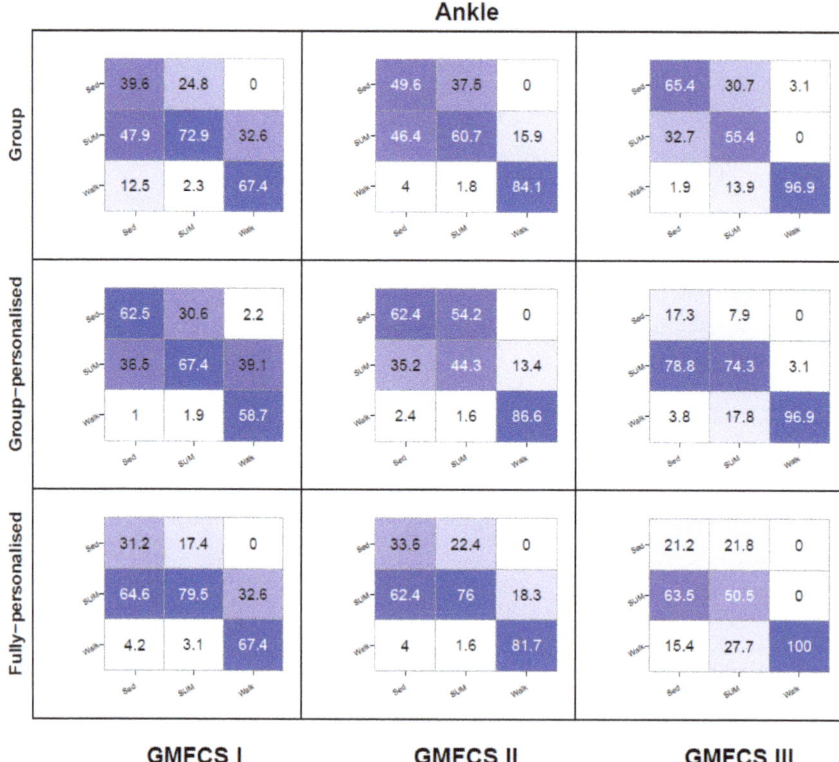

Figure 6. Ankle placement activity class recognition for group, group-personalized, and fully-personalized classification models during the simulated free-living trial. Columns represent observed (%); rows represent predictions (%); values on the diagonal represent correct predictions (%).

For the ankle placement, the recognition of SED for the group models was poor among children at GMFCS levels I and II (40–50%), and modest among children at GMFCS III (65%). For group-personalized models recognition was modest among children at GMFCS I and II (62%), but poor among children at GMFCS III (17%). For the fully-personalized models, recognition of SED was poor among all GMFCS levels (21–34%), with just under two-thirds of SED instances being misclassified as SUM. The recognition of SUM was good for group models among children at GMFCS I (73%), but generally poor among children at GMFCS levels II and III (55–61%), with the majority being misclassified as SED (31–38%). Group-personalized models exhibited a good recognition of SUM among children at GMFCS III (74%) and modest recognition among GMFCS I children (67%), but poor recognition among children at GMFCS II (44%). At GMFCS levels I and II, the misclassification of SUM as WALK occurred at rates of less than 2%, but at GMFCS III, this misclassification was 18%. Fully-personalized models exhibited good to very good recognition of SUM among children at GMFCS I and II (76–80%), but poor recognition among children at GMFCS III (51%), with misclassification occurring as either SED (22%) or WALK (28%). For all model types, the recognition of WALK was very good among children at GMFCS II (82–87%) and excellent among children at GMFCS III (97–100%). Among children at GMFCS I, the recognition of WALK was modest (59–67%), with 33–39% of WALK instances being misclassified as SUM.

Hip and Ankle

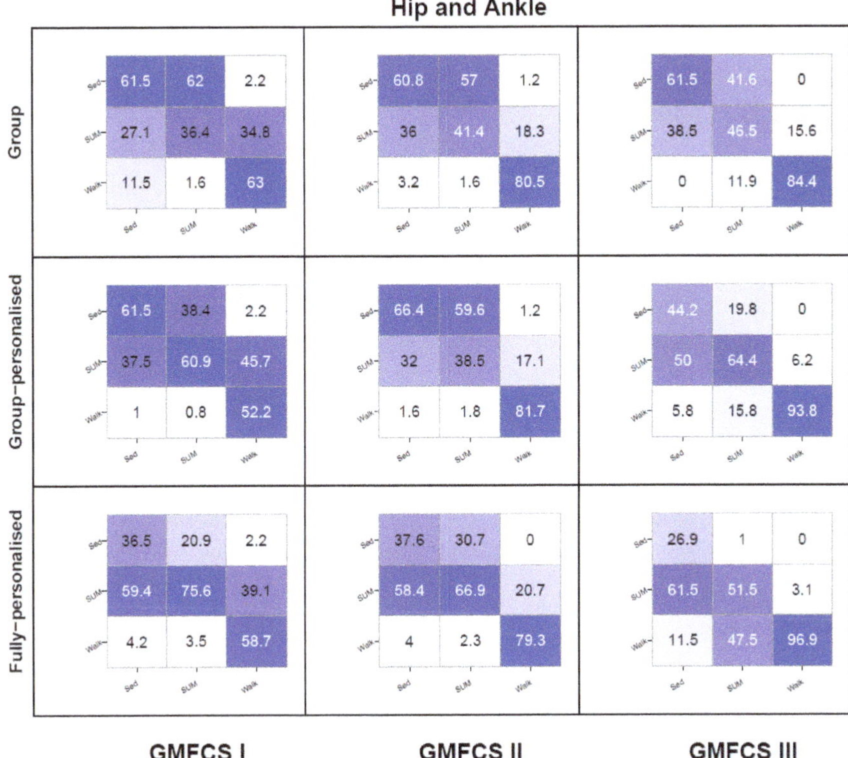

Figure 7. Combined hip and ankle placement activity class recognition for group, group-personalized, and fully-personalized classification models during the simulated free-living trial. Columns represent observed (%); rows represent predictions (%); values on the diagonal represent correct predictions (%).

For the combined hip and ankle (H + A) placement, the recognition of SED by group models was modest (61–62%) at all GMFCS levels. For group-personalized models, the recognition of SED was modest among children at GMFCS I and II (62–66%), but poor among children at GMFCS III (44%). At all GMFCS levels, the recognition of SED for the fully-personalized models was poor (27–38%). For each model type, between 27% and 59% of SED instances were misclassified as SUM. At all GMFCS levels, group models displayed poor recognition of SUM (36–47%), with the majority of misclassifications occurring as SED (42–62%). Group-personalized models exhibited modest recognition of SUM among children at GMFCS levels I and II (61–64%), but poor recognition among children at GMFCS III (39%). At GMFCS levels I and II, the majority of misclassification occurred as SED (38–60%), whilst for children at GMFCS III, misclassification occurred as both SED (20%) and WALK (16%). The recognition of SUM for fully-personalized models was good among GMFCS I children (76%), modest among GMFCS II children (67%), and poor among GMFCS III children (52%), with the majority being misclassified as WALK (48%). For group models, the recognition of WALK was very good among both children at GMFCS II and III (81–84%), but modest among children at GMFCS I (63%). For the group-personalized and fully-personalized models, the recognition of WALK was very good among children at GMFCS II (79–82%) and excellent among children at GMFCS III (94–97%), but poor among children at GMFCS I (52–59%).

4. Discussion

To the best of our knowledge, this is the first study to compare group, group-personalized, and fully-personalized activity classification models for children with CP. During cross-validation evaluation, fully-personalized RF activity classification models were more accurate than group and group-personalized models. Fully-personalized models were most accurate among children at GMFCS III, for whom the overall accuracy exceeded 95%, compared to group and group-personalized models, for which the overall accuracy was below 83%. Models trained on ankle data and all two- and three-placement combinations provided a greater overall accuracy greater than those trained on data from the hip or wrist. Irrespective of the model type or placement configuration, however, none of the models performed well when tested under conditions that replicated how activities are performed under real-world conditions.

When evaluated under laboratory conditions, fully-personalized classifiers exhibited a greater accuracy than group and group-personalized classifiers, with the largest performance differential being observed in children at GMFCS III. Furthermore, when children completed activities under simulated free-living conditions, the group-personalized and/or fully-personalized classifiers exhibited a greater accuracy than conventional group classifiers; although, none of the models generalized well and attained a high level of accuracy. These findings are consistent with the notion that the use of personalized classifiers to measure physical activity type and intensity may be advantageous among children with more severe motor impairments, where there is substantial heterogeneity in functional capacity. Notably, more than 90% of children classified at GMFCS III have bilateral CP with spastic, dyskinetic, ataxic, or hypotonic motor impairments and require assistive mobility devices for ambulation [33]. Conversely, most children classified at GMFCS I and II have unilateral CP, with only spastic motor impairment and can ambulate independently. Hence, the common practice of training one-size-fits-all group-based classifiers may not be useful in children with CP.

The fully-personalized, group-personalized, and group RF classifiers trained on laboratory-based activity trials displayed excellent recognition accuracy when cross-validated under laboratory conditions. However, when tested under conditions that replicated real-world conditions, overall accuracy decreased from 89–94% to 5–62%. The substantial decrease in accuracy when tested under simulated free-living conditions is indicative that the cross-validation performance under laboratory conditions is not reflective of a model's generalizability in real-world scenarios, particularly if the model is trained on data from choreographed activity trials. This supports the findings of prior studies conducted in typically developing children and healthy adults that reported substantial decreases in accuracy when classifiers trained on laboratory-based activity trials were implemented under real-world conditions [27–29]. Furthermore, prior studies have demonstrated that when classifiers are trained with free-living data, they have a much better generalizability to real-world conditions [34,35]. Future studies developing machine learning classification models for children with CP should therefore train models using accelerometer data collected under true free-living conditions, where children are allowed to naturally engage in physical activities that are representative of daily activity behaviors. Such studies should have sufficient participants from each GMFCS level to determine if models perform well in children with more severe impairments. Furthermore, to obtain ground-truth activities, video-based direct observation techniques that have been successfully implemented in both the current study and in free-living studies with children should be used to ensure the reliability and precision of activity type labeling [28].

When evaluated under simulated free-living conditions, group-personalized classifiers exhibited greater overall accuracy than the group or fully-personalized classifiers among children at GMFCS III. In contrast, fully-personalized classifiers exhibited greater accuracy than the group and group-personalized models among children at GMFCS levels I and II. The relatively poor performance of the fully-personalized models among children with more severe impairments under simulated free-living conditions was surprising and difficult to explain. When completing the structured activity trials in the laboratory, children functioning at GMFCS III performed sedentary and SUM activities with minimal movement and, subsequently, when the fully-personalized classifier was trained using data from a single individual, there may have been insufficient diversity in the accelerometer data to differentiate these two activity classes. Conversely, group-personalized classifiers trained on data from all GMFCS III participants displayed more diversity in the training data and therefore a greater discriminative ability, which was reflected in the better performance demonstrated during the simulated free-living evaluation. This finding, in combination with the poor generalizability of laboratory-trained classifiers, suggests that fully-personalized classifiers need to be trained on large quantities of accelerometer data collected from an individual completing a wide range of activities under free-living conditions. This, however, significantly increases the research burden placed on both researchers and participants. Alternatively, group-personalized classifiers trained for specific GMFCS classifications can be implemented as an off-the-shelf model; however, such models appear to be less accurate among children with less severe activity limitations.

Random Forest classifiers trained on ankle accelerometer data provided the best overall accuracy during the leave-one-subject-out cross-validation and simulated free-living activity trial. This finding is consistent with prior studies conducted in typically developing children and healthy adults [24,36–38]. The distinct rhythmic pattern of walking is easily captured by features from an ankle-worn accelerometer. Consequently, models trained on ankle data had the highest recognition of walking and it was the only placement to able achieve ≥ 95% walking recognition among children at GMFCS III under simulated free-living conditions. Additionally, classification models trained on ankle data provided better recognition of standing utilitarian movements, such as wiping down a counter or playing active video games. For researchers and clinicians interested in monitoring time spent walking, the ankle may be an ideal wear location. An accelerometer can be readily attached to the ankle-foot orthoses commonly worn by this patient group to assist in mobility [39,40]. Prior studies have established that children with CP have a high compliance with wearing ankle-mounted accelerometers, such as the StepWatch, as they do not cause undue irritation or discomfort [41–43].

Although the classifiers achieved a high degree of recognition accuracy in the laboratory-based evaluation, SED was frequently misclassified as SUM and vice versa during the simulated free-living trial. During the laboratory-based activity trials, sedentary activities consisted of lying down and quietly sitting and required little movement, while SUM activities such as wiping down a table and playing an active videogame required the children to be standing, while moving side to side or forward and backwards. In contrast, during the simulated free-living trial, children rarely sat quietly during sedentary activities and performed SUM activities with little or no lateral movement. The misclassification of sedentary and SUM during the simulated free-living trial may therefore be attributable, at least in part, to differences in the way that activities were performed in the structured laboratory trials and the simulated free-living evaluation. Regardless of how the activities are performed, a principle difference between the two activity classes is posture. When an accelerometer is placed on a body location that moves into a fixed anatomical plane in different postures, the changes in accelerometer orientation from tilt angle can be used to detect the posture with a high accuracy. Skotte et al. [44] and Edwardson et al. [45] have previously demonstrated that thigh-worn classifiers can differentiate sitting and standing, whereas Gjoreski et al. [46], Tang et al. [47], and Narayanan et al. [48] have shown that a two-placement combination of the thigh and hip/back results in an excellent recognition of lying, sitting, and standing. Prior studies have also demonstrated that thigh-worn monitors, such as activPAL and Uptimer, provide acceptable estimates of sitting and standing among children with CP who have mild motor impairments [49,50]. Although there may be optimal wear locations for a model to recognize certain activities [23], a consideration of the trade-off between activity recognition and wear compliance by children will dictate the optimal wear location for investigators.

The current study had several strengths. To the best of our knowledge, this is the first study to develop and test personalized classifiers in ambulant children with CP that account for significant heterogeneity in relation to movement impairment and functional capacity. Second, the study had sufficient numbers of children classified at GMFCS levels I, II, and III to evaluate classifier performance across the full spectrum of ambulatory ability. Third, classification models were tested under simulated free-living conditions. This allowed for an examination of the models' performance when activities were completed in sequence, rather than separate activity trials. Opposing these strengths were several limitations. First, although the classifiers were evaluated under simulated free-living conditions that were intended to replicate a real-world scenario, the tasks performed and brief duration did not fully replicate the activity performances of children with CP in true free-living contexts. Consequently, future studies should evaluate the performance of group and personalized classifiers under true free-living conditions. Second, the study did not include a thigh placement. The propensity for classifiers trained on hip and wrist accelerometer data to misclassify sitting and standing is well-documented [51–53] and the inclusion of posture-related features from a thigh-mounted accelerometer may have improved the recognition of SED and SUM in the simulated free-living evaluation trial. Third, the current study only trained RF classifiers and did not benchmark the performance with other supervised or unsupervised learning algorithms. RF classifiers were chosen in the current study because they are ensemble learning models which have been shown to provide accurate activity recognition in children with CP [22,54], as well as those with typical development [55,56]. As the aim of the current study was to evaluate the influence of personalization on the classification accuracy, it was important that models were trained using the same supervised learning algorithm. Future studies could examine the performance of other machine learning algorithms and benchmark the performance to the results observed in this study. The final group and group-personalized models with the annotated dataset and code for implementation are available at https://github.com/QUTcparg/Sensors_CP_PersonalisedModels.

5. Conclusions

In summary, when evaluated under laboratory conditions, group-personalized and fully-personalized RF activity classification models provide a more accurate recognition of physical activity in children with CP than "one-size-fits-all" group models. Personalized models yielded the greatest improvement in accuracy among children with the more severe motor impairments. When evaluated under simulated free-living conditions, personalized models exhibited a higher classification accuracy than conventional group models; however, the performance for all models declined substantially. Accordingly, future studies should evaluate the feasibility and comparative accuracy of group-personalized and fully-personalized activity classification models trained on accelerometer data collected under true free-living conditions.

Supplementary Materials: The following are available online at http://www.mdpi.com/1424-8220/20/14/3976/s1.

Author Contributions: Conceptualization, M.N.A. and S.G.T.; methodology, M.N.A., S.G.T., M.E.O. and E.B.; formal analysis, M.N.A. and S.G.T.; participant recruitment: E.B.; writing—original draft preparation, M.N.A. and S.G.T.; writing—review and editing, S.G.T., M.E.O., R.N.B. and E.B.; funding acquisition, S.G.T., M.E.O. and R.N.B. All authors have read and agreed to the published version of the manuscript.

Funding: This research was funded by the National Institutes of Health, grant number R21 HD086745-01A1.

Acknowledgments: We wish to thank the children and families for their participation and Denise K. Brookes for her contributions throughout the study.

Conflicts of Interest: The authors declare no conflicts of interest. The funders had no role in the design of the study; in the collection, analyses, or interpretation of data; in the writing of the manuscript; or in the decision to publish the results.

References

1. Novak, I.; Morgan, C.; Adde, L.; Blackman, J.; Boyd, R.N.; Brunstrom-Hernandez, J.; Cioni, G.; Damiano, D.; Darrah, J.; Eliasson, A.C.; et al. Early, accurate diagnosis and early intervention in cerebral palsy: Advances in diagnosis and treatment. *JAMA Pediatr.* **2017**, *171*, 897–907. [CrossRef]

2. Report of the Australian Cerebral Palsy Register Birth Years 1995–2012. 2018. Available online: https://cpregister.com/wp-content/uploads/2019/02/Report-of-the-Australian-Cerebral-Palsy-Register-Birth-Years-1995-2012.pdf (accessed on 18 June 2020).

3. Bjornson, K.; Fiss, A.; Avery, L.; Wentz, E.; Kerfeld, C.; Cicirello, N.; Hanna, S.E. Longitudinal trajectories of physical activity and walking performance by gross motor function classification system level for children with cerebral palsy. *Disabil Rehabil.* **2018**, 1–9. [CrossRef] [PubMed]

4. Nooijen, C.F.J.; Slaman, J.; Stam, H.J.; Roebroeck, M.E.; Van Den Berg-Emons, R.J. Inactive and sedentary lifestyles amongst ambulatory adolescents and young adults with cerebral palsy. *J. Neuroeng. Rehabil.* **2014**, *11*, 49. [CrossRef] [PubMed]

5. Ross, S.M.; Smit, E.; Yun, J.; Bogart, K.; Hatfield, B.; Logan, S.W. Updated national estimates of disparities in physical activity and sports participation experienced by children and adolescents with disabilities: NSCH 2016–2017. *J. Phys. Act. Health* **2020**, *17*, 443–455. [CrossRef]

6. Van Wely, L.; Becher, J.G.; Balemans, A.C.J.; Dallmeijer, A.J. Ambulatory activity of children with cerebral palsy: Which characteristics are important? *Dev. Med. Child Neurol.* **2012**, *54*, 436–442. [CrossRef]

7. Gorter, J.W.; Noorduyn, S.G.; Obeid, J.; Timmons, B.W. Accelerometry: A Feasible Method to Quantify Physical Activity in Ambulatory and Nonambulatory Adolescents with Cerebral Palsy. *Int. J. Pediatr.* **2012**, *2012*, 1–6. [CrossRef] [PubMed]

8. Bjornson, K.F.; Belza, B.; Kartin, D.; Logsdon, R.; McLaughlin, J.F. Ambulatory physical activity performance in youth with cerebral palsy and youth who are developing typically. *Phys. Ther.* **2007**, *87*, 248–257. [CrossRef]

9. Reedman, S.E.; Boyd, R.N.; Trost, S.G.; Elliott, C.; Sakzewski, L. Efficacy of participation-focused therapy on performance of physical activity participation goals and habitual physical activity in children with cerebral palsy: A randomized controlled trial. *Arch. Phys. Med. Rehabil.* **2019**, *100*, 676–686. [CrossRef]

10. Maher, C.A.; Williams, M.T.; Olds, T.; Lane, A.E. An internet-based physical activity intervention for adolescents with cerebral palsy: A randomized controlled trial. *Dev. Med. Child Neurol.* **2010**, *52*, 448–455. [CrossRef]

11. Boyd, R.N.; Davies, P.S.; Ziviani, J.; Trost, S.; Barber, L.; Ware, R.; Rose, S.; Whittingham, K.; Sakzewski, L.; Bell, K.; et al. PREDICT-CP: Study protocol of implementation of comprehensive surveillance to predict outcomes for school-aged children with cerebral palsy. *BMJ Open* **2017**, *7*, 1–19. [CrossRef]

12. Trost, S.G.; O'Neil, M. Clinical use of objective measures of physical activity. *Br. J. Sports Med.* **2014**, *48*, 178–181. [CrossRef]

13. Van Wely, L.; Balemans, A.C.J.; Becher, J.G.; Dallmeijer, A.J. Physical activity stimulation program for children with cerebral palsy did not improve physical activity: A randomised trial. *J. Physiother.* **2014**, *60*, 40–49. [CrossRef]

14. Mitchell, L.E.; Ziviani, J.; Boyd, R.N. A randomized controlled trial of web-based training to increase activity in children with cerebral palsy. *Dev. Med. Child Neurol.* **2016**, *58*, 767–773. [CrossRef]

15. Trost, S.G.; Mciver, K.L.; Pate, R.R. Conducting accelerometer-based activity assessments in field-based research. *Med. Sci. Sports Exerc.* **2005**, *37* (Suppl. 11), S531–S543. [CrossRef] [PubMed]

16. O'Neil, M.E.; Fragala-Pinkham, M.; Lennon, N.; George, A.; Forman, J.; Trost, S.G. Reliability and validity of objective measures of physical activity in youth with cerebral palsy who are ambulatory. *Phys. Ther.* **2016**, *96*, 37–45. [CrossRef] [PubMed]

17. O'Neil, M.E.; Fragala-Pinkham, M.; Forman, J.L.; Trost, S.G. Measuring reliability and validity of the ActiGraph GT3X accelerometer for children with cerebral palsy: A feasibility study. *J. Pediatr. Rehabil. Med.* **2014**, *7*, 233–240. [CrossRef] [PubMed]

18. Clanchy, K.M.; Tweedy, S.M.; Boyd, R.N.; Trost, S.G. Validity of accelerometry in ambulatory children and adolescents with cerebral palsy. *Eur. J. Appl. Physiol.* **2011**, *111*, 2951–2959. [CrossRef] [PubMed]

19. Mitchell, L.E.; Ziviani, J.; Boyd, R.N. Variability in measuring physical activity in children with cerebral palsy. *Med. Sci. Sports Exerc.* **2015**, *47*, 194–200. [CrossRef] [PubMed]

20. Ryan, J.; Walsh, M.; Gormley, J. Ability of RT3 accelerometer cut points to detect physical activity intensity in ambulatory children with cerebral palsy. *Adapt. Phys. Act. Q.* **2014**, *31*, 310–324. [CrossRef]

21. Trost, S.G.; Fragala-Pinkham, M.; Lennon, N.; O'Neil, M.E. Decision trees for detection of activity intensity in youth with cerebral palsy. *Med. Sci. Sports Exerc.* **2016**, *48*, 958–966. [CrossRef]

22. Ahmadi, M.N.; O'Neil, M.E.; Fragala-pinkham, M.; Lennon, N.; Trost, S.G. Machine learning algorithms for activity recognition in ambulant children and adolescents with cerebral palsy. *J. Neuroeng. Rehabil.* **2018**, *15*, 1–9. [CrossRef] [PubMed]

23. Chowdhury, A.K.; Tjondronegoro, D.; Chandran, V.; Trost, S.G. Physical activity recognition using posterior-adapted class-based fusion of multiaccelerometer data. *IEEE J. Biomed. Health Inform.* **2018**, *22*, 678–685. [CrossRef] [PubMed]

24. Mannini, A.; Intille, S.S.; Rosenberger, M.; Sabatini, A.M.; Haskell, W. Activity recognition using a single accelerometer placed at the wrist or ankle. *Med. Sci. Sports Exerc.* **2013**, *45*, 2193–2203. [CrossRef] [PubMed]

25. Lockhart, J.W.; Weiss, G.M. The benefits of personalized smartphone-based activity recognition models. In Proceedings of the 2014 SIAM International Conference on Data Mining, Philadelphia, PA, USA, 24–26 April 2014; p. 9.

26. Carcreff, L.; Paraschiv-Ionescu, A.; Gerber, C.N.; Newman, C.J.; Armand, S.; Aminian, K. A personalized approach to improve walking detection in real-life settings: Application to children with cerebral palsy. *Sensors* **2019**, *19*, 5316. [CrossRef] [PubMed]

27. Sasaki, J.E.; Hickey, A.M.; Staudenmayer, J.W.; John, D.; Kent, J.A.; Freedson, P.S. Performance of activity classification algorithms in free-living older adults. *Med. Sci. Sports Exerc.* **2016**, *48*, 941–949. [CrossRef]

28. Ahmadi, M.N.; Brookes, D.; Chowdhury, A.; Pavey, T.; Trost, S.G. Free-living evaluation of laboratory-based activity classifiers in preschoolers. *Med. Sci. Sports Exerc.* **2020**, *52*, 1227–1234. [CrossRef]

29. Bastian, T.; Maire, A.; Dugas, J.; Ataya, A.; Villars, C.; Gris, F.; Perrin, E.; Caritu, Y.; Doron, M.; Blanc, S.; et al. Automatic identification of physical activity types and sedentary behaviors from triaxial accelerometer: Laboratory-based calibrations are not enough. *J. Appl. Physiol.* **2015**, *118*, 716–722. [CrossRef]

30. Breiman, L. Random forests. *Mach. Learn.* **2001**, *45*, 5–32. [CrossRef]

31. Chowdhury, A.K.; Tjondronegoro, D.; Chandran, V.; Trost, S.G. Ensemble methods for classification of physical activities from wrist accelerometry. *Med. Sci. Sports Exerc.* **2017**, *49*, 1965–1973. [CrossRef]

32. Peng, H.; Long, F.; Ding, C. Feature selection based on mutual information: Criteria of max-dependency, max-relevance, and min-redundancy. *IEEE Trans. Pattern Anal. Mach. Intell.* **2005**, *27*, 1226–1238. [CrossRef] [PubMed]

33. Report of the Australian Cerebral Palsy Register, Birth Years 1993–2009. 2016. Available online: https://cpregister.com/wp-content/uploads/2018/05/ACPR-Report_Web_2016.pdf (accessed on 18 June 2020).

34. Kerr, J.; Patterson, R.E.; Ellis, K.; Godbole, S.; Johnson, E.; Lanckriet, G.; Staudenmayer, J. Objective assessment of physical activity: Classifiers for public health. *Med. Sci. Sports Exerc.* **2016**, *48*, 951–957. [CrossRef] [PubMed]

35. Rosenberg, D.; Godbole, S.; Ellis, K.; Di, C.; LaCroix, A.Z.; Natarajan, L.; Kerr, J. Classifiers for accelerometer-measured behaviors in older women. *Med. Sci. Sports Exerc.* **2017**, *49*, 610–616. [CrossRef] [PubMed]

36. Strath, S.J.; Kate, R.J.; Keenan, K.G.; Welch, W.A.; Swartz, A.M. Ngram time series model to predict activity type and energy cost from wrist, hip and ankle accelerometers: Implications of age. *Physiol. Meas.* **2015**, *36*, 2335–2351. [CrossRef]

37. Gjoreski, M.; Gjoreski, H.; Lustrek, M.; Gams, M. How accurately can your wrist device recognize daily activities and detect falls? *Sensors* **2016**, *16*, 800. [CrossRef]

38. Preece, S.J.; Goulermas, J.Y.; Kenney, L.P.J.; Howard, D. A comparison of feature extraction methods for the classification of dynamic activities from accelerometer data. *IEEE Trans. Biomed. Eng.* **2009**, *56*, 871–879. [CrossRef]

39. Wingstrand, M.; Hägglund, G.; Rodby-Bousquet, E. Ankle-foot orthoses in children with cerebral palsy: A cross sectional population based study of 2200 children. *BMC Musculoskelet. Disord.* **2014**, *15*, 327. [CrossRef] [PubMed]

40. Lobo, M.A.; Hall, M.L.; Greenspan, B.; Rohloff, P.; Prosser, L.A.; Smith, B.A. Wearables for pediatric rehabilitation: How to optimally design and use products to meet the needs of users. *Phys. Ther.* **2019**, *99*, 647–657. [CrossRef]

41. McDonald, C.M.; Widman, L.; Abresch, R.T.; Walsh, S.A.; Walsh, D.D. Utility of a step activity monitor for the measurement of daily ambulatory activity in children. *Arch. Phys. Med. Rehabil.* **2005**, *86*, 793–801. [CrossRef]

42. Clanchy, K.M.; Tweedy, S.M.; Boyd, R. Measurement of habitual physical activity performance in adolescents with cerebral palsy: A systematic review. *Dev. Med. Child Neurol.* **2011**, *53*, 499–505. [CrossRef] [PubMed]

43. Bjornson, K.F.; Zhou, C.; Stevenson, R.D.; Christakis, D. Relation of stride activity and participation in mobility-based life habits among children with cerebral palsy. *Arch. Phys. Med. Rehabil.* **2014**, *95*, 360–368. [CrossRef]

44. Skotte, J.; Korshøj, M.; Kristiansen, J.; Hanisch, C.; Holtermann, A. Detection of physical activity types using triaxial accelerometers. *J. Phys. Act. Health* **2014**, *11*, 76–84. [CrossRef] [PubMed]

45. Edwardson, C.L.; Rowlands, A.V.; Bunnewell, S.; Sanders, J.P.; Esliger, D.W.; Gorely, T.; O'Connell, S.; Davies, M.J.; Khunti, K.; Yates, T.E. Accuracy of posture allocation algorithms for thigh- and waist-worn accelerometers. *Med. Sci. Sports Exerc.* **2016**, *48*, 1085–1090. [CrossRef] [PubMed]

46. Gjoreski, H.; Luštrek, M.; Gams, M. Accelerometer placement for posture recognition and fall detection. In Proceedings of the 2011 7th International Conference on Intelligent Environments, Nottingham, UK, 25–28 July 2011; pp. 47–54.

47. Tang, Q.; John, D.; Thapa-Chhetry, B.; Arguello, D.J.; Intille, S. Posture and physical activity detection. *Med. Sci. Sports Exerc.* **2020**. [CrossRef]

48. Narayanan, A.; Stewart, T.; Mackay, L. A Dual-accelerometer system for detecting human movement in a free-living environment. *Med. Sci. Sports Exerc.* **2020**, *52*, 252–258. [CrossRef]

49. O'Donoghue, D.; Kennedy, N. Validity of an activity monitor in young people with cerebral palsy gross motor function classification system level I. *Physiol. Meas.* **2014**, *35*, 2307–2318. [CrossRef]

50. Pirpiris, M.; Graham, H.K. Uptime in children with cerebral palsy. *J. Pediatr. Orthop.* **2004**, *24*, 521–528. [CrossRef]

51. Grant, P.M.; Ryan, C.G.; Tigbe, W.W.; Granat, M.H. The validation of a novel activity monitor in the measurement of posture and motion during everyday activities. *Br. J. Sports Med.* **2006**, *40*, 992–997. [CrossRef] [PubMed]

52. Lyden, K.; Kozey Keadle, S.L.; Staudenmayer, J.W.; Freedson, P.S. Validity of two wearable monitors to estimate breaks from sedentary time. *Med. Sci. Sports Exerc.* **2012**, *44*, 2243–2252. [CrossRef] [PubMed]

53. Steeves, J.A.; Bowles, H.R.; Mcclain, J.J.; Dodd, K.W.; Brychta, R.J.; Wang, J.; Chen, K.Y. Ability of thigh-worn actigraph and activpal monitors to classify posture and motion. *Med. Sci. Sports Exerc.* **2015**, *47*, 952–959. [CrossRef]

54. Goodlich, B.I.; Armstrong, E.L.; Horan, S.A.; Baque, E.; Carty, C.P.; Ahmadi, M.N.; Trost, S.G. Machine learning to quantify habitual phyisical activity in children with cerebral palsy. *Dev. Med. Child Neurol.* **2020**. [CrossRef]

55. Trost, S.G.; Cliff, D.P.; Ahmadi, M.N.; Van Tuc, N.; Hagenbuchner, M. Sensor-enabled activity class recognition in preschoolers: Hip versus wrist data. *Med. Sci. Sports Exerc.* **2018**, *50*, 634–641. [CrossRef] [PubMed]

56. Ahmadi, M.N.; Pfeiffer, K.A.; Trost, S.G. Physical activity classification in youth using raw accelerometer data from the hip. *Meas. Phys. Educ. Exerc. Sci.* **2020**, *24*, 129–136. [CrossRef]

Review

Machine Learning Approaches for Activity Recognition and/or Activity Prediction in Locomotion Assistive Devices—A Systematic Review

Floriant Labarrière [1], Elizabeth Thomas [1], Laurine Calistri [2], Virgil Optasanu [3], Mathieu Gueugnon [4], Paul Ornetti [1,4,5] and Davy Laroche [1,4,*]

[1]   INSERM, UMR1093-CAPS, Université de Bourgogne Franche Comté, UFR des Sciences du Sport, F-21000 Dijon, France; floriant_labarriere@etu.u-bourgogne.fr (F.L.); elizabeth.thomas@u-bourgogne.fr (E.T.); paul.ornetti@chu-dijon.fr (P.O.)

[2]   PROTEOR, 6 rue de la Redoute, CS 37833, CEDEX 21078 Dijon, France; Laurine.Calistri@proteor.com

[3]   ICB, UMR 6303 CNRS, Université de Bourgogne Franche Comté 9 Av. Alain Savary, CEDEX 21078 Dijon, France; virgil.optasanu@u-bourgogne.fr

[4]   INSERM, CIC 1432, Module Plurithematique, Plateforme d'Investigation Technologique, CHU Dijon-Bourgogne, Centre d'Investigation Clinique, Module Plurithématique, Plateforme d'Investigation Technologique, 21079 Dijon, France; mathieu.gueugnon@chu-dijon.fr

[5]   Department of Rheumatology, Dijon University Hospital, 21079 Dijon, France

*   Correspondence: davy.laroche@chu-dijon.fr; Tel.: +33-380295665

Received: 25 September 2020; Accepted: 4 November 2020; Published: 6 November 2020

Abstract: Locomotion assistive devices equipped with a microprocessor can potentially automatically adapt their behavior when the user is transitioning from one locomotion mode to another. Many developments in the field have come from machine learning driven controllers on locomotion assistive devices that recognize/predict the current locomotion mode or the upcoming one. This review synthesizes the machine learning algorithms designed to recognize or to predict a locomotion mode in order to automatically adapt the behavior of a locomotion assistive device. A systematic review was conducted on the Web of Science and MEDLINE databases (as well as in the retrieved papers) to identify articles published between 1 January 2000 to 31 July 2020. This systematic review is reported in accordance with the Preferred Reporting Items for Systematic reviews and Meta-Analyses (PRISMA) guidelines and is registered on Prospero (CRD42020149352). Study characteristics, sensors and algorithms used, accuracy and robustness were also summarized. In total, 1343 records were identified and 58 studies were included in this review. The experimental condition which was most often investigated was level ground walking along with stair and ramp ascent/descent activities. The machine learning algorithms implemented in the included studies reached global mean accuracies of around 90%. However, the robustness of those algorithms seems to be more broadly evaluated, notably, in everyday life. We also propose some guidelines for homogenizing future reports.

Keywords: machine learning; locomotion; assistive devices; embedded sensors

---

1. Introduction

Healthy humans are easily able to adjust locomotor pattern to deal with multiple environments encountered in daily living situations such as stair ascent/descent, slope ascent/descent, obstacle clearance, walking on uneven floors, cross-slopes or different surfaces. Hence, with lower limb impairments such as unilateral lower limb amputation, it becomes challenging to deal with most of these environmental changes [1].

To handle this issue, intelligent devices such as the C-leg TM (OTTOBOCK, Berlin, Germany) or the Rheo knee (ÖSSUR, Reykjavík, Iceland) have been developed. These variable-damping prostheses, compared to mechanically passive prostheses, improved the smoothness of gait, and decreased hip work during level-ground walking [2]. Additional improvement was provided by a powered prosthesis which was reported to decrease the metabolic cost of transport when compared to a conventional passive prosthesis in similar conditions [3]. Prosthetic devices which passively or actively mimicked human actions were found to be of help. One historic example of such innovation was the energy return foot that reproduced foot behavior and improved the gait of patients with amputation. Other innovations in the attempt to create intelligent devices can be seen with some microprocessor-controlled prostheses with the ability to recognize the terrain being traversed (e.g., Genium OTTOBOCK, Berlin, Germany, Linx BLATCHFORD, Basingstoke, UK). It only stands to reason that the next step in this progression would be the development of devices with the ability to make predictions for automatic gait adjustments across multiple terrains.

Developments in these efforts have come in the form of intelligent controllers on locomotion assistive devices. In such devices, gait is regulated by a hierarchical three-level controller [4]. The highest-level controller is responsible for detecting the user-intent. The mid-level controller automatically switches the control law (e.g., the powered active transfemoral prosthesis developed by Vanderbilt University [5]) of the device in accordance with the high-level controller output. The low-level controller compares the desired state of the device to the sensed state and corrects it when needed. The detection of user-intent is done either by the user directly communicating his intentions to the device using a controller, or by automatic interpretation by an algorithm. Examples of the first are the control buttons found in the ReWalk TM exoskeleton (ARGO MEDICAL TECHNOLOGIES Ltd., Yokneam, Israel) or predefined body movements which allow the wearer of the Power KneeTM (ÖSSUR, Reykjavík, Iceland) to switch between locomotion modes. In this device, switching between locomotion modes requires the user to stop or to perform certain unnatural body movements. As opposed to these explicit methods, algorithm-based implicit methods interpret user intent. Such algorithm-based techniques allow smoother transitions by automatically switching between the control laws of the device. A more promising approach might be one based on machine learning algorithms. Such algorithms automatically detect user-intent by mapping sensor data to an associated locomotion task.

There are numerous studies in which machine learning has been used to adapt the behavior of orthotic/prosthetic devices to user locomotion mode. We performed a systematic review that identifies and summarizes such studies. Under the scope of this review, reports were selected if (1) body-worn sensors or sensors embedded in the devices were used (2) machine learning classifiers were able to identify the investigated locomotion modes of human volunteers. It covers essential technical details such as the pre-processing methods which were used, the specific Machine Learning algorithms which were employed, and the corresponding accuracies obtained. By the end of this review we aim to propose recommendations for future studies and some suggestions concerning the uniformization of the terms used to report results in the field.

2. Material and Methods

This systematic review, registered on PROSPERO (CRD42020149352), is reported in accordance with the Preferred Reporting Items for Systematic reviews and Meta-Analyses (PRISMA) guidelines [6].

2.1. Eligibility Criteria

This systematic review included peer-reviewed articles and patents focusing on the Machine Learning (ML) approach for classifying locomotion modes in volunteers wearing assistive devices (see below for definition of the included devices). For this purpose:

- The algorithms must be based on locomotion data collected from embedded sensors in the device or from body-worn sensors. Studies evaluating a previously developed ML-based pattern recognition algorithms were also included. We focused on Machine Learning methods that carried out classification for recognizing locomotion modes. Studies using a Machine Learning regression approach were excluded.

- The articles must be related to locomotion in various environments, e.g., level ground walking, stair ascent/descent, ramp ascent/descent, obstacle clearance, walking on a cross-slope, turning, walking on different surfaces, ... Studies were included if at least two locomotion modes were investigated.
- Only lower limb assistive devices such as exoskeletons, prostheses (for below or above knee amputation) or orthoses were considered.
- Studies were excluded if they met at least one of the following exclusion criteria: (1) non-human (robots or animals), (2) volunteers who are minors (under 18 years old), (3) studies focusing on volunteers equipped with an upper-limb device.

2.2. Information Sources

The PubMed and Web of Knowledge (including Web of Science core collection, Derwent Innovation Index, Russian Citation index, SciELO Citation Index) databases were searched on 31 July 2020. The two search strings used are given in the Supplementary Material. Published articles in English between 1 January 2000 and 31 July 2020 were included. Systematic reviews and meta-analyses were excluded. Conference papers were excluded if a corresponding published peer-reviewed article by the same authors had been included. Additional articles were included by further searching the references within the papers which were first identified by the search strategy described above.

2.3. Study Selection

The search strings were defined and validated by all authors. One person (FL) performed the initial search and removed the duplicates. Two main readers (DL, FL) independently screened the titles and the abstracts of all articles identified during the initial search. In case of disagreement, a third reader (LC) decided to include/exclude the article. Afterwards, the two readers (DL, FL) read the full text of the articles which had been picked from the previous step and checked them for eligibility using the criteria of our Modified QualSyst Tool which can be found in the Supplementary Material of this article. The process used to create the Modified QualSyst Tool can be found in the Section 2.4.1. Any disagreements on the eligibility of an article were resolved by the third person (LC).

2.4. Quality Assessment in Included Articles

The quality of the included articles was assessed with a dedicated QualSyst Tool [7] modified for the purposes of studying Machine Learning algorithms implemented on locomotion assistive devices. In the sections below, we provide further explanations of the score assignment for each article using this tool.

2.4.1. Creating the Modified QualSyst Tool

Our first step was to remove irrelevant items from the QualSyst Tool [7] (Criteria 3 and 5 to 12, e.g., blinding of investigators, of subjects, etc.). Next, we added items which are relevant to the implementation of Machine Learning algorithms such as analysis windows, selected features, evaluation method of the algorithm, etc. All items were validated by all the authors and the quality of included articles was assessed by the main readers (FL, DL). The final version of this Modified QualSyst Tool can be found in the Supplementary Material of this article.

2.4.2. Rating Articles Using the Modified QualSyst Tool

Twelve items were used for rating the articles. For each item, the article was rated with a score between 0 and 2 (with 2 indicating full supply of information, 1 a partial supply and 0 no information provided). Guidelines to allow consistent ratings across the included papers were created. These guidelines are provided in the Supplementary Material. The description of the twelve items were as follows:

- The first two items evaluated if the hypotheses and objectives of the study were sufficiently described and if the study design was appropriate.
- Item 3 evaluated if the volunteer characteristics were sufficiently described.
- Items 4 to 10 evaluated if the Machine Learning approach was sufficiently described to allow repeatability.
- Items 11 and 12 evaluated if the results were reported with enough details and if the conclusions were in accordance with them.

The score of each article was computed as the average of the 12 rated items. The maximum score possible for an article was 2. The score from 0–2 was transformed to a scale of 0–100% for ease of comprehension (0 indicating no information provided at all and 100 with maximum lucidity). More details on this scoring procedure and the guidelines used can be found in the Supplementary Material.

2.5. Synthesis of the Results

The following elements were extracted and grouped from the included studies:

- Investigated population (pathology and number of volunteers) and type of assistive device (above-knee prosthesis or below-knee prosthesis or orthosis or exoskeleton).
- The main elements of the experimental protocol are reported.

  - The studied locomotor activities along with the walking speed of the volunteers are given.
  - The 'Critical Timing' is reported. It is the latest moment when the behavior of the locomotion assistive device can be adapted to the new locomotion mode without disturbing the user.
  - The type of sensors used in each study along with the total number of measurement axes per sensor are reported.
  - Details on the machine learning algorithm implementation are also reported (online and/or offline implementation; forward prediction and/or backward recognition [8]).

- The signal processing techniques and Machine Learning algorithms used are reported as well:

  - This includes the type and length of the analysis windows.
  - The extracted features used for the analyses. If several configurations were tested, only the optimal configuration is given.
  - The machine learning algorithms are provided. Overall results of the machine learning algorithms are reported in terms of accuracy (A). So, if studies indicated the error rates (E), the corresponding mean overall accuracy was computed (A = 100—E in percent). For studies recruiting both healthy volunteers and patients, the reported accuracy of the machine learning algorithms corresponded to the patients (accuracy).

3. Results

3.1. Study Selection

The literature search produced 288 articles on PubMed and 1078 articles on Web of Science. Additionally, four studies were manually identified from references in the articles and added to the review. After removing the duplicates, there remained 1343 articles for screening. On the basis of titles and abstracts screening, 1267 articles were excluded from the review. Two authors independently read the full texts of the remaining 76 articles and checked them for eligibility. Finally, 58 articles were considered eligible to be included in this review. The PRISMA Flow Chart [6] is provided (Figure 1).

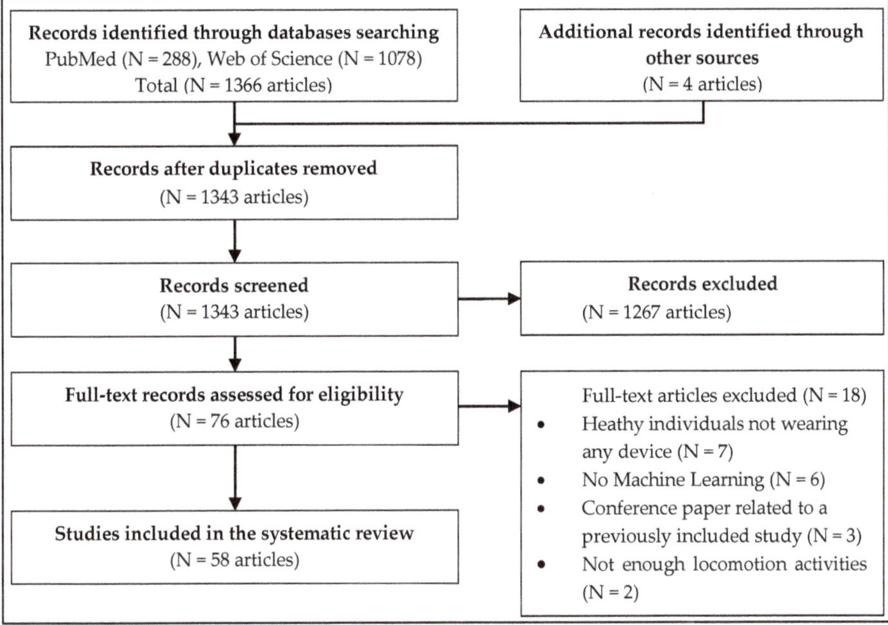

Figure 1. Preferred Reporting Items for Systematic reviews and Meta-Analyses (PRISMA) flow chart of the systematic review.

3.2. Quality of the Included Studies

The mean quality score of each study using the items of the Modified QualSyst Tool is provided in Table 1 and the detailed quality scores are presented in the Supplementary Material. The mean quality score was 68.4% +/− 13.4 for the articles.

Table 1. Quality assessment and recruited volunteers in the included studies.

Article	Quality Score	Groups (N)	Locomotion Assistive Device
Ai et al. 2017 [9]	70.8%	TT (4)/Healthy (1)	Ankle Prosthesis
Beil et al. 2018 [10]	90.9%	Healthy (10)	Exoskeleton
Chen et al. 2013 [11]	72.7%	TT (5)/Healthy (8)	Ankle Prosthesis
Chen et al. 2014 [12]	79.2%	TT (1)/Healthy (7)	Ankle Prosthesis
Chen et al. 2015 [13]	77.3%	TT (1)/Healthy (5)	Ankle Prosthesis
Du et al. 2012 [14]	75.0%	TF (9)	Ankle Knee Prosthesis
Du et al. 2013 [15]	45.8%	TF (4)	Ankle Knee Prosthesis
Feng et al. 2019 [16]	77.3%	TT (3)	Ankle Prosthesis
Godiyal et al. 2018 [17]	86.4%	TF (2)/Healthy (8)	Ankle Knee Prosthesis
Gong et al. 2018 [18]	86.4%	Healthy (1)	Orthosis
Gong et al. 2020 [19]	86.4%	Healthy (3)	Orthosis
Hernandez et al. 2012 [20]	37.5%	TF (1)	Ankle Knee Prosthesis
Hernandez et al. 2013 [21]	54.2%	Healthy (1)	Ankle Knee Prosthesis
Huang et al. 2009 [22]	81.8%	TF (2)/Healthy (8)	Ankle Knee Prosthesis
Huang et al. 2010 [23]	79.2%	TF (1)/Healthy (5)	Ankle Knee Prosthesis
Huang et al. 2011 [24]	83.3%	TF (5)	Ankle Knee Prosthesis
Kim et al. 2017 [25]	63.6%	Healthy (8)	Exoskeleton
Liu et al. 2016 [26]	70.8%	TF (1)/Healthy (6)	Ankle Knee Prosthesis

Table 1. *Cont.*

Article	Quality Score	Groups (N)	Locomotion Assistive Device
Liu et al. 2017 [27]	66.7%	TF (2)/Healthy (2)	Ankle Knee Prosthesis
Liu et al. 2017 [28]	63.6%	TF (2)/Healthy (3)	Ankle Knee Prosthesis
Long et al. 2016 [29]	83.3%	Healthy (3)	Exoskeleton
Mai et al. 2011 [30]	50.0%	TT (1)	Ankle Prosthesis
Mai et al. 2018a [31]	45.8%	TT (1)	Ankle Prosthesis
Mai et al. 2018b [32]	54.2%	TT (1)	Ankle Prosthesis
Miller et al. 2013 [33]	90.9%	TT (5)/Healthy (5)	Ankle Prosthesis
Moon et al. 2019 [34]	33.3%	Healthy (1)	Exoskeleton
Pew et al. 2017 [35]	66.7%	TT (5)	Ankle Prosthesis
Shell et al. 2018 [36]	70.8%	TT (3)	Ankle Prosthesis
Simon et al. 2017 [37]	66.7%	TF (6)	Ankle Knee Prosthesis
Spanias et al. 2014 [38]	54.2%	TF (4)	Ankle Knee Prosthesis
Spanias et al. 2015 [39]	54.2%	TF (6)	Ankle Knee Prosthesis
Spanias et al. 2016a [40]	62.5%	TF (8)	Ankle Knee Prosthesis
Spanias et al. 2016b [8]	58.3%	Healthy (2)	Ankle Knee Prosthesis
Spanias et al. 2017 [41]	58.3%	TF (3)	Ankle Knee Prosthesis
Spanias et al. 2018 [42]	62.5%	TF (8)	Ankle Knee Prosthesis
Stolyarov et al. 2017 [43]	79.2%	TF (6)	Ankle Knee Prosthesis
Su et al. 2019 [44]	77.3%	TF (1)/Healthy (10)	Ankle Knee Prosthesis
Tkach et al. 2013 [45]	62.5%	TT (5)	Ankle Prosthesis
Wang et al. 2013 [46]	66.7%	TT (1)	Ankle Prosthesis
Wang et al. 2018 [47]	79.2%	Healthy (22)	Exoskeleton
Woodward et al. 2016 [48]	91.7%	TF (6)	Ankle Knee Prosthesis
Xu et al. 2018 [49]	75.0%	TT (3)	Ankle Prosthesis
Young et al. 2013a [50]	66.7%	TF (4)	Ankle Knee Prosthesis
Young et al. 2013b [51]	79.2%	TF (6)	Ankle Knee Prosthesis
Young et al. 2013c [52]	62.5%	TF (4)	Ankle Knee Prosthesis
Young et al. 2014a [53]	66.7%	TF (6)	Ankle Knee Prosthesis
Young et al. 2014b [54]	75.0%	TF (8)	Ankle Knee Prosthesis
Young et al. 2016 [55]	75.0%	TF (8)	Ankle Knee Prosthesis
Zhang et al. 2011 [56]	70.8%	TF (1)/Healthy (1)	Ankle Knee Prosthesis
Zhang et al. 2013 [57]	66.7%	TF (4)	Ankle Knee Prosthesis
Zhang et al. 2019 [58]	63.6%	TF (3)/Healthy (6)	Ankle Knee Prosthesis
Zhang et al. 2019 [59]	59.1%	TF (3)/Healthy (6)	Ankle Knee Prosthesis
Zhang et al. 2012 [60]	62.5%	Healthy (1)	Ankle Knee Prosthesis
Zheng et al. 2013 [61]	86.4%	TT (1)	Ankle Prosthesis
Zheng et al. 2014 [62]	86.4%	TT (6)	Ankle Prosthesis
Zheng et al. 2016 [63]	75.0%	TT (6)	Ankle Prosthesis
Zheng et al. 2019 [64]	54.2%	TT (6)	Ankle Prosthesis
Zhou et al. 2019 [65]	54.2%	Healthy (3)	Exoskeleton

TT = Volunteer with a unilateral transtibial amputation, TF = Volunteer with a unilateral transfemoral amputation, N = Number of recruited volunteers.

3.3. Extracted Elements of the Included Studies

In this section, we summarize some of the key aspects of the extracted elements of the included studies.

3.3.1. Type of Assistive Device and Related Population

The type of assistive device used in each study and the related population are detailed in Table 1. Four types of devices were used in the included studies: prostheses for transfemoral amputation (i.e., above-knee prostheses), prostheses for transtibial amputation (i.e., below-knee prostheses), exoskeletons and orthoses.

- **Above-knee prostheses.** This was the largest group among the published studies (N = 32). Among these thirty-two studies, the recruited population were either patients with unilateral transfemoral amputation or knee disarticulation (N = 19). There were healthy volunteers and patients with transfemoral amputation or knee disarticulation (N = 10). Finally, there were healthy volunteers wearing an above-knee prosthesis with an L-shape adaptor (N = 3).

- **Below-knee prosthesis.** This was the second largest group in this review (N = 18). Among those eighteen studies, the recruited population were either patients with unilateral transtibial amputation (N = 13) or healthy volunteers or patients with unilateral transtibial amputation (N = 5).
- **Exoskeletons** and **orthoses.** This constituted the smallest group in this review (N = 6 and N = 2 respectively). Among those eight studies, the recruited population was always healthy volunteers wearing the assistive device.

3.3.2. Locomotion Activities and Walking Speed

The locomotion activities and walking speed investigated in each study are reported in Table 2.

The most representative experimental protocol investigated level ground walking along with stair and ramp ascent/descent activities (N = 43). Secondly, in some studies, level ground walking was investigated only with stair ascent and/or descent activities (N = 13). Among those fifty-six (43 + 13) studies, additional activities were also considered such as obstacle clearance (N = 6), turning (N = 2) or squatting (N = 1/58) for 'dynamic' activities and standing (N = 23/58) or sitting (N = 6/58) for static activities. The remaining two papers investigated level ground walking with cross slope walking (N = 1) and level ground walking with turning (N = 1).

In most studies, the walking speed was not provided (N = 33). One can assume that the volunteers walked at a self-selected speed in these thirty-three studies. Next, the volunteers were asked to walk at a self-selected speed in seventeen studies (N = 17). Finally, a small number of studies investigated different walking speeds: volunteers were asked to walk either at self-selected speed or at a slower or faster pace for different locomotion activities (N = 6). In the two remaining studies, recruited volunteers were asked to walk at a predefined speed of 0.7 m/s (N = 2).

Table 2. Sensors used for recognition and/or prediction of the locomotion activities investigated in the included studies.

Article	Locomotion Activities	Critical Timing	Speed	Sensors	Axes × Sensors	Offline/Online	Recognition/Prediction
Ai et al. 2017 [9]	LW, SA, SD, ST, SQ	NP	NP	EMG IMU	1×4 $3(A) \times 1$	Off	R
Beil et al. 2018 [10]	LW, SA, SD, Turns, ST	NA	SSS	Force Sensors IMU	3×7 $6(A, \alpha) \times 3$	Off	R
Chen et al. 2013 [11]	LW, SA, SD, OBS, ST, SIT	NA	NP	Capacitive Pressure	1×10 2×1	Off	R
Chen et al. 2014 [12]	LW, SA, SD, RA, RD	3	NP	IMU Pressure	$9(A, G, \alpha) \times 2$ 4×2	Off	R
Chen et al. 2015 [13]	LW, SA, SD, OBS, ST, SIT	NA	SSS	Pressure	4×1	Off	R
Du et al. 2012 [14]	LW, SA, SD, RA, RD	2	NP	EMG Load cell	1×9 6×1	Off	P
Du et al. 2013 [15]	LW, SA, SD, RA, RD	NP	NP	EMG Load cell	1×7 6×1	Off	P
Feng et al. 2019 [16]	LW, SA, SD, RA, RD	NA	NP	Load cell Angle Sensor	NP 1×1	Off	R
Godiyal et al. 2018 [17]	LW, SA, SD, RA, RD	NA	SSS	FMG Pressure	1×8 3×1	Off	R
Gong et al. 2018 [18]	LW, SA, SD, RA, RD, ST	NA	Imposed Speed	IMU	$9(A, G, \alpha) \times 2$	Off and On	R
Gong et al. 2020 [19]	LW, SA, SD, RA, RD, ST	NA	Imposed Speed	IMU	$9(A, G, \alpha) \times 2$	Off and On	R
Hernandez et al. 2012 [20]	LW, SA, SD, RA, RD, ST, SIT	NP	NP	Load cell EMG	6×1 1×7	Off	R
Hernandez et al. 2013 [21]	LW, SA, ST	2	NP	Load cell EMG	6×1 1×7	Off and On	P
Huang et al. 2009 [22]	LW, SA, SD, OBS, Turns, ST	NA	SSS	EMG Pressure	1×11 2×1	Off	R
Huang et al. 2010 [23]	LW, SA, SD, OBS	1	SSS	EMG Pressure	1×11 NP	Off	P
Huang et al. 2011 [24]	LW, SA, SD, RA, RD, OBS	2	SSS	EMG Load cell Pressure	1×11 6×1 NP	Off	P

Table 2. *Cont.*

Article	Locomotion Activities	Critical Timing	Speed	Sensors	Axes × Sensors	Offline/Online	Recognition/Prediction
Kim et al. 2017 [25]	LW, SA, SD, RA, RD	NA	NP	Joint angle IMU Load cell Pressure	1×4 $3(\alpha) \times 5$ 1×4 4×1	Off	R
Liu et al. 2016 [26]	LW, SA, SD, RA, RD	2	SSS	EMG Load cell IMU Laser	1×8 6×1 $6(A, G, \alpha) \times 1$ 1×1	Off and On	P
Liu et al. 2017 [27]	LW, SA, SD, RA, RD	NP	NP	EMG Load cell	1×7 6×1	Off and On	P
Liu et al. 2017 [28]	LW, SA, SD, RA, RD	NA	SSS, SL, F	IMU Pressure	$4(A, G) \times 1$ 2×2	Off	R
Long et al. 2016 [29]	LW, SA, SD, RA, RD	5	SSS	IMU Pressure	$3(\alpha) \times 4$ 3×2	Off and On	P
Mai et al. 2011 [30]	LW, SA, SD	NA	SSS, F	Load cell	1×12	Off	R
Mai et al. 2018a [31]	LW, SA, SD, RA, RD, ST	NP	NP	IMU	$9(A, G, \alpha) \times 2$	Off and On	R
Mai et al. 2018b [32]	LW, SA, SD, RA, RD	NP	NP	IMU	$8(3A, 3G, 2\alpha) \times 2$	Off and On	R
Miller et al. 2013 [33]	LW, SA, SD, RA, RD	NA	SSS, (SL, F) for LW	EMG Pressure	1×4 2×2	Off	R
Moon et al. 2019 [34]	LW, SA, SD	NP	NP	Motor Encoder Spring length	1×1 1×1	Off and On	R
Pew et al. 2017 [35]	LW, Turns	NP	SSS	Load cell	6×1	Off	P
Shell et al. 2018 [36]	LW, cross-slope	NP	NP	IMU	$5(3A, 2G) \times 1$	Off	R
Simon et al. 2017 [37]	LW, SA, SD, RA, RD, ST	3	SSS, (SL, F) for LW	Joint Angle Joint Velocity Motor Current IMU Load cell	1×2 1×2 1×2 $6(A, G, \alpha) \times 1$ 6×1	Off	P
Spanias et al. 2014 [38]	LW, SA, SD, RA, RD	2	NP	Joint Angle Joint Velocity Motor Current IMU Load cell EMG	1×2 1×2 1×2 $6(A, G, \alpha) \times 1$ 1×1 1×9	Off	P and R

Table 2. *Cont.*

Article	Locomotion Activities	Critical Timing	Speed	Sensors	Axes × Sensors	Offline/Online	Recognition/Prediction
Spanias et al. 2015 [39]	LW, SA, SD, RA, RD	2	SSS, SL, F	Joint Angle Joint Velocity Motor Current IMU Load cell EMG	1×2 1×2 1×2 $8(3A, 3G, 2\alpha) \times 1$ 6×1 1×4	Off	P
Spanias et al. 2016a [40]	LW, SA, SD, RA, RD	2	NP	Joint Angle Joint Velocity Motor Current IMU Load cell EMG	1×2 1×2 1×2 $6(A, G, \alpha) \times 1$ 1×1 1×9	Off	P
Spanias et al. 2016b [8]	LW, SA, SD, RA, RD, ST	3	NP	Joint Angle Joint Velocity Motor Current IMU Load cell	1×2 1×2 1×2 $10(3A, 3G, 4\alpha) \times 1$ 6×1	Off and On	P and R
Spanias et al. 2017 [41]	LW, SA, SD, RA, RD, ST	3	NP	Joint Angle Joint Velocity Motor Current IMU Load cell	1×2 1×2 1×2 $10(3A, 3G, 4\alpha) \times 1$ 6×1	Off and On	P and R
Spanias et al. 2018 [42]	LW, SA, SD, RA, RD, ST	3	NP	Joint Angle Joint Velocity Motor Current IMU Load cell EMG	1×2 1×2 1×2 $10(3A, 3G, 4\alpha) \times 1$ 6×1 1×8	Off and On	P and R
Stolyarov et al. 2017 [43]	LW, SA, SD, RA, RD	NP	SSS	IMU	$6(A, G) \times 1$	Off	P
Su et al. 2019 [44]	LW, SA, SD, RA, RD	NP	SSS	IMU	$6(A, G) \times 3$	Off	R
Tkach et al. 2013 [45]	LW, SA, SD, RA, RD	NP	SSS	IMU Joint Angle Joint Velocity Joint Current EMG	$3(A) \times 1$ 1×1 1×1 1×1 1×4	Off	R

Table 2. *Cont.*

Article	Locomotion Activities	Critical Timing	Speed	Sensors	Axes × Sensors	Offline/Online	Recognition/Prediction
Wang et al. 2013 [46]	LW, SA, SD, ST, SIT	NA	NP	Pressure	4×1	Off	R
Wang et al. 2018 [47]	LW, SA, SD, ST, SIT	2	NP	Joint Angles	1×6	Off and On	P
Woodward et al. 2016 [48]	LW, SA, SD, RA, RD	NP	NP	IMU Joint Angle Joint Velocity Joint Current Load cell	$7(3A, 3G, 1\alpha) \times 1$ 1×2 1×2 1×2 6×1	Off	P
Xu et al. 2018 [49]	LW, SA, SD, RA, RD, ST	4	NP	IMU Load cell	$9(A, G, \alpha) \times 1$ 1×1	Off and On	P
Young et al. 2013a [50]	LW, SA, SD, RA, RD	2	NP	IMU Joint Angle Joint Velocity Joint Current Load cell EMG	$6(A, G) \times 1$ 1×2 1×2 1×2 1×1 1×9	Off	P
Young et al. 2013b [51]	LW, SA, SD, RA, RD	2	NP	IMU Joint Angle Joint Velocity Joint Current Load cell	$6(A, G) \times 1$ 1×2 1×2 1×2 1×1	Off	P
Young et al. 2013c [52]	LW, SA, SD, RA, RD	2	NP	IMU Joint Angle Joint Velocity Joint Current Load cell EMG	$6(A, G) \times 1$ 1×2 1×2 1×2 1×1 1×7	Off	P
Young et al. 2014a [53]	LW, SA, SD, RA, RD	2	SSS	IMU Joint Angle Joint Velocity Joint Current Load cell	$6(A, G) \times 1$ 1×2 1×2 1×2 1×1	Off	P
Young et al. 2014b [54]	LW, SA, SD, RA, RD	2	SSS, (SL, F) for LW	IMU Joint Angle Joint Velocity Joint Current Load cell EMG	$6(A, G) \times 1$ 1×2 1×2 1×2 1×1 1×9	Off	P

Table 2. *Cont.*

Article	Locomotion Activities	Critical Timing	Speed	Sensors	Axes × Sensors	Offline/Online	Recognition/Prediction
Young et al. 2016 [55]	LW, SA, SD, RA, RD	2	SSS	IMU Joint Angle Joint Velocity Joint Current Load cell	6(A, G) × 1 1 × 2 1 × 2 1 × 2 1 × 1	Off	P
Zhang et al. 2011 [56]	LW, SA, SD, RA, RD	2	NP	Load cell EMG	6 × 1 1 × 11	Off and On	P
Zhang et al. 2013 [57]	LW, SA, SD, RA, RD, ST, SIT	NP	NP	IMU Load cell EMG	6(A, G) × 2 6 × 1 1 × 8	Off and On	P
Zhang et al. 2019 [58]	LW, SA, SD, RA, RD	NP	NP	Depth Camera IMU	224 × 171 3(α) × 1	Off	P
Zhang et al. 2019 [59]	LW, SA, SD, RA, RD	NP	NP	Depth Camera IMU	224 × 171 3(α) × 1	Off	P
Zhang et al. 2012 [60]	LW, SA, SD, ST	2	NP	IMU Laser Load cell EMG	6(A, G) × 1 1 × 1 6 × 1 1 × 7	Off and On	P
Zheng et al. 2013 [61]	LW, SA, SD, RA, RD, OBS, ST	NA	SSS	Capacitive Pressure	1 × 7 3 × 1	Off	R
Zheng et al. 2014 [62]	LW, SA, SD, RA, RD, ST	NA	SSS	Capacitive Pressure	1 × 6 3 × 1	Off	R
Zheng et al. 2016 [63]	LW, SA, SD, RA, RD, ST	NP	SSS	IMU Load cell Joint angle Pressure Capacitive	8(3A, 3G, 2α) × 2 1 × 1 1 × 1 4 × 1 1 × 6	Off	P

Table 2. *Cont.*

Article	Locomotion Activities	Critical Timing	Speed	Sensors	Axes × Sensors	Offline/Online	Recognition/Prediction
Zheng et al. 2019 [64]	LW, SA, SD, RA, RD, ST	NP	NP	IMU Load cell Joint angle Pressure	$8(3A, 3G, 2\alpha) \times 2$ 1×1 1×1 4×1	Off	P
Zhou et al. 2019 [65]	LW, SA, SD	3	NP	IMU Load cell Joint angle	$9(A, G, \alpha) \times 2$ 1×2 1×1	Off and On	P

Locomotion Activities: LW = Level-ground Walking, SA = Stairs Ascent, SD = Stairs Descent, RA = Ramp Ascent, RD = Ramp Descent, ST = Standing, SIT = Sitting, SQ = Squatting, OBS = Obstacle clearance. Critical Timing: NP = Not Provided, NA = Not Applicable, 1 = 200 ms before the prosthesis foot off of the ground for all transitions, 2 = for transitions from level ground walking to any other locomotion mode, the critical timing was defined either at the prosthesis foot off of the ground or at mid-swing and for transitions from any locomotion mode to level ground walking, the critical timing was defined at prosthesis foot contact on level ground walking or at mid-stance, 3 = at foot off of the previous locomotion mode for all transitions,4 = For level ground walking to stairs ascent or stairs descent transitions, the critical timing was defined either at the prosthesis foot off of the ground or at prosthesis foot contact on the stairs. For any other transitions, the critical timing was defined either at the prosthesis foot contact on the new locomotion mode or at the first prosthesis foot off of the new locomotion mode,5 = the critical timing occurred at foot contact of the contralateral leg of the exoskeleton. Speed: NP = Not Provided, SSS = Self-Selected Speed, SL = Slow, F = Fast, (SL, F) for LW = Slower and faster paces tested only for level-ground walking. Sensors: EMG = ElectroMyoGraphs, IMU = Inertial Motion Unit, FMG = Force MyoGraph. Axes × Sensors: The number of measurement axes and the number of sensors is reported. For IMUs, the signals used are specified with A = Accelerometer, G = Gyroscope, α = Inclination. For instance, $9(A, G, \alpha) \times 2$ means that 2 IMUs were used and for each IMU the 3D accelerations, the 3D rotational speed and the 3D orientation were extracted. Offline/Online: Off = Offline, On = Online. Recognition/Prediction: R = Recognition, P = Prediction.

3.3.3. Identifying the Critical Timing

The Critical Timings used in each study are provided in Table 2.

Among the studies focusing on ankle-knee or ankle-foot prostheses (N = 50), most investigated the transitions between locomotion modes (N = 39). Several definitions of critical timing were used. We describe these definitions below:

Firstly, a study (N = 1) conducted by Huang et al. [23] in 2010 defined the critical timing as 200 ms before the prosthesis foot off of the ground for all transitions. Figure 2 illustrates the critical timing used in Huang et al. [23] for both level ground walking to stair ascent and stair descent to level ground walking transitions.

Secondly, some studies (N = 15) (in Huang et al. 2011 [24] for example) chose the critical timings at well-defined gait events (e.g., Foot-Off and Foot Contact): for transitions from level ground walking to any other locomotion mode, the critical timing was defined at the prosthesis foot off of the ground and for transitions from any locomotion mode to level ground walking, the critical timing was defined at prosthesis foot contact on level ground.

Thirdly, some studies (N = 5) (in Spanias et al. [42] for example) attempted to delay the critical timing in order to improve the locomotion mode prediction. Here, for transitions from level ground walking to any other locomotion mode, the critical timing was defined 90 ms after a gait event, such as the prosthesis foot off, mid-swing, prosthesis foot contact or mid-stance.

Finally, in a recent study (N = 1) conducted by Xu et al. [49] in 2018 defined the definitions of critical timings were altered based on the transition type and on the transitioning leg. As a result, the critical timing was delayed when the amputated leg was the leading leg for the transition. For level ground walking to stair ascent or stairs descent transitions, the critical timing was defined either at the last prosthesis foot off of the ground or at the first prosthesis foot contact on the stairs. For any other transitions, the critical timing was defined either at the first prosthesis foot contact on the new locomotion mode or at the first prosthesis foot off of the new locomotion mode.

The other studies did not investigate the transitions (N = 11) or did not report the critical timings used in the study (N = 17).

Among the studies focusing on orthoses or exoskeletons (N = 8), only three studies investigated the transitions between locomotion modes. In Long et al. [29], the critical timing occurred at foot contact of the contralateral leg of the exoskeleton. In Wang et al. [47], the critical timing occurred at foot contact of the ipsilateral leg in the new locomotion mode. Finally, in Zhou et al. [65], the critical timing occurred at mid-swing when the leg wearing the exoskeleton led the transition. It occurred at the last foot off of the ground for transitions from level ground to any other locomotion mode and finally for transitions from any locomotion mode to level ground walking it was at the first foot off of the ground. The remaining studies either did not investigate the transitions (N = 4) or did not report the critical timings used in the study (N = 1).

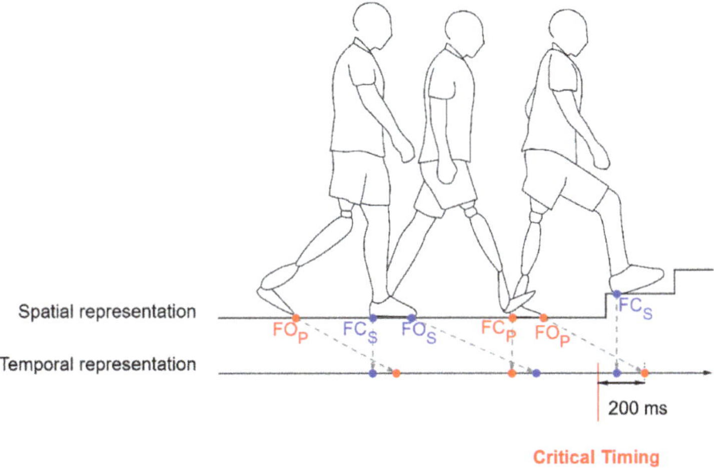

A—Level walking to stair ascent

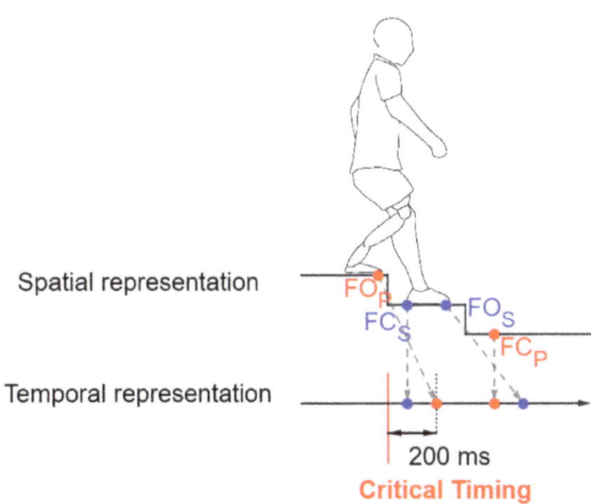

B—Stair descent to level walking

Figure 2. Example of the critical timings used in Huang et al. [23].

Panel A represents a patient with amputation in his transition from level walking to stair ascent. The superior part of this panel is a spatial representation of the patient motion. The line below is the temporal representation of the foot contact events. A dashed line maps the spatial representation to the temporal representation. For the spatial representation, the points refer to the spatial coordinates where the foot will hit/leave the ground. The temporal axis details the Foot Contact (FC) and Foot Off (FO) gait events for both sides. Critical timing is defined 200 ms prior to the prosthesis Foot Off event according to the Huang et al. Study [23]. The blue points are associated with the sound leg (Index S) and the red points are associated with the prosthesis side (index P). The panel B uses the same representation for patient from level walking to stair descent.

3.3.4. Online/Offline Implementation of Machine Learning Algorithm for Prediction of the Upcoming Locomotion Mode or Recognition of the Current Locomotion Mode

Information regarding the type of implementation of the Machine Learning algorithms is provided in Table 2: recognition and/or prediction algorithm and online and/or offline implementation.

The Machine Learning algorithms developed in the studies included in this systematic review were designed either to predict the upcoming locomotion mode (N = 30) or to recognize the current locomotion mode (N = 24). Some studies developed a Locomotion Mode Recognition system with adaptive strategies (N = 4). A forward predictor identified the upcoming locomotion mode while a backward estimator recognized the current locomotion mode. The backward estimator was used to label new data and the forward predictor could be updated with these newly labeled data.

Most of algorithms were trained and evaluated offline (N = 40) with a few which were trained offline and evaluated online (N = 18).

3.3.5. Data Type and Sensors Used

The details concerning the sensors used in the studies are provided in Table 2.
Sensors used in the included studies were of four types:

- Kinematic data were measured with sensors such as Inertial Motion Units (IMUs) (N = 36), or angle encoders (N = 21).
- Kinetic data such as interaction force between the device and the user were measured with load cell (N = 31). Ground reaction force was measured with foot insoles (N = 17) and torque at the joint was measured with motor current sensors (N = 14) or by measuring the length of a spring (N = 1).
- Physiological data were measured with sensors such as Electromyographs (EMG) (N = 21), Capacitive Sensing Systems (CSS) (N = 4) or Forcemyographs (FMG) (N = 1).
- Extrinsic data such as the distance between the user and an upcoming obstacle were measured with laser distance meters (N = 2) or with depth cameras (N = 2).

3.3.6. Analysis Windows

The details concerning the analysis windows used in each study are provided in Table 3.
Three types of analysis windows can be distinguished: sliding (N = 30), unique (N = 7) or multiple (N = 19) windows. The first method consisted of using a sliding analysis window by defining a window length and a window increment. The windows can therefore overlap. For the unique and multiple methods, the analysis window(s) was (were) defined either by a starting or ending point and a fixed window length (N = 21) or by both end points with a variable window length (N = 5). The remaining studies (N = 2) did not provide any information concerning analysis windows.

Table 3. Preprocessing techniques, Machine Learning algorithms and reported accuracies of the included studies. The details of the preprocessing techniques (windows and features) and the machine learning algorithms used in each study are reported along with the corresponding accuracy. If several configurations were tested, only the optimal configuration is reported.

Article	Analysis Windows			Sensors	Features	Algorithm	Accuracy
	Type	Number	Length				
Ai et al. 2017 [9]	Sliding	NA	250	EMG Mech	WT DTW	SVM	97.9
Beil et al. 2018 [10]	Sliding	NA	300	Mech	Raw data	HMM	92.8
Chen et al. 2013 [11]	Sliding	NA	150	Capacitive	Mean, Max, Min, RMS	LDA	94.54
Chen et al. 2014 [12]	Sliding	NA	160	Pressure Mech	Mean, Max, Min, SD, RMS, WL, CORR Mean, Max, Min, SD, RMS, WL, ZC, CORR	LR	98.2
Chen et al. 2015 [13]	Multiple	4	200	Pressure	SD, AR	LDA	98.4
Du et al. 2012 [14]	Sliding	NA	150	EMG Mech	MAV, SSC, WL, ZC Mean, Max, Min	LDA	98
Du et al. 2013 [15]	Sliding	NA	160	EMG Mech	MAV, SSC, WL, ZC Mean, Max, Min	EBA	92.5
Feng et al. 2019 [16]	Unique	1	Gait Cycle	Mech	Raw Data	CNN	92.1
Godiyal et al. 2018 [17]	Unique	1	Stance	FMG	Mean, Max, Min, SD, RMS, WL, SSC, MAD	LDA	96.1
Gong et al. 2018 [18]	Sliding	NA	250	Mech	Mean, Max, Min, SD, MAD	ANN	97.8
Gong et al. 2020 [19]	Sliding	NA	250	Mech	Mean, Max, Min, SD, MAD	ANN	98.4
Hernandez et al. 2012 [20]	Sliding	NA	150	EMG Mech	MAV, SSC, WL, ZC Mean, Max, Min	SVM	NP
Hernandez et al. 2013 [21]	Sliding	NA	160	EMG Mech	MAV, SSC, WL, ZC Mean, Max, Min	SVM	99.9
Huang et al. 2009 [22]	Sliding	NA	140	EMG	MAV, SSC, WL, ZC	LDA	95.5
Huang et al. 2010 [23]	Multiple	3	100	EMG	MAV, SSC, WL, ZC	LDA	NR
Huang et al. 2011 [24]	Sliding	NA	150	EMG Mech	MAV, SSC, WL, ZC Mean, Max, Min	SVM	100
Kim et al. 2017 [25]	Unique	1	FC contro to FC ipsi	Mech	Custom values	DT	99.1
Liu et al. 2016 [26]	Sliding	NA	50	EMG Mech	MAV, SSC, WL, ZC Mean, Max, Min, SD	LDA	~98

Table 3. *Cont.*

Article	Analysis Windows			Sensors	Features	Algorithm	Accuracy
	Type	Number	Length				
Liu et al. 2017 [27]	Sliding	NA	160	EMG Mech	MAV, SSC, WL, ZC Mean, Max, Min, SD	EBA/LIFT	94.3
Liu et al. 2017 [28]	Unique	1	800	Mech	ICC	HMM	95.8
Long et al. 2016 [29]	NP	NP	NP	Mech	WT	SVM	98.4
Mai et al. 2011 [30]	Unique	1	Stance	Mech	Mean, Force Changing Rate, Force Ratio	ANN	98.5
Mai et al. 2018a [31]	Sliding	NA	100	Mech	Mean, Max, Min, SD, Diff	SVM	NP
Mai et al. 2018b [32]	Sliding	NA	100 pts	Mech	Mean, Max, Min, SD, Diff	SVM	99.4
Miller et al. 2013 [33]	Multiple	3	200/300/100	EMG	MAV, SSC, WL, ZC, SD	SVM	98.5
Moon et al. 2019 [34]	Sliding	NA	NP	Mech	Raw data	ANN	NP
Pew et al. 2017 [35]	NP	NP	NP	Mech	NP	KNN	93.8
Shell et al. 2018 [36]	Sliding	NA	150	Mech	Mean, SD, Max, Min	LDA	78
Simon et al. 2017 [37]	Multiple	2	300	Mech	WT	DBN	99.6
Spanias et al. 2014 [38]	Multiple	2	300	EMG Mech	MAV, SSC, WL, ZC, AR Mean, Max, Min, SD, IV, FV	LDA	~ 96
Spanias et al. 2015 [39]	Multiple	8	300	EMG Mech	MAV, SSC, WL, ZC, AR Mean, Max, Min, SD, IV, FV	DBN	~ 99
Spanias et al. 2016a [40]	Multiple	2	300	EMG Mech	MAV, SSC, WL, ZC, AR Mean, Max, Min, SD	DBN	NR
Spanias et al. 2016b [8]	Multiple	8	300	Mech	Mean, Max, Min, SD, IV, FV	DBN	96.7
Spanias et al. 2017 [41]	Multiple	8	300	Mech	Mean, Max, Min, SD, IV, FV	DBN	98.8
Spanias et al. 2018 [42]	Multiple	4	300	EMG Mech	MAV, SSC, WL, ZC, AR Mean, Max, Min, SD, IV, FV	DBN	95.97
Stolyarov et al. 2017 [43]	Unique	1	FF to FO	Mech	Mean, Max, Min, SD	LDA	94.1
Su et al. 2019 [44]	Unique	1	490	Mech	Raw Data	CNN	89.2
Tkach et al. 2013 [45]	Multiple	3	250	EMG Mech	MAV, SSC, WL, ZC Mean, SD	LDA	96
Wang et al. 2013 [46]	Multiple	4	200	Mech	Range, AR, CORR	LDA	99.01
Wang et al. 2018 [47]	Sliding	NA	50 pts	Mech	Raw Data	LSTM	95

Table 3. *Cont.*

Article	Analysis Windows			Sensors	Features	Algorithm	Accuracy
	Type	Number	Length				
Woodward et al. 2016 [48]	Multiple	2	300	Mech	Mean, Max, Min, SD, IV, FV	ANN	98.9
Xu et al. 2018 [49]	Sliding	NA	250	Mech	Mean, Max, Min, SD, Diff	QDA	93.2
Young et al. 2013a [50]	Multiple	8	300	EMG Mech	MAV, SSC, WL, ZC, AR Mean, Max, Min, SD	DBN	~98.2
Young et al. 2013b [51]	Multiple	8	300	Mech	Mean, Max, Min, SD	DBN	~98
Young et al. 2013c [52]	Multiple	2	300	EMG Mech	MAV, SSC, WL, ZC, AR Mean, Max, Min, SD	LDA	86.4
Young et al. 2014a [53]	Multiple	2	250	Mech	Mean, Max, Min, SD	LDA	~99
Young et al. 2014b [54]	Multiple	8	300	EMG Mech	MAV, SSC, WL, ZC, AR Mean, Max, Min, SD	DBN	~99
Young et al. 2016 [55]	Multiple	8	300	Mech	Mean, Max, Min, SD, IV, FV	DBN	~99
Zhang et al. 2011 [56]	Sliding	NA	150	EMG Mech	MAV, SSC, WL, ZC Mean, Max, Min	LDA	>97
Zhang et al. 2013 [57]	Sliding	NA	150	EMG Mech	MAV, SSC, WL, ZC Mean, Max, Min	SVM	95
Zhang et al. 2019 [58]	Sliding	NA	600	Depth Camera	Raw data	CNN + HMM	96.4
Zhang et al. 2019 [59]	Sliding	NA	733	Depth Camera	Raw data	CNN	94.9
Zhang et al. 2012 [60]	Sliding	NA	160	EMG Mech	MAV, SSC, WL, ZC Mean, Max, Min	LDA	97.6
Zheng et al. 2013 [61]	Sliding	NA	250	Capacitive	Mean, Max, Min, SD, sum(abs(diff(X))), mean(diff(X)), sum(abs(X)), Std(abs(diff(X))), CORR	QDA	95
Zheng et al. 2014 [62]	Sliding	NA	250	Capacitive	Mean, Max, Min, SD, sum(abs(diff(X))), mean(diff(X)), sum(abs(X)), Std(abs(diff(X))), CORR	QDA	95.1

Table 3. *Cont.*

Article	Analysis Windows			Sensors	Features	Algorithm	Accuracy
	Type	Number	Length				
Zheng et al. 2016 [63]	Sliding	NA	250	Capacitive Mech	Mean, Max, Min, SD, sum(abs(diff(X))), sum(abs(X)) Mean, Max, Min, SD	SVM	95.8
Zheng et al. 2019 [64]	Sliding	NA	250	Mech	Mean, Max, SD	SVM	92.7
Zhou et al. 2019 [65]	Sliding	NA	150	Mech	Mean, Max, Min, SD, RMS	SVM	>90

Analysis Windows: Type: Three types of analysis windows are used: sliding windows, multiple windows, unique window. **Number:** When using multiple windows, the number of windows is reported. NA = Not Applicable (for unique and sliding windows). NP = Not Provided. **Length:** The window length is reported in ms. When the window length is variable, both the beginning and the end of the window(s) are reported: Gait Cycle = the data of the complete gait cycle are extracted, Stance = The data of the stance phase of the tested side are extracted, FF to FO = the data from Foot Flat to Foot Off (of the tested side) are extracted, FC contro to FC ipsi = the data from the Foot Contact of the contralateral side to the Foot Contact of the ipsilateral side are extracted. If the acquisition frequency was not reported, the window length is reported in terms of point numbers. NP = Not Provided. **Sensors:** EMG = ElectroMyoGraphs, FMG = ForceMyoGraph, Mech = Mechanical sensors (e.g., IMU, joint angle, joint rotational speed, data from load cell, etc.), for more details refer to Table 3. **Features:** NP = Not Provided, WT = Wavelet Transform, DTW = Dynamic Time Wrapping, Max = Maximum value, Min = Minimum Value, IV = Initial Value, FV = Final Value, RMS = Root Mean Square, SD = Standard Deviation, WL = Waveform Length, ZC = Zero Crossings, SSC = Slope Sign Change, MAD = Mean Absolute Deviation, Diff = Differential Values, AR = autocorrelation coefficients of an autoregressive model (the number of coefficients and the order of the model are not reported), CORR = Correlation between signals, ICC = Intraclass Correlation Coefficients. **Algorithm:** (by order of appearance) SVM = Support Vector Machine, HMM = Hidden Markov Model, LDA = Linear Discriminant Analysis, LR = Logistic Regression, EBA = Entropy Based Algorithm, CNN = Convolutional Neural Network, ANN = Artificial Neural Network, DT = Decision Tree, LIFT = Learning From Testing Data, KNN = K-Nearest Neighbor, DBN = Dynamic Bayesian Network, LSTM = Long-Short Term Memory network, QDA = Quadratic Discriminant Analysis. **Accuracy:** NP = Not Provided, NR = Not Reported (in Huang et al. [23] and in Spanias et al. [40], the influence of simulated noise on the EMG was tested, the reported accuracies were lower and were not comparable to other studies), A '~' sign means that results were obtained from reading graphs un the paper.

3.3.7. Features

The detailed features and domains used in each study can be found in Table 3.

Two main domains of features have been investigated in the included studies: time-domain (e.g., mean, minimum, maximum, standard deviation, etc.) (N = 48) and time-frequential domain features (e.g., coefficients of the wavelet transform) (N = 1). One study compared the performances of machine learning algorithm using either time-domain features or time-frequency domain features [9].

The remaining studies did not provide any information concerning the features used (N = 1) or used the temporal data measured by the sensors and did not extract any features (N = 7).

3.3.8. Machine Learning Algorithms and Their Accuracies

Details on the machine learning algorithms used in studies and their reported accuracies are presented in Table 3.

Most of the studies used the classical pattern recognition algorithms (Bishop 2006 [66]) which are available. Three algorithms were implemented more often than others: Linear Discriminant Analysis (LDA) (N = 29), Support Vector Machines (SVM) (N = 17) and Dynamic Bayesian Network (DBN) (N = 10). Other algorithms were investigated a few times. Quadratic Discriminant Analysis (QDA) (N = 8) was used either with data from Capacitive Sensing Systems (N = 4) or from Inertial Motion Units (N = 4). Small Artificial Neural Networks (ANN) with 1 or 2 hidden layers were used to recognize the current locomotion mode (N = 6) or to predict the upcoming locomotion mode (N = 1). Convolutional Neural Networks (CNN) (N = 4) have started to be applied more recently for locomotion mode classification (since 2019). CNNs were essentially used to avoid feature selection: all studies using CNNs did not extract any feature and instead fed raw sensor data into the algorithm. Other algorithms were used only once (K-Nearest Neighbors—KNN and Long Short-Term Memory neural networks - LSTM) or twice (Decision Tree—DT and Hidden Markov Model—HMM).

Some less typical adaptive algorithms were also sometimes used. Learning From Testing data (LIFT) and Entropy Based Algorithm (EBA) were each used twice and Transductive SVM was used once [15,27].

4. Discussion

This systematic review included 58 articles implementing Machine Learning classifiers designed to identify the locomotion mode of assistive device user. Such algorithms were generally implemented as high-level controllers able to automatically adapt the behavior of lower limb prostheses, exoskeletons, or orthoses. We used the PubMed and Web of Science core collection databases for finding our references. This was done because most medical related literature (including biomedical engineering) can be found in these two databases. In addition, we performed an extensive search through the references of the papers from the aforementioned databases. As we were focusing on medical literature, we did not include Scopus as one of the databases for this review. This may have led to a very small number of papers that have not been included in this review.

Accuracy and the robustness (e.g., stable performance in the face of long-term use) of the algorithm were the variables most often used to report the results from studies investigating locomotion on different terrains. The influence of (1) sensors, (2) analysis windows and features, (3) machine learning algorithms on the accuracy and on the robustness of the locomotion mode classifiers are discussed below. It should be noted that the accuracies reported in this review are those which were supplied in each paper. Since each study was conducted with different circumstances such as number of subjects and conditions tested, accuracies can be compared within each study but cannot be compared between studies with precision.

4.1. Influence of Sensor Choice

Several sensors have been used to build locomotion mode classifiers. The choices of these sensors may influence the accuracy and the robustness of the classifiers. More details are provided in the sections below.

4.1.1. Algorithm Accuracy

Among the included studies the three most used sensors were Inertial Motion Units (IMU) (N = 36, see Table 2), load cells (N = 31, Table 2) and electromyographs (EMG) (N = 21, Table 2).

Firstly, IMUs measure the acceleration and the rotational speed along three orthogonal axes. For example, Stolyarov et al. [43] classified level-ground walking (LW), stair ascent (SA), stair descent (SD), ramp ascent (RA) and ramp descent (RD) with LDA. They showed that including trajectory information of the prosthesis increased the averaged accuracy compared to using only the accelerations and rotational speeds (from 80.9% to 94.1%). They suggested using filtering techniques to reduce drift (e.g., Kalman filters, particle filters, etc.). These researchers also brought up the point that the performance of the classification algorithms might be reduced when applied to gait at slow walking speed. Other researchers demonstrating the capacity of IMUs for the detection of locomotion mode were Zhou et al. [65]. They were able with the SVM to classify three locomotion modes (LW, SA, SD) with the exclusive use of IMU data. They achieved above 90% accuracy using orientation information. The signals combining acceleration, rotational speed and orientation were directly extracted from the IMUs (MPU 9250, Ivensense®—the filter technique was not reported in the data sheet of the sensor).

However, these studies suggested that the algorithm performances could increase when fusing IMUs signals with other sensors signals. Thus, in most studies using IMUs, information from this sensor was fused with measurements from other sensors (see below).

Secondly, load cells measured the interaction force between the device and the user. For example, Huang et al. [24] classified five locomotion modes (LW, SA, SD, RA, RD) with LDA and SVM by using only a 6 degrees of freedom (DOF) load cell mounted on the prosthetic pylon of an above-knee prosthesis. The phase-dependent strategy achieved 85 to 95% accuracy during stance phase (Initial Double Limb Stance (DS1), Single Limb Stance (SS) and Terminal Double Limb Stance (DS2)) but the accuracies dropped to 50–60% during swing (SW) phase for both LDA and SVM classifiers. Similar drops in accuracy were reported when using only plantar pressure measurements [13,46]. According to the authors [24], the low classification accuracies in the swing phase were almost certainly due to low forces/moments generated during swing phase.

Thirdly, EMG signals measured from the residual limb were reported to contain useful information for locomotion mode predictions in early studies. Indeed, for example, Huang et al. [24] and Miller et al. [33] achieved classification of five locomotion modes (LW, SA, SD, RA, RD) using EMG signals measured in the residual limb of patients with transfemoral and transtibial unilateral amputation respectively. LDA and SVM classifiers were used in both studies. For volunteers with transfemoral amputation [24], the SVM achieved an accuracy of above 90% for all phases. The LDA algorithm achieved similar accuracies in the stance phase but a slightly lower accuracy of 85% in the swing phase. For volunteers with transtibial amputation [33], both LDA and SVM algorithms achieved around 98% accuracy. Many researchers have pointed out that the EMG signals suffer from disturbances especially because of shifts in electrode position when donning and doffing a prosthesis for example. Miller et al. [33] reported a mean loss in accuracy of 15.8% and 23.1% for LDA and SVM classifiers when the medial gastrocnemius electrode was shifted. Both studies concluded that EMG signals could be helpful for classifying locomotion modes as long as the signals are not disturbed. Several studies have provided suggestions for reducing these problems. They are discussed in the 'Algorithm robustness' Section 4.1.2 below.

Finally, sensor fusion has been proven to significantly increase accuracies of locomotion mode classifiers [24,54]. For example, Huang et al. [24] observed an increase in accuracy by combining EMG and load cell data instead of using either only EMG data or only load cell data (accuracy increase of up to 5.9% for an SVM classifier). Since then, data from different sensors have been fused together to reach higher accuracies. In another example, Young et al. [54] used 13 mechanical sensors (IMU, load cell, position, velocity and torque at knee and ankle joints) and recorded EMG signals from 9 muscles of the residual limb of volunteers with a transfemoral amputation. A DBN algorithm predicting upcoming locomotion modes reached 99% accuracy for steady-state steps and 88% accuracy for transitional steps.

4.1.2. Algorithm Robustness

Sensors measurement noise over time can affect the performances of locomotion mode classifiers. To achieve reliable behavior of locomotion assistive device for long-term use, the influence of such noise should be considered. Techniques implemented to take into account sensors noise are discussed here.

EMG signals were mostly reported to be disturbed by environmental noise, electrode conductivity changes, shifts in electrode position or even loss of electrode contact [67,68]. Three techniques have been used to cope with such disturbances. The first one aims at training ML algorithm with several electrode displacement configurations [33]. The second one consists of building a sensor fault detection system so that disturbed EMG channel are removed if detected as noisy [23,40]. The third one uses an adaptive framework so that ML algorithm can be updated when EMG signals are disturbed [42]. The latter adaptive algorithm also included a sensor fault detection system. Alternatively, according to some researchers [62,63], capacitive sensing systems, measuring the gap change between the residual limb and the prosthetic socket [63], could eventually replace EMG signals since such sensors appear to be robust to donning and doffing an ankle-knee prosthesis and to load bearing changes [62].

4.2. Influence of Analysis Windows

In this section, we will discuss the influence of the analysis window configuration on the accuracy of locomotion mode prediction.

Among the included studies, sliding (N = 30, Table 3) and multiple (N = 19, Table 3) analysis windows were the preferred configurations. While the implementation of sliding windows requires the building of one classifier per gait phase, the implementation of multiple windows is performed by building one classifier per analysis window [50–55]. The number of classifiers depends on the number of gait phases for sliding windows and depends on the number of windows for multiple windows. In the case of sliding windows, Chen et al. [12] observed that the number of phases for phase-dependent classification significantly influences algorithm accuracy. As a result, using four gait phases (DS1, SS, DS2, SW) increased the accuracy of both LDA and QDA compared to when using only two phases (Stance, Swing). As the sliding window method generally involves a longer portion of the gait phase in question, the data to be classified are generally more variable.

Several studies reported that the length of analysis windows had a significant impact on algorithm performances for multiple window [53,54] and sliding window [22] configurations. Young et al. [53,54], using multiple windows, showed that there was an optimal window length (between 200 and 300 ms) for classification accuracies using mechanical and EMG data for both steady state and transitional data. The same was found in a study using sliding windows, where the length of the window but not its increments were found to affect algorithm performances [22]. For online implementation however smaller window increment ensures a faster response time since locomotion mode classification is performed more often.

More recently, some researchers did not use analysis windows which ended at classic gait events like foot contact but instead allowed for a delay in the termination of the analysis window. For example, Simon et al. [37,69] had an analysis window which ended 90ms after foot contact or foot off. This delay increased the accuracy of a DBN algorithm and did not affect the stability of the users of a powered above-knee prosthesis.

4.3. Influence of Features

The features set used in each study was highly dependent on the sensors used.

For EMG signals, two types of features were tested: (1) time-domain features and (2) time-frequential domain features. The most commonly used time-domain features were mean absolute value, waveform length, number of zero crossings, number of slope sign changes (N = 21) and the coefficients of autoregressive models (N = 8). For time-frequential domain features, the coefficients of the wavelet transform of EMG signals were used once [9]. Ai et al. [9] compared LDA and SVM performances when using time-domain features or time-frequency domain features. Both algorithms reached higher accuracies with time domain features for one volunteer with below-knee amputation, e.g., in the case of the SVM 91.9% with time-domain features vs. 82.3% with time-frequency features. Additionally, time-domain features were easier and faster to compute [9].

A large number of studies (N = 48, Table 3) used mechanical sensors (IMU, load cells, encoders, pressure insoles, etc.). The most representative feature (N = 34, Table 3) set was a combination of the following time-domain features: mean, maximum, minimum and standard deviation. Initial and final values were also sometimes added to the feature set (N = 8, Table 3).

Finally, several feature reduction techniques were sometimes used to find the minimal feature set necessary for successful classification and to avoid overfitting (N = 14): Wrapper techniques such as Sequential Forward Selection (SFS) and Selection Backward Selection (SBS) were used to pick the features having the highest impact on the classification accuracy [39] (N = 8). Such methods are time consuming [18]. Zhang et al. [57] compared the processing time taken by two wrapper methods and a filter method. The filter method was found to be faster compared to wrapper methods (84 s for the filter method vs. 1978 s for SBS).

4.4. Influence of Machine Learning Algorithm

4.4.1. On Accuracy

A variety of ML algorithms were used in the included studies. The most frequently used algorithms were LDA (N = 29, Table 3), SVM (N = 19, Table 3) and DBN (N = 10, Table 3). Also, CNNs were used to avoid features selection (N = 4, Table 3).

LDA is easy to implement since no hyperparameters need to be tuned [48,70]. This algorithm is fast (1.29 ms [48], 0.078 ms with parallelization [32]) and not prone to overfitting [9]. For these reasons, this algorithm is often used as a baseline for performance comparisons between several algorithms [32,42]. More importantly, in some studies, LDA obtained accuracies similar to neural networks [48] and to SVM [33].

Even though, hyperparameters such as kernel parameter and the penalty factor need to be tuned for SVM [16], optimization techniques (e.g., grid search [9], particle swarm optimization [29]) have been found in some studies to reach slightly better performances than LDA [9,24] or QDA [62].

One of the first researchers to use DBNs were Young et al. in 2013 [50,51]. By adding past information to those of the current state, the DBN was able to obtain higher classification accuracies than LDA [54] (88% vs. 85% for transitional accuracies for DBN and LDA respectively). The DBN, unlike LDA with uniform priors, take transitional probabilities into account (e.g., in stair ascent mode, the next mode is more likely to be stair ascent or level ground walking).

Finally, CNNs were recently used in a few studies [16,44,58,59]. For example, Zhang et al. [58,59] used depth-images with a depth-camera coupled with an IMU mounted on the prosthetic pylon of an above-knee prosthesis. CNNs, known to perform well when handling image datasets are often used to avoid manual feature selection. CNNs were also used in the case of non-image data, e.g., IMU data [44] or load cell data [16]. All four studies using CNNs reported an accuracy above 89% but none of those studies implemented the designed CNN online.

The most common mistake was misclassification between ramp ascent and level ground walking modes [50]. Grouping ramp ascent and level walking classes were reported to improve the performances of locomotion mode classifiers [50]. Such a technique is relevant when the control laws (impedance in [43,50]) are similar for both modes. Zhang et al. [59] evaluated the influence of such errors (misclassifications between level walking and incline walking) on the stability of the user of an above-knee prosthesis using angular momentum and a subjective questionnaire. It was observed that the effect of the errors depends on the type of error, the error duration, and the gait phase where the error occurred. Errors were considered critical if the stability of prosthesis users was disturbed. This appears to be a good criterion for evaluating the importance of errors when designing a locomotion mode classifier.

4.4.2. On Robustness

Very few studies have evaluated the performances of locomotion mode classifiers for long term use. Adaptive frameworks have been proposed to deal with EMG disturbances [42] or to achieve stable performances for long term use [27]. For example, Spanias et al. [42], designed a forward predictor and a backward estimator. The forward predictor is an ML algorithm designed to predict the upcoming locomotion mode of an assistive device user. The backward estimator is an ML algorithm designed to recognize the current locomotion mode. The latter algorithm was used to label new data. Then, the newly labelled data were incorporated into the training set and then used to update the forward predictor parameters. Spanias et al. [42] used this framework to deal with EMG disturbances. The adaptive algorithm learned to reincorporate disturbed EMG channels over time. The adaptive algorithm was reported to perform significantly better than a non-adaptive algorithm. In another example, Liu et al. [27] evaluated the performance of adaptive algorithms compared to a non-adaptive algorithm across multiple session within a single experimental day. After donning and doffing the prosthesis, the adaptive algorithms were reported to update classifiers boundaries and to recover initial accuracy whereas the performances of the non-adaptive algorithms gradually decreased. To sum up, adaptive frameworks seem to be a promising solution to achieve long-term locomotion mode classification.

4.5. Propositions for Future Work

This systematic review included 58 articles published between 1 January 2000 and 31 July 2020. All 58 articles implemented ML-based locomotion mode classifiers designed for users of lower limb assistive devices. As can be seen from Table 3, classification accuracies under the tested conditions were almost always very high, hence indicating good progress in the attempts to construct more intelligent prosthetic devices. Nevertheless, there is always room for improvement. We try here to propose some recommendations concerning the research reports in the field as well as suggestions for moving forward with the implementation of these devices in the daily lives of the patients.

4.5.1. Homogenization of Reports

We will start first with the question of terms that are used in the field. This is not a trivial matter as the homogenization of terms would increase the understanding between researchers and hence speed up progress. There is much confusion around the use of terms recognition and prediction. The two terms are used in an interchangeable manner across studies but do not refer to the same goal. We propose that classifying the locomotion mode before the critical timing can be considered as a prediction task while a classification made after the critical timing can be considered as a recognition task. For recall, the critical timing is the latest moment when the behavior of the locomotion assistive device can be adapted to the new locomotion mode without disturbing the user. A more discriminating use of the two terms, recognition and prediction, would ease the comprehension of the studies.

The report of accuracies also suffers from a similar lack of precision. While many reports have distinguished between accuracies during steady state and the transitional step, several have not. Adding together the success obtained in steady state with that which is obtained in transitional steps is misleading, as the errors made in the latter tend to be higher. We therefore propose that there should be a systematic distinction of accuracies for these two modes.

4.5.2. Recommendations for Generalization to Daily Life Conditions

The review shows that significant progress has been made in the efforts to ease the use of prosthetic devices across multiple terrains. Nevertheless, some obvious steps are necessary to move ahead with ensuring the comfortable use of these devices in the daily lives of the patients.

An obvious thing to add on the list would be the inclusion of more daily life conditions for testing the devices. Examples of these would be different angles of approaches towards stairs or slope [9,28], different staircases [18] or load bearing changes [51]. A good extension for many of the studies included in this review would be a test of the algorithms outside the laboratory. Only a very small number of studies managed to take this step. For example, the work of Zhang et al. [58,59] evaluated a CNN classifier with data acquired both indoors and outdoors. Such studies are to be encouraged.

Another important condition to be included, to make the prosthetic devices more usable in daily life, would be the integration of multiple speeds in the study. Once again, very few researchers have investigated this condition. One researcher who has taken a step in this direction is Liu et al., 2017 [28].

A third variation which is not often taken into consideration is the transitioning leg which is used when entering a new terrain. While subjects tend to use one leg more than the other when crossing into new conditions, the side used is not always identical and subjects can change the transitioning leg. A handful of studies such as one by Zhou et al. [65] have taken this into account. They reported better accuracies when the locomotion mode classifier was trained with data from both transitioning legs. This may be a simple condition to include in more of the future studies in the field.

We turn here, from a discussion of conditions to be tested, to comments on how to decrease the burden of developing an algorithm which is tuned to each patient. The process of gathering data for the purpose of training the ML algorithm for each patient can be long and burdensome. A few researchers have provided recommendations on how to reduce the difficulty of this step. For example, Zhang et al. [60] proposed an automatic training method through environmental sensing. A radar distance meter coupled with an IMU helped to sense the environment and to automatically label the acquired data. Automatic labelling could also be achieved with depth cameras [58,59]. Another step in this direction has been the use of subject independent models which could potentially reduce the amount of training data needed. Efforts of this type have been made by Young et al. [52] and Spanias [8]. It seems that the addition of this step to future investigations of predictive or recognition algorithms would provide the additional bonus of reduced training time for the patient.

Supplementary Materials: The following are available online at http://www.mdpi.com/1424-8220/20/21/6345/s1, S1: Detailed Search Strategy, S2: Quality Assessment of the studies, Table S2: Detailed quality scores of the included studies, Table S3: EMG used in the studies, S4: References of the supplementary material.

Author Contributions: Conceptualization, D.L. and E.T.; methodology, D.L., F.L., M.G.; validation, D.L., E.T., F.L. and P.O.; investigation, D.L., F.L., L.C.; resources, F.L. and L.C.; data curation, F.L., V.O.; writing—original draft preparation, F.L.; writing—review and editing, D.L., E.T., L.C., V.O., P.O., M.G.; visualization, F.L.; supervision, F.L. and E.T.; project administration, D.L.; funding acquisition, D.L., E.T. and L.C. All authors have read and agreed to the published version of the manuscript. Authorship must be limited to those who have contributed substantially to the work reported.

Funding: This research was funded by Conseil Régional de Bourgogne Franche Comté and supported by university Hospital of Dijon. F.L. is supported by a PhD grant from Conseil Régional de Bourgogne Franche Comté.

Acknowledgments: Authors want to thank all staff of the university of Burgundy and University Hospital of Dijon for their help for funding and during the manuscript preparation. Authors want to thank Lucile Antonini for her help for the Figure 2.

Conflicts of Interest: The authors declare no conflict of interest.

References

1. Goujon-Pillet, H.; Drevelle, X.; Bonnet, X.; Villa, C.; Martinet, N.; Sauret, C.; Bascou, J.; Loiret, I.; Djian, F.; Rapin, N.; et al. APSIC: Training and fitting amputees during situations of daily living. *IRBM* **2014**, *35*, 60–65. [CrossRef]

2. Johansson, J.L.; Sherrill, D.M.; Riley, P.O.; Bonato, P.; Herr, H. A Clinical Comparison of Variable-Damping and Mechanically Passive Prosthetic Knee Devices. *Am. J. Phys. Med. Rehabil.* **2005**, *84*, 563–575. [CrossRef]

3. Au, S.K.; Weber, J.A.; Herr, H.M. Powered Ankle—Foot Prosthesis Improves Walking Metabolic Economy. *IEEE Trans. Robot.* **2009**, *25*, 51–66. [CrossRef]

4. Tucker, M.R.; Olivier, L.; Pagel, A.; Bleuler, H.; Bouri, M.; Lambercy, O.; Millán, J.D.R.; Riener, R.; Vallery, H.; Gassert, R. Control strategies for active lower extremity prosthetics and orthotics: A review. *J. Neuroeng. Rehabil.* **2015**, *12*, 1. [CrossRef]

5. Sup, F.; Varol, H.A.; Mitchell, J.; Withrow, T.J.; Goldfarb, M. Preliminary Evaluations of a Self-Contained Anthropomorphic Transfemoral Prosthesis. *IEEE/ASME Trans. Mechatron.* **2009**, *14*, 667–676. [CrossRef] [PubMed]

6. Moher, D.; Liberati, A.; Tetzlaff, J.; Altman, D.G. Preferred Reporting Items for Systematic Reviews and Meta-Analyses: The PRISMA Statement. *PLoS Med.* **2009**, *6*, 7. [CrossRef] [PubMed]

7. Kmet, L.M.; Lee, R.C.; Cook, L.S.; Alberta Heritage Foundation for Medical Research. *Standard Quality Assessment Criteria for Evaluating Primary Research Papers from a Variety of Fields*; Alberta Heritage Foundation for Medical Research: Edmonton, AB, Canada, 2004.

8. Spanias, J.; Simon, A.M.; Perreault, E.J.; Hargrove, L.J. Preliminary results for an adaptive pattern recognition system for novel users using a powered lower limb prosthesis. In Proceedings of the 2016 38th Annual International Conference of the IEEE Engineering in Medicine and Biology Society (EMBC), Orlando, FL, USA, 17–20 August 2016; Volume 2016, pp. 5083–5086.

9. Ai, Q.; Zhang, Y.; Qi, W.; Liu, Q.; Chen, K. Research on Lower Limb Motion Recognition Based on Fusion of sEMG and Accelerometer Signals. *Symmetry* **2017**, *9*, 147. [CrossRef]

10. Beil, J.; Ehrenberger, I.; Scherer, C.; Mandery, C.; Asfour, T. Human Motion Classification Based on Multi-Modal Sensor Data for Lower Limb Exoskeletons. In Proceedings of the 2018 IEEE/RSJ International Conference on Intelligent Robots and Systems (IROS), Madrid, Spain, 1–5 October 2018; pp. 5431–5436.

11. Chen, B.; Zheng, E.; Fan, X.; Liang, T.; Wang, Q.; Wei, K.; Wang, L. Locomotion Mode Classification Using a Wearable Capacitive Sensing System. *IEEE Trans. Neural Syst. Rehabil. Eng.* **2013**, *21*, 744–755. [CrossRef] [PubMed]

12. Chen, B.; Zheng, E.; Wang, Q.; Wang, L. A new strategy for parameter optimization to improve phase-dependent locomotion mode recognition. *Neurocomputing* **2015**, *149*, 585–593. [CrossRef]

13. Chen, B.; Wang, X.; Huang, Y.; Wei, K.; Wang, Q. A foot-wearable interface for locomotion mode recognition based on discrete contact force distribution. *Mechatronics* **2015**, *32*, 12–21. [CrossRef]

14. Du, L.; Zhang, F.; Liu, M.; Huang, H. Toward Design of an Environment-Aware Adaptive Locomotion-Mode-Recognition System. *IEEE Trans. Biomed. Eng.* **2012**, *59*, 2716–2725. [CrossRef]

15. Du, L.; Zhang, F.; He, H.; Huang, H. Improving the performance of a neural-machine interface for prosthetic legs using adaptive pattern classifiers. In Proceedings of the 2013 35th Annual International Conference of the IEEE Engineering in Medicine and Biology Society (EMBC), Osaka, Japan, 3–7 July 2013; Volume 2013, pp. 1571–1574.

16. Feng, Y.; Chen, W.; Wang, Q. A strain gauge based locomotion mode recognition method using convolutional neural network. *Adv. Robot.* **2019**, *33*, 254–263. [CrossRef]

17. Godiyal, A.K.; Mondal, M.; Joshi, S.D.; Joshi, D. Force Myography Based Novel Strategy for Locomotion Classification. *IEEE Trans. Human-Mach. Syst.* **2018**, *48*, 648–657. [CrossRef]

18. Gong, C.; Xu, D.; Zhou, Z.; Vitiello, N.; Wang, Q. Real-Time on-Board Recognition of Locomotion Modes for an Active Pelvis Orthosis. In Proceedings of the 2018 IEEE-RAS 18th International Conference on Humanoid Robots (Humanoids), Beijing, China, 6–9 November 2018; pp. 346–350.

19. Gong, C.; Xu, D.; Zhou, Z.; Vitiello, N.; Wang, Q. BPNN-Based Real-Time Recognition of Locomotion Modes for an Active Pelvis Orthosis with Different Assistive Strategies. *Int. J. Hum. Robot.* **2020**, *17*, 2050004. [CrossRef]

20. Hernandez, R.; Zhang, F.; Zhang, X.; Huang, H.; Yang, Q. Promise of a low power mobile CPU based embedded system in artificial leg control. In Proceedings of the 2012 Annual International Conference of the IEEE Engineering in Medicine and Biology Society, San Diego, CA, USA, 28 August–1 September 2012; Volume 2012, pp. 5250–5253.

21. Hernandez, R.; Yang, Q.; Huang, H.; Zhang, F.; Zhang, X. Design and implementation of a low power mobile CPU based embedded system for artificial leg control. In Proceedings of the 2013 35th Annual International Conference of the IEEE Engineering in Medicine and Biology Society (EMBC), Osaka, Japan, 3–7 July 2013; Volume 2013, pp. 5769–5772.

22. Huang, H.; Kuiken, T.A.; Lipschutz, R.D. A Strategy for Identifying Locomotion Modes Using Surface Electromyography. *IEEE Trans. Biomed. Eng.* **2009**, *56*, 65–73. [CrossRef]

23. Huang, H.; Zhang, F.; Sun, Y.L.; He, H. Design of a robust EMG sensing interface for pattern classification. *J. Neural Eng.* **2010**, *7*, 056005. [CrossRef] [PubMed]

24. Huang, H.; Zhang, F.; Hargrove, L.J.; Dou, Z.; Rogers, D.R.; Englehart, K.B. Continuous Locomotion-Mode Identification for Prosthetic Legs Based on Neuromuscular–Mechanical Fusion. *IEEE Trans. Biomed. Eng.* **2011**, *58*, 2867–2875. [CrossRef]

25. Kim, H.; Shin, Y.J.; Kim, J. Kinematic-based locomotion mode recognition for power augmentation exoskeleton. *Int. J. Adv. Robot. Syst.* **2017**, *14*, 172988141773032. [CrossRef]

26. Liu, M.; Wang, D.; Huang, H.H. Development of an Environment-Aware Locomotion Mode Recognition System for Powered Lower Limb Prostheses. *IEEE Trans. Neural Syst. Rehabil. Eng.* **2015**, *24*, 434–443. [CrossRef]

27. Liu, M.; Zhang, F.; Huang, H. (Helen) an Adaptive Classification Strategy for Reliable Locomotion Mode Recognition. *Sensors* **2017**, *17*, 2020. [CrossRef]

28. Liu, Z.; Lin, W.; Geng, Y.; Yang, P. Intent pattern recognition of lower-limb motion based on mechanical sensors. *IEEE/CAA J. Autom. Sin.* **2017**, *4*, 651–660. [CrossRef]

29. Long, Y.; Du, Z.-J.; Wang, W.-D.; Zhao, G.-Y.; Xu, G.-Q.; He, L.; Mao, X.-W.; Dong, W. PSO-SVM-Based Online Locomotion Mode Identification for Rehabilitation Robotic Exoskeletons. *Sensors* **2016**, *16*, 1408. [CrossRef]

30. Mai, A.; Commuri, S. Gait identification for an intelligent prosthetic foot. In Proceedings of the 2011 IEEE International Conference on Control Applications (CCA), Denver, CO, USA, 28–30 September 2011; pp. 1341–1346.

31. Mai, J.; Xu, D.; Li, H.; Zhang, S.; Tan, J.; Wang, Q. Implementing a SoC-FPGA Based Acceleration System for On-Board SVM Training for Robotic Transtibial Prostheses. In Proceedings of the 2018 IEEE International Conference on Real-time Computing and Robotics (RCAR), Kandima, Maldives, 1–5 August 2018; Institute of Electrical and Electronics Engineers (IEEE): Kandima, Maldives; pp. 150–155.

32. Mai, J.; Chen, W.; Zhang, S.; Xu, D.; Wang, Q. Performance analysis of hardware acceleration for locomotion mode recognition in robotic prosthetic control. In Proceedings of the 2018 IEEE International Conference on Cyborg and Bionic Systems (CBS), Shenzhen, China, 25–27 October 2018; pp. 607–611.

33. Miller, J.D.; Beazer, M.S.; Hahn, M.E. Myoelectric Walking Mode Classification for Transtibial Amputees. *IEEE Trans. Biomed. Eng.* **2013**, *60*, 2745–2750. [CrossRef]

34. Moon, D.-H.; Kim, D.; Hong, Y.-D. Development of a Single Leg Knee Exoskeleton and Sensing Knee Center of Rotation Change for Intention Detection. *Sensors* **2019**, *19*, 3960. [CrossRef]

35. Pew, C.; Klute, G.K. Turn Intent Detection for Control of a Lower Limb Prosthesis. *IEEE Trans. Biomed. Eng.* **2017**, *65*, 789–796. [CrossRef]

36. Shell, C.E.; Klute, G.K.; Neptune, R.R. Identifying classifier input signals to predict a cross-slope during transtibial amputee walking. *PLoS ONE* **2018**, *13*, e0192950. [CrossRef]

37. Simon, A.M.; Ingraham, K.A.; Spanias, J.A.; Young, A.J.; Finucane, S.B.; Halsne, E.G.; Hargrove, L.J. Delaying Ambulation Mode Transition Decisions Improves Accuracy of a Flexible Control System for Powered Knee-Ankle Prosthesis. *IEEE Trans. Neural Syst. Rehabil. Eng.* **2016**, *25*, 1164–1171. [CrossRef]

38. Spanias, J.A.; Perreault, E.J.; Hargrove, L.J. A strategy for labeling data for the neural adaptation of a powered lower limb prosthesis. In Proceedings of the 2014 36th Annual International Conference of the IEEE Engineering in Medicine and Biology Society, Chicago, IL, USA, 26–30 August 2014; Volume 2014, pp. 3090–3093.

39. Spanias, J.A.; Simon, A.M.; Ingraham, K.A.; Hargrove, L.J. Effect of additional mechanical sensor data on an EMG-based pattern recognition system for a powered leg prosthesis. In Proceedings of the 2015 7th International IEEE/EMBS Conference on Neural Engineering (NER), Montpellier, France, 22–24 April 2015; pp. 639–642.

40. Spanias, J.A.; Perreault, E.J.; Hargrove, L.J. Detection of and Compensation for EMG Disturbances for Powered Lower Limb Prosthesis Control. *IEEE Trans. Neural Syst. Rehabil. Eng.* **2015**, *24*, 226–234. [CrossRef]

41. Spanias, J.A.; Simon, A.M.; Hargrove, L.J. Across-user adaptation for a powered lower limb prosthesis. In Proceedings of the 2017 International Conference on Rehabilitation Robotics (ICORR), London, UK, 17–20 July 2017; Volume 2017, pp. 1580–1583.

42. Spanias, J.A.; Simon, A.M.; Finucane, S.B.; Perreault, E.J.; Hargrove, L.J. Online adaptive neural control of a robotic lower limb prosthesis. *J. Neural Eng.* **2018**, *15*, 016015. [CrossRef]

43. Stolyarov, R.; Burnett, G.; Herr, H.M. Translational Motion Tracking of Leg Joints for Enhanced Prediction of Walking Tasks. *IEEE Trans. Biomed. Eng.* **2018**, *65*, 763–769. [CrossRef]

44. Su, B.; Wang, J.; Liu, S.-Q.; Sheng, M.; Jiang, J.; Xiang, K. A CNN-Based Method for Intent Recognition Using Inertial Measurement Units and Intelligent Lower Limb Prosthesis. *IEEE Trans. Neural Syst. Rehabil. Eng.* **2019**, *27*, 1032–1042. [CrossRef] [PubMed]

45. Tkach, D.C.; Hargrove, L.J. Neuromechanical sensor fusion yields highest accuracies in predicting ambulation mode transitions for trans-tibial amputees. In Proceedings of the 2013 35th Annual International Conference of the IEEE Engineering in Medicine and Biology Society (EMBC), Osaka, Japan, 3–7 July 2013; Institute of Electrical and Electronics Engineers (IEEE): Osaka, Japan; Volume 2013, pp. 3074–3077.

46. Wang, X.; Wang, Q.; Zheng, E.; Wei, K.; Wang, L. A Wearable Plantar Pressure Measurement System: Design Specifications and First Experiments with an Amputee. In *Intelligent Autonomous Systems 12*; Lee, S., Cho, H., Yoon, K.-J., Lee, J., Eds.; Springer: Berlin/Heidelberg, Germany, 2013; Volume 194, pp. 273–281.

47. Wang, C.; Wu, X.; Ma, Y.; Wu, G.; Luo, Y. A Flexible Lower Extremity Exoskeleton Robot with Deep Locomotion Mode Identification. *Complexity* **2018**, *2018*, 5712108. [CrossRef]

48. Woodward, R.B.; Spanias, J.; Hargrove, L. User intent prediction with a scaled conjugate gradient trained artificial neural network for lower limb amputees using a powered prosthesis. In Proceedings of the 2016 38th Annual International Conference of the IEEE Engineering in Medicine and Biology Society (EMBC), Orlando, FL, USA, 17–20 August 2016; Volume 2016, pp. 6405–6408.

49. Xu, D.; Feng, Y.; Mai, J.; Wang, Q. Real-Time On-Board Recognition of Continuous Locomotion Modes for Amputees with Robotic Transtibial Prostheses. *IEEE Trans. Neural Syst. Rehabil. Eng.* **2018**, *26*, 2015–2025. [CrossRef]

50. Young, A.J.; Simon, A.; Hargrove, L.J. An intent recognition strategy for transfemoral amputee ambulation across different locomotion modes. In Proceedings of the 2013 35th Annual International Conference of the IEEE Engineering in Medicine and Biology Society (EMBC) Osaka, Japan, 3–7 July 2013; Volume 2013, pp. 1587–1590.

51. Young, A.J.; Simon, A.M.; Fey, N.P.; Hargrove, L.J. Intent Recognition in a Powered Lower Limb Prosthesis Using Time History Information. *Ann. Biomed. Eng.* **2013**, *42*, 631–641. [CrossRef] [PubMed]

52. Young, A.J.; Simon, A.M.; Fey, N.P.; Hargrove, L.J. Classifying the intent of novel users during human locomotion using powered lower limb prostheses. In Proceedings of the 2013 6th International IEEE/EMBS Conference on Neural Engineering (NER), San Diego, CA, USA, 6–8 November 2013; pp. 311–314.

53. Young, A.J.; Simon, A.M.; Hargrove, L.J. A Training Method for Locomotion Mode Prediction Using Powered Lower Limb Prostheses. *IEEE Trans. Neural Syst. Rehabil. Eng.* **2013**, *22*, 671–677. [CrossRef]

54. Young, A.J.; Kuiken, T.A.; Hargrove, L.J. Analysis of using EMG and mechanical sensors to enhance intent recognition in powered lower limb prostheses. *J. Neural Eng.* **2014**, *11*, 056021. [CrossRef]

55. Young, A.J.; Hargrove, L.J. A Classification Method for User-Independent Intent Recognition for Transfemoral Amputees Using Powered Lower Limb Prostheses. *IEEE Trans. Neural Syst. Rehabil. Eng.* **2015**, *24*, 217–225. [CrossRef]

56. Zhang, F.; Disanto, W.; Ren, J.; Dou, Z.; Yang, Q.; Huang, H. A Novel CPS System for Evaluating a Neural-Machine Interface for Artificial Legs. In Proceedings of the 2011 IEEE/ACM Second International Conference on Cyber-Physical Systems, Chicago, IL, USA, 12–14 April 2011; pp. 67–76.

57. Zhang, F.; Huang, H. Source Selection for Real-Time User Intent Recognition toward Volitional Control of Artificial Legs. *IEEE J. Biomed. Health Inform.* **2012**, *17*, 907–914. [CrossRef]

58. Zhang, K.; Zhang, W.; Xiao, W.; Liu, H.; De Silva, C.W.; Fu, C. Sequential Decision Fusion for Environmental Classification in Assistive Walking. *IEEE Trans. Neural Syst. Rehabil. Eng.* **2019**, *27*, 1780–1790. [CrossRef] [PubMed]

59. Zhang, K.; Xiong, C.; Zhang, W.; Liu, H.; Lai, D.; Rong, Y.; Fu, C. Environmental Features Recognition for Lower Limb Prostheses toward Predictive Walking. *IEEE Trans. Neural Syst. Rehabil. Eng.* **2019**, *27*, 465–476. [CrossRef]

60. Zhang, X.; Wang, D.; Yang, Q.; Huang, H. An automatic and user-driven training method for locomotion mode recognition for artificial leg control. In Proceedings of the 2012 Annual International Conference of the IEEE Engineering in Medicine and Biology Society, San Diego, CA, USA, 28 August–1 September 2012; Volume 2012, pp. 6116–6119.

61. Zheng, E.; Wang, L.; Luo, Y.; Wei, K.; Wang, Q. Non-contact capacitance sensing for continuous locomotion mode recognition: Design specifications and experiments with an amputee. In Proceedings of the 2013 IEEE 13th International Conference on Rehabilitation Robotics (ICORR), Seattle, WA, USA, 24–26 June 2013; Volume 2013, pp. 1–6.

62. Zheng, E.; Wang, L.; Wei, K.; Wang, Q. A Noncontact Capacitive Sensing System for Recognizing Locomotion Modes of Transtibial Amputees. *IEEE Trans. Biomed. Eng.* **2014**, *61*, 2911–2920. [CrossRef]

63. Zheng, E.; Wang, Q. Noncontact Capacitive Sensing-Based Locomotion Transition Recognition for Amputees With Robotic Transtibial Prostheses. *IEEE Trans. Neural Syst. Rehabil. Eng.* **2016**, *25*, 161–170. [CrossRef]

64. Zheng, E.; Wang, Q.; Qiao, H. Locomotion Mode Recognition with Robotic Transtibial Prosthesis in Inter-Session and Inter-Day Applications. *IEEE Trans. Neural Syst. Rehabil. Eng.* **2019**, *27*, 1836–1845. [CrossRef]

65. Zhou, Z.; Liu, X.; Jiang, Y.; Mai, J.; Wang, Q. Real-time onboard SVM-based human locomotion recognition for a bionic knee exoskeleton on different terrains. In Proceedings of the 2019 Wearable Robotics Association Conference (WearRAcon), Scottsdale, AZ, USA, 25–27 March 2019; pp. 34–39.

66. Bishop, C.M. *Pattern Recognition and Machine Learning*; Springer: Berlin, Germany, 2006.

67. Hargrove, L.J.; Englehart, K.; Hudgins, B. A Comparison of Surface and Intramuscular Myoelectric Signal Classification. *IEEE Trans. Biomed. Eng.* **2007**, *54*, 847–853. [CrossRef]
68. Parker, P.; Englehart, K.; Hudgins, B. Myoelectric signal processing for control of powered limb prostheses. *J. Electromyogr. Kinesiol.* **2006**, *16*, 541–548. [CrossRef]
69. Simon, A.M.; Spanias, J.A.; Ingraham, K.A.; Hargrove, L.J. Delaying ambulation mode transitions in a powered knee-ankle prosthesis. In Proceedings of the 2016 38th Annual International Conference of the IEEE Engineering in Medicine and Biology Society (EMBC), Orlando, FL, USA, 17–20 August 2016; Volume 2016, pp. 5079–5082.
70. Fisher, R.A. The use of multiple measurements in taxonomic problems. *Ann. Eugen.* **1936**, *7*, 179–188. [CrossRef]

Publisher's Note: MDPI stays neutral with regard to jurisdictional claims in published maps and institutional affiliations.

Letter

Estimation of Relative Hand-Finger Orientation Using a Small IMU Configuration

Zhicheng Yang [1,2,*], Bert-Jan F. van Beijnum [1], Bin Li [2], Shenggang Yan [2] and Peter H. Veltink [1]

[1]   Department of Biomedical Signals Systems, Technical Medical Centre, University of Twente, 7500 AE Enschede, The Netherlands; b.j.f.vanbeijnum@utwente.nl (B.-J.F.v.B.); P.H.Veltink@utwente.nl (P.H.V.)

[2]   School of Marine Science and Technology, Northwestern Polytechnical University, Xi'an 710072, China; libin_cme@nwpu.edu.cn (B.L.); yshgang@nwpu.edu.cn (S.Y.)

*   Correspondence: z.yang-1@utwente.nl

Received: 30 June 2020; Accepted: 17 July 2020; Published: 19 July 2020

Abstract: Relative orientation estimation between the hand and its fingers is important in many applications, such as virtual reality (VR), augmented reality (AR) and rehabilitation. It is still quite a big challenge to do the estimation by only exploiting inertial measurement units (IMUs) because of the integration drift that occurs in most approaches. When the hand is functionally used, there are many instances in which hand and finger tips move together, experiencing almost the same angular velocities, and in some of these cases, almost the same accelerations are measured in different 3D coordinate systems. Therefore, we hypothesize that relative orientations between the hand and the finger tips can be adequately estimated using 3D IMUs during such designated events (DEs) and in between these events. We fused this extra information from the DEs and IMU data with an extended Kalman filter (EKF). Our results show that errors in relative orientation can be smaller than five degrees if DEs are constantly present and the linear and angular movements of the whole hand are adequately rich. When the DEs are partially available in a functional water-drinking task, the orientation error is smaller than 10 degrees.

Keywords: relative orientation estimation; IMU; magnetometer-free

---

1. Introduction

Hand-finger movement tracing is useful in many areas, such as virtual reality (VR), augmented reality (AR), ergonomic assessment and especially medical applications [1–5]. People who suffered from stroke or injury of the spinal cord need an effective rehabilitation therapy for recovery of body functions, including hand function. In a hospital, therapists evaluate the hand function through some traditional assessments such as the Fugl–Meyer or Jebsen–Taylor hand function assessment [6,7]. Currently, the results may be subjective and dependent on the therapist. Therefore, it is essential to provide a quantitative and understandable measurement to make the therapist's diagnosis more objective. Several sensory systems can be used to trace hand motion, which can be categorized as camera-based, glove-based, magnetic actuator-based and inertial measurement unit (IMU)-based. Camera-based systems can be divided into two different types. One uses high-speed cameras to trace markers attached to body segments, which is quite accurate and often used as the reference [8]. However, occlusion problems will influence its accuracy and the distance between cameras and hands needs to be below a few meters in order to accurately measure hand and finger orientations. Because of these problems, you need many cameras (6 to 12). The other camera-based system traces objects, including their orientations, by exploiting depth maps to reconstruct the object [9,10]. Its advantage is that no finger or hand attachments are needed, making it friendly to users. However, this system also suffers from the occlusion problem and only allows hand movements to be evaluated if they occur in the vicinity of the cameras [11,12]. Besides, it requires a powerful processor to process the images. Glove-based sensor systems exploit varying sensors, such as resistive-bend sensors and optical-fiber sensors on the glove, transducing finger movement into corresponding signals to estimate relative orientations between hand and finger segments [13,14]. It has the benefit of a low price. However, the glove needs to be well attached for the measurement and requires thorough calibration before utilization. The magnetic actuator-based system also has two types, active actuation and passive actuation. The first one deploys active magnetic actuators on the finger tip and receivers on the dorsal side of the hand [15]. It has high accuracy and no occlusion issue. However, it requires different frequencies for each degree of freedom (DoF) of the actuator, which often needs equipment such as multiple power signal sources and a high-speed processor. This affects the complexity of the system and its physical dimensions. The second one uses magnets as passive sources, magnets are placed on the finger tips while magnetic sensors are worn on the wrist [16]. It has the benefit of having a simple structure and a low cost. However, it is difficult to distinguish the fields of different magnets, since only the sum of the fields are measured, especially when the magnets get close. The IMU-based system utilizes inertial sensors to trace the hand [17–19]. Compared with previous methods, it can provide raw data including angular velocity and acceleration. Orientation can be estimated by fusing the raw data. This operation suffers from drift, since it involves integration operations. However, this drift can be compensated using magnetometer data, which is easily accessible since it is often embedded in IMU systems. However, magnetometers are used in this solution and are therefore vulnerable to external magnetic disturbances, such as indoor iron surroundings [20,21]. Thomas and Wolfgang et al. proposed magnetometer-free methods for the joint angle estimation [22–25]. However, such methods assume the rotation is restricted to two DoFs because of the anatomy constraint [23]. Thus, they cannot be applied to flexible joints, such as the metacarpophalangeal joint (MCP) of the thumb. Relative orientations between the hand and its fingers are important for the reconstruction of hand-finger movement, which is essential information for AR, VR and rehabilitation.

Our goal is to estimate relative 3D orientations between finger tips and the dorsal side of the hand with only IMUs, essentially getting rid of magnetic disturbance by not using magnetometers. In order to reduce the integration drift, we exploit information during the daily life rather than using the biomechanical constraint. The information is based on the assumption that there are many instances in which hand and finger tips move together, experiencing almost the same angular velocities and accelerations represented in different 3D coordinate systems. The method was verified with a small sensor configuration: one sensor on the dorsal side of hand, and one on the most distal finger segment of interest.

2. Methods

In order to estimate the relative orientation, the information from the gyroscope and accelerometer and extra information during DEs need to be combined in an optimal way. Therefore, an extended Kalman filter (EKF) was introduced to estimate 3D relative orientations between the dorsal side of the hand and finger tips, assuming angular velocities and accelerations are the same, but just represented in a different coordinate system. The process model is based on integrating relative angular velocity, the measurement model is mainly based on the information during the DE. The quality of the DE is considered in the measurement variance. When the DE is available with small variance, we trust the measurement model more; otherwise, we trust the process model more. Thus, the information from process and measurement models is optimally fused to estimate relative 3D orientations during functional hand and finger movements.

2.1. Sensor Model

The gain error and non-orthogonality error are assumed to be time-invariant and can be obtained through sensor calibration; thus, the outputs of calibrated gyroscope can be expressed as

$$
\begin{cases}
y^h_{gyr,h} = \omega^h_h + b_h + \zeta_h \\
y^f_{gyr,f} = \omega^f_f + b_f + \zeta_f
\end{cases}
\tag{1}
$$

where $y^h_{gyr,h}$ and $y^f_{gyr,f}$ are gyroscope outputs on the hand and finger tip in their own frames. $b_x (x = h, f)$ is the slowly varying offset. $\zeta_x (x = h, f)$ is Gaussian noise.

For the calibrated accelerometer, the outputs on the hand and finger tips are

$$
\begin{cases}
y^h_{acc,h} = a^h_h + g^h + \eta^h_h \\
y^f_{acc,f} = a^f_f + g^f + \eta^f_f
\end{cases}
\tag{2}
$$

where g is the gravity, and η^h_h and η^f_f are Gaussian noise.

2.2. Process Model

The process model is based on integrating the relative angular velocity between the hand and its fingers in its own frame. We choose the quaternion $q_{hf} = \begin{bmatrix} q_0 & q_1 & q_2 & q_3 \end{bmatrix}^T$ that expresses relative orientation from a finger tip to the dorsal side of the hand as the state vector $x = q_{hf}$. The relative orientation x_k is updated as

$$
x_k = x_{k-1} \otimes \begin{bmatrix} 1 & \tfrac{1}{2}\omega_k d_t \end{bmatrix} + m
\tag{3}
$$

where m is the process error. ω_k is the relative angular velocity between the hand and fingers; $\otimes$ represents the multiplication operator between two quaternions.

$$
\omega_k = (\omega^h_h)_k - x_{k-1} \otimes (\omega^f_f)_k \otimes x^*_{k-1}
\tag{4}
$$

where ω^h_h and ω^f_f are hand and finger angular velocities.

2.3. Measurement Model

The measurement update of EKF is based on the DE. During the DE, the hand and fingers share the same angular velocity in different coordinate frames

$$
\omega^h_h = q_{hf} \otimes \omega^f_f \otimes q^*_{hf}
\tag{5}
$$

where $\omega^y_x (x = h, f, y = h, f)$ is the angular velocity of an object in frame x expressed in the coordinate frame of object y. h represents the hand and f represents the finger tip. Combining Equations (1) and (5), we find:

$$
\begin{aligned}
y^h_{gyr,h} &= q_{hf} \otimes y^f_{gyr,f} \otimes q^*_{hf} + b_h - q_{hf} \otimes b_f \otimes q^*_{hf} + \\
&\quad \zeta_h - q_{hf} \otimes \zeta_f \otimes q^*_{hf} \\
&= q_{hf} \otimes y^f_{gyr,f} \otimes q^*_{hf} + d_{gyr}
\end{aligned}
\tag{6}
$$

where the combined error of gyroscope d_{gyr} is

$$
d_{gyr} = (b_h - q_{hf} \otimes d_f \otimes q^*_{hf}) + (\zeta_h - q_{hf} \otimes \zeta_f \otimes q^*_{hf})
\tag{7}
$$

Unlike the angular velocity, accelerations at different positions are different, which can be expressed as

$$
\begin{aligned}
a_h^h &= q_{hf} \otimes a_f^f \otimes q_{hf}^* + \omega_h^h \times (\omega_h^h \times r_{fh}^h) + \dot{\omega}_h^h \times r_{fh}^h \\
&= q_{hf} \otimes a_f^f \otimes q_{hf}^* + (\lfloor \omega_h^h \rfloor_\times \lfloor \omega_h^h \rfloor_\times + \lfloor \dot{\omega}_h^h \rfloor_\times) r_{fh}^h
\end{aligned}
\tag{8}
$$

where $a_x^y (x = h, f, y = h, f)$ is the acceleration of object in frame x relative to frame y. $\dot{\omega}_h^h$ is the hand angular acceleration in its own frame. r_{fh}^h is the position vector between hand and fingers in the hand frame. $\lfloor \rfloor_\times$ denotes a skew-symmetric matrix.

$$
\lfloor a \rfloor_\times =
\begin{bmatrix}
0 & -a_z & a_y \\
a_z & 0 & -a_x \\
-a_y & a_x & 0
\end{bmatrix}
\tag{9}
$$

If the second term $(\lfloor \omega_h^h \rfloor_\times \lfloor \omega_h^h \rfloor_\times + \lfloor \dot{\omega}_h^h \rfloor_\times) r_{fh}^h$ is relatively small compared with the first term $q_{hf} \otimes a_f^f \otimes q_{hf}^*$, then Equation (8) can be approximated as the following equation:

$$
a_h^h \approx q_{hf} \otimes a_f^f \otimes q_{hf}^*
\tag{10}
$$

Combining Equations (2) and (8), we find:

$$
y_{acc,h}^h = q_{hf} \otimes y_{acc,f}^f \otimes q_{hf}^* + (\lfloor \omega_h^h \rfloor_\times \lfloor \omega_h^h \rfloor_\times + \lfloor \dot{\omega}_h^h \rfloor_\times) r_{fh}^h + \eta_c
\tag{11}
$$

where the combined error η_c can be expressed as

$$
\eta_c = \eta_h^h - q_{hf} \otimes \eta_f^f \otimes q_{hf}^*
\tag{12}
$$

Finally, an overall relation between hand and fingers based on Equations (6) and (11) is

$$
\begin{cases}
y_{gyr,h}^h = q_{hf} \otimes y_{gyr,f}^f \otimes q_{hf}^* + d_{gyr} \\
y_{acc,h}^h = q_{hf} \otimes y_{acc,f}^f \otimes q_{hf}^* + (\lfloor \omega_h^h \rfloor_\times \lfloor \omega_h^h \rfloor_\times + \lfloor \dot{\omega}_h^h \rfloor_\times) r_{fh}^h + \eta_c
\end{cases}
\tag{13}
$$

Subsequently, we can get the measurement model based on the sensor model and quaternion constraint

$$
y_k = f(x_k) + v
\tag{14}
$$

where y and f can be expressed as

$$
y_k = \begin{bmatrix} (y_{acc,h}^h)^T & (y_{gyr,h}^h)^T & 0 \end{bmatrix}^T
\tag{15}
$$

$$
f(x) =
\begin{bmatrix}
x_k \otimes y_{acc,f}^f \otimes x_k^* \\
x_k \otimes y_{gyr,f}^f \otimes x_k^* \\
q_0^2 + q_1^2 + q_2^2 + q_3^2 - 1
\end{bmatrix}
\tag{16}
$$

As shown in Equations (3) and (16), the process and measurement model are both nonlinear with respect to x_k. In order to update the covariance matrix for x_k, linearization is performed and the Jacobian matrix F and H for process and measurement model are calculated; the details can be found in the Appendix A.

2.4. Uncertainty Error Variance

In order to assess the relative confidence in the measurement model (based on our DE assumptions) and the process model, the measurement variance is determined. According to the assumption that a hand and finger share approximately the same angular velocity and acceleration based on Equation (13), the differences in angular velocity and acceleration between the hand and fingers measured by the IMU determine the measurement variance. From Equation (7), the error is related to the offset error, the white noise and relative orientation. d_{gyr} can be expressed with following equation from Equation (6).

$$d_{gyr} = y_{gyr,h}^{h} - q_{hf} \otimes y_{gyr,f}^{f} \otimes q_{hf}^{*} \tag{17}$$

We approximate the distribution of d_{gyr} as Gaussian distribution with zero mean and standard deviation $\sigma_g \begin{bmatrix} 1 & 1 & 1 \end{bmatrix}$ (rad/s)

$$\sigma_g = \left\| y_{gyr,h}^{h} - q_{hf} \otimes y_{gyr,f}^{f} \otimes q_{hf}^{*} \right\|_2 \tag{18}$$

For Equation (13), the error d_{acc} can be expressed with the following equation:

$$d_{acc} = (\lfloor \omega_h^h \rfloor_\times \lfloor \omega_h^h \rfloor_\times + \lfloor \dot{\omega}_h^h \rfloor_\times) r_{fh}^h + \eta \tag{19}$$

We can express the error in another format from Equation (11).

$$d_{acc} = y_{acc,h}^{h} - q_{hf} \otimes y_{acc,f}^{f} \otimes q_{hf}^{*} \tag{20}$$

Similarly to the gyroscope, we assume the error d_{acc} has an approximate Gaussian distribution with zero mean while its standard deviation $\sigma_a \begin{bmatrix} 1 & 1 & 1 \end{bmatrix}$ is

$$\sigma_a = \left\| y_{acc,h}^{h} - q_{hf} \otimes y_{acc,f}^{f} \otimes q_{hf}^{*} \right\|_2 \tag{21}$$

Based on the Gaussian approximation, as described in Equations (17) and (20), it is essential to know the rotation quaternion q_{hf} before we get the variance. However, q_{hf} is the variable we try to estimate which is also unknown. As we assume there is no or a slow orientation change between the hand and finger tips, the estimated relative orientation at time $k - 1$ is used as the true relative orientation at time k.

$$q_{hf,k} = \hat{q}_{hf,k-1} \tag{22}$$

where $q_{hf,k}$ is the "true" rotation quaternion we use to estimate the variance at time k. $\hat{q}_{hf,k-1}$ is the estimated rotation quaternion at time $k - 1$. The measurement covariance is determined as

$$R_m = \begin{bmatrix} \sigma_g I_{3\times3} & 0 & 0 \\ 0 & \sigma_a I_{3\times3} & 0 \\ 0 & 0 & 0 \end{bmatrix} \tag{23}$$

The initial value for the state vector of relative orientations x_k was set as $\begin{bmatrix} 1 & 0 & 0 & 0 \end{bmatrix}^T$.

3. Experiments

3.1. Experiment Setup

The sensor system includes three IMUs fixed on the most distal segments of the thumb and index finger and the dorsal side of the hand, as shown in Figure 1. MPU9250 (InvenSense) was chosen for the IMU, which contains a tri-axis accelerometer and tri-axis gyroscope (it also contains a tri-axis magnetometer, which was not used in the current study). All IMUs were sampled synchronously; the sample frequencies of gyroscope and accelerometer were 200 Hz and 100 Hz respectively. All the data were collected by a master micro-controller (Atmel XMEGA) and then transmitted to the PC via a USB connection. Prior to the experiment, the accelerometer was calibrated based on local gravity; the gyroscope was calibrated based on the calibrated accelerometer [26]. An optical Vicon system with eight cameras was used to perform 3D orientation reference measurements. For this purpose, three optical markers were attached to each IMU. The sampling frequency of Vicon was 100 Hz.

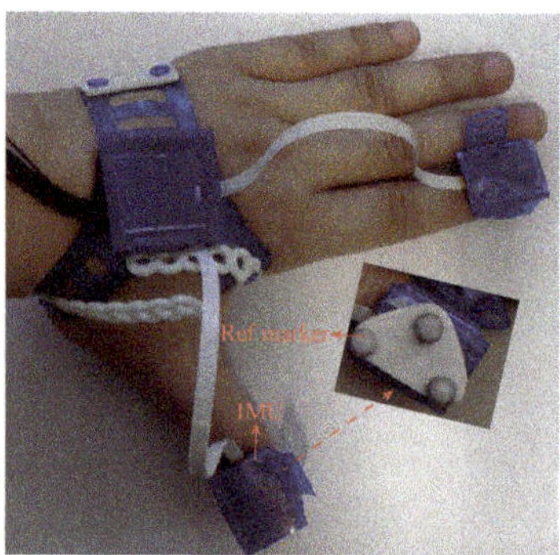

Figure 1. IMUs on the dorsal side of the hand and fingertips. The inset shows the cluster of optical markers used on top of each IMU for reference measurement of segment orientations using the optical VICON system. Every cluster contains three markers, which determine a 3D coordinate frame.

3.2. Alignment of the IMU and Reference Marker Frame for the Validation Experiment

For evaluation of the IMU-based 3D relative orientation estimation using the optical system, it is essential to calibrate the relative orientation between the sensor and marker-based reference frame. Here, we used the accelerometer for this marker system's IMU calibration. Holding the system static, we obtained the gravity in the IMU frame from the accelerometer readings. Meanwhile, we obtained the orientation from the global Vicon frame to marker frame q_{mg}. Gravity in marker frame is $q_{mg} \otimes g \otimes q_{mg}^*$, where g is gravity in global Vicon frame (z-axis of global Vicon system was vertical upward; gravity in this frame was $g = \begin{bmatrix} 0 & 0 & -g \end{bmatrix}$; g is the local gravity value). When we have at least two poses, we obtain more than two vectors expressed in the marker frame and IMU frame respectively, which is enough to determine the relative orientation between the IMU and marker frame.

3.3. Sensor to Segment Calibration

Before the experiment, IMU errors were calibrated according to D Tedaldi et al.'s and WT Fong et al.'s research [26,27], including sensitivity error, offset error, non-orthogonal error and misalignment between the accelerometer and gyroscope. After the IMU was fixed on the hand and fingers, the relative orientations between IMU sensors and body segments were calibrated. An accelerometer was used to achieve the alignment by exploiting static accelerometer measures of gravity. When we held our hand sequentially horizontally and vertically, we obtained the 3D relative orientation between two frames. More details can be found in Kortier et al.'s research [28].

3.4. Synchronization of Vicon and IMU System

In this experiment, the two measurement systems were synchronized by recording the sensed responses of an induced impact at the start and end of each experiment. At the start and end of every experiment, we hit the IMU on a desk, resulting in an acceleration peak measured by the IMU system and a minimum vertical position of the Vicon markers simultaneously, which was used to synchronize the two systems.

3.5. Protocols for the Experiment

In order to demonstrate the feasibility of our approach, an experimental part was designed to estimate the accuracy of the algorithm compared with the optical system. Our feasibility experiment involved three participants. The protocol was reviewed, approved and conducted under the auspices of the Ethics Committee EEMCS, Univerisity of Twente. The following tasks were performed:

Task1: Movements and rotations of the hand, while not varying relative orientations between hand and fingers: IMUs were fixed on fingers and the dorsal side of the hand. Then, the participant did the pronation and supination movements with the arm while the axis of pronation and supination was continuously changing. The orientation was changed over approximately $160°$ around the rotation axis; see Figure 2. Furthermore, we varied the angular velocity by performing these cyclical movements with varying repetition rate of pronation and supination (60, 120, 240 cycles/min), with the help of a metronome. This was done in order to test the performance of the algorithm under different conditions. During the process, the subject was asked to close the hand and not change the relative orientations between the hand and fingers, while displacing or rotating the hand.

Task2: Simple functional task. The subject was asked to place the hand on the desk; then rise the hand and grasp a cup; subsequently drink some water and place the cup back; and finally place the hand on the original position. The illustration of the movement can be seen in Figure 3.

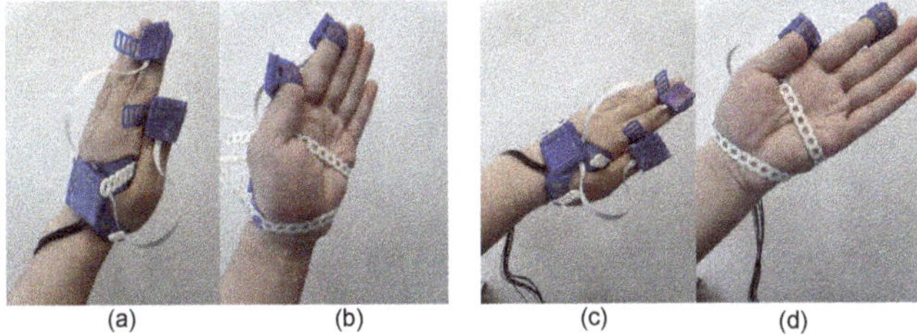

Figure 2. Movement for task 1: rotations of the hand, while not varying relative orientations between the hand and fingers. Subfigure (**a,b**) are a set of pronation and supination movements. Subfigure (**c,d**) are another set of pronation and supination but with a different rotation axis. During this task, we did the pronation and supination movements with different rotation axes.

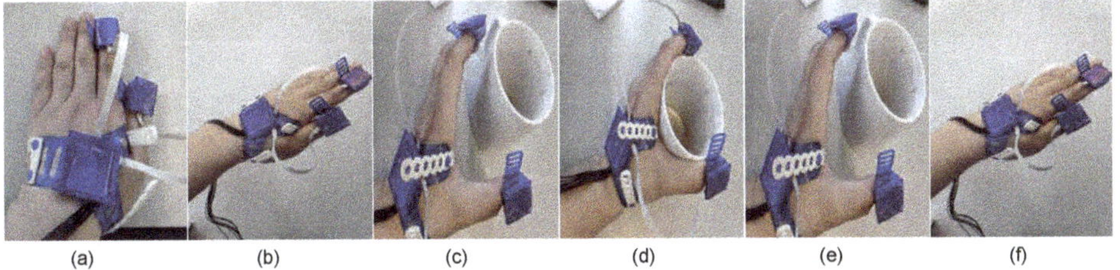

Figure 3. Movement for task 2: Simple functional task. The task can be divided into several phases. (**a**) Put the hand static on the desk; (**b**) raise the hand; (**c**) grasp the cup; (**d**) drink the water; (**e**) release the hand; (**f**) withdraw the hand.

For task 1, the orientation reference was directly derived from the IMUs, because the relative orientation was imposed by the hand, and therefore, known and not varying. For task 2, the reference measurement was performed using the optical VICON system (software version 2.8.2).

4. Results

4.1. Movements and Rotations of the Hand, While Not Varying Relative Orientations between the Hand and Finger (Task 1)

The error angle used was the arccos of the first component of quaternion error q_{err} [29]:

$$q_{err} = q_{est}^{-1} \otimes q_{ref} = \begin{bmatrix} 1 & \frac{1}{2}\theta_{err} \end{bmatrix} \tag{24}$$

where q_{est} was the estimated relative orientation and q_{ref} was the orientation reference. We obtained more than two independent vectors from the gyroscope, accelerometer or both from 3D movements. The error angle estimated when DE is available is shown in Figure 4. The orientation error is smallest with the gyroscope and accelerometer, while the orientation error is largest with accelerometer data only.

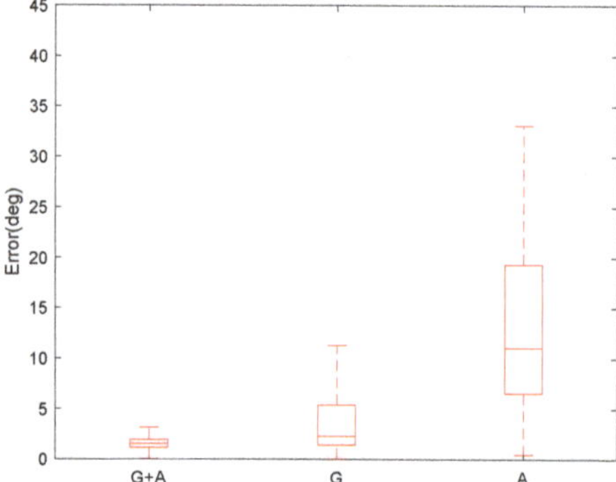

Figure 4. Estimated orientation error $|\theta_{err}|$ with gyroscope and accelerometer (values under 99.3 percent coverage are shown in the boxplot figures). "G", "A" and "G+A" represent estimated results based on gyroscope, accelerometer and gyroscope plus accelerometer respectively.

Influence of Repetition Rate of Movement

The estimation may be influenced by the repetition rate of movements. Figures 5 and 6 show the relation between the norm of gyroscope or accelerometer on thte hand and finger for several repetition rates. Ideally, the gyroscope output norms $\|y_{gyr,h}\|$, $\|y_{gyr,f}\|$ should be equal for the measurement update and for the accelerometer. The differential output norms cause estimation errors, as shown in Equation (13). For the accelerometer, the different output norms $\|\|y_{acc,h}\| - \|y_{acc,f}\|\|$ were 29.3 m/s^2, 66.4 m/s^2 and 370.2 m/s^2 under the repetition rates 60, 120 and 240 beats/min respectively. Meanwhile, the correspondingly differential output norms of gyroscope were 2.2 rad/s, 2.7 rad/s and 4.4 rad/s. As shown in Figure 7b,c, the estimated orientation error based on the accelerometer became larger when the repetition rate increased, while orientation error based on gyroscope changed little when the repetition rate increased. As shown in Figure 7a, the estimated result based on the gyroscope and accelerometer trusted the gyroscope more than the accelerometer because it contained less error; thus, it was also insensitive to repetition rate.

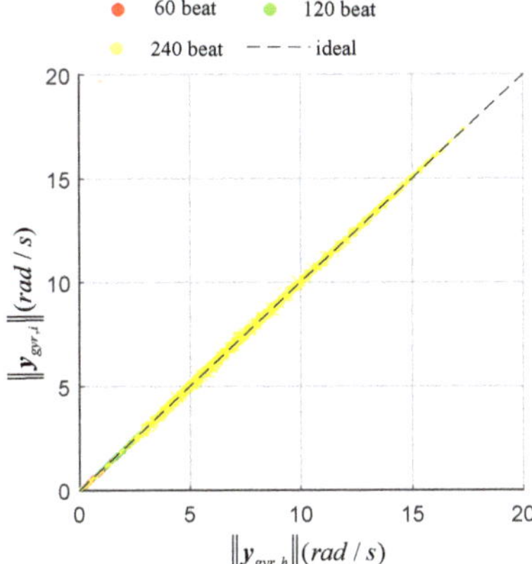

Figure 5. Relation of output norms between gyroscopes on the dorsal side of the hand and finger tip with different repetition rates.

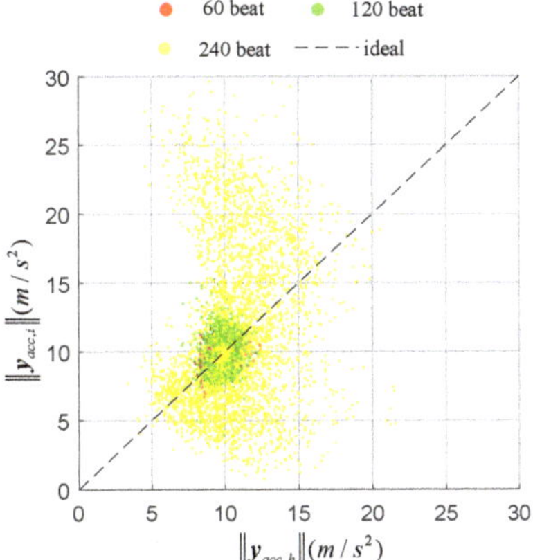

Figure 6. Relation of output norms between accelerometers on the dorsal side of the hand and finger tip with different repetition rates.

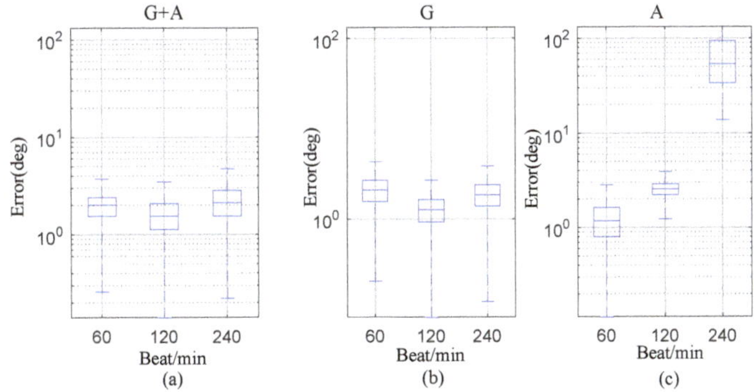

Figure 7. Estimation error $|\theta_{err}|$ with different repetition rates (values under 99.3 percent coverage are shown in the boxplot figures). Subfigures (**a–c**) are estimations with gyroscope plus accelerometer, and gyroscope and accelerometer individually.

4.2. Simple Functional Task (Task 2)

According to Figure 3, the whole process was divided into several phases; the estimated orientation errors based on the optical system in different phases are shown in Figure 8. The quaternion-based orientation estimated by IMU system and optical can be seen in Figure A3 in the Appendix B. The error during the drinking part was relatively low because the cub imposed a constant relative orientation on the hand and fingers and the whole hand moved with varying position and orientation, as shown in Figure 8. Since the angular velocity and acceleration norms were close to each other, the standard deviations of measurement noise σ_a and σ_g were small, as shown in subfigure (b); the measurement model was trusted relatively more relative to the process model under said condition. For the other phases of this functional task, there were bigger differences between gyroscope and accelerometer norms on the hand and fingers; thus σ_a and σ_g were bigger; the trust in the process model was relatively high. A good estimation of relative orientation was achieved by choosing a suitable standard deviation for the process error (see Figure 8c). The results of other two participants can be seen in Figures A1 and A2 in the Appendix B.

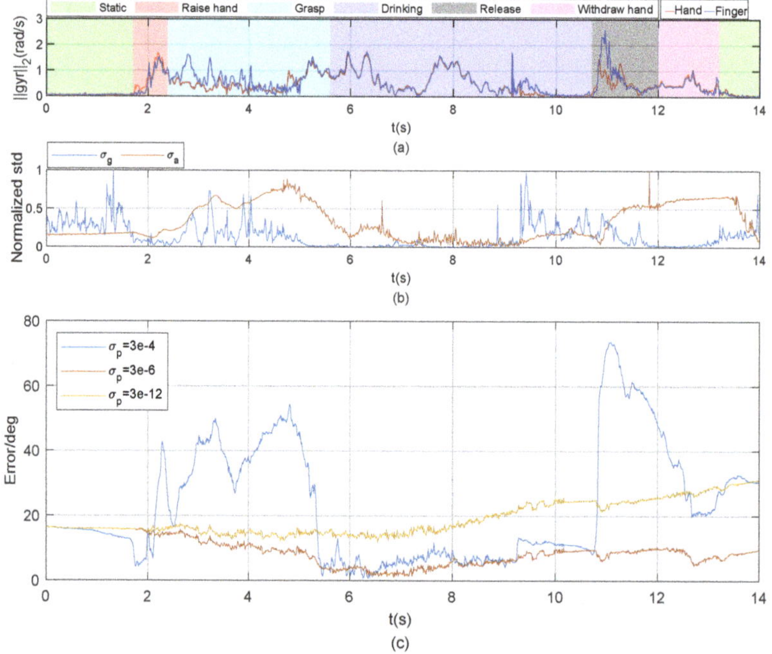

Figure 8. Relative orientation between hand and thumb during the water-drinking process. Subfigure (**a**) shows the output norms of the two gyroscopes (on the hand and finger tip respectively). Subfigure (**b**) shows the normalized SDs σ_a and σ_g from Equations (18) and (21). Larger σ_a and σ_g mean larger measurement error. The EKF trusts the process model more and the measurement model less when σ_a and σ_g are larger. Subfigure (**c**) shows the estimated results with different SDs of the process model. The variance of process error Q was determined as $\sigma_p I_4$.

5. Discussion

We proposed and evaluated an IMU-based setup for estimating 3D relative orientation between hand and finger tips. Compared with the IMU-based data glove system described by Salchov-Homer et al. [19] and Kortier et al. [28], we reduced the number of IMUs as much as possible and avoided magnetic disturbance, but still obtained comparable precision of estimated orientation.

In reference [19], the orientation error magnitude is approximately five to ten degrees. In our research, the orientation error is related to the movement quality. When the hand and fingers move together, the median orientation error can be smaller than five degrees. For the water-drinking experiment, the estimated error is less than ten degrees when hand and fingers approximately move together, but around ten degrees during the rest periods. In our view, this is a promising method for the hand finger orientation estimation with a small IMU configuration which can be used if rich whole-hand movements occur and the change of relative orientations between hand and finger tips is regular and relatively small. Standard deviations σ_g and σ_a can be used to assess whether such DEs regularly apply during a specify movement.

Most previous IMU-based systems [17,28] for finger orientation estimation usually require a magnetometer to reduce the drift caused by the gyroscope, which will suffer from the magnetic disturbance problem in indoor environments. To our knowledge, in order to remove magnetic disturbance but still suppress the drift, a biomechanical model is additionally used in methods described in the literature (e.g., [17,28]). We have not applied additional information from a bioimechanical model in our current study, although this additional information could be applied. However, it should be noted that finger movements are usually assumed to be restricted to two DoFs while using biomechanical constraints. In construct, our method can be implemented without biomechanical constraints and can be applied to estimating three-DoF-relative orientation during 3D hand movements without such biomechanical assumptions.

For the result in task 1, the relative orientation estimation is less sensitive to an increase of repetition rate of the same movement when using gyroscopes or gyroscopes plus an accelerometer than the accelerometer only. That is because as the difference among the accelerometer signals from the hand and finger becomes larger, the non-gravitational acceleration caused by increasing angular velocity or angular acceleration becomes relatively more important compared to the gravity component.

Position estimation only based on inertial sensors is quite challenging and limited by integration drift. Our further research will concentrate on relative position estimation based on IMUs combined with sensing the magnetic field of a magnetic source. For this to be feasible, an adequate estimate of relative orientation is required, so the 3D magnetic field measurement can be expressed in the coordinate system of the magnetic source. This is an essential first step in estimating relative positions. In this research, only one healthy participant was involved since we are mainly concentrating on verifying the performance of the algorithm. Subsequently, the proposed relative orientation and position estimation methods for the hand and finger using a small sensing configuration need to be evaluated in healthy subjects and patients during more complex daily tasks, in order to assess the applicability in clinical and daily-life settings. To make the system more friendly to users, the system could be wireless in the future.

6. Conclusions

In conclusion, IMUs can be used to estimate the relative orientation between the hand and fingers without using magnetometers. Compared with previous systems, we only exploit IMUs on finger tips and the dorsal side of the hand rather than having IMUs on every segment. The performance is dependent on how well the hand and fingers move together, which influences the accuracy of the estimate. The median value of estimation error can be smaller than five degrees when IMUs are on our hand and fingers if their relative orientation is not variant over time, while the object or hand is moving. During the water-drinking task, the estimation error can be smaller than 10 degrees during periods when the hand and fingers approximately move together, which may be adequately accurate to provide useful information to clinicians when judging.

Author Contributions: Conceptualization, Z.Y. and P.H.V.; methodology, Z.Y., P.H.V., B.-J.F.v.B., S.Y. and B.L.; validation, Z.Y.; formal analysis, Z.Y.; writing—original draft preparation, Z.Y. and P.H.V.; writing—review and editing, Z.Y., P.H.V., B.-J.F.v.B., S.Y. and B.L.; supervision, P.H.V. All authors have read and agreed to the published version of the manuscript.

Funding: This research received funding from China Scholarship Council (CSC).

Acknowledgments: The authors would like to thank Roessingh Research and Development (Enschede, Netherlands) for sharing the gait laboratory, and the laboratory manager, Leendert Schaake, for helping with the optical system and the processing of data. Thanks to A. Droog and G.J.W. Wolterink from Biomedical Signals and Systems, University of Twente for providing the inertial sensor setup and 3D printed coat for inertial sensors.

Conflicts of Interest: The authors declare no conflict of interest

Appendix A

By linearizing the nonlinear function of the process model and measurement model, we obtained the Jacobian matrixes F and H respectively, which were used in EKF for the covariance update. Based on Equation (3):

$$F = \frac{\partial x_k}{\partial x_{k-1}} = \frac{1}{2} \begin{bmatrix} a_{11}\omega_2 & a_{12}\omega_2 & a_{13}\omega_2 & a_{14}\omega_2 \\ a_{21}\omega_2 & a_{22}\omega_2 & a_{23}\omega_2 & a_{24}\omega_2 \\ a_{31}\omega_2 & a_{32}\omega_2 & a_{33}\omega_2 & a_{34}\omega_2 \\ a_{41}\omega_2 & a_{42}\omega_2 & a_{43}\omega_2 & a_{44}\omega_2 \end{bmatrix} d_t$$

$$+ \frac{1}{2} \begin{bmatrix} 0 & -\omega_{1x} & -\omega_{1y} & -\omega_{1z} \\ \omega_{1x} & 0 & \omega_{1z} & -\omega_{1y} \\ \omega_{1y} & -\omega_{1z} & 0 & \omega_{1x} \\ \omega_{1z} & \omega_{1y} & -\omega_{1x} & 0 \end{bmatrix} d_t + I_{4\times4} \tag{A1}$$

$$\begin{cases} a_{11} = 2\begin{bmatrix} q_0q_1 & q_0q_2 & q_0q_3 \end{bmatrix} & a_{12} = \begin{bmatrix} 1+2q_1^2 & 2q_1q_2 & 2q_1q_3 \end{bmatrix} \\ a_{13} = \begin{bmatrix} 2q_1q_2 & 1+2q_2^2 & 2q_2q_3 \end{bmatrix} & a_{14} = \begin{bmatrix} 2q_1q_3 & 2q_2q_3 & 1+2q_3^2 \end{bmatrix} \\ a_{21} = -\begin{bmatrix} 1+2q_0^2 & 2q_0q_3 & -2q_0q_2 \end{bmatrix} & a_{22} = -2\begin{bmatrix} q_0q_1 & q_1q_3 & -q_1q_2 \end{bmatrix} \\ a_{23} = \begin{bmatrix} -2q_0q_2 & -2q_2q_3 & 1+2q_2^2 \end{bmatrix} & a_{24} = -\begin{bmatrix} 2q_0q_3 & 1+2q_3^2 & -2q_2q_3 \end{bmatrix} \\ a_{31} = -\begin{bmatrix} -2q_0q_3 & 1+2q_0^2 & 2q_0q_1 \end{bmatrix} & a_{32} = -\begin{bmatrix} -2q_1q_3 & 2q_0q_1 & 1+2q_1^2 \end{bmatrix} \\ a_{33} = -2\begin{bmatrix} -q_2q_3 & q_0q_2 & q_1q_2 \end{bmatrix} & a_{34} = \begin{bmatrix} 1+2q_3^2 & -2q_0q_3 & -2q_1q_3 \end{bmatrix} \\ a_{41} = -\begin{bmatrix} 2q_0q_2 & -2q_0q_1 & 1+2q_0^2 \end{bmatrix} & a_{42} = \begin{bmatrix} -2q_1q_2 & 1+2q_1^2 & -2q_0q_1 \end{bmatrix} \\ a_{43} = -\begin{bmatrix} 1+2q_2^2 & -2q_1q_2 & q_0q_2 \end{bmatrix} & a_{44} = -2\begin{bmatrix} q_2q_3 & -q_1q_3 & q_0q_3 \end{bmatrix} \end{cases} \tag{A2}$$

Based on Equation (16):

$$H = \frac{\partial f}{\partial x_k} = \begin{bmatrix} \dfrac{\partial(x_k \otimes y_{acc,f}^f \otimes x_k^*)}{\partial x_k} \\ \dfrac{\partial(x_k \otimes y_{gyr,f}^f \otimes x_k^*)}{\partial x_k} \\ \dfrac{\partial(q_0^2+q_1^2+q_2^2+q_3^2)}{\partial x_k} \end{bmatrix}$$

$$= 2 \begin{bmatrix} b_{11}y_{acc,f}^f & b_{12}y_{acc,f}^f & b_{13}y_{acc,f}^f & b_{14}y_{acc,f}^f \\ b_{21}y_{acc,f}^f & b_{22}y_{acc,f}^f & b_{23}y_{acc,f}^f & b_{24}y_{acc,f}^f \\ b_{31}y_{acc,f}^f & b_{32}y_{acc,f}^f & b_{33}y_{acc,f}^f & b_{34}y_{acc,f}^f \\ b_{11}y_{gyr,f}^f & b_{12}y_{gyr,f}^f & b_{13}y_{gyr,f}^f & b_{14}y_{gyr,f}^f \\ b_{21}y_{gyr,f}^f & b_{22}y_{gyr,f}^f & b_{23}y_{gyr,f}^f & b_{24}y_{gyr,f}^f \\ b_{31}y_{gyr,f}^f & b_{32}y_{gyr,f}^f & b_{33}y_{gyr,f}^f & b_{34}y_{gyr,f}^f \\ q_0 & q_1 & q_2 & q_3 \end{bmatrix} \tag{A3}$$

$$\begin{cases} \boldsymbol{b}_{11} = \begin{bmatrix} q_0 & q_3 & -q_2 \end{bmatrix} & \boldsymbol{b}_{12} = \begin{bmatrix} q_1 & q_2 & q_3 \end{bmatrix} \\ \boldsymbol{b}_{13} = \begin{bmatrix} -q_2 & q_1 & -q_0 \end{bmatrix} & \boldsymbol{b}_{14} = \begin{bmatrix} -q_3 & q_0 & q_1 \end{bmatrix} \\ \boldsymbol{b}_{21} = \begin{bmatrix} -q_3 & q_0 & q_1 \end{bmatrix} & \boldsymbol{b}_{22} = \begin{bmatrix} q_2 & -q_1 & q_0 \end{bmatrix} \\ \boldsymbol{b}_{23} = \begin{bmatrix} q_1 & q_2 & q_3 \end{bmatrix} & \boldsymbol{b}_{24} = \begin{bmatrix} -q_0 & -q_3 & q_2 \end{bmatrix} \\ \boldsymbol{b}_{31} = \begin{bmatrix} q_2 & -q_1 & q_0 \end{bmatrix} & \boldsymbol{b}_{32} = \begin{bmatrix} q_3 & -q_0 & -q_1 \end{bmatrix} \\ \boldsymbol{b}_{33} = \begin{bmatrix} q_0 & q_3 & -q_2 \end{bmatrix} & \boldsymbol{b}_{34} = \begin{bmatrix} q_1 & q_2 & q_3 \end{bmatrix} \end{cases} \tag{A4}$$

Appendix B

In this section, three figures are shown. Figures A1 and A2 are the results of task 2 from other two participants. Compared with the first participant, the drinking phase (see subfigure (d) of Figure 3) is replaced as displacing, which means, we did not put the cup to the mouth but to another position on the desk. Figure A3 is the estimation of relative orientation between the hand and fingers based on the IMU and optical system; based on this, we obtained the orientation error in subfigure (c) in Figure 8.

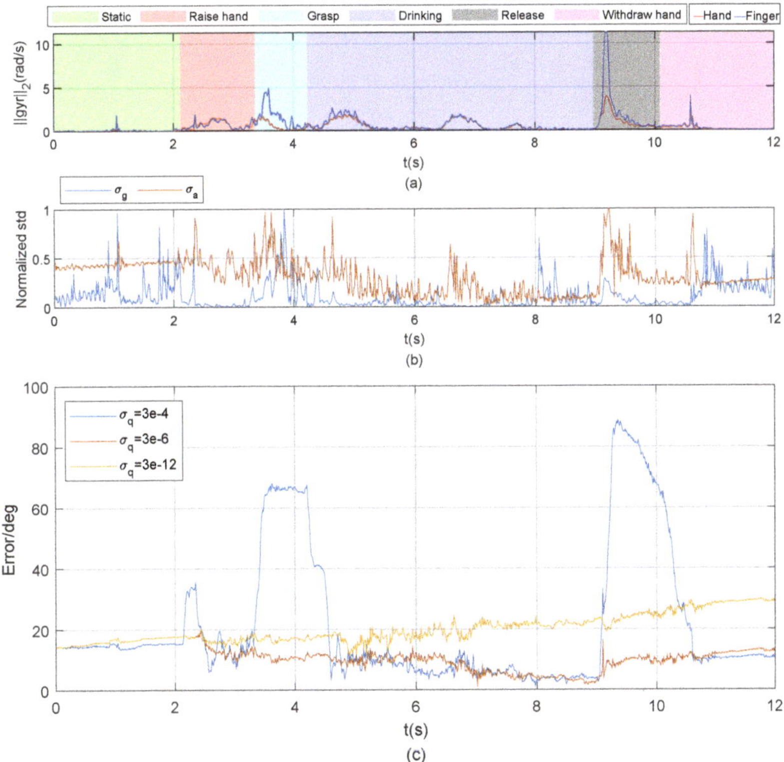

Figure A1. Relative orientation between the hand and thumb during the water-drinking process. Subfigure (**a**) shows the output norms of the two gyroscopes (on the hand and finger tip respectively). Subfigure (**b**) shows the normalized SDs σ_a and σ_g from Equations (18) and (21). Larger σ_a and σ_g mean larger measurement error. The EKF trusts the process model more and the measurement model less when σ_a and σ_g are larger. Subfigure (**c**) shows the estimated results with different SDs of the process model. The variance of process error Q was determined as $\sigma_p I_4$.

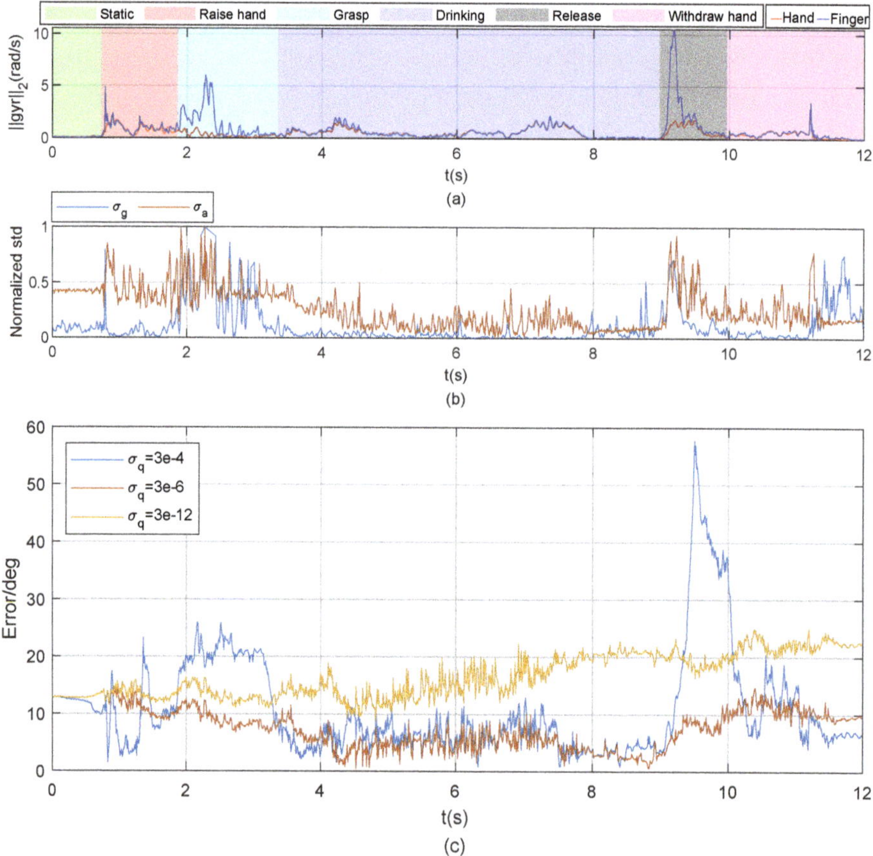

Figure A2. Relative orientation between the hand and thumb during the water-drinking process. Subfigure (**a**) shows the output norms of the two gyroscopes (on the hand and finger tip respectively). Subfigure (**b**) shows the normalized SDs σ_a and σ_g from Equation (18) and (21). Larger σ_a and σ_g mean larger measurement error. The EKF trusts the process model more and the measurement model less when σ_a and σ_g are larger. Subfigure (**c**) shows the estimated results with different SDs of the process model. The variance of process error Q was determined as $\sigma_p I_4$.

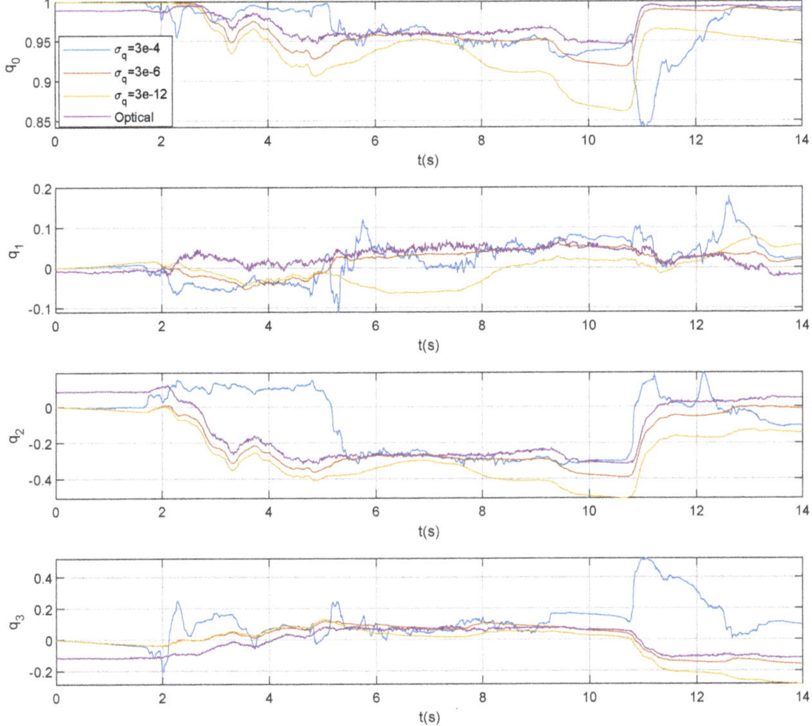

Figure A3. Estimation based on relative orientation between the hand and thumb based on IMU system and optical system. Orientations are expressed based on quaternion; based on these results, we obtained the orientation error in subfigure (c) of Figure 8. σ_p has the same meaning as in Figure 8.

References

1. Oka, K.; Sato, Y.; Koike, H. Real-time fingertip tracking and gesture recognition. *IEEE Comput. Graph. Appl.* **2002**, *22*, 64–71. [CrossRef]
2. Tunik, E.; Saleh, S.; Adamovich, S.V. Visuomotor discordance during visually-guided hand movement in virtual reality modulates sensorimotor cortical activity in healthy and hemiparetic subjects. *IEEE Trans. Neural Syst. Rehabil. Eng.* **2013**, *21*, 198–207. [CrossRef] [PubMed]
3. Vignais, N.; Miezal, M.; Bleser, G.; Mura, K.; Gorecky, D.; Marin, F. Innovative system for real-time ergonomic feedback in industrial manufacturing. *Appl. Ergon.* **2013**, *44*, 566–574. [CrossRef] [PubMed]
4. Szturm, T.; Peters, J.F.; Otto, C.; Kapadia, N.; Desai, A. Task-specific rehabilitation of finger-hand function using interactive computer gaming. *Arch. Phys. Med. Rehabil.* **2008**, *89*, 2213–2217. [CrossRef] [PubMed]
5. Ahmad, N.; Ghazilla, R.A.R.; Khairi, N.M.; Kasi, V. Reviews on various inertial measurement unit (IMU) sensor applications. *Int. J. Signal Process. Syst.* **2013**, *1*, 256–262. [CrossRef]
6. Jebsen, R.H.; Taylor, N.; Trieschmann, R.B.; Trotter, M.J.; Howard, L.A. An objective and standardized test of hand function. *Arch. Phys. Med. Rehabil.* **1969**, *50*, 311–319.
7. Platz, T.; Pinkowski, C.; van Wijck, F.; Kim, I.H.; Di Bella, P.; Johnson, G. Reliability and validity of arm function assessment with standardized guidelines for the Fugl-Meyer Test, Action Research Arm Test and Box and Block Test: a multicentre study. *Clin. Rehabil.* **2005**, *19*, 404–411. [CrossRef]
8. Kapur, A.; Virji-Babul, N.; Tzanetakis, G.; Driessen, P.F. Gesture-Based Affective Computing on Motion Capture Data. In Proceedings of the International Conference on Affective Computing and Intelligent Interaction, Beijing, China, 22–24 October 2005; Springer: Berlin/Heidelberg, Germany, 2005; Volume 3784, pp. 1–7.

9. Guna, J.; Jakus, G.; Pogačnik, M.; Tomažič, S.; Sodnik, J. An analysis of the precision and reliability of the leap motion sensor and its suitability for static and dynamic tracking. *Sensors* **2014**, *14*, 3702–3720. [CrossRef]

10. Smeragliuolo, A.H.; Hill, N.J.; Disla, L.; Putrino, D. Validation of the Leap Motion Controller using markered motion capture technology. *J. Biomech.* **2016**, *49*, 1742–1750. [CrossRef]

11. Lu, W.; Tong, Z.; Chu, J. Dynamic hand gesture recognition with leap motion controller. *IEEE Signal Process. Lett.* **2016**, *23*, 1188–1192. [CrossRef]

12. Mohandes, M.; Aliyu, S.; Deriche, M. Arabic sign language recognition using the leap motion controller. In Proceedings of the 2014 IEEE 23rd International Symposium on Industrial Electronics (ISIE), Istanbul, Turkey, 1–4 June 2014; pp. 960–965.

13. Borghetti, M.; Sardini, E.; Serpelloni, M. Sensorized Glove for Measuring Hand Finger Flexion for Rehabilitation Purposes. *IEEE Trans. Instrum. Meas.* **2013**, *62*, 3308–3314. [CrossRef]

14. Chen, S.; Lou, Z.; Chen, D.; Jiang, K.; Shen, G. Polymer-Enhanced Highly Stretchable Conductive Fiber Strain Sensor Used for Electronic Data Gloves. *Adv. Mater. Technol.* **2016**, *1*, 1600136. [CrossRef]

15. Chen, K.Y.; Patel, S.N.; Keller, S. Finexus. In *CHI 2016*; Kaye, J., Druin, A., Lampe, C., Morris, D., Hourcade, J.P., Eds.; The Association for Computing Machinery: New York, NY, USA, 2016; pp. 1504–1514. [CrossRef]

16. Ma, Y.; Mao, Z.H.; Jia, W.; Li, C.; Yang, J.; Sun, M. Magnetic Hand Tracking for Human-Computer Interface. *IEEE Trans. Magn.* **2011**, *47*, 970–973. [CrossRef]

17. Lin, B.S.; Hsiao, P.C.; Yang, S.Y.; Su, C.S.; Lee, I.J. Data Glove System Embedded With Inertial Measurement Units for Hand Function Evaluation in Stroke Patients. *IEEE Trans. Neural Syst. Rehabil. Eng.* **2017**, *25*, 2204–2213. [CrossRef] [PubMed]

18. Chang, H.T.; Chang, J.Y. Sensor Glove based on Novel Inertial Sensor Fusion Control Algorithm for 3D Real-time Hand Gestures Measurements. *IEEE Trans. Ind. Electron.* **2019**, 1. [CrossRef]

19. Salchow-Hömmen, C.; Callies, L.; Laidig, D.; Valtin, M.; Schauer, T.; Seel, T. A Tangible Solution for Hand Motion Tracking in Clinical Applications. *Sensors* **2019**, *19*, 208. [CrossRef]

20. Seel, T.; Ruppin, S. Eliminating the Effect of Magnetic Disturbances on the Inclination Estimates of Inertial Sensors. *IFAC-PapersOnLine* **2017**, *50*, 8798–8803. [CrossRef]

21. Madgwick, S.O.; Harrison, A.J.; Vaidyanathan, R. Estimation of IMU and MARG orientation using a gradient descent algorithm. In Proceedings of the 2011 IEEE International Conference on Rehabilitation Robotics, Zurich, Switzerland, 29 June–1 July 2011; pp. 1–7.

22. Teufl, W.; Miezal, M.; Taetz, B.; Fröhlich, M.; Bleser, G. Validity, Test-Retest Reliability and Long-Term Stability of Magnetometer Free Inertial Sensor Based 3D Joint Kinematics. *Sensors* **2018**, *18*, 1980. [CrossRef]

23. Seel, T.; Schauer, T.; Raisch, J. Joint axis and position estimation from inertial measurement data by exploiting kinematic constraints. In Proceedings of the IEEE International Conference on Control Applications (CCA), Dubrovnik, Croatia, 3–5 October 2012; IEEE: Piscataway, NJ, USA, 2012; pp. 45–49. [CrossRef]

24. Laidig, D.; Lehmann, D.; Bégin, M.A.; Seel, T. Magnetometer-free realtime inertial motion tracking by exploitation of kinematic constraints in 2-dof joints. In Proceedings of the 2019 41st Annual International Conference of the IEEE Engineering in Medicine and Biology Society (EMBC), Berlin, Germany, 23–27 July 2019; pp. 1233–1238.

25. Lehmann, D.; Laidig, D.; Seel, T. Magnetometer-free motion tracking of one-dimensional joints by exploiting kinematic constraints. *Proc. Autom. Med. Eng.* **2020**, *1*, 027.

26. Tedaldi, D.; Pretto, A.; Menegatti, E. A robust and easy to implement method for IMU calibration without external equipments. In Proceedings of the 2014 IEEE International Conference on Robotics and Automation (ICRA), Hong Kong, China, 31 May–5 June 2014; pp. 3042–3049.

27. Fong, W.; Ong, S.; Nee, A. Methods for in-field user calibration of an inertial measurement unit without external equipment. *Meas. Sci. Technol.* **2008**, *19*, 085202. [CrossRef]

28. Kortier, H.G.; Sluiter, V.I.; Roetenberg, D.; Veltink, P.H. Assessment of hand kinematics using inertial and magnetic sensors. *J. NeuroEng. Rehabil.* **2014**, *11*, 70. [CrossRef] [PubMed]

29. Shepperd, S.W. Quaternion from rotation matrix. *J. Guid. Control* **1978**, *1*, 223–224. [CrossRef]

sensors

Systematic Review

Wearable Sensor-Based Real-Time Gait Detection: A Systematic Review

Hari Prasanth [1,2,†]**, Miroslav Caban** [3,4,†]**, Urs Keller** [4]**, Grégoire Courtine** [5,6,7,8]**, Auke Ijspeert** [3]**, Heike Vallery** [2,9,*] **and Joachim von Zitzewitz** [4]

[1] ONWARD, Building 32, Hightech Campus, 5656 AE Eindhoven, The Netherlands; hari.prasanth@onwd.com

[2] Faculty of Mechanical, Maritime and Materials Engineering, Delft University of Technology, Mekelweg 2, 2628 CD Delft, The Netherlands

[3] Institute of Bioengineering, École Polytechnique Fédérale de Lausanne (EPFL), 1015 Lausanne, Switzerland; miroslav.caban@epfl.ch (M.C.); auke.ijspeert@epfl.ch (A.I.)

[4] ONWARD, EPFL Innovation Park Building C, 1015 Lausanne, Switzerland; urs.keller@onwd.com (U.K.); joachim.vonzitzewitz@onwd.com (J.v.Z.)

[5] Center for Neuroprosthetics and Brain Mind Institute, School of Life Sciences, Swiss Federal Institute of Technology (EPFL), 1015 Lausanne, Switzerland; gregoire.courtine@epfl.ch

[6] Department of Neurosurgery, Lausanne University Hospital (CHUV) and University of Lausanne (UNIL), 1011 Lausanne, Switzerland

[7] Department of Clinical Neuroscience, Lausanne University Hospital (CHUV) and University of Lausanne (UNIL), 1011 Lausanne, Switzerland

[8] Defitech Center for Interventional Neurotherapies (.NeuroRestore), CHUV/UNIL/EPFL, 1011 Lausanne, Switzerland

[9] Department of Rehabilitation Medicine, Erasmus MC, 3000 CA Rotterdam, The Netherlands

[*] Correspondence: h.vallery@tudelft.nl

[†] These authors contributed equally to this work.

Citation: Prasanth, H.; Caban, M.; Keller, U.; Courtine, G.; Ijspeert, A.; Vallery, H.; von Zitzewitz, J. Wearable Sensor-Based Real-Time Gait Detection: A Systematic Review. *Sensors* **2021**, *21*, 2727. https://doi.org/10.3390/s21082727

Academic Editor: Jochen Klenk

Received: 28 February 2021
Accepted: 8 April 2021
Published: 13 April 2021

Publisher's Note: MDPI stays neutral with regard to jurisdictional claims in published maps and institutional affiliations.

Abstract: Gait analysis has traditionally been carried out in a laboratory environment using expensive equipment, but, recently, reliable, affordable, and wearable sensors have enabled integration into clinical applications as well as use during activities of daily living. Real-time gait analysis is key to the development of gait rehabilitation techniques and assistive devices such as neuroprostheses. This article presents a systematic review of wearable sensors and techniques used in real-time gait analysis, and their application to pathological gait. From four major scientific databases, we identified 1262 articles of which 113 were analyzed in full-text. We found that heel strike and toe off are the most sought-after gait events. Inertial measurement units (IMU) are the most widely used wearable sensors and the shank and foot are the preferred placements. Insole pressure sensors are the most common sensors for ground-truth validation for IMU-based gait detection. Rule-based techniques relying on threshold or peak detection are the most widely used gait detection method. The heterogeneity of evaluation criteria prevented quantitative performance comparison of all methods. Although most studies predicted that the proposed methods would work on pathological gait, less than one third were validated on such data. Clinical applications of gait detection algorithms were considered, and we recommend a combination of IMU and rule-based methods as an optimal solution.

Keywords: wearable sensor; real-time gait detection; gait analysis; insole pressure sensors; inertial measurement unit; pathological gait; gait rehabilitation; assistive device

1. Introduction

1.1. Motivation

Traditionally, performing gait analysis required a dedicated laboratory and expensive equipment, which has limited its scope of applications. Recent advancements in technology have led to reliable, affordable, and wearable sensors for gait analysis that enable its use outside of a laboratory environment and during activities of daily living. One of its primary uses has been in diagnosing walking impairment in people with gait disabilities [1–4] and inspires control mechanisms of exoskeletons [5,6] and prostheses [7], among other applications [8–12]. More specifically, real-time gait analysis has proven essential in applications necessitating real-time control such as exoskeletons and prostheses, as well as gait rehabilitation involving Functional Electrical Stimulation (FES) [13,14] or Epidural Electrical Stimulation (EES) [15].

Physiological human gait is a quasi-periodic, synergistic process involving the timely actuation of several lower-limb muscles, well-coordinated by neurons in the brain and the spinal cord [5]. Gait disorders and disabilities can arise due to various reasons including amputation of lower limbs, neurological diseases such as Parkinson's, cerebral palsy and Huntington's, as well as through stroke or paralysis following an injury to the brain or the spinal cord. This review is particularly motivated by the use of real-time gait analysis in an on-going clinical study (STIMO, ClinicalTrials.gov, NCT02936453): a First-in-Man study to confirm the safety and feasibility of a closed-loop EES in combination with overground robot assisted rehabilitation training for patients with chronic incomplete spinal cord injury (SCI) [16].

We first carried out a systematic review and meta analysis across four major scientific databases (Scopus, Web of Science, Cochrane and PubMed) to identify the current state of the art in wearable sensor-based real-time gait analysis. We then extracted studies that focused on pathological gait and analyzed the most common sensors and techniques used in clinical applications.

1.2. Previous Reviews

Before moving into the details of our study, we will briefly discuss our analysis of existing review articles from literature, the results of which are summarized in Table 1. It can be noted from the table that reviews done so far are either specific to a particular category of gait detection method, are not systematic, lack coverage across major citation databases or focus only on wearable sensing. Furthermore, throughout the literature, we noticed the synonymous usage of the terminologies gait detection, gait event detection, and gait phase detection. Although we appreciate the specific difference in terminologies, for the sake of brevity, we will use gait detection to imply gait event and/or gait phase detection.

Among the 11 review studies analyzed, only four of them were systematic reviews. Taborri et al. [17] performed a systematic review on wearable and non-wearable sensors used in gait detection. The study identified various wearable sensors such as inertial measurement units (IMU), insole pressure sensors (IPS), electromyography (EMG) and electroneurogram (ENG), and non-wearable sensors such as opto-electronic systems, force plates, and ultrasonic sensors. The study, however, was limited to sensors and did not provide any review of gait detection methods. Panebianco et al. [18] performed a systematic review covering PubMed, Scopus, and Web of Science. Although the search keywords were not restrictive to rule-based methods, all 17 of the studies involved were limited to rule-based methods. Caldasa et al. [19] performed a systematic review across major databases such as Web of Science, ScienceDirect, IEEE, PubMed/MEDLINE, SCOPUS, CINAHL, and Cochrane, thereby ensuring an exhaustive coverage. However, the review was limited to only artificial intelligence-based gait detection methods using inertial measurements, resulting in only 22 studies that met the acceptance criteria. Chen et al. [20] performed a systematic review focusing particularly on quantifiable gait measures and tangible evaluation techniques that are based on wearable sensors, particularly inertial measurement units (IMU). The study also includes a review of nonlinear analysis techniques such as phase portrait, Poincaré map, Lyapunov exponent (for gait stability assessment), and elliptical Fourier analysis (for gait complexity assessment). The study, however, did not report any real-time gait analysis methods. It can be noticed from these systematic reviews that they are either limited to review of sensors or to a specific category of gait detection method or did not consider real-time gait detection methods. Finally, none of these studies presented information about pathological aspects of wearable sensor-based gait detection.

To the best of the authors' knowledge, there does not exist a systematic literature review that identifies various wearable sensing options, real-time gait analysis methods, and presents pathological aspects of wearable sensor-based gait detection. We therefore performed a systematic review and meta-analysis across the four major scientific databases mentioned earlier, following the PRISMA (Preferred Reporting Items for Systematic Reviews and Meta-Analyses) standard [21], covering 1262 articles, as will be discussed in the next sections.

Table 1. Previous reviews: gaps identified from existing review articles from literature, covering wearable sensor-based gait detection.

References	Focus of Review	Database Covered	Gaps Identified in Existing Reviews	Number of Articles Included
Song et al. [8]	Health sensing techniques with a particular focus on smartphone sensing	Not specified	Not a systematic review, no review of gait detection methods	-
Shull et al. [22]	Clinical impact of wearable sensing	MEDLINE, Science Citation Index Expanded, CINAHL, Cochrane	Not a systematic review, no review of gait detection methods	76
López-Nava and Muñoz-Meléndez [23]	Review on inertial sensors and sensor fusion methods for human motion analysis,	ACM Digital Library, IEEE Xplore, PubMed, ScienceDirect, Scopus, Taylor Francis Online, Web of Science, Wiley Online Library	Not a systematic review, no review of gait detection methods, review limited to inertial sensors	37
Novak and Riener [24]	Sensor fusion methods in wearable robotics	Not specified	Not a systematic review, no review of gait detection methods	-
Vu et al. [25]	Gait event detection methods applicable specifically for prosthetic devices	Scopus, ScienceDirect, Google Scholar	Not a systematic review, review restricted to one category of rehabilitation devices	87
Rueterbories et al. [26]	Review of sensor configurations and placements, and a brief review of gait detection methods	Not specified	Not a systematic review, gait detection methods were reviewed very briefly	-
Perez-Ibarra et al. [27]	Brief review comparing gait event detection methods, sensors used, placement of sensors and subjects involved	Not specified	Brief review, as a subset of the article	18
Taborri et al. [17]	Wearable and non-wearable sensors used in gait detection	Scopus, Google Scholar, PubMed	No review of gait detection methods	72
Caldasa et al. [19]	Artificial intelligence-based gait event detection methods using inertial measurements	Web of Science, ScienceDirect, IEEE, PubMed/MEDLINE, Scopus, CINAHL, Cochrane	Review was limited to only one type of sensor and one type of gait detection algorithm	22
Panebianco et al. [18]	Rule-based methods	PubMed, Scopus and Web of Science	Review was limited to only one category of gait detection algorithm	17
Chen et al. [20]	Quantifiable gait measures and tangible evaluation techniques that are based on wearable sensors	PubMed, IEEE Xplore, ACM Digital Library, EBSCO and Cochrane Library	No review of real-time gait analysis methods	35

1.3. Structure of the Report

The remainder of the article is organized into three sections. In Section 2, we describe the methods followed in setting up the systematic review. In Section 3, we present and discuss the results regarding: the search results in general in Section 3.1, gait events and gait phases in Section 3.2, wearable sensors in Section 3.3, algorithms used for wearable sensor-based real-time gait analysis in Section 3.4, and interpretations towards clinical applications in Section 3.5. Finally, in Section 4, we present the conclusions.

2. Method: Setting up the Review

2.1. Choice of Databases

Haddaway et al. [28] classified scientific literature databases into two categories: Academic Citation Database (ACDB) and Academic Citation Search Engine (ACSE). ACDBs include the traditional Boolean string-based search engines such as Scopus, Web of Science and PubMed, while ACSEs include Google scholar and semantic/natural language-based search engines such as Microsoft Academic Search and Semantic Scholar. We first explored both categories before making a choice.

We considered nine of the most popular ACDBs in our selection process: CINAHL, EBSCO, ACM digital library, IEEE Xplore, Science Direct, Scopus, Web of Science, Cochrane, and PubMed. CINAHL, EBSCO, and ACM digital libraries were not included in the study because of a lack of access to them. We could not include IEEE Xplore since it limits the number of search terms to 15, which is noticeably lower than the number of keywords used in this study (see Table 2). However, this should not impact the comprehensiveness of our study, since IEEE articles are already indexed in Scopus and Web of Science. Science Direct was not included, as a recent update in their search keyword input framework limited us from inputting all our keywords, thereby making it unsuitable for our systematic review. However, this should not impact the comprehensiveness of our study since Science Direct and Scopus share the same database [29] and come from the same parent company (Elsevier). Finally, we decided not to include ACSEs in the review primarily because of deficient repeatability and reproducibility of search results, among other factors [28,30–32]. The remaining databases were thus Scopus, Web of Science, Cochrane and PubMed. ACDBs such as Scopus and Web of Science use a selective procedure to safeguard against low-quality or low-impact material being indexed [33], while Cochrane and PubMed are expected to add more clinically relevant studies.

2.2. Choice of Keywords for Search

To establish an appropriate search phrase, a pre-search was carried out first, collecting a list of keywords used by gait analysis researchers. In an attempt to find an optimum keyword-combination from the list, we analyzed these keyword-combinations by taking the conducted search results (from Scopus) to VOS-viewer [34], a metadata analysis software. VOS-viewer performs clustering of search results based on title, abstract and keywords of corresponding articles and illustrates the results graphically as shown in Figure 1. This gives us a bigger picture of the nature of articles returned by the search engine for the corresponding choice of keywords. The size of each node indicates the relative relevance (based on the frequency of occurrence of keywords) of that topic among the list of articles returned by the search query. This procedure was iterated and refined several times before arriving at the final search phrase listed in Table 2. We believe that it made the decision-making less subjective and biased. Although this study is not limited to any particular wearable sensor, we included the keywords 'IMU' and 'insole' (see Table 2) explicitly so as not to miss out articles related to these two types of sensors, while also retaining the word 'sensor' in the search phrase to make it inclusive for every other type of wearable sensors.

Table 2. Keyword combination used for search in Scopus database which resulted in 697 articles (see Figure 2). The same keyword combination was used in the other databases as well, except adapting syntax to individual search engines.

realtime OR "real time" OR online
AND
gait OR walking OR locomotion OR "lower limb" OR "lower body" OR leg OR "lower extremity"
AND
analysis OR detection OR evaluation OR assessment OR estimation OR reconstruction OR tracking
AND
wearable OR portable OR mobile
AND
sensor OR "inertial measurement unit" OR accelerometer OR IMU OR gyroscope OR insole OR in-sole

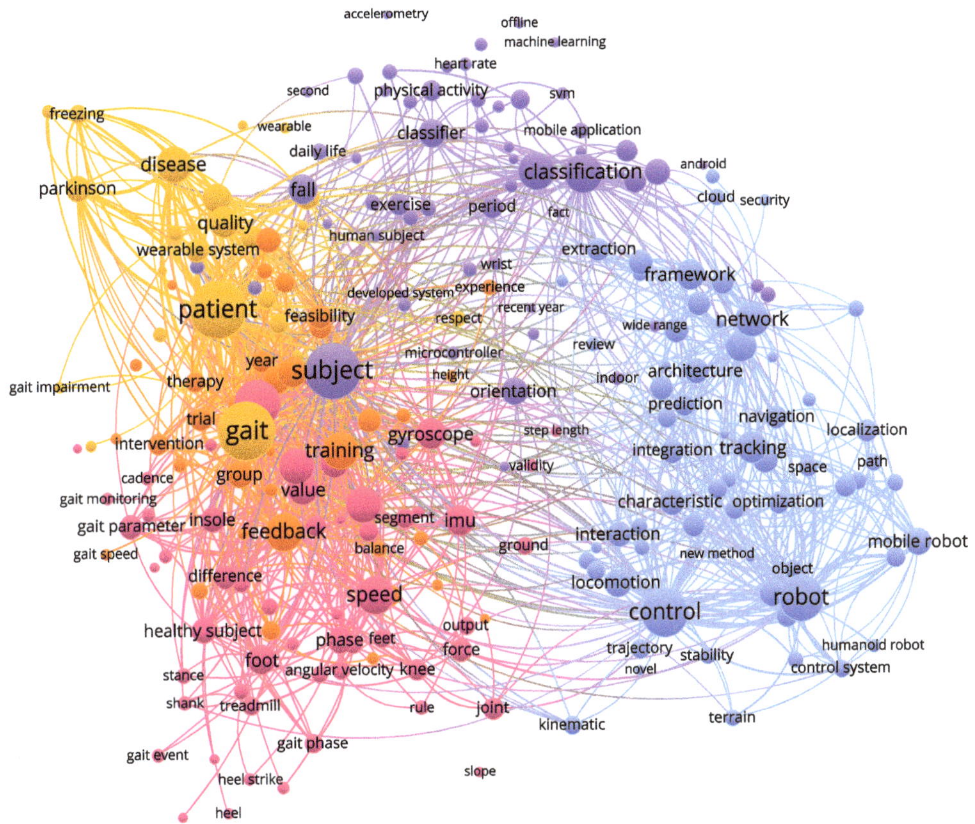

Figure 1. Word cloud showing most frequently occurring keywords present in search results for a search query in Scopus (for instance, see Table 2), visualized by the software VOS-Viewer [35]. The size of each node indicates the relative relevance (based on the frequency of occurrence of keywords) of that topic among the list of articles returned by the search query. Connections between nodes were not used in refining the search terms.

2.3. Carrying Out the Review in a Systematic Manner

In this study, we followed the PRISMA standard for systematic review. A total of 1262 articles were part of the review, including 1221 articles retrieved from the four databases (see Figure 2) and an additional 41 articles added later from bibliographies of the former. Figure 3 shows the PRISMA flow diagram illustrating the screening procedure followed. The screening procedure was performed independently by the two lead authors of our study. Studies for which the authors had difference of opinion on their exclusion (such as studies that were at the border line of the exclusion criteria) were mutually discussed and decided upon. Although assessment of paper quality is not mandated by the PRISMA standard, the authors' choice of databases ensured that low-quality and low-impact material was not considered.

We classify the most commonly used gait features into intra-stride (within stride) and inter-stride (between strides) features, and into temporal and spatial features. Intra-stride features are of higher granularity, looking at the gait in more detail, while inter-stride features are of lower granularity. The choice of each depends on the application. Examples of each category are listed in Table 3.

In this review, we focus on intra-stride temporal gait features (ISTGFs), the rationale being that inter-stride features can be obtained from related intra-stride features (e.g., stride duration and cadence can be obtained from temporal information of consecutive heel strikes). We therefore believe that detecting sufficient ISTGFs is sufficient for insightful gait analysis in practice. In addition, ISTGFs are necessary for real-time control applications such as in the clinical study [16] that has served as a particular motivation for this review. Hence, studies not estimating the ISTGFs (such as indoor localization algorithms, gait reconstruction methods and activity classification methods) were excluded from this review.

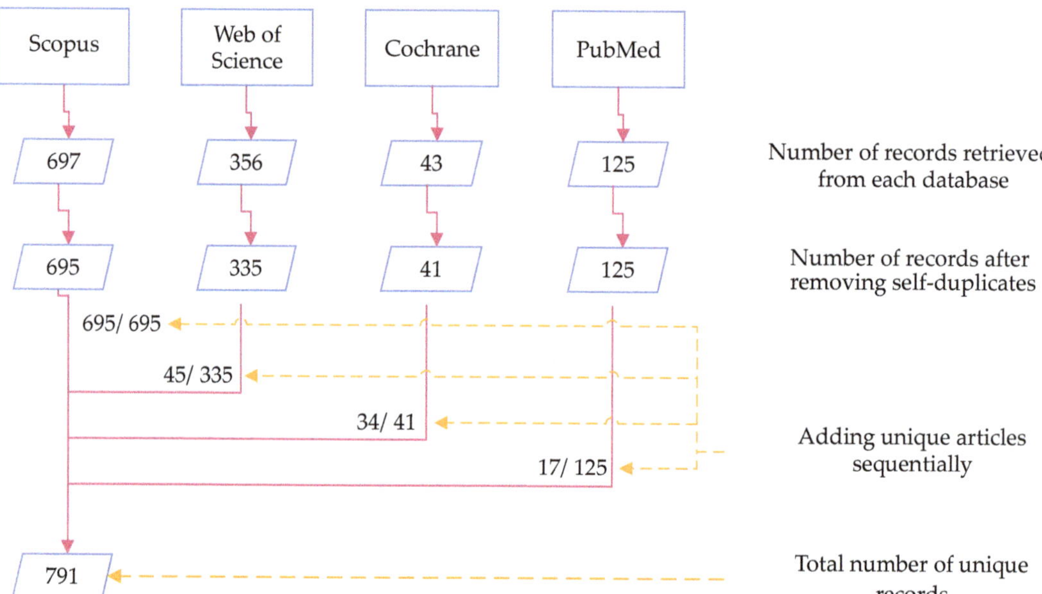

Figure 2. Collecting unique articles from the databases was carried out sequentially, starting with Scopus where 697 articles were extracted and of which 695 unique records were identified. Out of the 335 unique records identified from Web of Science, 290 articles already appeared in the results from Scopus and hence the remaining 45 unique records were added. Similarly, 34 from Cochrane and 17 from PubMed were added to the list of unique records.

Table 3. Classification of the most commonly used gait features into intra-stride and inter-stride as well as into temporal, spatial, and spatio-temporal features. The scope of this review is primarily limited to the intra-stride temporal gait features (ISTGFs) highlighted in blue.

Features	Intra-Stride Features	Inter-Stride Features
Temporal	Gait events Gait phases Step duration Swing/stance duration	Stride duration Cadence
Spatial	Step length	Stride length
Spatio-temporal	Joint angles Segment angles, segment positions Joint torques Ground reaction force Centre of pressure	

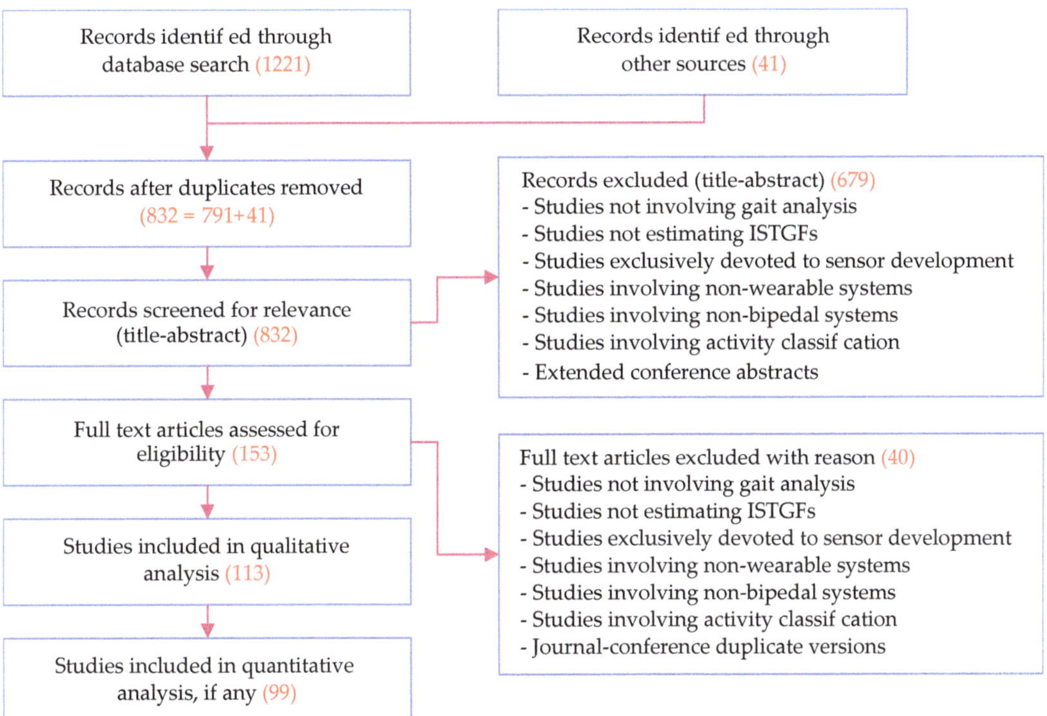

Figure 3. PRISMA flow diagram illustrating the screening procedure. Reasons for exclusion and the number of articles retrieved at each stage are indicated in red. In addition, 832 articles were left after removing duplicates. Out of those, 679 articles were eliminated through title-abstract screening, based on a set of exclusion criteria as listed in the PRISMA flow diagram. The remaining 153 articles qualified for full-text screening, of which 40 were excluded and the remaining 113 qualified for full-text review. Out of these, 99 articles were also used for quantitative analysis. ISTGF—intra-stride temporal gait feature.

Other major reasons for exclusion are listed in the PRISMA flow diagram (see Figure 3). Studies not involving bipedal systems, studies not involving wearable systems, and studies devoted purely to (wearable) sensor development are directly excluded. When both a conference version of an article and its extended journal version appeared in our search results, the conference version was excluded. In a rare observation, we noted two sets of nearly duplicate conference publications from the same set of authors [7,36–38]. In this case, only the latest ones were considered for further review. Finally, if an article was found to compare, list, or review multiple gait event detection methods introduced in other studies, the original studies were included in the review rather than the former.

3. Results and Discussion

3.1. Search Results

The systematic review resulted in the identification of 1262 studies, as indicated in Figure 3. After removal of duplicates, we were left with 832 unique studies. Out of these, 113 and 99 qualified for qualitative and quantitative analysis respectively. Studies that underwent qualitative analysis influenced our discussions while quantitative analysis resulted in the extraction of metadata that was presented throughout the paper. The 14 studies not included in the quantitative analysis did not contain all the necessary metadata and therefore did not contribute directly to the metrics presented.

Review studies cited in Section 1.2 have an average of 30 full-text articles per review, with a maximum of 76 by Shull et al. [22]. The present work is thus one of the most comprehensive reviews on the subject.

3.2. Gait Events and Gait Phases

Irrespective of the type of wearable sensors used and the type of real-time gait analysis methods followed, here we will briefly discuss the two major ISTGFs from literature: gait events and gait phases.

Researchers followed different terminologies for various gait events. Some authors prefer to use initial contact (IC) (or sometimes touch down) instead of being more specific as to whether the contact is with heel strike (HS) or with toe strike (TS). Although HS is most often the obvious initial contact in unimpaired gait, it is not necessarily the case with impaired gait. For instance, initial contact in the case of toe walking could be TS instead of HS. Similarly, some authors prefer end contact or foot off instead of using the more explicit terminologies: toe off (TO) or heel off (HO). On the other hand, for gait phases, researchers tend to use consistent terminology to decompose stance and swing: loading response, mid-stance, terminal stance, pre-swing, initial swing, mid-swing, and terminal swing [6,39–42].

A detailed count of gait events and gait phases used in the resulting studies is shown in Figure 4a,b respectively. TO and HS are the most widely identified gait events irrespective of the type of sensor used. A total of 42 studies detected TO while 45 detected HS, suggesting the high relevance and ease with which these events can be identified from gait signals. Among the gait phases, swing (22 studies) was the most widely identified gait phase followed by mid-stance (17 studies).

3.3. Sensors

In order to have an overview of relevant wearable sensors available on the market, a survey of off-the-shelf devices was conducted. Wearable sensors identified include primarily IMUs, insole pressure sensors (IPS), electromyography (EMG) sensors, goniometers, inclinometers, electromagnetic trackers, and stretch sensors. However, only three main types of wearable sensors could be identified among the 99 studies that featured in the quantitative analysis of our review: IMU, IPS, and a combination of the two. The distribution of sensors used is shown in Figure 5a and discussed in the following sections. A check was performed on the Scopus database to ensure that our explicit addition of search terms regarding IMU and IPS were not heavily biasing the results. In fact, only 98 additional studies were found compared to the 725 identified without the search terms, and we therefore conclude that their explicit addition was not responsible for the dominance of these sensor types.

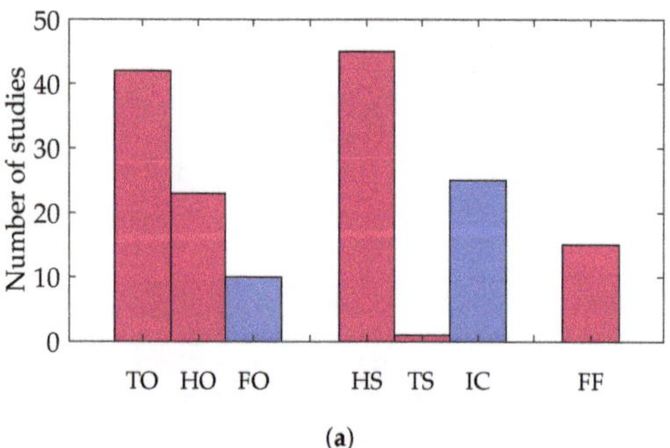
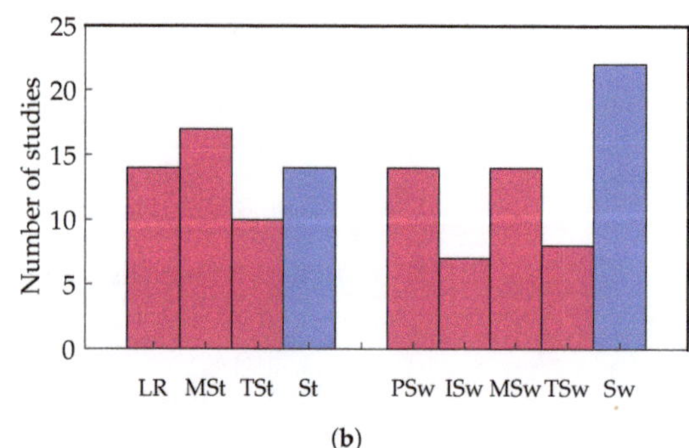

Figure 4. (**a**) Distribution of studies based on the detected gait events; (**b**) distribution of studies based on the gait phases identified. Gait events/phases reported with greater (temporal) specificity are shown in magenta while gait events/phases reported with less specificity are shown in blue. For instance, initial contact (IC, blue) is not specific as to whether the contact is with the heel or toe while heel strike (HS, magenta) and toe strike (TS, magenta). TO—toe off, HO—heel off, FO—foot off, HS—heel strike, TS—toe strike, IC—initial contact, FF—foot flat, LR—loading response, MSt—mid-stance, TSt—terminal stance, St—stance, PSw—pre-swing, ISw—initial swing, MSw—mid-swing, TSw—terminal swing and Sw—swing.

3.3.1. Inertial Measurement Units

Inertial measurement units (IMUs) are sensors combining accelerometers and gyroscopes to measure linear acceleration and angular velocity of the body to which it is attached. Optionally, it also comes with a magnetometer that can estimate the magnetic north based on the earth's magnetic field and is sometimes called an inertial-magnetic measurement unit. As shown in Figure 5a, we notice that IMUs are the most widely used sensors with 77% of studies using it either alone (67%), or in combination with IPSs (10%).

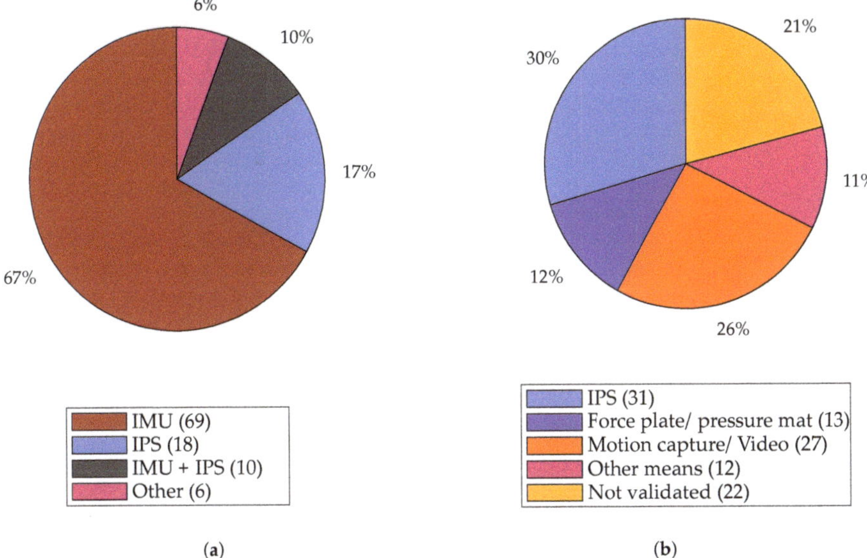

Figure 5. (**a**) Distribution of studies based on the type of wearable sensors used; (**b**) distribution of studies based on the type of sensors used for ground-truth validation of IMU-based gait analysis. Absolute number of studies in each category is listed within parentheses. IMU—inertial measurement unit, IPS—insole pressure sensor, EMG—electromyography sensor.

Appropriate sensor placement often simplifies or even eliminates any calibration required for gait detection algorithms. Gyroscopes are invariant to translation in position [39] since the angular velocity of a rigid body is the same at any point along the body (assuming the orientation of the sensor remains the same with respect to the body segment). They are also unaffected by gravity and are less prone to noise. Accelerometers, on the other hand, are reported to be more noisy, subject to the influence of gravity and sensitive to both position and orientation. The sensitivity to sensor orientation is typically avoided by considering the norm of acceleration instead of acceleration along individual axes. The influence of gravity is often used to estimate the orientation of the sensor with respect to the earth frame of reference (in combination with additional constraints such as the Earth's magnetic field).

With IMUs, various possibilities for sensor placement exist and researchers have tried a number of approaches for gait-related studies, placing IMUs on different body segments or combinations thereof. The approaches are quantified in Figure 6. Among the studies which used IMUs, the shank was the most widely preferred lower-body segment for gait analysis (with 39 studies) closely followed by the foot (with 38 studies).

Although these numbers give us a better understanding of the preference followed in literature, they alone do not necessarily tell us whether these segments are the ones that provide the richest information of gait or if preferring these segments over others makes it easy to identify ISTGFs from gait data. Some researchers attempted to give a clearer answer to these questions. Li et al. [43] compares IMU signals from the thigh, the shank and the foot based on what they call, the "energy of acceleration," which is the norm of raw acceleration minus gravity. They argue that the "energy of acceleration" (when inspected graphically) appears to be relatively more "stable" (i.e., constant) in the foot compared to the other two body segments and hence recommend IMU placement at the foot. Mazilu et al. [12], in the context of freezing of gait, reports 98% or more detection performance for all three body segments, suggesting that the question of optimal sensor placement is irrelevant in the context of freezing of gait. Jasiewicz et al. [44] report that, upon comparison of three rule-based gait event detection methods, the method based on foot angular velocity and linear acceleration was significantly more accurate than that of the method based on the shank for spinal cord injured (SCI) subjects. Taborri et al. [45] made a similar observation for a hidden Markov model (HMM)-based classifier. They report that the accuracy of a HMM-based classifier for gait event detection was better when the angular velocity of the foot was used rather than of the thigh or the shank. The relatively high preference for the foot could also be justified by the results from neuro-behavioral experiments suggesting that limb endpoints are the primary variables used to coordinate locomotion in animals and humans [15]. Bejarano et al. [46] analyzed four signals—linear acceleration in forward and vertical direction, angular velocity, and segment angle normal to the sagittal plane—for the thigh, the shank, and the foot. Acceleration components were discarded (after a preliminary investigation) due to noise and vibrations while the root mean square error between each cycle (as well as the average) was computed for the other two signals. For both angular velocity as well as segment angle, sensors placed on the shank were identified with noticeably low root mean square error and hence the authors recommended using the shank as the preferred location for IMU placement.

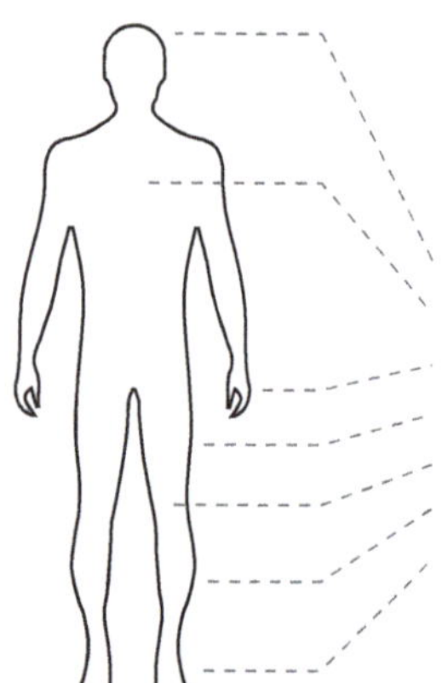

	Number of Studies									
	Total	Single Placement	Placement Combinations							
Head	2	2								
Trunk	10	3								
Upper Limb	2	1								
Thigh	20	5								
Knee	2	1								
Shank	39	23								
Foot	38	25								
			1	1	5	4	4	2	1	1

Figure 6. Number of studies using inertial measurement units (IMUs) that placed the sensor(s) on specific anatomical locations. Single placement contains studies where sensor(s) were placed only in one anatomical location. Placement combinations' columns indicate studies where sensor(s) were placed in more than one location. Each relevant location is marked by a shaded cell and the number of studies using this combination is indicated at the bottom of the column. The total indicates the sum of studies where the sensor(s) were placed on that given anatomical location.

Much like the inter-segment IMU placement problem just discussed, a user could also place the IMU at a different position and orientation within a given body segment, each time it is attached to the body. Such intra-segment differences may result in undesirable variations across data sets and across subjects. This is typically avoided by using a mount/socket so that the sensor falls into the same location every time it is inserted. Anwary et al. [47] suggests that the optimal location for IMU placement on the foot is the medial arch followed by the Achilles tendon.

Raw signals from IMUs are noisy, particularly the accelerometer signals, and thus filters are widely used. Meta-analysis on preprocessing filters used in the case of IMUs revealed that 39 out of 69 studies used at least one preprocessing filter, 31 of which used a low-pass filter among which 15 used the Butterworth low-pass filter. Note that preprocessing filters add to the latency in data processing, which is undesirable in a real-time system.

The orientation measured by an IMU is often useful in gait analysis. One way to estimate the orientation is by integrating the angular velocity from the gyroscope. However, due to gyroscopic bias, such an approach is prone to drift from numerical integration. Under static conditions, accelerometers can be used to estimate the inclination with respect to the gravity vector, while magnetometers can be used to estimate orientation with respect to the Earth's magnetic field (magnetic North). Since acceleration measurements from accelerometers are prone to noise, estimating the orientation outside of static conditions is not accurate from instantaneous sensor data. Magnetometers, on the other hand, are sensitive to external magnetic fields. Sensor fusion methods combine the information from accelerometers, gyroscopes, and magnetometers (or a subset of these) to provide a better estimate of the orientation of the sensor. Kalman filter-based and complementary filter-based methods are the two most popular sensor fusion methods used for estimating orientation from IMUs. Casamassima et al. [48] compared these two methods based on accuracy, computational cost, and energy efficiency and concluded that the Kalman filter-based method was their preferred choice. Two of the most popular methods using the complementary filter are proposed by Mahony et al. [49] and Madgwick et al. [50]. Cirillo et al. [51] observed that a version of the extended Kalman filter-based method that offers similar performance to the two took approximately one order of magnitude more time to process a sample in the MATLAB/Simulink environment and two orders of magnitude more time in an embedded system environment. Overall, Kalman filter-based methods are known to be more accurate but computationally demanding, while complementary filter-based methods are known to be computationally light and fairly accurate [52].

Within the context of walking, accuracy of orientation (and position) estimation using IMUs can be enhanced using additional constraints from the foot–ground interaction. The zero velocity update (ZUPT) algorithm or one of its variants are typically used to compensate drift. The algorithm exploits the fact that, during a part of the stance phase, the stance foot is quasi-static. During this moment, the linear and angular velocity of the foot is assumed to be zero and the drift errors due to integration are reset. Yang et al. [53] estimated the stance duration from thresholds set on both angular velocity and acceleration, which helped in correctly applying the ZUPT. Skog et al. [54] compared four different detectors to identify the zero velocity interval—"acceleration moving variance detector, the acceleration magnitude detector, the angular rate energy (ARE) detector, and a generalized likelihood ratio test detector, referred to as the SHOE"—and concluded that both ARE and SHOE performed with very high accuracy. Inspired by this, Refs. [48,55] used a threshold on the ARE to estimate the ZUPT interval by hypothesizing that the IMU is stationary when ARE is below the threshold. Other variants of ZUPT are also attempted [56,57]. In [57], the foot inclination angle, obtained by integrating the gyroscope signal, was reset to zero during the stance phase based on input from an IPS.

Often, sensors regarded as the gold standard, providing ground-truth information, are used to validate results obtained from IMU-based gait analysis. Figure 5b illustrates the distribution of those sensors used in validating IMU-based gait analysis. Among the studies which used IMU-based gait analysis, it can be observed that IPSs are the most widely used sensors for validating the results with 31 studies compared to 27 studies using motion capture/video. In addition, they are the only wearable sensor to be used for validation.

3.3.2. Insole Pressure Sensors

An insole pressure sensor (IPS) measures the pressure distribution at the foot, which is widely used to estimate the COP along with other gait parameters such as step count, duration of the gait cycle, swing duration, stance duration [58], and foot–ground interaction events (such as HS or TO). IPSs are available in different variants based on optoelectronic sensors, force-sensing resistors (FSRs), capacitive sensors, and piezoelectric sensors, which are based on polyvinylidene difluoride (PVDF) films [59]. PVDF films lack durability, although they are reliable and inexpensive. FSRs, on the other hand, are highly durable, flexible, and inexpensive. The reliability of FSRs is low when estimating the magnitude of force in real-time, but FSRs perform well in detecting the temporal information such as the instant of application of force, as shown by [59], and hence are good candidates for real-time gait event detection. Delgado-Gonzalo et al. [60] report that IPSs have a short lifespan, although the claim is not adequately validated. We noticed in our review that 38 out of 59 studies are from 2014 or later, possibly suggesting that IPSs have become more reliable over the years.

IPSs are the second most widely used wearable sensor with 57% of the studies using it either for sensing (28 studies, see Figure 5a) or validation (31 studies, see Figure 5b). It is the only wearable sensor to be used in validation studies, which arguably makes it the gold standard in wearable sensing [61]. IPSs are often considered as an alternative to force plates in validation studies due to several advantages including cost factor, wearability, and unconstrained movement that allows natural gait in both indoor and outdoor environments. Despite these advantages, there are some constraints to consider. IPSs are typically placed inside the shoe and are thus subject to pressure between it and the foot, which can lead to non-zero pressure readings even when the foot is in swing phase [42,59]. Although IPSs are comparable to force plates when it comes to estimation of temporal features, using them for real-time ground reaction force estimation is not recommended since it takes a considerably longer time to reach the set value compared to a force plate [59].

Unlike IMUs, sensor placement is not a challenging problem for IPS. While the user could place the IMU anywhere within the body segment of interest, IPSs are almost always placed in the subject's shoe, making them fall into the same position with respect to the foot. The traditional approach to place FSRs within an IPS has been to place them at specific hotspots such as the heel, toe, first, and fifth metatarsals. Such IPSs require the correct foot size of the subject so that FSRs are aligned with the correct hotspots. Senanayake et al. [42] reported errors in measurement owing to subjects involving varying foot sizes (6–11) while the IPS was at a fixed size (eight). Lin et al. [62] reported robustness against this offset, caused by the size mismatch of the IPS with the foot, by using the derivative of pressure signals. The authors used an array of 48 pressure sensors, giving a better resolution than the conventional approaches, which place a few FSRs at carefully chosen hotspots. The authors of [63,64] followed a similar approach using a pressure signal from the IPS and its first derivative while using an IPS with 64 optoelectronic sensors. With IPSs getting better in resolution, the approach is shifting towards packing as many sensors within the insole as possible so as to collect data throughout the feet and identify the hotspots not at the hardware end, but later at the software end during signal processing. When used in real-time, this demands more bandwidth for communication and computational power to process the additional information.

In contrast to IMUs where 39 out of 69 studies at least used a preprocessing filter, only six out of 28 studies related to IPS used any sort of preprocessing/filtering. Instead, these studies relied directly on the raw signal from the IPS, likely contributing to shorter latency, an advantage when it comes to real-time systems.

3.3.3. Combination of IPS and IMU

A new kind of product that is emerging in the wearable sensor market is an IPS combined with an IMU, such as Moticon Science from Moticon GmbH, Munich, Germany, Stridalyzer from Retisense, Bangalore, India and Arion Wearable from ATO-GEAR, Eindhoven, The Netherlands. Such a set up allows combining the advantages of both types of sensors. Ten out of 52 studies used a combination of IPS and IMU (in addition to the studies that used IPS separately for validating IMU-based gait detection results).

Depending on the product, it is possible that the position of the IMU is fixed relative to the IPS, thereby minimizing the errors caused by differences in IMU placements between and within segments as well as across data sets and subjects.

3.3.4. Other Wearable Sensors

Other wearable sensors used for gait detection include Electromyography (EMG) sensors, rotary encoders, laser range finders, flex sensors, and capacitive shank orthosis. Other than EMG sensors, all the other sensors measure kinematics. EMG sensors, on the other hand, measure electrical activity in muscles, which gives them an inherent advantage that signals appear earlier than the corresponding movement from muscle activation [5]. Fleischer et al. [65] reports that EMG signals appear 20–80 ms before the resulting contraction begins. This should contribute to early sensing and hence decrease latency in control. Farmer et al. [66] presented an auto-correlation model that takes EMG signals as input to predict the ankle angle, which is claimed to predict around 100 ms in advance. Despite this advantage, which is particularly important in real-time gait detection, it is interesting to note that only one out of the 99 studies used EMG sensors. This could partly be because of usability constraints: the skin is typically prepared by shaving body hairs and applying abrasive gel to increase signal-to-noise ratio, and the sensor is taped to the skin in order to keep the contact constant and reduce motion artifacts. It could also be owed to the fact that EMG signals require more preprocessing/filtering and EMG signals of persons with certain impairments (primarily neurological deficits) can be weak and hard to interpret. Furthermore, EMG-related parameters are subject-dependent and can change regularly due to varying conditions of the skin and body state, such as sweat. Correct sensor placement is also non-trivial and requires some training because the sensor should be placed as close as possible to the belly of the appropriate muscle. This approach may be less appropriate for lay users thus limiting its application. Evaluation of EMG patterns are mostly done using classification algorithms and less often using physiological models [65].

3.4. Real-Time Gait Analysis

We classify various gait analysis methods identified from literature into three main categories: time domain-based, frequency domain-based, and time-frequency domain-based. Table 4 shows various real-time gait analysis methods used by researchers based on this classification, some of which are discussed in greater detail in the subsequent sections.

In summary, we observe that the rule-based methods are the most popular, with a majority of the studies using it, likely due to their simplicity and intuitiveness compared to other computationally expensive solutions. Phase portraits and adaptive oscillators are among the limited number of methods noted for continuous gait phase estimation. Wavelet transforms are seen as more suitable for fast motion transitions, and the method may serve as a better candidate in gait phase estimation than adaptive oscillators.

Performance of the different methods is not compared quantitatively since evaluation criteria varied from one study to another, which makes an objective comparison difficult. For instance, a study that was intended for impaired gait but was tested only with unimpaired subject can present better results that need not translate to impaired subjects.

3.4.1. Rule-Based Methods

Rule-based methods are the most widely used gait detection technique, employed by 63 out of the 99 studies. The wider adoption of the method could be attributed to their simplicity, intuitiveness, and less computational complexity involved (and hence less latency in the real-time processing).

Table 4. Classification of studies based on the type of gait analysis methods used; the number of studies which followed a type of method is listed in the table. Note that a few studies were counted in more than one category when those studies involved more than one method.

Domain	Algorithm	Number of Studies	
	Rule-based methods	63	
	Fuzzy inference system (FIS)	4	
Time domain	Machine learning (ML)	19	92
	Phase portrait (PP)	1	
	Other	5	
Frequency domain	Adaptive oscillator (AO)	4	5
	Spectral analysis	1	
Time-frequency domain	Wavelet transform (WT)	3	4
	Empirical mode decomposition	1	

One way of implementing a rule-based method is by setting a threshold on the raw or processed signal from the IMU (for instance, on the acceleration, angular velocity, segment orientation angle, or joint angle). Rule-based methods often employ multiple rules built on conditional statements (typically, if-else logic) that are connected using inequality constraints or logical AND/OR operators.

Sometimes, threshold-based techniques are replaced with peak detection techniques. One disadvantage of peak detection is that the presence of a peak can be confirmed only after both the rising edge and the falling edge appear. This may introduce a delay in gait event/phase detection depending on which part of the peak the event/phase temporally overlaps with. Maqbool et al. [7] followed such an approach wherein the shank angular velocity in the sagittal plane is used with a window of 80 ms before confirming the peak, while Ref. [37] additionally used accelerometer signals.

Instead of using predetermined thresholds, Ref. [58] proposed an adaptive threshold-based method which automatically computes and updates the threshold in real-time. This is done through what is called the "dynamics of sensor data", defined as a function of linear acceleration and angular rate, averaged over the last five data samples. The adaptive threshold is used to distinguish between swing and stance phase.

Rule-based approaches are also popular in the case of IPSs , with several studies using threshold or peak detection based approaches on IPSs, either to distinguish between stance phase and swing phase or between multiple gait phases [57,67–70]. Lin et al. [62] set a threshold on the first derivative of the pressure sensor data in identifying HS and TO and reports that it makes the detection robust against spurious signals, offset variation between IPSs and between-subject variations. Hanlon et al. [71] used a similar approach of setting a threshold on the derivative of the pressure sensor while additionally using a threshold on the accelerometer data along with its first and second derivative.

Rule-based methods are often implemented as finite-state machines (FSM). Pappas et al. [57] reported an FSM that considers four states: swing phase, stance phase, HS, and TO. Seven transitions were defined between these four states based on input from the IPS and the foot pitch angle. The IPS used three FSRs, one each on the heel, and first and fourth metatarsals. The FSRs were used as foot switches to identify if and when weight was applied at these hotspots. It was the only study that considered both stroke and spinal cord injured subjects. The latter were able to walk short distances with or without crutches, but no ASIA impairment scale (AIS) score was mentioned. The study reported above 99% detection reliability for both unimpaired and pathological gait, with detection delay always less than 90 ms. The method is often considered as a benchmark in literature for gait event detection.

Lambrecht et al. [72] used the same to benchmark the performance of three versions of peak detection-based FSM implemented by them. The three methods differed in their input signals, which were chosen from: shank angular velocity, shank segment angle, ankle joint angle, heel linear velocity, toe linear velocity, shank position, foot angular velocity, and foot angle. The methods were otherwise identical in that the state transitions were defined from TO to mid-swing to IC to foot-flat to HO and back to TO. Although the study reported better performance, it is to be noted that the data were extracted using motion capture and is therefore hard to replicate using a wearable sensor (such as linear velocity and position). Hence, the corresponding FSMs may not be easily transferable to a system based on wearable sensing. Furthermore, a direct quantitative comparison between [57] and [72] would be questionable since the data-set used by the former involved both unimpaired and impaired gait while the one used by the latter only involved unimpaired subjects walking on treadmill.

3.4.2. Fuzzy Inference System

An advanced version of the rule-based technique is the fuzzy inference system (FIS). Instead of using thresholds to specify binary states (true or false scenarios) to decide state membership, an FIS fuzzifies the input variable and provides a continuous map between input and output variables based on a systematically designed rule base. González et al. [73] fuzzified the input from pressure sensors placed at the heel, the hallux, and the first and fifth metatarsals into fuzzy variables and defined a rule base whose outputs are gait phases. Senanayake et al. [42] followed a similar approach of fuzzifying four FSR variables while also using the knee angle, obtained from IMUs at the thigh and the shank, as the fifth fuzzy variable. However, the use of additional sensors appears to be counterproductive since a quick comparison of the latencies reported by both approaches reveals that the former (latency less than 77 ms) performed better than the latter (latency less than 300 ms).

One disadvantage of FIS is that it requires the state membership functions to be set by the user [74] and then be adapted each time to a new subject or data set for optimal performance, similarly to what is done with thresholds as discussed in Section 3.4.1. The adaptive neuro-fuzzy inference system (ANFIS) provides a workaround which combines the benefits of artificial neural networks (ANN) and FIS by letting the nonlinear membership functions be learned through the neural network, provided that sufficient training data sets are available. Lauer et al. [74] combined ANFIS with a subtractive clustering method to identify state membership functions followed by a supervisory control system (if-then rules) to prevent gait events from being identified in the reverse order. The subtractive clustering algorithm provided a quick method of estimating the minimal number of clusters required, and these clusters formed the initial shape of the state membership functions. A similar two-level approach using FIS with a supervisory function was also implemented by [13].

3.4.3. Machine Learning

Machine learning (ML) methods are the second most widely used gait analysis technique, with 19 out of 99 studies using them. ML approaches have been gaining popularity in recent years as 18 out of 19 studies are from 2012 or later. The Hidden Markov model (HMM) is the most favored with nine out of 19 studies using this approach. Abaid et al. [3] used angular velocity of the foot in the sagittal plane as input to the HMM, while Ref. [45] used angular velocity of the thigh, the shank and the foot in the sagittal plane. In both cases, FSR based IPSs were used for creating a labelled data set, which is necessary for training the model. Chen et al. [75] used inputs from both IPSs and accelerometers and used a third order fast Fourier transform followed by a principal component analysis for feature generation which was then fed to a support vector machine classifier to identify the gait phases. The study considered five ISTGFs and reported a 97.26% success rate for initial contact. Overall, machine learning techniques reported noticeably high accuracy with 10 out of the 15 studies (which reported at least some quantitative metrics) reporting above 91% accuracy. It is also interesting to note that 11 out of 19 studies detected at least four ISTGFs, which is much higher compared to most of the rule-based methods discussed in Section 3.4.1.

3.4.4. Phase Portrait

Among all the algorithms that were listed in Table 4, only a few studies used continuous gait phase estimation methods. The idea of continuous gait phase estimation is to have a variable keeping track of the progress of gait, continuously and bounded within the gait cycle. Quintero et al. [76] estimated continuous gait phase in real-time from the phase portrait of the hip angle and its derivative. Here, the phase portrait angle of the hip is considered as the continuous gait phase variable—placing the hip angle on the horizontal axis and its derivative on the vertical axis, a phase portrait angle is the angle subtended on the horizontal axis by the line joining the origin to a point on the phase portrait. The hip was chosen based on a more extensive study (although offline) carried out by [77], which reports that the phase angle obtained from the phase portrait of the hip is linearly and monotonically increasing, and bounded, even under perturbations. The phase portrait was scaled by a factor estimated by the ratio of difference in maximum phase angle and minimum phase angle to the difference in the first derivative of the same, so as to improve the monotonicity and linearity. These properties were further improved by filtering, at the expense of some delay. Although the method performed well in unimpaired subjects [76] and the offline analysis reported robustness to perturbations [77], it remains to be seen how the method would work with pathological gait in real-time.

3.4.5. Adaptive Oscillators

An adaptive oscillator (AO) is a frequency domain method that can synchronize to any periodic or pseudo-periodic input signal without any preprocessing [78]. Yan et al. [79] used peak detection to identify a gait event, based on the occurrence of a desired bio-mechanical event (e.g., max hip flexion angle, heel strike). This is used to mark the initialization of a new gait cycle, following which continuous phase estimation of the current gait cycle is carried out using adaptive oscillators. Chen et al. [6] developed a robust adaptive oscillator-based gait phase estimation which is reported to be working robustly even for abnormal gait. HMM was used for gait event detection which in turn was fed to AOs, instead of feeding the entire gait signal continuously. The robustness, according to the authors, is due to the fact that gait events were the only information needed to achieve synchronization which minimized the influence of gait abnormality on the algorithm.

3.4.6. Wavelet Transform

Wavelet transform (WT) is a time-frequency domain method that uses basis functions localized in both the time and frequency domain, through so-called wavelets analogous to sinusoids in a Fourier transform (see Figures A1 and A2 in Appendix A for a detailed description of WT). Features that are identifiable in the frequency domain can therefore be localized in the time domain, for example characteristic high-frequency content during heel strike. Aminian et al. [80] reported that TO and HS events consist of combined features that can be well resolved in the time-frequency domain. They identified distinctive features in the shank angular velocity involving some medium- and high-frequency content with sharp characteristic peaks. A discrete wavelet transform (DWT) with fifth-order Coiflet wavelets was used to enhance gait events in the signal, thereby enabling easier identification of global maxima corresponding to the gait events. This identification was followed by customized rules that found specific peaks in the time domain to confirm TO and HS. Coiflet wavelets were chosen because they resemble characteristic peaks observed in the angular velocity signal. Although the study reported accurate temporal estimation of TO and HS, it should be noted that it was only tested on unimpaired subjects and implemented for offline analysis.

3.5. Towards Clinical Applications

3.5.1. Sensor and Algorithm Choice

The motivation of most studies to develop real-time gait analysis techniques was to apply them in gait rehabilitation. Although many of the studies anticipated their proposed methods to work on pathological gait, less than one-third of studies included impaired/pathological gait for validation (31 studies out of 99). When considering clinical applications, the algorithm performance is only one of equally important requirements. The approach must also use sensors that are durable, easy to manipulate, and require a relatively low setup time—in other words, the approach should be simple and user-friendly. Finally, the algorithm must be validated on a sufficient sample of the target population in order to to take into account the unique characteristics of the target impairment and to account for the higher inter-subject variability present in impaired gait.

In Sections 3.3 and 3.4, we concluded that IMUs are the most common sensor and that rule-based methods are the most common algorithm type, when considered individually. Machine learning approaches are the next most common, under which we combine the hidden Markov model, support vector machine, and Bayesian approaches amongst others. In Table 5, we consider studies that were validated at least on unimpaired subjects (84 out of 99), since we deem these more relevant when it comes to applications, compared to studies that were not validated on subjects at all. The studies are listed according to the method used. We then quantify the number of studies using IMUs for each algorithm type and notice that IMUs remain dominant for both rule-based and machine learning approaches with 75% and 87%, respectively. This does not come as a surprise as they are good candidates for clinical applications: they can easily be placed/removed on/from the relevant body segment and avoid mechanical stress relatively well during typical use. IPSs, the second most popular sensors, are more cumbersome as they must be placed in the shoe and taken out for recharging. For persons with impaired hand-function, this is a point that cannot be neglected. For certain gait impairments where orthopaedic insoles are prescribed, the additional sensorized insole may cause discomfort or provide skewed signals. Lastly, they are subject to repeated mechanical stress, making them considerably less durable.

Improved usability and versatility can be observed with rule-based approaches compared to machine learning. In fact, only 11% of studies using a rule-based method required the placement of more than one IMU per leg, in contrast to 53% of machine learning approaches. This means applications using the rule-based methods typically require a less complex sensor setup, which would be preferable to the end-user. Furthermore, in 84% of studies using rule-based methods, the approach could be used independently for one leg or the other, compared to 47% for machine learning. This means that rule-based methods can be more easily tailored to specific use-cases, such as providing unilateral assistance.

Validation of rule-based and machine learning methods was done on 485 and 138 unimpaired subjects, respectively, which is an average of roughly nine subjects per study for both categories. Impaired and unimpaired gait, however, can vary significantly and thus these numbers speak little towards clinical applications. As stated previously, less than one third of studies were found to validate on impaired subjects, which mirrors findings by Perez et al. [27] that not many real-time gait detection algorithms are validated on populations with gait impairments. Despite this fact, literature and our previous observations are pointing towards IMUs and rule-based methods as primary candidates for clinical applications.

3.5.2. Impaired Gait Considerations

IMUs and rule-based algorithms are the preferred option amongst the studies that validated on impaired subjects. These studies are listed exhaustively in Table 6 and the combination amounts to 67%. We categorized the target impairments into two classes based on how they were presented in the respective studies—first, generally diminished ambulatory function, where gait is impaired due to general degeneration of the locomotor system, such as with Parkinson's disease, osteoarthritis, Huntington's disease, diplegic cerebral palsy (CP), ageing, or spinal cord injury (SCI). Second, unilateral loss of ambulatory function, where gait is impaired on one side of the body, such as amputation, stroke, and hemiplegic CP.

Table 5. Distribution of gait detection techniques for studies that validated on unimpaired subjects. Details on the usage of inertial measurement units (IMU) are presented together with the total number of unimpaired subjects the algorithms types were validated on. ML—Machine learning.

Algorithm	Total Number of Studies	Number of Studies that Used IMU	Number of Studies That Used More than One IMU per Leg	Number of Studies Where the Proposed Method Can Work Independently on Either One of the Legs	Total Number of Unimpaired Subjects	References
Rule-based method	51	38	4	32	485	[7,14,27,37,39,44,46,55, 57,59,60,62,68,69,71,72, 79,81–111]
Fuzzy inference system	3	0	0	0	14	[42,73,112]
Hidden Markov model	8	7	3	2	70	[3,6,40,45,64,113,114]
Support vector machine	2	1	1	0	30	[75,115]
Bayesian	2	2	2	2	18	[116,117]
Other ML methods	3	3	2	3	20	[118–120]
Phase portrait	1	1	0	1	1	[76]
Lookup table	1	1	0	1	1	[121]
Other time domain methods	4	2	1	1	42	[122–125]
Adaptive oscillators	4	1	1	0	29	[6,79,83,126]
Wavelet transform	3	3	0	3	61	[80,127,128]

For generally diminished ambulatory function, rule-based algorithms can exploit the fact that gait features typically become less prominent but are not lost. This means that gait remains periodic and features on which rules can be built exist. Applications seeking this category should focus primarily on understanding the unique gait features of the target population. Behboodi et al. [86] used such an approach for gait detection in children with diplegic CP. The authors circumvented the lack of an identifiable heel strike in equinus gait by using angular velocity at the shank, which still shows characteristic peaks, valleys, and zero-crossings.

On the other hand, unilateral loss of ambulatory function typically means that gait features remain on the unimpaired side but are lost on the other. A gait event detection method must thus be extended to additionally capture the irregularities on the impaired side. One approach is presented by Perez et al. [27], who derived a rule-based algorithm from eight other studies, but corrected detection rules that would fail with impaired gait. The authors claim that, whereas normal gait is regular and smooth, the thresholds typically used to detect gait events are tricked by the irregularities found in neurological impaired gait.

Depending on the impairment, assistive devices such as walking frames or orthoses and training devices such as body-weight support systems or exoskeletons can enable or enhance locomotion. The prominence of gait features can then depend on the assistive device. For example, while considering complete SCI patients with functional electrical stimulation (FES), Skelly et al. [13] chose to perform gait event detection using a fuzzy inference system as it can specifically accommodate for the relatively large step-to-step variability observed in FES gait. In Jasiewicz et al. [44], for example, algorithms were significantly under-performing during gaits exhibited when using walking aids. Similarly, with active neuroprostheses such as EES, the stimulation itself can substantially modify the gait pattern [16].

Table 6. Algorithm types categorized with respect to the impairment of subjects on which they were validated. Impairments are categorized based on how they were characterized in the respective study. The number of impaired and unimpaired subjects involved in the study suggest the reliability and popularity of the given algorithmic approach for that specific impairment. Note that some studies (such as [63]) are listed more than once in the table depending on whether they employed more than one category of impaired subjects. FIS—Fuzzy inference system, ANFIS—Adaptive neuro fuzzy inference system, HMM—Hidden Markov model.

Impairment	Algorithm Type	Sensor Type	References	Number of Impaired Subjects	Number of Unimpaired Subjects
Parkinson's disease	Rule-based method	IMU	[55]	16	12
	Rule-based method	IMU	[104]	5	15
	Wavelet transform	IMU	[127]	48	40
Osteoarthritis	Support vector machine	IMU + IPS	[75]	14	10
Huntington's disease	HMM	IMU	[63]	10	0
Cerebral palsy	Rule-based method	IMU	[86]	5	7
	Rule-based method	IPS	[92]	3	8
	ANFIS	EMG	[74]	8	0
	HMM	IMU	[113]	10	10
Spinal cord injury	Rule-based method	IMU	[44]	14	26
	FIS	IPS	[13]	3	0
Elderly	Spectral analysis	IMU	[129]	92	0
	HMM	IMU	[63]	10	0
Amputee	Rule-based method	IMU	[7]	1	8
	Rule-based method	IMU	[37]	1	9
	Rule-based method	IMU + IPS	[69]	3	5
	Rule-based method	IPS	[110]	1	1
Stroke	Rule-based method	IMU	[130]	2	0
	Rule-based method	IMU + IPS	[131]	1	0
	Rule-based method	IMU	[132]	1	0
	Rule-based method	IMU	[90]	4	10
	Rule-based method	IMU	[133]	1	0
	Rule-based method	IMU	[104]	4	15
	Rule-based method	IMU	[134]	6	0
	Rule-based method	IMU	[97]	10	22
	Rule-based method	IMU	[135]	2	0
	Rule-based method	IMU	[27]	1	1
	HMM	IMU	[63]	10	0
Unspecified Hemiplegia/ Hemiparesis	Rule-based method	IMU	[14]	10	10
	HMM	IMU	[3]	10	10

Performing algorithm validations with impaired subjects is essential for advancing clinical applications, but, in doing so, patient safety should not be neglected. All studies in Table 6 besides three, [75,133,135], explicitly state having received ethics approval and obtained informed consent. However, only nine out of 27 mention any safety considerations in their text and only one study has a dedicated section. We recommend that a section on safety always be included in future studies, covering a basic risk analysis and mitigation put in place, and documenting potentially hazardous system failures during experiments. Some example considerations would be preventing skin irritation, ensuring that sensors do not fall off while walking, or verifying that sensors can be securely manipulated by the target population. Such a section will help the community accelerate meaningful development of sensors and algorithms for gait detection and build trust towards their use in clinical applications.

4. Conclusions

In the present work, we performed a comprehensive systematic review and provided a broad overview on wearable sensors and methods used in real-time gait analysis. We performed meta-data analysis and identified trends among researchers such as the most sought-after gait events, body segments for IMU placement, and sensor types for ground truth validation of IMU-based gait detection methods. Studies that validated on subjects with impaired gait were then extracted and sensors and methods for clinical applications discussed.

Based on popularity in our findings, we recommend performing gait detection in a subject by using a rule-based method to determine toe off and heel strike. We propose that an IMU is placed on either the foot or shank, and insole pressure sensors are used as ground-truth for validation. When investigating new gait detection methods for clinical use, it is crucial that they are evaluated on their target population and across relevant conditions such as using various walking aids.

One of the limitations of the present review is that the performance of gait detection methods could not be compared quantitatively due to the heterogeneity of the metrics used across the algorithm types. Therefore, we suggest that future studies report performance metrics that would allow benchmarking with respect to comparable methods.

In our review method, we did not consider clinical applications explicitly. However, a subset of the reviewed publications revealed that the algorithm performance can be heavily influenced by gait impairments up to a point where the impairment dictates the algorithm choice. We believe it is a very relevant direction worth further systematic investigation.

Real-time gait detection using wearable sensors provides an unprecedented means to deliver clinical interventions for people with gait impairments. As opposed to traditional gait detection equipment, wearable sensors can inherently be used in an ambulatory setting, and, compared to offline gait analysis, real-time gait detection can be integrated in closed-loop control. The right combination of sensor and gait detection method thus enables the development of assistive devices that have the potential to increase the effectiveness of rehabilitation and improve the lives of people with ambulatory deficits.

Author Contributions: H.P. and M.C. performed conceptualization, methodology, data curation, investigation, prepared visualizations, and wrote the manuscript. U.K. reviewed and edited the manuscript. G.C. and A.I. provided the resources, and reviewed and edited the manuscript. H.V. and J.v.Z. performed project administration, supervised, reviewed, edited, and approved the manuscript. H.V. additionally provided resources and J.v.Z. additionally secured funding. All authors have read and agreed to the published version of the manuscript.

Funding: This research was partially funded by the European Union's Horizon 2020 research and innovation program Grant No. 779963 "EUROBENCH".

Institutional Review Board Statement: Not applicable.

Informed Consent Statement: Not applicable.

Data Availability Statement: Not applicable.

Acknowledgments: The authors thank Niek Borgers for participating in early phase conceptualization of the systematic review. The authors thank Fang Lyu for her valuable suggestions and reviewing the manuscript.

Conflicts of Interest: H.P., M.C., U.K., G.C., and J.Z. work at ONWARD, which is developing a therapy to improve rehabilitation of people with spinal cord injury. The therapy may involve the use of gait detection algorithms.

Appendix A. Background of Wavelet Transform Method

One way to extract gait cadence is through frequency spectrum analysis. The two prominent peaks in the frequency spectrum analysis of gait signals (e.g., angular velocity of the foot) correspond to stride frequency and step frequency (first and second harmonics). Cadence can be evaluated from one of the two peaks. However, to extract cadence in real-time, we need to perform a Fourier transform (FT) over a window of data streamed continuously, the so-called short-time Fourier transform.

STFT suffers from poor resolution in the time-frequency domain [11]. To improve the resolution in one domain, the resolution in the other domain must be compromised. For instance, to get better resolution of frequency content, we need to provide it with a larger (time) window of data, thereby reducing the resolution in the time domain, resulting in output cadence evaluated over a larger interval (see Figure A1). Hence, STFT can be considered as a suitable technique for long duration, steady-state walking (quasi-periodic with no abrupt changes), but not for fast motion transitions [11].

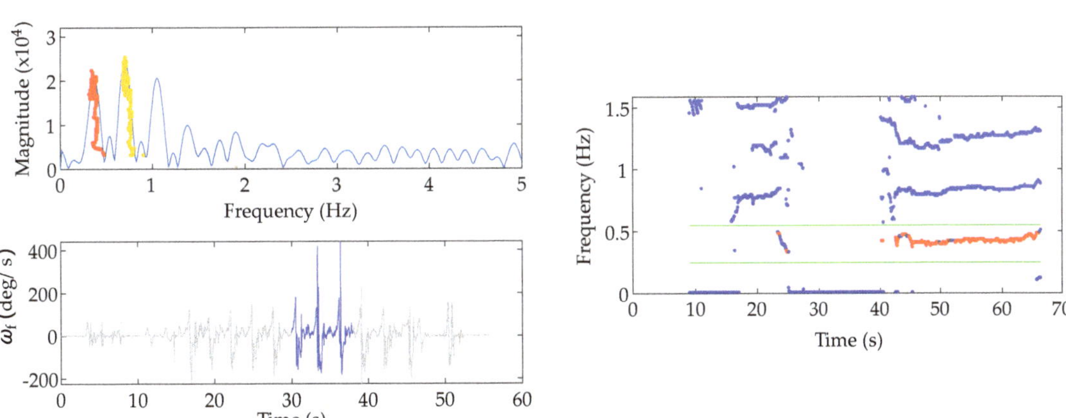

Figure A1. An illustration of the STFT (short-time Fourier transform) implementation in real-time. In the lower left plot, the sagittal plane angular velocity (ω_f) of the foot is shown in grey. The window of sample used during the snapshot (at around time $t = 38\,$s) is highlighted in blue. The upper left plot shows the frequency spectrum corresponding to time $t = 38\,$s. The red and yellow dots represent the paths traced over the duration by the peaks corresponding to the first harmonic (stride frequency) and the second harmonic (step frequency), respectively. In the right plot, the time-frequency domain output (of the entire data set) from an offline implementation of STFT is shown in blue dots while the corresponding output from the implementation in real-time (for the first harmonic) is shown in the red dots. Note that the sampling interval is slightly irregular in the real-time implementation due to computational complexity associated with STFT. It can be observed that there is a clear separation between the first, second, and third harmonics (the green lines border the first harmonics, which corresponds to the stride frequency).

The constraint of the resolution trade-off can be overcome with a wavelet transform (WT). Much like an FT, a WT also decomposes the signal based on a set of basis functions. While these basis functions are sinusoids in the case of FTs, the basis functions in the case of WTs are wavelets. The difference is that, while sinusoids only differ in their frequencies, wavelets are localized both in the time and the frequency domain. Therefore, unlike STFT, which uses windows of fixed size in time domain, a WT uses windows of varying size in the time-frequency domain, increasing the time resolution with high frequency signals and vice versa [136,137].

There are two variants of discrete wavelet transform (DWT). The first variant, where 'discrete' implies discretized time-frequency domain while otherwise being identical to continuous wavelet transform (CWT), is described in [137] and implemented in [11].

Han et al. [11] used this version of DWT to extract step frequency/cadence and reports that WT performs better than STFT in estimating step frequency when fast motion changes occur. The second variant, which is filter bank-based, is described in [136] and implemented in [138]. Klingbeil et al. [138] carried out step detection using this version of DWT with a Daubechies wavelet. Here, the signal is split into the so-called detail levels and then reconstructed with detail levels spread across 0.8 Hz to 3.2 Hz (assuming most of the gait signal is in this range), followed by step detection using thresholds. The MATLAB implementations of both version of DWT are available, but the former is called by the name *cwt* (because of its resemblance to CWT), while the latter is called by the name *dwt*. Note that, although Refs. [11,138] carried out the study in real-time, these studies were limited to step detection and step frequency, but not ISTGF detection. Our own implementation of wavelet transform for gait detection, based on *cwt*, is shown in Figure A2.

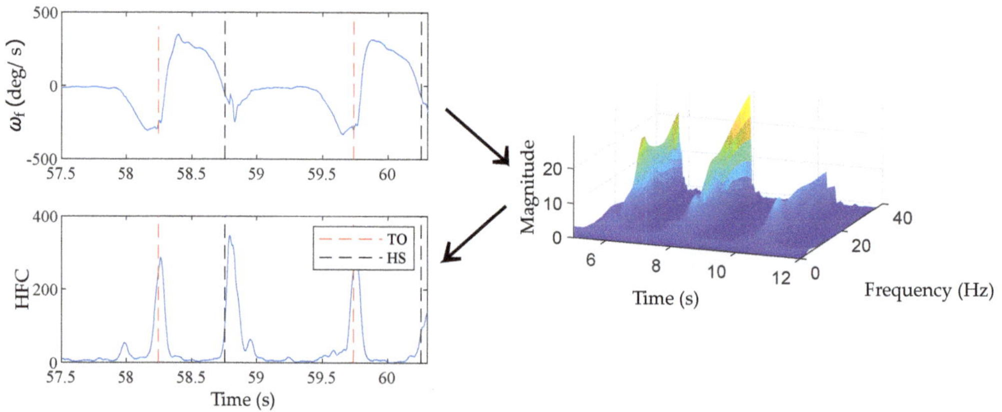

Figure A2. An implementation of wavelet transform for gait event detection. Sagittal plane angular velocity (ω_f) of the foot (top left plot), which is in time domain, was first transformed into time-frequency domain using wavelet transform (right plot). Then, the high-frequency region was condensed into time domain (bottom left plot) by evaluating the cross section area of the magnitude of wavelet transform. A real-time implementation of the method was limited by computational complexity. HFC—high frequency content.

One limitation of WT is the high computational complexity required for such an implementation, which we also noticed in our own implementation presented in Figure A2. Since WTs involve comparing the signal to wavelets, the transform at the beginning and end of the signal are less reliable because the wavelet cannot be overlapped completely with the extreme ends of the signal. This leads to the so-called cone of influence, a region outside of which the result of wavelet transform is no longer reliable. This is usually of low significance in offline analysis since the entire signal is available for analysis at once; however, in real-time analysis, the cone of influence becomes more important. This is because what we are interested in for every iteration is the transform of the latest sample of data, which is precisely where the reliability is poor.

References

1. Luo, J.; Tang, J.; Xiao, X. Abnormal Gait behavior detection for elderly based on enhanced Wigner-Ville analysis and cloud incremental SVM learning. *J. Sens.* **2016**, *2016*. [CrossRef]
2. Chen, M.; Huang, B.; Xu, Y. Intelligent shoes for abnormal gait detection. In Proceedings of the 2008 IEEE International Conference on Robotics and Automation, Pasadena, CA, USA, 19–23 May 2008; pp. 2019–2024.
3. Abaid, N.; Cappa, P.; Palermo, E.; Petrarca, M.; Porfiri, M. Gait detection in children with and without hemiplegia using single-axis wearable gyroscopes. *PLoS ONE* **2013**, *8*, e73152. [CrossRef] [PubMed]

4. Tay, A.; Yen, S.; Lee, P.; Wang, C.; Neo, A.; Phan, S.; Yogaprakash, K.; Liew, S.; Au, W. Freezing of Gait (FoG) detection for Parkinson Disease. In Proceedings of the 2015 10th Asian Control Conference (ASCC), Kota Kinabalu, Malaysia, 31 May–3 June 2015; pp. 1–6.

5. Duan, P.; Li, S.; Duan, Z.; Chen, Y. Bio-Inspired Real-Time Prediction of Human Locomotion for Exoskeletal Robot Control. *Appl. Sci.* **2017**, *7*, 1130. [CrossRef]

6. Chen, G.; Qi, P.; Guo, Z.; Yu, H. Gait-Event-Based Synchronization Method for Gait Rehabilitation Robots via a Bioinspired Adaptive Oscillator. *IEEE Trans. Biomed. Eng.* **2017**, *64*, 1345–1356. [CrossRef] [PubMed]

7. Maqbool, H.F.; Husman, M.A.B.; Awad, M.I.; Abouhossein, A.; Iqbal, N.; Dehghani-Sanij, A.A. A real-time gait event detection for lower limb prosthesis control and evaluation. *IEEE Trans. Neural Syst. Rehabil. Eng.* **2017**, *25*, 1500–1509. [CrossRef] [PubMed]

8. Song, L.; Wang, Y.; Yang, J.J.; Li, J. Health sensing by wearable sensors and mobile phones: A survey. In Proceedings of the IEEE 16th International Conference on e-Health Networking, Applications and Services (Healthcom), Natal, Brazil, 15–18 October 2014; pp. 453–459.

9. Derawi, M.; Bours, P. Gait and activity recognition using commercial phones. *Comput. Secur.* **2013**, *39*, 137–144. [CrossRef]

10. Schneider, O.S.; MacLean, K.E.; Altun, K.; Karuei, I.; Wu, M. Real-time gait classification for persuasive smartphone apps: Structuring the literature and pushing the limits. In Proceedings of the 2013 International Conference on Intelligent User Interfaces, Santa Monica, CA, USA, 19–22 March 2013; pp. 161–172. [CrossRef]

11. Han, D.; Renaudin, V.; Ortiz, M. Smartphone based gait analysis using STFT and wavelet transform for indoor navigation. In Proceedings of the 2014 International Conference on Indoor Positioning and Indoor Navigation (IPIN), Busan, Korea, 27–30 October 2014; pp. 157–166.

12. Mazilu, S.; Hardegger, M.; Zhu, Z.; Roggen, D.; Troster, G.; Plotnik, M.; Hausdorff, J.M. Online detection of freezing of gait with smartphones and machine learning techniques. In Proceedings of the 2012 6th International Conference on Pervasive Computing Technologies for Healthcare (PervasiveHealth), San Diego, CA, USA, 21–24 May 2012; pp. 123–130.

13. Skelly, M.M.; Chizeck, H.J. Real-time gait event detection for paraplegic FES walking. *IEEE Trans. Neural Syst. Rehabil. Eng.* **2001**, *9*, 59–68. [CrossRef]

14. Rueterbories, J.; Spaich, E.G.; Andersen, O.K. Gait event detection for use in FES rehabilitation by radial and tangential foot accelerations. *Med. Eng. Phys.* **2014**, *36*, 502–508. [CrossRef]

15. Wenger, N.; Moraud, E.M.; Raspopovic, S.; Bonizzato, M.; DiGiovanna, J.; Musienko, P.; Morari, M.; Micera, S.; Courtine, G. Closed-loop neuromodulation of spinal sensorimotor circuits controls refined locomotion after complete spinal cord injury. *Sci. Transl. Med.* **2014**, *6*, 255ra133. [CrossRef]

16. Wagner, F.B.; Mignardot, J.B.; Le Goff-Mignardot, C.G.; Demesmaeker, R.; Komi, S.; Capogrosso, M.; Rowald, A.; Seáñez, I.; Caban, M.; Pirondini, E.; et al. Targeted neurotechnology restores walking in humans with spinal cord injury. *Nature* **2018**, *563*, 65–71. [CrossRef] [PubMed]

17. Taborri, J.; Palermo, E.; Rossi, S.; Cappa, P. Gait partitioning methods: A systematic review. *Sensors* **2016**, *16*, 66. [CrossRef] [PubMed]

18. Panebianco, G.P.; Bisi, M.C.; Stagni, R.; Fantozzi, S. Analysis of the performance of 17 algorithms from a systematic review: Influence of sensor position, analysed variable and computational approach in gait timing estimation from IMU measurements. *Gait Posture* **2018**, *66*, 76–82. [CrossRef] [PubMed]

19. Caldas, R.; Mundt, M.; Potthast, W.; de Lima Neto, F.B.; Markert, B. A systematic review of gait analysis methods based on inertial sensors and adaptive algorithms. *Gait Posture* **2017**, *57*, 204–210. [CrossRef] [PubMed]

20. Chen, S.; Lach, J.; Lo, B.; Yang, G.Z. Toward pervasive gait analysis with wearable sensors: A systematic review. *IEEE J. Biomed. Health Inform.* **2016**, *20*, 1521–1537. [CrossRef]

21. Moher, D.; Shamseer, L.; Clarke, M.; Ghersi, D.; Liberati, A.; Petticrew, M.; Shekelle, P.; Stewart, L.A. Preferred reporting items for systematic review and meta-analysis protocols (PRISMA-P) 2015 statement. *Syst. Rev.* **2015**, *4*, 1. [CrossRef] [PubMed]

22. Shull, P.B.; Jirattigalachote, W.; Hunt, M.A.; Cutkosky, M.R.; Delp, S.L. Quantified self and human movement: a review on the clinical impact of wearable sensing and feedback for gait analysis and intervention. *Gait Posture* **2014**, *40*, 11–19. [CrossRef]

23. López-Nava, I.H.; Muñoz-Meléndez, A. Wearable inertial sensors for human motion analysis: A review. *IEEE Sens. J.* **2016**, *16*, 7821–7834. [CrossRef]

24. Novak, D.; Riener, R. A survey of sensor fusion methods in wearable robotics. *Robot. Auton. Syst.* **2015**, *73*, 155–170. [CrossRef]

25. Vu, H.T.T.; Dong, D.; Cao, H.L.; Verstraten, T.; Lefeber, D.; Vanderborght, B.; Geeroms, J. A review of gait phase detection algorithms for lower limb prostheses. *Sensors* **2020**, *20*, 3972. [CrossRef]

26. Rueterbories, J.; Spaich, E.G.; Larsen, B.; Andersen, O.K. Methods for gait event detection and analysis in ambulatory systems. *Med. Eng. Phys.* **2010**, *32*, 545–552. [CrossRef]

27. Perez-Ibarra, J.C.; Siqueira, A.A.; Krebs, H.I. Real-time identification of gait events in impaired subjects using a single-IMU foot-mounted device. *IEEE Sens. J.* **2019**, *20*, 2616–2624. [CrossRef]

28. Haddaway, N.R.; Collins, A.M.; Coughlin, D.; Kirk, S. The Role of Google Scholar in Evidence Reviews and Its Applicability to Grey Literature Searching. *PLoS ONE* **2015**, *10*, e0138237. [CrossRef] [PubMed]

29. Tober, M. PubMed, ScienceDirect, Scopus or Google Scholar—Which is the best search engine for an effective literature research in laser medicine? *Med. Laser Appl.* **2011**, *26*, 139–144. [CrossRef]

30. Boeker, M.; Vach, W.; Motschall, E. Google Scholar as replacement for systematic literature searches: Good relative recall and precision are not enough. *BMC Med. Res. Methodol.* **2013**, *13*, 131. [CrossRef]

31. Bramer, W.M.; Giustini, D.; Kramer, B.M. Comparing the coverage, recall, and precision of searches for 120 systematic reviews in Embase, MEDLINE, and Google Scholar: A prospective study. *Syst. Rev.* **2016**, *5*, 39. [CrossRef] [PubMed]

32. Fagan, J.C. An evidence-based review of academic web search engines, 2014-2016: Implications for librarians' practice and research agenda. *Inf. Technol. Libr.* **2017**, *36*, 7–47. [CrossRef]

33. de Winter, J.C.F.; Zadpoor, A.A.; Dodou, D. The expansion of Google Scholar versus Web of Science: A longitudinal study. *Scientometrics* **2014**, *98*, 1547–1565. [CrossRef]

34. Van Eck, N.J.; Waltman, L. VOS: A new method for visualizing similarities between objects. In *Advances in Data Analysis*; Springer: Berlin/Heidelberg, Germany, 2007; pp. 299–306. [CrossRef]

35. Van Eck, N.J.; Waltman, L. Software survey: VOSviewer, a computer program for bibliometric mapping. *Scientometrics* **2010**, *84*, 523–538. [CrossRef]

36. Maqbool, H.F.; Husman, M.A.B.; Awad, M.I.; Abouhossein, A.; Dehghani-Sanij, A.A. Real-time gait event detection for transfemoral amputees during ramp ascending and descending. In Proceedings of the 2015 37th Annual International Conference of the IEEE Engineering in Medicine and Biology Society (EMBC), Milan, Italy, 25–29 August 2015; pp. 4785–4788.

37. Maqbool, H.F.; Husman, M.A.; Awad, M.I.; Abouhossein, A.; Iqbal, N.; Dehghani-Sanij, A.A. Stance Sub-phases Gait Event Detection in Real-Time for Ramp Ascent and Descent. In *Converging Clinical and Engineering Research on Neurorehabilitation II*; Springer International Publishing: Berlin/Heidelberg, Germany, 2017; pp. 191–196. [CrossRef]

38. Maqbool, H.F.; Husman, M.A.B.; Awad, M.I.; Abouhossein, A.; Mehryar, P.; Iqbal, N.; Dehghani-Sanij, A.A. Real-time gait event detection for lower limb amputees using a single wearable sensor. In Proceedings of the 2016 IEEE 38th Annual International Conference of the Engineering in Medicine and Biology Society (EMBC), Orlando, FL, USA, 16–20 August 2016; pp. 5067–5070.

39. Behboodi, A.; Wright, H.; Zahradka, N.; Lee, S. Seven phases of gait detected in real-time using shank attached gyroscopes. In Proceedings of the 2015 37th Annual International Conference of the IEEE Engineering in Medicine and Biology Society (EMBC), Milan, Italy, 25–29 August 2015; pp. 5529–5532.

40. Chen, G.; Salim, V.; Yu, H. A novel gait phase-based control strategy for a portable knee-ankle-foot robot. In Proceedings of the 2015 IEEE International Conference on Rehabilitation Robotics (ICORR), Singapore, 11–14 August 2015; pp. 571–576.

41. Cai, V.A.D.; Ibanez, A.; Granata, C.; Nguyen, V.T.; Nguyen, M.T. Transparency enhancement for an active knee orthosis by a constraint-free mechanical design and a gait phase detection based predictive control. *Meccanica* **2017**, *52*, 729–748. [CrossRef]

42. Senanayake, C.M.; Senanayake, S.A. Computational intelligent gait-phase detection system to identify pathological gait. *IEEE Trans. Inf. Technol. Biomed.* **2010**, *14*, 1173–1179. [CrossRef]

43. Li, F.; Liu, G.; Liu, J.; Chen, X.; Ma, X. 3D Tracking via Shoe Sensing. *Sensors* **2016**, *16*, 1809. [CrossRef]

44. Jasiewicz, J.M.; Allum, J.H.; Middleton, J.W.; Barriskill, A.; Condie, P.; Purcell, B.; Li, R.C.T. Gait event detection using linear accelerometers or angular velocity transducers in able-bodied and spinal-cord injured individuals. *Gait Posture* **2006**, *24*, 502–509. [CrossRef]

45. Taborri, J.; Rossi, S.; Palermo, E.; Patanè, F.; Cappa, P. A novel HMM distributed classifier for the detection of gait phases by means of a wearable inertial sensor network. *Sensors* **2014**, *14*, 16212–16234. [CrossRef]

46. Bejarano, N.C.; Ambrosini, E.; Pedrocchi, A.; Ferrigno, G.; Monticone, M.; Ferrante, S. An adaptive real-time algorithm to detect gait events using inertial sensors. In Proceedings of the XIII Mediterranean Conference on Medical and Biological Engineering and Computing 2013, Seville, Spain, 25–28 September 2013; pp. 1799–1802. [CrossRef]

47. Anwary, A.R.; Yu, H.; Vassallo, M. Optimal foot location for placing wearable IMU sensors and automatic feature extraction for gait analysis. *IEEE Sens. J.* **2017**, *18*, 2555–2567. [CrossRef]

48. Casamassima, F.; Ferrari, A.; Milosevic, B.; Ginis, P.; Farella, E.; Rocchi, L. A wearable system for gait training in subjects with Parkinson's disease. *Sensors* **2014**, *14*, 6229–6246. [CrossRef]

49. Mahony, R.; Hamel, T.; Pflimlin, J.M. Nonlinear complementary filters on the special orthogonal group. *IEEE Trans. Autom. Control.* **2008**, *53*, 1203–1218. [CrossRef]

50. Madgwick, S.O.; Harrison, A.J.; Vaidyanathan, R. Estimation of IMU and MARG orientation using a gradient descent algorithm. In Proceedings of the 2011 IEEE International Conference on Rehabilitation Robotics, Zurich, Switzerland, 29 June–1 July 2011; pp. 1–7.

51. Cirillo, A.; Cirillo, P.; De Maria, G.; Natale, C.; Pirozzi, S.; Fourati, H.; Belkhiat, D. A comparison of multisensor attitude estimation algorithms. In *Multisensor Attitude Estimation: Fundamental Concepts and Applications*; CRC Press: Boca Raton, FL, USA, 2016; pp. 529–540.

52. Higgins, W.T. A comparison of complementary and Kalman filtering. *IEEE Trans. Aerosp. Electron. Syst.* **1975**, *AES-11*, 321–325. [CrossRef]

53. Yang, L.; Ye, S.; Wang, Z.; Huang, Z.; Wu, J.; Kong, Y.; Zhang, L. An error-based micro-sensor capture system for real-time motion estimation. *J. Semicond.* **2017**, *38*, 105004. [CrossRef]

54. Skog, I.; Nilsson, J.O.; Händel, P. Evaluation of zero-velocity detectors for foot-mounted inertial navigation systems. In Proceedings of the 2010 International Conference on Indoor Positioning and Indoor Navigation, Zurich, Switzerland, 15–17 September 2010; pp. 1–6.

55. Ferrari, A.; Ginis, P.; Hardegger, M.; Casamassima, F.; Rocchi, L.; Chiari, L. A mobile Kalman-filter based solution for the real-time estimation of spatio-temporal gait parameters. *IEEE Trans. Neural Syst. Rehabil. Eng.* **2016**, *24*, 764–773. [CrossRef] [PubMed]

56. Anacleto, R.; Figueiredo, L.; Almeida, A.; Novais, P. Localization system for pedestrians based on sensor and information fusion. In Proceedings of the 17th International Conference on Information Fusion (FUSION), Salamanca, Spain, 7–10 July 2014; pp. 1–8.

57. Pappas, I.P.; Popovic, M.R.; Keller, T.; Dietz, V.; Morari, M. A reliable gait phase detection system. *IEEE Trans. Neural Syst. Rehabil. Eng.* **2001**, *9*, 113–125. [CrossRef]

58. Van Nguyen, L.; La, H.M. Real-time human foot motion localization algorithm with dynamic speed. *IEEE Trans.-Hum.-Mach. Syst.* **2016**, *46*, 822–833. [CrossRef]

59. Harle, R.; Taherian, S.; Pias, M.; Coulouris, G.; Hopper, A.; Cameron, J.; Lasenby, J.; Kuntze, G.; Bezodis, I.; Irwin, G.; et al. Towards real-time profiling of sprints using wearable pressure sensors. *Comput. Commun.* **2012**, *35*, 650–660. [CrossRef]

60. Delgado-Gonzalo, R.; Hubbard, J.; Renevey, P.; Lemkaddem, A.; Vellinga, Q.; Ashby, D.; Willardson, J.; Bertschi, M. Real-time gait analysis with accelerometer-based smart shoes. In Proceedings of the 2017 39th Annual International Conference of the IEEE Engineering in Medicine and Biology Society (EMBC), Jeju, Korea, 11–15 July 2017.

61. Stöggl, T.; Martiner, A. Validation of Moticon's OpenGo sensor insoles during gait, jumps, balance and cross-country skiing specific imitation movements. *J. Sport. Sci.* **2017**, *35*, 196–206. [CrossRef]

62. Lin, F.; Wang, A.; Zhuang, Y.; Tomita, M.R.; Xu, W. Smart insole: A wearable sensor device for unobtrusive gait monitoring in daily life. *IEEE Trans. Ind. Inform.* **2016**, *12*, 2281–2291. [CrossRef]

63. Mannini, A.; Trojaniello, D.; Della Croce, U.; Sabatini, A.M. Hidden Markov model-based strategy for gait segmentation using inertial sensors: Application to elderly, hemiparetic patients and Huntington's disease patients. In Proceedings of the 2015 37th Annual International Conference of the IEEE Engineering in Medicine and Biology Society (EMBC), Milan, Italy, 25–29 August 2015; pp. 5179–5182.

64. Crea, S.; De Rossi, S.M.; Donati, M.; Reberšek, P.; Novak, D.; Vitiello, N.; Lenzi, T.; Podobnik, J.; Munih, M.; Carrozza, M.C. Development of gait segmentation methods for wearable foot pressure sensors. In Proceedings of the 2012 Annual International Conference of the IEEE Engineering in Medicine and Biology Society (EMBC), San Diego, CA, USA, 28 August–1 September 2012; pp. 5018–5021.

65. Fleischer, C.; Hommel, G. A human–exoskeleton interface utilizing electromyography. *IEEE Trans. Robot.* **2008**, *24*, 872–882. [CrossRef]

66. Farmer, S.; Silver-Thorn, B.; Voglewede, P.; Beardsley, S.A. Within-socket myoelectric prediction of continuous ankle kinematics for control of a powered transtibial prosthesis. *J. Neural Eng.* **2014**, *11*, 056027. [CrossRef]

67. Mazhar, O.; Bari, A.Z.; Faudzi, A.A.M. Real-time gait phase detection using wearable sensors. In Proceedings of the 2015 10th Asian Control Conference (ASCC), Kota Kinabalu, Malaysia, 31 May–3 June 2015; pp. 1–4.

68. Li, J.; Zhou, X.; Li, C.; Li, W.; Zhang, H.; Gu, H. A Real-Time Gait Phase Detection Method for Prosthesis Control. In *Assistive Robotics, Proceedings of the 18th International Conference on CLAWAR 2015*; National Natural Science Foundation of China: Beijing, China, 2016; pp. 577–584. [CrossRef]

69. Goršič, M.; Kamnik, R.; Ambrožič, L.; Vitiello, N.; Lefeber, D.; Pasquini, G.; Munih, M. Online phase detection using wearable sensors for walking with a robotic prosthesis. *Sensors* **2014**, *14*, 2776–2794. [CrossRef] [PubMed]

70. Yang, J.; Chen, X.; Guo, H.; Zhang, Q. Implementation of omnidirectional lower limbs rehabilitation training robot. In Proceedings of the 2007 International Conference on Electrical Machines and Systems (ICEMS), Seoul, Korea, 8–11 October 2007; pp. 2033–2036.

71. Hanlon, M.; Anderson, R. Real-time gait event detection using wearable sensors. *Gait Posture* **2009**, *30*, 523–527. [CrossRef]

72. Lambrecht, S.; Harutyunyan, A.; Tanghe, K.; Afschrift, M.; De Schutter, J.; Jonkers, I. Real-Time Gait Event Detection Based on Kinematic Data Coupled to a Biomechanical Model. *Sensors* **2017**, *17*, 671. [CrossRef] [PubMed]

73. González, I.; Fontecha, J.; Hervás, R.; Bravo, J. An ambulatory system for gait monitoring based on wireless sensorized insoles. *Sensors* **2015**, *15*, 16589–16613. [CrossRef]

74. Lauer, R.T.; Smith, B.T.; Betz, R.R. Application of a neuro-fuzzy network for gait event detection using electromyography in the child with cerebral palsy. *IEEE Trans. Biomed. Eng.* **2005**, *52*, 1532–1540. [CrossRef]

75. Chen, W.; Xu, Y.; Wang, J.; Zhang, J. Kinematic analysis of human gait based on wearable sensor system for gait rehabilitation. *J. Med Biol. Eng.* **2016**, *36*, 843–856. [CrossRef]

76. Quintero, D.; Lambert, D.J.; Villarreal, D.J.; Gregg, R.D. Real-Time continuous gait phase and speed estimation from a single sensor. In Proceedings of the 2017 IEEE Conference on Control Technology and Applications (CCTA), Maui, HI, USA, 27–30 August 2017; pp. 847–852.

77. Villarreal, D.J.; Poonawala, H.A.; Gregg, R.D. A robust parameterization of human gait patterns across phase-shifting perturbations. *IEEE Trans. Neural Syst. Rehabil. Eng.* **2017**, *25*, 265–278. [CrossRef]

78. Righetti, L.; Buchli, J.; Ijspeert, A.J. Dynamic hebbian learning in adaptive frequency oscillators. *Phys. Nonlinear Phenom.* **2006**, *216*, 269–281. [CrossRef]

79. Yan, T.; Parri, A.; Garate, V.R.; Cempini, M.; Ronsse, R.; Vitiello, N. An oscillator-based smooth real-time estimate of gait phase for wearable robotics. *Auton. Robot.* **2017**, *41*, 759–774. [CrossRef]

80. Aminian, K.; Najafi, B.; Büla, C.; Leyvraz, P.F.; Robert, P. Spatio-temporal parameters of gait measured by an ambulatory system using miniature gyroscopes. *J. Biomech.* **2002**, *35*, 689–699. [CrossRef]

81. Anwary, A.R.; Yu, H.; Vassallo, M. Gait quantification and visualization for digital healthcare. *Health Policy Technol.* **2020**, *9*, 204–212. [CrossRef]

82. Li, Q.; Liao, X.; Gao, Z. Indoor Localization with Particle Filter in Multiple Motion Patterns. In Proceedings of the 2020 IEEE Wireless Communications and Networking Conference (WCNC), Seoul, Korea, 25–28 May 2020; pp. 1–6.

83. Singh, Y.; Kher, M.; Vashista, V. Intention detection and gait recognition (IDGR) system for gait assessment: A pilot study. In Proceedings of the 2019 28th IEEE International Conference on Robot and Human Interactive Communication (RO-MAN), New Delhi, India, 14–18 October 2019; pp. 1–6.

84. Hua, R.; Wang, Y. Monitoring insole (MONI): A low power solution toward daily gait monitoring and analysis. *IEEE Sens. J.* **2019**, *19*, 6410–6420. [CrossRef]

85. Hu, H.; Fang, K.; Guan, H.; Wu, X.; Chen, C. A Novel Control Method of A Soft Exosuit with Plantar Pressure Sensors. In Proceedings of the 2019 IEEE 4th International Conference on Advanced Robotics and Mechatronics (ICARM), Toyonaka, Japan, 3–5 July 2019; pp. 581–586.

86. Behboodi, A.; Zahradka, N.; Wright, H.; Alesi, J.; Lee, S. Real-time detection of seven phases of gait in children with cerebral palsy using two gyroscopes. *Sensors* **2019**, *19*, 2517. [CrossRef]

87. Benson, L.C.; Clermont, C.A.; Watari, R.; Exley, T.; Ferber, R. Automated accelerometer-based gait event detection during multiple running conditions. *Sensors* **2019**, *19*, 1483. [CrossRef] [PubMed]

88. Ji, Q.; Yang, L.; Li, W.; Zhou, C.; Ye, X. Real-time gait event detection in a real-world environment using a laser-ranging sensor and gyroscope fusion method. *Physiol. Meas.* **2018**, *39*, 125003. [CrossRef]

89. Figueiredo, J.; Felix, P.; Costa, L.; Moreno, J.C.; Santos, C.P. Gait event detection in controlled and real-life situations: repeated measures from healthy subjects. *IEEE Trans. Neural Syst. Rehabil. Eng.* **2018**, *26*, 1945–1956. [CrossRef] [PubMed]

90. Wang, L.; Sun, Y.; Li, Q.; Liu, T. Estimation of step length and gait asymmetry using wearable inertial sensors. *IEEE Sens. J.* **2018**, *18*, 3844–3851. [CrossRef]

91. Hwang, T.H.; Reh, J.; Effenberg, A.O.; Blume, H. Real-time gait analysis using a single head-worn inertial measurement unit. *IEEE Trans. Consum. Electron.* **2018**, *64*, 240–248. [CrossRef]

92. Pitale, J.T.; Bolte, J.H. A heel-strike real-time auditory feedback device to promote motor learning in children who have cerebral palsy: A pilot study to test device accuracy and feasibility to use a music and dance-based learning paradigm. *Pilot Feasibility Stud.* **2018**, *4*, 1–7. [CrossRef]

93. Gonçalves, H.R.; Moreira, R.; Rodrigues, A.; Minas, G.; Reis, L.P.; Santos, C.P. Real-time tool for human gait detection from lower trunk acceleration. In Proceedings of the World Conference on Information Systems and Technologies, Naples, Italy, 27–29 March 2018; pp. 9–18. [CrossRef]

94. Khoo, I.H.; Marayong, P.; Krishnan, V.; Balagtas, M.; Rojas, O.; Leyba, K. Real-time biofeedback device for gait rehabilitation of post-stroke patients. *Biomed. Eng. Lett.* **2017**, *7*, 287–298. [CrossRef]

95. Piriyakulkit, S.; Hirata, Y.; Ozawa, H. Real-time gait event recognition for wearable assistive device using an imu on thigh. In Proceedings of the 2017 IEEE International Conference on Cyborg and Bionic Systems (CBS), Beijing, China, 17–19 October 2017; pp. 314–318.

96. Ahmadi, A.; Destelle, F.; Unzueta, L.; Monaghan, D.S.; Linaza, M.T.; Moran, K.; O'Connor, N.E. 3D human gait reconstruction and monitoring using body-worn inertial sensors and kinematic modeling. *IEEE Sens. J.* **2016**, *16*, 8823–8831. [CrossRef]

97. Bejarano, N.C.; Ambrosini, E.; Pedrocchi, A.; Ferrigno, G.; Monticone, M.; Ferrante, S. A novel adaptive, real-time algorithm to detect gait events from wearable sensors. *IEEE Trans. Neural Syst. Rehabil. Eng.* **2015**, *23*, 413–422. [CrossRef]

98. Kim, B.H.; Jo, S. Real-time motion artifact detection and removal for ambulatory BCI. In Proceedings of the 3rd International Winter Conference on Brain-Computer Interface, Gangwon, Korea, 12–14 January 2015; pp. 1–4.

99. Alahakone, A.U.; Senanayake, S.A.; Senanayake, C.M. Smart wearable device for real time gait event detection during running. In Proceedings of the 2010 IEEE Asia Pacific Conference on Circuits and Systems (APCCAS), Kuala Lumpur, Malaysia, 6–9 December 2010; pp. 612–615.

100. Félix, P.; Figueiredo, J.; Santos, C.P.; Moreno, J.C. Adaptive real-time tool for human gait event detection using a wearable gyroscope. In Proceedings of the 20th International Conference on Climbing Walking Robots Support Technologies for Mobile Machines (CLAWAR), Porto, Portugal, 11–13 September 2017; pp. 1–9.

101. Lee, J.K.; Park, E.J. Quasi real-time gait event detection using shank-attached gyroscopes. *Med. Biol. Eng. Comput.* **2011**, *49*, 707–712. [CrossRef]

102. Šprdlík, O.; Hurák, Z. Inertial gait phase detection: Polynomial nullspace approach. *IFAC Proc. Vol.* **2006**, *39*, 375–380. [CrossRef]

103. Allseits, E.; Lučarević, J.; Gailey, R.; Agrawal, V.; Gaunaurd, I.; Bennett, C. The development and concurrent validity of a real-time algorithm for temporal gait analysis using inertial measurement units. *J. Biomech.* **2017**, *55*, 27–33. [CrossRef] [PubMed]

104. Chang, H.C.; Hsu, Y.L.; Yang, S.C.; Lin, J.C.; Wu, Z.H. A wearable inertial measurement system with complementary filter for gait analysis of patients with stroke or Parkinson's disease. *IEEE Access* **2016**, *4*, 8442–8453. [CrossRef]

105. Kang, D.W.; Choi, J.S.; Kim, H.S.; Oh, H.S.; Seo, J.W.; Lee, J.W.; Tack, G.R. Wireless gait event detection system based on single gyroscope. In Proceedings of the 6th International Conference on Ubiquitous Information Management and Communication, New York, NY, USA, 20–22 February 2012; pp. 1–4. [CrossRef]

106. Catalfamo, P.; Ghoussayni, S.; Ewins, D. Gait event detection on level ground and incline walking using a rate gyroscope. *Sensors* **2010**, *10*, 5683–5702. [CrossRef] [PubMed]

107. Zhou, H.; Ji, N.; Samuel, O.W.; Cao, Y.; Zhao, Z.; Chen, S.; Li, G. Towards real-time detection of gait events on different terrains using time-frequency analysis and peak heuristics algorithm. *Sensors* **2016**, *16*, 1634. [CrossRef]

108. Gouwanda, D.; Gopalai, A.A. A robust real-time gait event detection using wireless gyroscope and its application on normal and altered gaits. *Med. Eng. Phys.* **2015**, *37*, 219–225. [CrossRef]

109. Figueiredo, J.; Ferreira, C.; Costa, L.; Sepúlveda, J.; Reis, L.P.; Moreno, J.C.; Santos, C.P. Instrumented insole system for ambulatory and robotic walking assistance: First advances. In Proceedings of the 2017 IEEE International Conference on Autonomous Robot Systems and Competitions (ICARSC), Coimbra, Portugal, 26–28 April 2017; pp. 116–121.

110. Zhang, F.; DiSanto, W.; Ren, J.; Dou, Z.; Yang, Q.; Huang, H. A novel CPS system for evaluating a neural-machine interface for artificial legs. In Proceedings of the 2011 IEEE/ACM Second International Conference on Cyber-Physical Systems, Chicago, IL, USA, 12–14 April 2011; pp. 67–76.

111. Sabatini, A.M.; Martelloni, C.; Scapellato, S.; Cavallo, F. Assessment of walking features from foot inertial sensing. *IEEE Trans. Biomed. Eng.* **2005**, *52*, 486–494. [CrossRef]

112. Chinimilli, P.T.; Qiao, Z.; Sorkhabadi, S.M.R.; Jhawar, V.; Fong, I.H.; Zhang, W. Automatic virtual impedance adaptation of a knee exoskeleton for personalized walking assistance. *Robot. Auton. Syst.* **2019**, *114*, 66–76. [CrossRef]

113. Taborri, J.; Scalona, E.; Palermo, E.; Rossi, S.; Cappa, P. Validation of inter-subject training for hidden Markov models applied to gait phase detection in children with cerebral palsy. *Sensors* **2015**, *15*, 24514–24529. [CrossRef]

114. Mannini, A.; Genovese, V.; Sabatini, A.M. Online decoding of hidden Markov models for gait event detection using foot-mounted gyroscopes. *IEEE J. Biomed. Health Inform.* **2013**, *18*, 1122–1130. [CrossRef]

115. Nutakki, C.; Mathew, R.J.; Suresh, A.; Vijay, A.R.; Krishna, S.; Babu, A.S.; Diwakar, S. Classification and Kinetic Analysis of Healthy Gait using Multiple Accelerometer Sensors. *Procedia Comput. Sci.* **2020**, *171*, 395–402. [CrossRef]

116. Martinez-Hernandez, U.; Mahmood, I.; Dehghani-Sanij, A.A. Simultaneous Bayesian recognition of locomotion and gait phases with wearable sensors. *IEEE Sens. J.* **2017**, *18*, 1282–1290. [CrossRef]

117. Meng, L.; Martinez-Hernandez, U.; Childs, C.; Dehghani-Sanij, A.A.; Buis, A. A practical gait feedback method based on wearable inertial sensors for a drop foot assistance device. *IEEE Sens. J.* **2019**, *19*, 12235–12243. [CrossRef]

118. Yang, J.; Huang, T.H.; Yu, S.; Yang, X.; Su, H.; Spungen, A.M.; Tsai, C.Y. Machine learning based adaptive gait phase estimation using inertial measurement sensors. In Proceedings of the 2019 Design of Medical Devices Conference, Minneapolis, MN, USA, 15–18 April 2019.

119. Vu, H.T.T.; Gomez, F.; Cherelle, P.; Lefeber, D.; Nowé, A.; Vanderborght, B. ED-FNN: A new deep learning algorithm to detect percentage of the gait cycle for powered prostheses. *Sensors* **2018**, *18*, 2389. [CrossRef] [PubMed]

120. Li, H.; Derrode, S.; Pieczynski, W. Lower limb locomotion activity recognition of healthy individuals using semi-Markov model and single wearable inertial sensor. *Sensors* **2019**, *19*, 4242. [CrossRef] [PubMed]

121. Grimmer, M.; Holgate, M.; Holgate, R.; Boehler, A.; Ward, J.; Hollander, K.; Sugar, T.; Seyfarth, A. A powered prosthetic ankle joint for walking and running. *Biomed. Eng. Online* **2016**, *15*, 37–52. [CrossRef] [PubMed]

122. Zhang, H.; Tay, M.O.; Suar, Z.; Kurt, M.; Zanotto, D. Regression models for estimating kinematic gait parameters with instrumented footwear. In Proceedings of the 2018 7th IEEE International Conference on Biomedical Robotics and Biomechatronics (Biorob), Enschede, The Netherlands, 26–29 August 2018; pp. 1169–1174.

123. Teufl, W.; Lorenz, M.; Miezal, M.; Taetz, B.; Fröhlich, M.; Bleser, G. Towards inertial sensor based mobile gait analysis: Event-detection and spatio-temporal parameters. *Sensors* **2019**, *19*, 38. [CrossRef] [PubMed]

124. Alvarez, D.; González, R.C.; López, A.; Alvarez, J.C. Comparison of step length estimators from weareable accelerometer devices. In Proceedings of the 28th Annual International Conference of the IEEE Engineering in Medicine and Biology Society, EMBS'06, New York, NY, USA, 30 August–3 September 2006; pp. 5964–5967. [CrossRef]

125. Afzal, M.R.; Lee, H.; Yoon, J.; Oh, M.K.; Lee, C.H. Development of an augmented feedback system for training of gait improvement using vibrotactile cues. In Proceedings of the 2017 14th International Conference on Ubiquitous Robots and Ambient Intelligence (URAI), Jeju, Korea, 28 June–1 July 2017; pp. 818–823.

126. Crea, S.; Manca, S.; Parri, A.; Zheng, E.; Mai, J.; Lova, R.M.; Vitiello, N.; Wang, Q. Controlling a robotic hip exoskeleton with noncontact capacitive sensors. *IEEE/ASME Trans. Mech.* **2019**, *24*, 2227–2235. [CrossRef]

127. Aich, S.; Pradhan, P.M.; Chakraborty, S.; Kim, H.C.; Kim, H.T.; Lee, H.G.; Kim, I.H.; Joo, M.i.; Jong Seong, S.; Park, J. Design of a Machine Learning-Assisted Wearable Accelerometer-Based Automated System for Studying the Effect of Dopaminergic Medicine on Gait Characteristics of Parkinson's Patients. *J. Healthc. Eng.* **2020**, *2020*. [CrossRef] [PubMed]

128. Zhang, Z.; Wu, J.; Huang, Z. Wearable sensors for realtime accurate hip angle estimation. In Proceedings of the 2008 IEEE International Conference on Systems, Man and Cybernetics, Singapore, 12–15 October 2008; pp. 2932–2937.

129. Waugh, J.L.; Huang, E.; Fraser, J.E.; Beyer, K.B.; Trinh, A.; McIlroy, W.E.; Kulić, D. Online learning of gait models from older adult data. *IEEE Trans. Neural Syst. Rehabil. Eng.* **2019**, *27*, 733–742. [CrossRef]

130. Wang, F.C.; Li, Y.C.; Wu, K.L.; Chen, P.Y.; Fu, L.C. Online Gait Detection with an Automatic Mobile Trainer Inspired by Neuro-Developmental Treatment. *Sensors* **2020**, *20*, 3389. [CrossRef]

131. Kwon, J.; Park, J.H.; Ku, S.; Jeong, Y.; Paik, N.J.; Park, Y.L. A soft wearable robotic ankle-foot-orthosis for post-stroke patients. *IEEE Robot. Autom. Lett.* **2019**, *4*, 2547–2552. [CrossRef]

132. Aguirre-Ollinger, G.; Narayan, A.; Reyes, F.A.; Cheng, H.J.; Yu, H. An Integrated Robotic Mobile Platform and Functional Electrical Stimulation System for Gait Rehabilitation Post-Stroke. In Proceedings of the International Conference on NeuroRehabilitation, Pisa, Italy, 16–20 October 2018; pp. 425–429. [CrossRef]
133. Kotiadis, D.; Hermens, H.J.; Veltink, P.H. Inertial gait phase detection for control of a drop foot stimulator. *Med Eng. Phys.* **2010**, *32*, 287–297. [CrossRef] [PubMed]
134. Seel, T.; Werner, C.; Raisch, J.; Schauer, T. Iterative learning control of a drop foot neuroprosthesis—Generating physiological foot motion in paretic gait by automatic feedback control. *Control. Eng. Pract.* **2016**, *48*, 87–97. [CrossRef]
135. Negård, N.O.; Schauer, T.; Kauert, R.; Raisch, J. An FES-assisted gait training system for hemiplegic stroke patients based on inertial sensors. *IFAC Proc. Vol.* **2006**, *39*, 315–320. [CrossRef]
136. Rioul, O.; Vetterli, M. Wavelets and signal processing. *IEEE Signal Process. Mag.* **1991**, *8*, 14–38. [CrossRef]
137. Lee, D.T.; Yamamoto, A. Wavelet analysis: Theory and applications. *Hewlett Packard J.* **1994**, *45*, 44–44.
138. Klingbeil, L.; Wark, T.; Bidargaddi, N. Efficient transfer of human motion data over a wireless delay tolerant network. In Proceedings of the 2007 3rd International Conference on Intelligent Sensors, Sensor Networks and Information, Melbourne, Australia, 3–6 December 2007; pp. 583–588.

Article

Whole-Body Movements Increase Arm Use Outcomes of Wrist-Worn Accelerometers in Stroke Patients

Gerrit Ruben Hendrik Regterschot [1,*], Ruud W. Selles [1,2], Gerard M. Ribbers [1,3] and Johannes B. J. Bussmann [1]

[1] Department of Rehabilitation Medicine, Erasmus University Medical Center Rotterdam, P.O. Box 2040, 3000 CA Rotterdam, The Netherlands; r.selles@erasmusmc.nl (R.W.S.); g.ribbers@erasmusmc.nl (G.M.R.); j.b.j.bussmann@erasmusmc.nl (J.B.J.B.)

[2] Department of Plastic and Reconstructive Surgery, Erasmus University Medical Center Rotterdam, P.O. Box 2040, 3000 CA Rotterdam, The Netherlands

[3] Rijndam Rehabilitation, Westersingel 300, 3015 LJ Rotterdam, The Netherlands

* Correspondence: g.r.h.regterschot@erasmusmc.nl

Abstract: Wrist-worn accelerometers are often applied to measure arm use after stroke. They measure arm movements during all activities, including whole-body movements, such as walking. Whole-body movements may influence clinimetric properties of arm use measurements—however, this has not yet been examined. This study investigates to what extent arm use measurements with wrist-worn accelerometers are affected by whole-body movements. Assuming that arm movements during whole-body movements are non-functional, we quantify the effect of whole-body movements by comparing two methods: Arm use measured with wrist-worn accelerometers during all whole-body postures and movements (P&M method), and during sitting/standing only (sit/stand method). We have performed a longitudinal observational cohort study with measurements in 33 stroke patients during weeks 3, 12, and 26 poststroke. The P&M method shows higher daily paretic arm use outcomes than the sit/stand method ($p < 0.001$), the mean difference increased from 31% at week three to 41% at week 26 ($p < 0.001$). Differences in daily paretic arm use between methods are strongly related to daily walking time (r = 0.83–0.92). Changes in the difference between methods are strongly related to changes in daily walking time (r = 0.89). We show that not correcting arm use measurements for whole-body movements substantially increases arm use outcomes, thereby threatening the validity of arm use outcomes and measured arm use changes.

Keywords: stroke; upper extremity; arm use; upper limb performance; accelerometer; sensor; walking; rehabilitation

Citation: Regterschot, G.R.H.; Selles, R.W.; Ribbers, G.M.; Bussmann, J.B.J. Whole-Body Movements Increase Arm Use Outcomes of Wrist-Worn Accelerometers in Stroke Patients. *Sensors* **2021**, *21*, 4353. https://doi.org/10.3390/s21134353

Academic Editor: Giuseppe Vannozzi

Received: 31 May 2021
Accepted: 20 June 2021
Published: 25 June 2021

Publisher's Note: MDPI stays neutral with regard to jurisdictional claims in published maps and institutional affiliations.

1. Introduction

In approximately 80% of the cases, a stroke leads to impairments in arm function in terms of muscle strength, voluntary control, coordination, and range of motion [1]. In-clinic assessment of arm function after stroke is often assumed to indicate arm use in daily life, i.e., the activities a person does with the arm in daily life. However, studies indicate that arm function and arm use are different constructs and need to be measured separately after stroke [2].

Wrist-worn accelerometers are often applied to measure arm use after stroke. For example, wrist-worn accelerometers have been used to assess arm use during rehabilitation poststroke [3–5], and to compare arm use levels between stroke patients and healthy subjects [6,7]. Wrist-worn accelerometers have also been applied to compare the arm use levels between the paretic and nonparetic arm after stroke [6,8]. Furthermore, studies explored the relationship between paretic arm use measured with wrist-worn accelerometers, arm function, and arm capacity after stroke [3,9]. Moreover, the change in arm use after stroke measured with wrist-worn accelerometers and the potential moderating role of psychological factors have been investigated [4].

217

Wrist-worn accelerometers measure arm use by recording the movements of the arms during all daily activities [3–5,7,9–12]. Thus, arm movements due to whole-body movements (e.g., walking, cycling, wheelchair transport, vehicle transport) influence arm use measurements with wrist-worn accelerometers [13,14]. However, arm movements due to whole-body movements are conceptually different from arm use during activities as eating with knife and fork, combing hair, and drinking. Arm movements due to whole-body movements are primarily non-functional, and therefore, they should ideally not be recorded as arm use. Quantifying non-functional arm movements due to whole-body movements as part of functional arm use is potentially problematic, since it may affect the clinimetric properties (e.g., validity, sensitivity to change, reliability) of arm use measurements with wrist-worn accelerometers [13]. For instance, when a patient walks, the non-functional arm movements due to walking will be measured as arm use by wrist-worn accelerometers, which may significantly increase the arm use outcomes of wrist-worn accelerometers, thereby reducing the validity of the arm use measurements. Similarly, an increase in the daily amount of walking after stroke may influence the change in daily arm use measured with wrist-worn accelerometers, thereby threatening the validity of the measured arm use changes.

These potential problems have been identified by previous studies, and different methods have already been proposed to correct arm use measurements with wrist-worn accelerometers for the effect of whole-body movements. These methods are measuring arm movements during sitting and standing [13,15], or calculating a ratio outcome between arms [14]. However, these methods are not widely adopted in the research field, because (1) they require more complex sensor set-ups and/or signal analyses, and (2) the effect of whole-body movements on arm use measurements with wrist-worn accelerometers is still unclear and has not yet been quantified. Most studies report arm use measurements with wrist-worn accelerometers without correcting for the effect of whole-body movements, which may affect the clinimetric properties of arm use outcomes. Therefore, studies quantifying the effect of whole-body movements on cross-sectional and longitudinal arm use measurements with wrist-worn accelerometers are urgently needed to determine the necessity of correcting arm use measurements regarding the effect of whole-body movements.

The present study quantifies the effect of whole-body movements on cross-sectional and longitudinal arm use measurements with wrist-worn accelerometers after stroke. Assuming that all arm movements during whole-body movements are non-functional, we quantified the effect of whole-body movements by comparing the arm use outcomes of two measurement methods: (1) Arm use outcomes measured with wrist-worn accelerometers during all whole-body postures and movements (P&M method), and (2) arm use outcomes measured with wrist-worn accelerometers during only sitting and standing periods (sit/stand method) [13,15]. The difference between the arm use outcomes of these two methods is the effect of whole-body movements on arm use measurements with wrist-worn accelerometers. We hypothesized that (1) whole-body movements, especially walking, increase arm use outcomes of wrist-worn accelerometers and the size of the effect depends on the amount of walking, and (2) the positive effect of walking on arm use measurements with wrist-worn accelerometers increases with time poststroke as a result of an increase in the daily amount of walking after stroke.

2. Materials and Methods

2.1. Participants

The present study was a longitudinal observational cohort study and part of another study investigating the change in objectively measured arm use during the first six months after stroke [15]. When designing and reporting the present study, we followed the STROBE recommendations for observational studies [16]. In the present study, we aimed to include at least 28 stroke patients, since this sample size can detect a medium effect (Cohen's d = 0.50) with an alpha of 0.05 and a power of 0.80. Included were patients admitted to Rijndam Rehabilitation (Rotterdam, The Netherlands) after an ischemic or hemorrhagic stroke that suffered from a paretic arm or leg (defined as National Institutes of Health Stroke Scale (NIHSS) 5A/B or 6A/B 4 $\geq$ score > 0). Inclusion criteria were (i) 18 years or older, (ii) Mini-Mental State Examination (MMSE) >19, (iii) able to sit at least 30 min with back support. Excluded were patients who were more than three weeks after stroke when admitted to the rehabilitation clinic. The study was performed between September 2016 and September 2019. The study was conducted in accordance with the Declaration of Helsinki. The study was approved by the Medical Ethics Committee of Erasmus MC University Medical Center Rotterdam in The Netherlands (MEC-2015-687), and all participants provided written informed consent.

2.2. Procedures

A researcher performed arm use measurements at 3 weeks, 12 weeks, and 26 weeks poststroke, assessed arm function (Fugl-Meyer upper extremity assessment) and stroke severity (National Institutes of Health Stroke Scale (NIHSS) [17,18]) and collected demographic data. At three weeks after stroke, all patients were inpatient at Rijndam Rehabilitation and received standard poststroke treatment. At Rijndam Rehabilitation, the arm-hand therapy after stroke consists of the Concise Arm and Hand Rehabilitation Approach in Stroke (CARAS) [19,20]. At week 12 poststroke, some individuals were still at the rehabilitation center, while at week 26 poststroke, all patients were at home. The arm use measurements at home were performed by the same researcher.

2.3. Arm Use Measurements

In this study, we used an arm use monitor developed and validated for the measurement of arm use in stroke patients [13]. The arm use monitor consists of three accelerometers (Activ8 Activity Monitor, Activ8; 30 $\times$ 32 $\times$ 10 mm; 20 g). One accelerometer was attached to each wrist to measure arm movement intensity (see Figure 1), and one accelerometer was attached to the front side of the nonparetic thigh to recognize body postures and movements (lying, sitting, standing, walking, cycling, running). The applied accelerometers measured with a sampling frequency of 12.5 Hz [13]. The sensors on the wrists converted acceleration data to movement counts with 1.6 Hz resolution [13], and stored these data in epochs of 30 s (per epoch 48 samples). The sensor on the thigh converted acceleration data to movement counts and body postures/movements with 1.6 Hz resolution [13], and stored these data in epochs of 30 s (per epoch 48 samples). The recognition of body postures and movements (lying, sitting, standing, walking, cycling, running) by the sensor on the thigh is based on (1) the orientation of the sensor compared to gravity, and (2) the intensity of the movement (in movement counts) [13,21]. An Activ8 sensor on the thigh provides an accurate recognition of whole-body postures and movements in stroke patients with an accuracy ranging from 82 to 100% [21].

During weeks 3, 12, and 26 poststroke, patients wore the three accelerometers for seven consecutive days. The wrist-worn sensors were attached with wristbands and were taken off during the night and during water activities (e.g., showering, swimming). The leg sensor was worn for seven consecutive days and attached with anti-allergic, water-resistant skin tape. The data of the sensors were downloaded on a PC for further processing and analysis after each measurement period of one week.

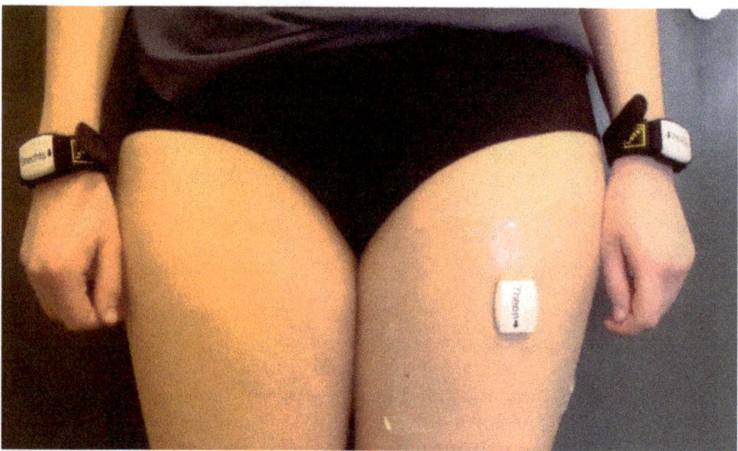

Figure 1. Participants wore three accelerometers: One accelerometer on each wrist and one accelerometer on the upper leg of the nonparetic side of the body.

2.4. Analysis of Sensor Data

To process and analyze the sensor data, we developed an algorithm in R [22] using RStudio (version 1.2.50001, RStudio, Inc., Boston, MA, USA). Firstly, the data of the sensors were time-synchronized based on the timestamps. We only analyzed waking hours from 7 a.m. to 10 p.m. Non-wear periods were excluded from further analysis and were defined as zero movement counts measured for at least one hour. Per participant, a measurement week was included in the analysis when at least two valid measurement days were available. A valid measurement day was defined as a day with at least ten hours of data of the whole sensor configuration.

In this study, we assumed that all arm movements during whole-body movements (e.g., walking, cycling, wheelchair transport, vehicle transport) are non-functional and conceptually different from arm use (e.g., combing hair, drinking, tooth brushing). Based on this assumption, we quantified the effect of whole-body movements on arm use measurements with wrist-worn accelerometers by comparing the arm use outcomes of two measurement methods: (1) Arm use measured with wrist-worn accelerometers during all whole-body postures and movements (P&M method), and (2) arm use measured with wrist-worn accelerometers during only sitting and standing periods (sit/stand method) [13,15]. The difference between the arm use outcomes of these two methods is the effect of whole-body movements on arm use measurements with wrist-worn accelerometers. Per valid measurement day, we calculated the arm use outcomes described below.

P&M method:

1. Paretic arm use: Calculated by summing the movement counts of the sensor on the paretic arm over all 30 s epochs.
2. Ratio between arms: Calculated as the paretic arm use during all whole-body postures and movements divided by the nonparetic arm use during all whole-body postures and movements.
3. Nonparetic arm use: Calculated by summing the movement counts of the sensor on the nonparetic arm over all 30 s epochs.

Sit/stand method:

1. Paretic arm use: Calculated by summing the movement counts of the sensor on the paretic arm over all 30 s epochs of which the posture was sitting or standing. An epoch was classified as sitting or standing when at least 90% of the 48 samples of the leg sensor were classified as sitting or standing.

2. Ratio between arms: Calculated as the paretic arm use during sitting and standing divided by the nonparetic arm use during sitting and standing.
3. Nonparetic arm use: Calculated by summing the movement counts of the sensor on the nonparetic arm over all 30 s epochs classified as sitting or standing.

Next, for each week (weeks 3, 12, and 26 poststroke), we calculated a mean daily value for each arm use outcome measure by averaging over valid measurement days.

2.5. Statistical Analysis

We performed the statistical analysis in R [22] using RStudio (version 1.2.50001, RStudio, Inc., Boston, MA, USA).

Differences in arm use outcomes between the P&M method and the sit/stand method were investigated at all time points (week 3, week 12, week 26 poststroke) by using Bland and Altman plots [23]. For the Bland and Altman plots, we calculated the mean difference in arm use outcome between the two methods (D), the SD of the differences in arm use outcome between the two methods (SDdiff), and the limits of agreement (LOA) as: LOA = D $\pm$ 1.96*SDdiff.

We applied Generalized Estimating Equation (GEE) [24] to test whether the P&M method and the sit/stand method differ significantly in cross-sectional and longitudinal arm use outcomes. In the GEE analysis, we included time (three levels: 3, 12, and 26 weeks), method (two levels: P&M method, sit/stand method), and the interaction time*method as factors. We used the Generalized Estimating Equation package ('geepack' package [25]) with as settings a Gaussian data distribution and an exchangeable correlation structure. Statistical significance was set at $p < 0.05$. For significant effects, posthoc comparisons were performed using the Estimated Marginal Means package ('emmeans' package) and by applying a Bonferroni correction [26].

To investigate whether differences in arm use outcomes between the methods are related to walking, we calculated Spearman's rank correlation coefficients between the daily walking time and the difference in arm use outcome between the methods at each time point (week 3, week 12, week 26 poststroke). To examine whether changes in the daily amount of walking after stroke are related to differences in longitudinal arm use outcomes between the methods, we calculated Spearman's rank correlation coefficients between the change in daily walking time from week 3 to week 26 poststroke and the change in arm use outcome difference between the methods from week 3 to week 26 poststroke. Correlations were interpreted as follows: Very weak when $0.00 < r < 0.25$; weak when $0.25 \leq r \leq 0.49$; moderate when $0.50 \leq r \leq 0.69$; strong when $0.70 \leq r \leq 0.89$; very strong when $0.90 \leq r \leq 1.00$ [27].

3. Results

We included 33 stroke patients (26 males, 7 females). Table 1 shows the patient characteristics. The arm use data of three measurement weeks (weeks 3, 12, 26) were available from 18 patients, while from the other patients' arm use data of only two measurement weeks were available. At week three poststroke, arm use data were missing in three participants as a result of a technical failure of the sensor system or non-wear of the system. At week 12 poststroke, arm use data were missing in five patients, due to a technical failure of the sensor system, non-wear of the system, or participant unavailability for the measurement. Arm use data were missing at week 26 in seven participants because of study dropout, a technical failure, or non-wear of the sensor system.

Table 1. Patient characteristics (n = 33). Data are reported as mean ± SD [minimal value, maximal value] unless otherwise stated.

Age in Years	55.9 ± 9.2 [37–75]
Gender	26 males, 7 females
Affected body side	12 left side, 21 right side
Dominant side affected	11 (33%)
Admitted to rehabilitation clinic in weeks poststroke	1.6 ± 0.7 [0.4–3.0]
Discharge from a rehabilitation clinic in weeks poststroke	10.5 ± 4.7 [3.7–20.3]
NIHSS [a] values week 12 poststroke	2.1 ± 2.7 [0–11]
Fugl-Meyer upper extremity assessment:	
week 3 poststroke	25.5 ± 20.6 [4–64]
week 12 poststroke	41.2 ± 21.8 [4–64]
week 26 poststroke	51.2 ± 16.8 [9–64]

[a] National Institutes of Health Stroke Scale.

The daily monitor wearing time did not change over time ($p = 0.73$; Figure 2A). Daily sitting and standing time decreased from week 3 to 12 and from week 3 to 26 (Figure 2B). Daily walking time increased from week 3 to 12 and from week 3 to 26 (Figure 2C).

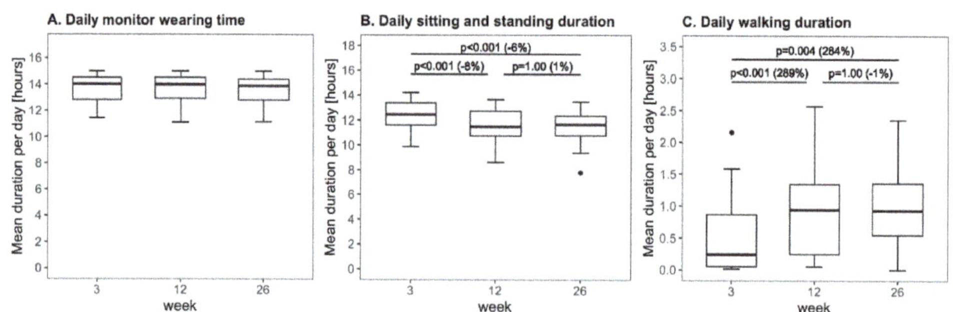

Figure 2. Boxplots of the daily monitor wearing time, daily sitting, and standing time, and daily walking time measured with the sensor system. The percentage between brackets represents the change in median value between time points.

The P&M method showed higher paretic arm use, the ratio between arms, and nonparetic arm use outcomes than the sit/stand method at all time points ($p < 0.001$; Figures 3 and 4). The mean difference in paretic arm use outcome between the methods increased over time ($p < 0.001$) from 31% at week 3 to 40% at week 12 and 41% at week 26 poststroke (Figure 4). The mean difference in ratio outcome between the methods did not change over time ($p = 0.16$) and was 8–9% at the different time points (Figure 4). The mean difference in nonparetic arm use outcome between the methods increased over time ($p < 0.001$) from 17% at week 3 to 30% at week 12 and 32% at week 26 poststroke (Figure 4).

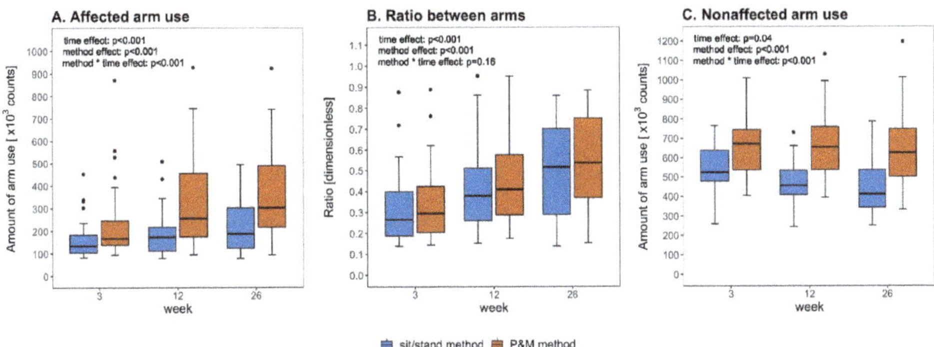

Figure 3. Boxplots showing the arm use outcomes of the P&M method and the sit/stand method at week 3, 12, and 26 poststroke. *p*-values of the GEE analyses are included in the plots.

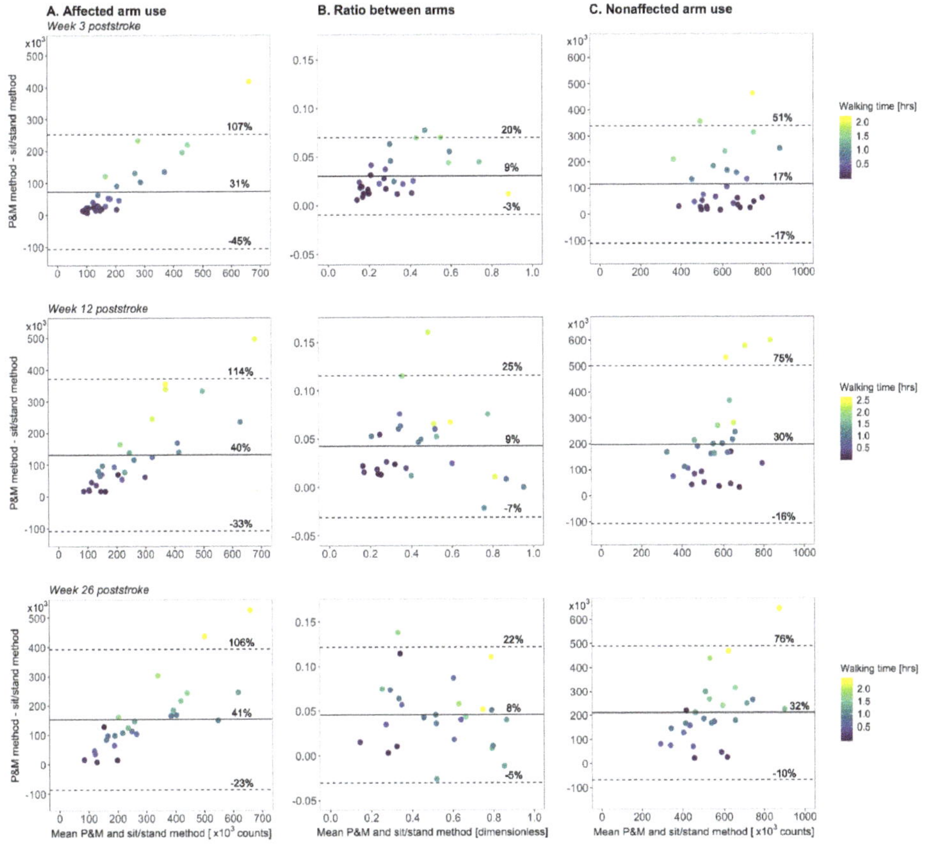

Figure 4. Bland and Altman plots showing the mean arm use outcome of the P&M method and the sit/stand method versus the difference in arm use outcome between the methods. The horizontal line indicates the mean difference between methods (D), and the dashed horizontal lines represent the limits of agreement (LOA). The D and LOA are expressed as a percentage of the mean arm use outcome of the P&M method. Bland and Altman plots are shown for week 3 poststroke (upper row), week 12 poststroke (middle row), and week 26 poststroke.

The differences in paretic and nonparetic arm use outcomes between the methods were strongly related to very strongly related to the daily walking time at all time points (r = 0.83–0.92; Figure 5), indicating a significant positive effect of walking on cross-sectional arm use measurements with the P&M method. The difference in ratio outcomes between the methods and the daily walking time were strongly related at week 3 poststroke (r = 0.70; Figure 5), but very weakly to weakly related at week 12 and week 26 poststroke (r = 0.22–0.33; Figure 5).

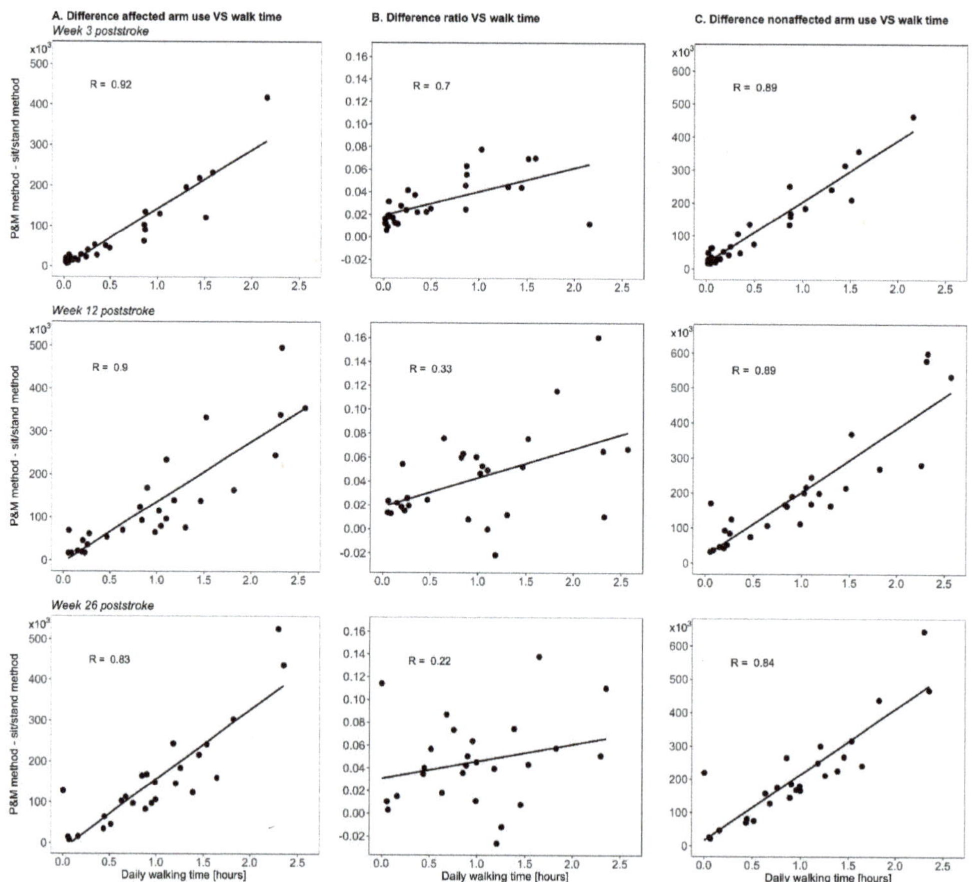

Figure 5. Scatterplots showing the daily walking time versus the difference in arm use outcome between the P&M method and the sit/stand method. Scatterplots are shown for week 3 poststroke (upper row), week 12 poststroke (middle row), and week 26 poststroke.

The increase in paretic and nonparetic arm use differences between the methods from week 3 to week 26 was strongly related to very strongly related to the increase in daily walking time poststroke (r = 0.89–0.90; Figure 6), indicating a significant positive effect of walking on longitudinal arm use measurements with the P&M method. The change in the ratio differences between the methods from week 3 to week 26 was moderately related to the change in daily walking time (r = 0.64; Figure 6).

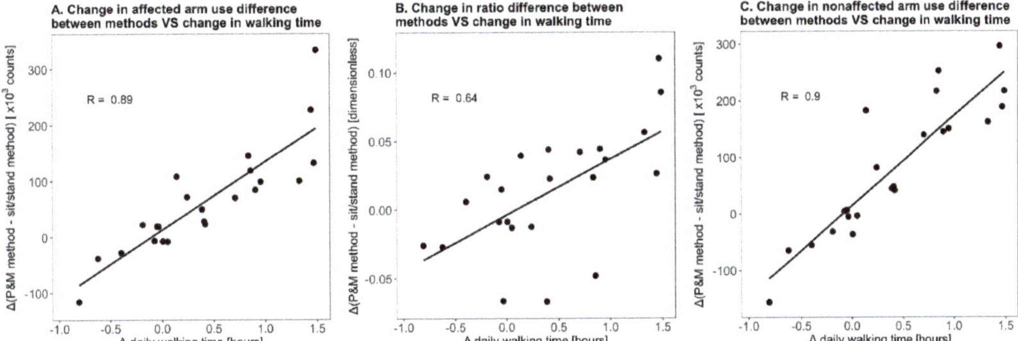

Figure 6. Scatterplots showing the change in daily walking time from week 3 to week 26 poststroke versus the change in arm use outcome difference between the P&M method and the sit/stand method from week 3 to week 26 poststroke.

4. Discussion

Results of this study confirm our hypotheses by showing that whole-body movements increase cross-sectionally measured arm use outcomes of wrist-worn accelerometers by 8–41% if arm use data are not corrected for whole-body movements. We found that the size of the effect of whole-body movements on arm use measurements depends largely on the amount of walking. Since the daily amount of walking increased from week 3 to week 26 after stroke, the average effect of whole-body movements on paretic arm use outcomes increased from 31% at week 3 to 41% at week 26 poststroke when not correcting for whole-body movements. These findings indicate that not correcting arm use data for whole-body movements may threaten the validity of arm use outcomes and of measured changes in arm use over time.

The positive effect of walking on arm use measurements with wrist-worn accelerometers can be explained by the fact that wrist-worn accelerometers measure arm use by recording all arm movements. This includes non-functional arm movements as a result of the center of mass displacement during walking, which are measured as arm use by wrist-worn accelerometers, and which consequently increase the arm use outcomes of wrist-worn accelerometers. Since most patients increased in daily walking time from week 3 to week 26 poststroke (Figure 2), the positive effect of walking on paretic and nonparetic arm use measurements increased from week 3 to week 26 after stroke.

The positive effect of walking on the ratio between arms is less clear. Our data suggest that walking has a larger absolute effect on nonparetic arm use outcomes than on paretic arm use outcomes (Figure 4A,C)—possibly because of more arm sway of the nonparetic arm during walking—and that as a result, the ratio between arms is only slightly higher during walking than during sitting/standing (Figure 4B). This would explain (1) why the positive effect of whole-body movements on the ratio between arms did not substantially change over time (Figure 4B), (2) why changes in daily walking time were not strongly related to changes in the ratio difference between the two methods (Figure 6B), and (3) why daily walking duration showed relatively weak associations with the difference in ratio outcome between the methods (Figure 5B).

A noteworthy observation was that at all time points, the differences in paretic arm use outcomes between the P&M method, and the sit/stand method were larger at higher paretic arm use levels (Figure 4A). This can be explained by the fact that individuals with higher paretic arm use levels spend more time walking during the day (see colored data in Figure 4A), resulting in larger differences in paretic arm use outcomes between the two methods. The relationship that we found between paretic arm use levels and daily walking time is in line with other studies showing a relationship between the disability level of the paretic arm and walking performance after stroke [28].

Since whole-body movements, especially walking, greatly affect arm use measurements with wrist-worn accelerometers and threaten the validity of arm use outcomes, it is important to correct arm use measurements for this effect. The use ratio between arms was proposed by a previous study [14] to correct for the effect of whole-body movements. However, our results demonstrate that the ratio between arms cannot fully correct for the effect of whole-body movements, since whole-body movements increased the ratio between arms on average by 8–9% at all time points (Figure 4B). To correct arm use outcomes for the effect of whole-body movements, we propose to measure arm use by recording arm movements with wrist-worn accelerometers during only sitting and standing periods. This practical and simple method avoids the effect of walking and provides accurate arm use measurements in stroke patients [13]. The sensor configuration of this method currently consists of two wrist-worn accelerometers combined with an accelerometer on the upper leg to detect whole-body postures and movements. To foster the clinical application of this method, we are currently developing a minimal sensing solution by enabling the detection of whole-body postures and movements based on wrist-worn accelerometers instead of using an accelerometer on the upper leg.

Our study may have consequences for interpreting the results of other studies that did not correct arm use measurements for the effect of whole-body movements. The arm use outcomes of these studies may be affected by whole-body movements, especially walking. For example, a recent study applied wrist-worn accelerometers to measure arm use during the first 12 weeks after stroke without correcting for the effect of whole-body movements [4]. Results showed that the mean daily paretic arm use in 29 stroke patients increased by approximately 85% from about 2.6 h in week 2 poststroke to almost 5 h in week 12 after stroke. Since the study did not correct arm use measurements for whole-body movements, it is possible that whole-body movements, such as walking, have affected the reported arm use changes. This is plausible since we found a comparable (approximately 75%) increase in the mean daily paretic arm use from week 3 to week 12 after stroke when not correcting for whole-body movements, but a much smaller increase (19%) when correcting for whole-body movements. This example shows that whole-body movements may have affected the arm use outcomes of studies that did not correct for such an effect. A potential effect of whole-body movements should be taken into consideration when interpreting the findings of these studies.

Several limitations may have influenced the outcomes of this study. First, the relatively small sample size and the single recruitment site may limit the generalizability of our results. However, the sample size of the present study (n = 33) is larger than the required sample size (n = 28) that we estimated a priori based on a statistical power analysis (see Methods section). Second, at each time point in the study, there were missing data due to technical issues, non-wear of the system, or unavailability of participants. To avoid that missing data affected the outcomes of the study, we applied generalized estimating equations that can handle missing data appropriately [24]. Third, the accuracy of the detection of sitting, standing, and walking by the leg accelerometer is not perfect (approximately 90–95% [21]). However, since the detection accuracy is very high, it is unlikely that misclassification has affected the findings of the present study. Fourth, we did not consider the effect of dominance on the paretic arm use measurements, since previous research indicated that the difference in daily use between the dominant and non-dominant arm in healthy adults is very small [10]. Fifth, the analysis in the present study assumed that all arm movements during whole-body movements are non-functional. Although this assumption might not be fully correct, previous research has shown that measuring arm movements with wrist-worn accelerometers during only sitting and standing periods provides very accurate arm use outcomes in stroke patients [13], thereby supporting the validity of our assumption.

5. Conclusions

This study shows that whole-body movements increase cross-sectionally measured arm use outcomes of wrist-worn accelerometers with 8–41% if not correcting arm use data for whole-body movements. We found that the size of the effect of whole-body movements on arm use measurements depends largely on the amount of walking. Since the daily amount of walking increased from week 3 to week 26 poststroke, the average effect of whole-body movements on paretic arm use outcomes increased from 31% at week 3 to 41% at week 26 poststroke when not correcting for whole-body movements. These findings indicate that not correcting arm use data for whole-body movements may threaten the validity of arm use outcomes and of measured changes in arm use over time. To correct arm use measurements with wrist-worn accelerometers for the effect of whole-body movements and specifically walking, we propose a practical and valid solution that measures arm use during only sitting and standing periods with wrist-worn accelerometers and an accelerometer on the upper leg [13].

Author Contributions: Conceptualization, G.R.H.R., R.W.S. and J.B.J.B.; methodology, G.R.H.R., R.W.S. and J.B.J.B.; software, G.R.H.R.; validation, G.R.H.R.; formal analysis, G.R.H.R.; investigation, G.R.H.R., R.W.S., G.M.R. and J.B.J.B.; resources, G.R.H.R.; data curation, G.R.H.R.; writing—original draft preparation, G.R.H.R.; writing—review and editing, R.W.S., J.B.J.B., G.R.H.R. and G.M.R.; visualization, G.R.H.R.; supervision, J.B.J.B., R.W.S., G.M.R.; project administration, J.B.J.B. and G.M.R.; funding acquisition, J.B.J.B., R.W.S. and G.M.R. All authors have read and agreed to the published version of the manuscript.

Funding: This research was funded by the ZonMW Innovative Medical Devices Initiative program (title: "PROFITS—Precision profiling to improve long-term outcome after stroke"; project number 104003008) and Rijndam Rehabilitation. The collaboration project is co-funded by the PPP Allowance made available by Health~Holland, Top Sector Life Sciences & Health, to stimulate public-private partnerships.

Institutional Review Board Statement: The study was conducted according to the guidelines of the Declaration of Helsinki, and approved by the Ethics Committee of Erasmus MC University Medical Center Rotterdam (MEC-2015-687; date of approval: 28-09-2016).

Informed Consent Statement: Informed consent was obtained from all subjects involved in the study.

Data Availability Statement: The data presented in this study are available on request from the corresponding author. The data are not publicly available due to privacy/ethical restrictions.

Conflicts of Interest: The authors declare no conflict of interest. The funders had no role in the design of the study; in the collection, analyses, or interpretation of data; in the writing of the manuscript, or in the decision to publish the results.

References

1. Kwakkel, G.; Veerbeek, J.M.; Wegen, E.E.H.V.; Wolf, S.L. Constraint-induced movement therapy after stroke. *Lancet Neurol.* **2015**, *14*, 224–234. [CrossRef]
2. Michielsen, M.E.; De Niet, M.; Ribbers, G.; Stam, H.J.; Bussmann, J.B. Evidence of a logarithmic relationship between motor capacity and actual performance in daily life of the paretic arm following stroke. *J. Rehabil. Med.* **2009**, *41*, 327–331. [CrossRef] [PubMed]
3. Doman, C.A.; Waddell, K.J.; Bailey, R.R.; Moore, J.L.; Lang, C.E. Changes in Upper-Extremity Functional Capacity and Daily Performance During Outpatient Occupational Therapy for People with Stroke. *Am. J. Occup. Ther.* **2016**, *70*, 1–11. [CrossRef]
4. Waddell, K.J.; Strube, M.J.; Tabak, R.G.; Haire-Joshu, D.; Lang, C.E. Upper Limb Performance in Daily Life Improves Over the First 12 Weeks Poststroke. *Neurorehabilit. Neural Repair* **2019**, *33*, 836–847. [CrossRef]
5. Lang, C.E.; Wagner, J.M.; Edwards, D.F.; Dromerick, A.W. Upper Extremity Use in People with Hemiparesis in the First Few Weeks After Stroke. *J. Neurol. Phys. Ther.* **2007**, *31*, 56–63. [CrossRef] [PubMed]
6. Michielsen, M.E.; Selles, R.W.; Stam, H.J.; Ribbers, G.; Bussmann, J.B. Quantifying Nonuse in Chronic Stroke Patients: A Study Into Paretic, Nonparetic, and Bimanual Upper-Limb Use in Daily Life. *Arch. Phys. Med. Rehabil.* **2012**, *93*, 1975–1981. [CrossRef]
7. Bailey, R.R.; Klaesner, J.W.; Lang, C.E. Quantifying Real-World Upper-Limb Activity in Nondisabled Adults and Adults with Chronic Stroke. *Neurorehabilit. Neural Repair* **2015**, *29*, 969–978. [CrossRef] [PubMed]
8. de Niet, M.; Bussmann, J.B.; Ribbers, G.; Stam, H.J. The Stroke Upper-Limb Activity Monitor: Its Sensitivity to Measure Hemiplegic Upper-Limb Activity During Daily Life. *Arch. Phys. Med. Rehabil.* **2007**, *88*, 1121–1126. [CrossRef]
9. Chin, L.F.; Hayward, K.; Soh, A.J.A.; Tan, C.M.; Wong, C.J.R.; Loh, J.W.; Loke, G.J.H.; Brauer, S. An accelerometry and observational study to quantify upper limb use after stroke during inpatient rehabilitation. *Physiother. Res. Int.* **2019**, *24*, e1784. [CrossRef]

10. Bailey, R.R.; Lang, C.E. Pt Upper-limb activity in adults: Referent values using accelerometry. *J. Rehabil. Res. Dev.* **2013**, *50*, 1213–1222. [CrossRef]
11. Moore, S.A.; Da Silva, R.; Balaam, M.; Brkic, L.; Jackson, D.; Jamieson, D.; Ploetz, T.; Rodgers, H.; Shaw, L.; Van Wijck, F.; et al. Wristband Accelerometers to motiVate arm Exercise after Stroke (WAVES): Study protocol for a pilot randomized controlled trial. *Trials* **2016**, *17*, 1–9. [CrossRef]
12. Held, J.P.O.; Luft, A.R.; Veerbeek, J.M. Encouragement-Induced Real-World Upper Limb Use after Stroke by a Tracking and Feedback Device: A Study Protocol for a Multi-Center, Assessor-Blinded, Randomized Controlled Trial. *Front. Neurol.* **2018**, *9*, 1–11. [CrossRef]
13. Fanchamps, M.; Selles, R.; Stam, H.; Bussmann, J. Development and validation of a clinically applicable arm use monitor for people after stroke. *J. Rehabil. Med.* **2018**, *50*, 705–712. [CrossRef]
14. Uswatte, G.; Foo, W.L.; Olmstead, H.; Lopez, K.; Holand, A.; Simms, L.B. Ambulatory Monitoring of Arm Movement Using Accelerometry: An Objective Measure of Upper-Extremity Rehabilitation in Persons with Chronic Stroke. *Arch. Phys. Med. Rehabilitation* **2005**, *86*, 1498–1501. [CrossRef] [PubMed]
15. Regterschot, G.R.H.; Bussmann, J.B.J.; Fanchamps, M.H.J.; Meskers, C.G.M.; Ribbers, G.M.; Selles, R.W. Objectively measured arm use in daily life improves during the first six months poststroke: A longitudinal observational cohort study. *J. NeuroEng. Rehabil.* **2021**, *18*, 1–10. [CrossRef] [PubMed]
16. Von Elm, E.; Altman, D.G.; Egger, M.; Pocock, S.J.; Gotzsche, P.C.; Vandenbroucke, J.P. STROBE Initiative. The strengthening the reporting of observational studies in epidemiology (STROBE) statement: Guidelines for reporting observational studies. *Lancet* **2007**, *370*, 1453–1457. [CrossRef]
17. Brott, T.; Adams, H.P.; Olinger, C.P.; Marler, J.R.; Barsan, W.G.; Biller, J.; Spilker, J.; Holleran, R.; Eberle, R.; Hertzberg, V. Measurements of acute cerebral infarction: A clinical examination scale. *Stroke* **1989**, *20*, 864–870. [CrossRef] [PubMed]
18. Kwah, L.K.; Diong, J. National Institutes of Health Stroke Scale (NIHSS). *J. Physiother.* **2014**, *60*, 61. [CrossRef]
19. Franck, J.A.; Smeets, R.J.E.M.; Seelen, H.A.M. Changes in arm-hand function and arm-hand skill performance in patients after stroke during and after rehabilitation. *PLoS ONE* **2017**, *12*, e0179453. [CrossRef]
20. Franck, J.A.; Halfens, J.; Smeets, R.; Seelen, H. Concise Arm and Hand Rehabilitation Approach in Stroke (CARAS): A practical and evidence-based framework for clinical rehabilitation management. *Open J. Occup. Ther.* **2015**, *3*, 10. [CrossRef]
21. Fanchamps, M.H.J.; Horemans, H.L.D.; Ribbers, G.M.; Stam, H.J.; Bussmann, J.B.J. The Accuracy of the Detection of Body Postures and Movements Using a Physical Activity Monitor in People after a Stroke. *Sensors* **2018**, *18*, 2167. [CrossRef]
22. R Core Team. *R: A Language and Environment for Statistical Computing*; R Foundation for Statistical Computing: Vienna, Austria, 2018; Available online: https://www.r-project.org/ (accessed on 4 November 2020).
23. Bland, J.M.; Altman, D.G. Statistical methods for assessing agreement between two methods of clinical measurement. *Lancet* **1986**, *1*, 307–310. [CrossRef]
24. Salazar, A.; Ojeda, B.; Dueñas, M.; Fernández, F.; Failde, I. Simple generalized estimating equations (GEEs) and weighted generalized estimating equations (WGEEs) in longitudinal studies with dropouts: Guidelines and implementation in R. *Stat. Med.* **2016**, *35*, 3424–3448. [CrossRef]
25. Højsgaard, S.; Halekoh, U.; Yan, J.; Ekstrøm, C. Generalized Estimating Equation Package. CRAN Repository, 2019. Available online: https://cran.r-project.org (accessed on 4 November 2020).
26. Lenth, R.; Singmann, H.; Love, J.; Buerkner, P.; Herve, M. Estimated Marginal Means, Aka Least-Squares Means. CRAN Repository, 2020. Available online: https://cran.r-project.org (accessed on 4 November 2020).
27. Domholdt, E. *Physical Therapy Research: Principles and Applications*; Saunders: Philadelphia, PA, USA, 2000.
28. Desrosiers, J.; Malouin, F.; Bourbonnais, D.; Richards, C.L.; Rochette, A.; Bravo, G. Arm and leg impairments and disabilities after stroke rehabilitation: Relation to handicap. *Clin. Rehabil.* **2003**, *17*, 666–673. [CrossRef]

Article

Classification of Neurological Patients to Identify Fallers Based on Spatial-Temporal Gait Characteristics Measured by a Wearable Device

Yuhan Zhou [1,*], Rana Zia Ur Rehman [2], Clint Hansen [3], Walter Maetzler [3], Silvia Del Din [2], Lynn Rochester [2,4], Tibor Hortobágyi [1] and Claudine J. C. Lamoth [1,*]

[1] Department of Human Movement Sciences, University Medical Center Groningen, University of Groningen, 9713 AV Groningen, The Netherlands; t.hortobagyi@umcg.nl

[2] Translational and Clinical Research Institute, Faculty of Medical Sciences, Newcastle University, Newcastle Upon Tyne NE4 5PL, UK; Rana.zia-ur-Rehman@newcastle.ac.uk (R.Z.U.R.); silvia.del-din@newcastle.ac.uk (S.D.D.); lynn.rochester@newcastle.ac.uk (L.R.)

[3] Department of Neurology, University Hospital Schleswig-Holstein, Campus Kiel, 24105 Kiel, Germany; c.hansen@neurologie.uni-kiel.de (C.H.); w.maetzler@neurologie.uni-kiel.de (W.M.)

[4] Newcastle Upon Tyne Hospitals NHS Foundation Trust, Newcastle Upon Tyne NE7 7DN, UK

* Correspondence: y.zhou01@umcg.nl (Y.Z.); c.j.c.lamoth@umcg.nl (C.J.C.L.)

Received: 17 June 2020; Accepted: 21 July 2020; Published: 23 July 2020

Abstract: Neurological patients can have severe gait impairments that contribute to fall risks. Predicting falls from gait abnormalities could aid clinicians and patients mitigate fall risk. The aim of this study was to predict fall status from spatial-temporal gait characteristics measured by a wearable device in a heterogeneous population of neurological patients. Participants ($n = 384$, age 49–80 s) were recruited from a neurology ward of a University hospital. They walked 20 m at a comfortable speed (single task: ST) and while performing a dual task with a motor component (DT1) and a dual task with a cognitive component (DT2). Twenty-seven spatial-temporal gait variables were measured with wearable sensors placed at the lower back and both ankles. Partial least square discriminant analysis (PLS-DA) was then applied to classify fallers and non-fallers. The PLS-DA classification model performed well for all three gait tasks (ST, DT1, and DT2) with an evaluation of classification performance Area under the receiver operating characteristic Curve (AUC) of 0.7, 0.6 and 0.7, respectively. Fallers differed from non-fallers in their specific gait patterns. Results from this study improve our understanding of how falls risk-related gait impairments in neurological patients could aid the design of tailored fall-prevention interventions.

Keywords: gait analysis; machine learning; inertial measurement units; neurological disorders; falls

1. Introduction

Falls are one of the leading causes of mortality, morbidity, and make up a substantial portion of health care costs worldwide [1]. Falls have a multifactorial origin and usually involve multiple interrelated intrinsic, as well as extrinsic factors [2]. There has been a plentitude of reviews aimed at the epidemiology of fall risk in the aging population. Identified risk factors in the aging population are reduced lower extremity strength, sarcopenia, dizziness, vision impairments, a decline in cognitive function, higher prevalence of comorbidities, polypharmacy, DBI drug use (cumulative anticholinergic and sedative exposure), depression, and extrinsic factors, such as poor lighting in the house, loose rugs, and slippery surface [2–5]. Healthy people can adapt easily to environmental perturbations, such as recovering from slipping or tripping or walking on uneven surfaces. With aging and/or age-related pathology, the ability to adapt to environmental perturbation while walking is diminished. Yet, the most consistent predictors of falls are impairments gait and balance disorders [6,7]. Moreover, various neurological disorders further increase the risk of falls by deteriorating specific nervous system functions contributing to gait and balance [8]. Therefore, the incidence of falls is high in neurological patients; compared with healthy subjects, neurological patients had a 49% increased risk of falling within 20 months [9,10]. Patients with Parkinson's Disease have falls at least once during their disease journey [11].Therefore, detecting gait impairments by wearable devices, such as inertial measurement units (IMUs), during walking could help clinicians identify patients prone to falling [12]. IMUs are attractive alternatives to laboratory motion analysis systems due to their small size, light weight, portability, low cost, and their simple use in the real world.

Spatial-temporal gait variables derived from IMU recordings are outcome parameters for the prediction of falls in patient groups. Fallers compared with non-fallers revealed higher standard deviations and coefficients of variation of stride time, swing time, stance time, and percentage stance time [13]. There exists a strong inter-relationship between cognition and gait control, since gait and cognitive function share cortical areas and several neurotransmitters [14]. An additional cognitive task while walking in clinical populations or older adults, might stress the system by competing for cortical resources, placing the patient at an increased fall risk while performing dual tasks [15,16]. Therefore, the classification accuracy of fallers and non-fallers might be more accurate when walking while performing a dual task. In a previous study, spatial-temporal gait variables obtained during dual task increased classification performance of fallers and non-fallers in 377 older adults compared to single task walking from 0.57 to 0.67 as quantified by the Area Under the receiver operating Curve (AUC) [17].

Different computational approaches, such as supervised machine learning approaches artificial neural network (ANN), K-nearest neighbors (KNN) or support vector machine (SVM), have become popular for the classification of fall risk using different types of tests and/or activities performed by healthy older persons or distinct group of patients [18,19]. Based on time-frequency domain features, different activities were classified using ANN, KNN, quadratic support vector machine (QSVM), and ensemble bagged tree (EBT), the classification accuracy of 85.8%, 91.8%, 96.1%, and 97.7% was obtained for fall detection, respectively [18]. Using a functional movement test, including walking and sit-to-stand with data from foot force sensors, different KNN-based classifiers were compared by classification accuracy of falls for older adults [19]. Although accuracy was reported to be 100% for local mean pseudo nearest neighbor method, the number of subjects included was small and was relatively healthy without any neurological or orthopedic condition that would affect their gait pattern.

Although these studies have successfully classified falls only for general older populations, or just focused on one specific neurologic disorder, the identification of gait impairments for the classification of falls in a more heterogeneous neurological population based on spatial-temporal variables requires a different approach [20,21].

Since the deconstruction of gait into clinically observable spatial-temporal components, such as shorter steps or longer strides, could assist clinicians in having a gait assessment for falls under direct clinical observation, there are advantages using these gait variables derived from IMUs to classify falls. For example, computational approaches can perform automated analyses of multivariate datasets and can deal with interdependency (collinearity) among gait variables from IMU, such as walking speed, mean stride times, and variability in stride times. However, many of these popular computational methods in previous studies with a relatively small sample size will result in overfitting due to the structure of spatial-temporal gait data. Alternatively, multivariate partial least square (PLS) regression or discriminant analysis (DA) analysis can be applied. PLS is a technique that combines features from principal component analysis and multiple regression and is not impeded by collinearity among variables. Besides, partial least square discriminant analysis (PLS-DA) is suitable for gait data in which classes (faller vs. non-fallers) are predicted from a relatively large set of independent (gait) variables with relatively few observations [22]. Clinically, the results of such a computational approach can assist clinicians in interpreting gait performance of patients and use it as a prevention tool that can identify patients with high fall risk. By extracting those gait features that include the information to distinguish classes, tailored intervention programs to reduce the probability of future falls can be developed [23].

The aim of the present study was to establish a quantitative model to classify fallers and non-fallers using spatial-temporal gait characteristics and to identify the specific gait characteristics that contributed to the classification model to target to mitigate falls risk. Considering the strong interrelation of many spatial-temporal gait variables, we hypothesized that, based on a subset of general gait features, fallers can be classified from non-fallers even in a heterogeneous group of neurological patients, and the classification accuracy will be improved while walking with an additional dual task.

2. Materials and Methods

2.1. Participants

Participants (n = 384, age range: 49–80 years) with neurological disorders were recruited from three neurology wards of the University Hospital of Tubingen between September 2014 and April 2015 [24]. The distribution of the major neurological disorders was: 19% Parkinson's disease (8% of fallers), 19% stroke (5% of fallers), 11% epilepsy (4% of fallers), 10% pain syndromes (3% of fallers), 9% multiple sclerosis (4% of fallers), 7% central nervous system tumor (2% of fallers), 6% vertigo (2% of fallers), 6% dementia (2% of fallers), and 6% meningitis/encephalitis (1% of fallers) (see Reference [24] for demographics). Participants were included if they were able to walk 20 m with or without walking aid. Exclusion criteria were: inability to give informed consent, a falling frequency of more than one fall per week, and impaired cognition (Mini-Mental State Examination (MMSE) score ≤10 [24]). Participants were classified as fallers if they had fallen at least once during a two-year period before recruitment. The ethics committee of the medical faculty of the University of Tübingen approved the study (No. 356/2014BO2), and all participants gave written informed consent prior to participation. The investigation was carried out following the rules of the Declaration of Helsinki of 1975, revised in 2013.

2.2. Procedure

Participants were instructed to walk 20 m at a comfortable speed (Single Task; ST), with a dual task (DT) containing mainly a motor component (walking and checking boxes on a paper sheet, DT1) and with a cognitive task (serial 7 s subtraction, DT2) [25]. A complete gait dataset was available for 349 of the 384 participants for ST, wherein 274 participants performed DT1 and 306 participants performed DT2. Table 1 shows participants' demographics for the two groups.

Table 1. Demographics of participants for the single task (ST) a motor dual tasks (DT1) and a cognitive dual task (DT2).

Tasks	Non-Fallers		Fallers		
	ST and DT1	DT2	ST	DT1	DT2
No. Males	115	115	88	41	64
No. Females	75	73	71	43	54
No. Total	190	188	159	84	118
Age, years	61.6 ± 12.2	61.5 ± 12.2	65.0 ± 12.7	61.8 ± 12.5	65.0 ± 12.5
Height, m	1.73 ± 0.1	1.73 ± 0.1	1.70 ± 0.1	1.71 ± 0.1	1.72 ± 0.1
Weight, kg	82.04 ± 16.25	82.04 ± 16.2	76.31 ± 14.87	75.97 ± 15.56	77.07 ± 14.61
BMI, kg/m^2	27.22 ± 4.79	27.25 ± 4.8	26.08 ± 4.34	25.8 ± 4.33	26.02 ± 3.97

Values are mean ± SD, BMI = body mass index, ST = walking at a comfortable speed without an additional task, DT1 = walking and checking boxes on a paper sheet, DT2 = serial 7 s subtraction.

2.3. Data Collection

An IMU-based wearable sensor system was attached with straps around the middle part of the foot around the shoe, to collect data during walking. The IMU system had 3D accelerometers (± 8 g), 3D gyroscopes ($\pm 2000°$/s), and 3D magnetometers (± 1.3 Gs), resulting in nine degrees of freedom (Rehawatch, Hasomed, Magdeburg, Germany) [26].

In each walking task, the following 27 gait variables were extracted from the accelerometer signals: mean and standard deviation of stride duration (s), stride length (m), stride velocity (m/s), number of steps (n), percent of stance (%), stance time (s), percent of swing (%), swing time (s), symmetry stance phase, symmetry swing phases, single support phase (s), ankle dorsiflexion at heel strike (°), plantar flexion at toe-off (°), circumduction of gait (cm), percent of gait cycle time variability (%), and percent of gait cycle spatial variability (%) [26]. In certain combinations, these variables are sensitive to aging and neurodegenerative diseases [10,12,27].

2.4. Statistical Analysis

The data set obtained from ST, DT1, and DT2 were analyzed separately. First, we evaluated if the 27 gait variables correlated linearly or non-linearly with each other, determining the choice of the subsequently used classification method. Partial Least Square Discriminate Analysis (PLS-DA) was applied to classify fallers from non-fallers.

The 27 gait outcomes were the independent variables, and the dependent variable was the classification of participants as non-fallers (Class 0) and fallers (Class 1). The PLS-DA model results in a data dimensionality reduction (Latent Variables; LVs) and provides parameters to evaluate the quality of the prediction and classification of the model. Besides, the Variable of Importance (VIP) gives information about the contribution of individual gait variables to the model. The VIP is calculated as:

$$VIP_j = \sqrt{p\sum_{k=1}^{N} \frac{\left[SS_k\left(\frac{w_{kj}}{\|w_k\|^2}\right)\right]}{\sum_{k=1}^{N}(SS)_k}}. \tag{1}$$

p is the total number of gait variables in the model. N is the number of LVs in the PLS-DA model, k represents the exact component of LV, ss_k explains the sum of the variance of the LVs, w_{kj} quantifies the contribution of variable j according to the kth LV, and w_k is the contribution of the kth LV.

Variables with a VIP >1 indicate a significant contribution of the variable to the classification model. The non-standardized data were tested for normality with the Shapiro-Wilk normality test in R. Since not all variables were normal distributed, a non-parametric Mann–Whitney–Wilcoxon test in R programming was applied [28] to test if the VIP variables >1 were significantly different between non-fallers and fallers.

The standard of the goodness of the model was addressed by the PLS-DA model parameters Q^2, R^2X and R^2Y. The values of these parameters need to be higher than zero in order to have an acceptable classification model. To avoid a too complex model with poor predictability (the problem of overfitting), leave one out cross-validation method was used with PLS-DA classification for assessing the classification model building [29].

Classification performance evaluation of the PLS-DA model was assessed by the receiver operating characteristic curve (ROC curve) and the Area Under the Curve (AUC). In addition, for the non-fallers group and fallers group, their corresponding true positive rate (sensitivity) and true negative rate (specificity) were calculated based on the confusion matrix.

3. Results

3.1. Classification of Fallers during Single and Dual Task Walking

The first five LVs for ST explained 61.4% variance of the original gait variables. Additional LVs did not explain substantially more of the variability of the spatial-temporal gait variables. The quality of the model based on five LVs was good as indicated by $Q^2 = 0.026$, $R^2X = 0.22$, $R^2Y = 0.61$.

PLS-DA classified participants into fallers and non-fallers for ST with AUC = 0.77 (Figure 1A). Figure 1B shows the classification results matrix. In the non-fallers group, the true positive rate and true negative rate of the non-fall group are 84% and 76%, respectively. In the fallers group, the true positive and negative rate, respectively, was 60% and 72%. Note that the AUC for each model is shown in Figure 1A directly, and the true positive rate is directly presented in the diagonal square of the confusion matrix in Figure 1B–D. The true negative rate was also calculated based on the confusion matrix but not directly show in the figures and was presented in Table A1 in the Appendix A.

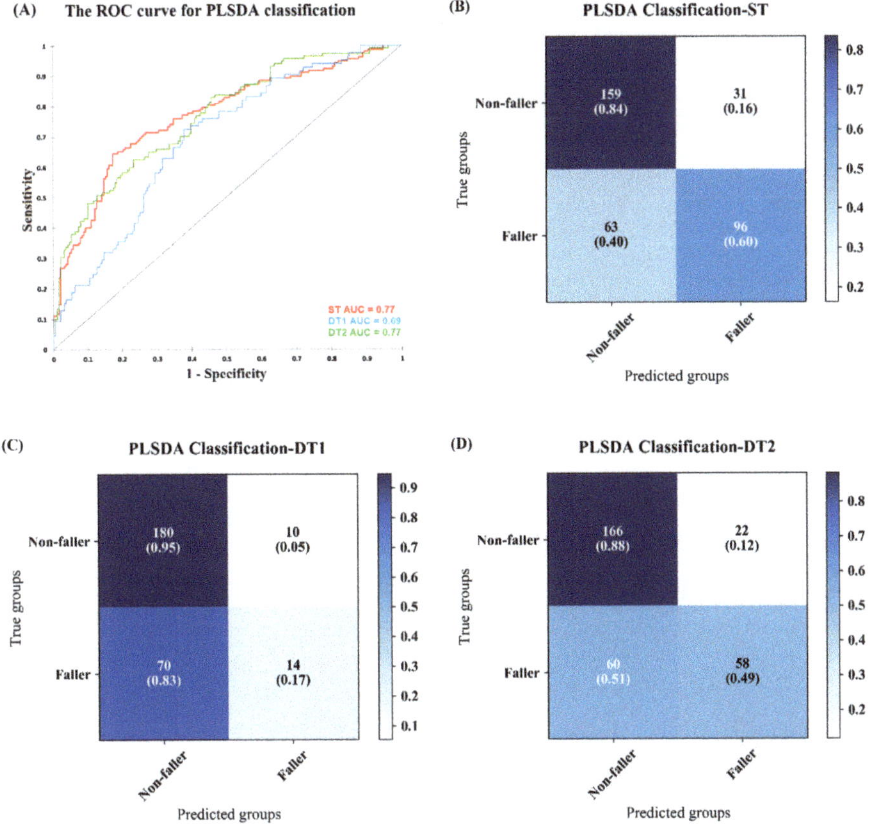

Figure 1. (**A**) shows the receiver operating characteristic (ROC) curves for partial least square discriminant analysis (PLS-DA) classification, based on ST (yellow), DT1 (green), and DT2 (blue) gait variables. (**B–D**) shows the classification matrix. The x-axis represents the participants in the predicted groups and the y-axis shows the participants in the original groups. The dark blue means more participants were assigned in this group. The numbers of participants and the percentages they occurred in the original group are shown in the squares and braces. DT1 = walking and checking boxes on a paper sheet; DT2 = serial 7 s subtraction.

For DT1 and DT2, the first five LVs were also selected and explained 58.6% and 59.7% variance, respectively. The goodness of the models of DT1 and DT2 was good as indicated by a $Q^2 = 0.018$, $R^2X = 0.063$, and $R^2Y = 0.18$ for DT1, and for DT2 of $Q^2 = 0.054$, $R^2X = 0.220$, and $R^2Y = 0.6$.

Classification of fallers and non-fallers based on DT1 and DT2 gait data obtained an AUC = 0.69 and an AUC = 0.77, respectively. The ROC curve for DT1 and DT2 is shown in Figure 1A. According to the confusion matrix in Figure 1C–D, in terms of DT1, the true positive rate and true negative rate of the non-faller group are 95% and 58%, respectively. While the faller group obtained 17% true positive and 72% true negative rate. For DT2, the true positive rate and specificity of the non-faller group are 88% and 72%, respectively, while the faller group obtained 49% true positive rate and 73% true negative rate.

3.2. Identified Gait Variables

Gait variables that contributed most to the classification model were identified by VIP scores with a value >1 (see Figure 2A–C).

For ST, the fallers could be distinguished from non-fallers by higher number of steps, lower mean stride velocity, stride length, and ankle dorsiflexion at heel strike with associated larger mean stance time, stride duration, and ankle plantar flexion at toe-off (for all $p < 0.05$) (Figure 3A).

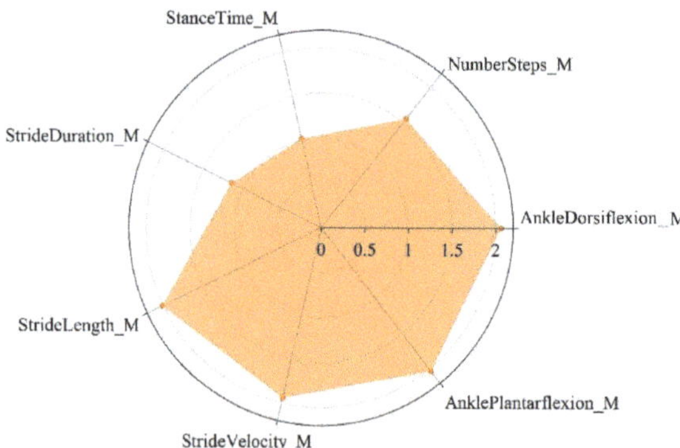

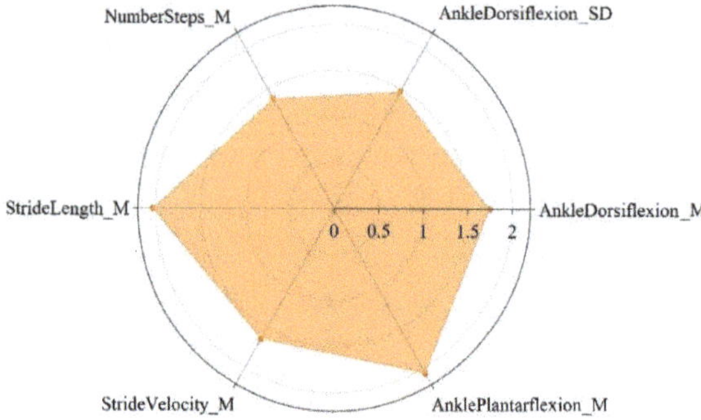

Figure 2. *Cont.*

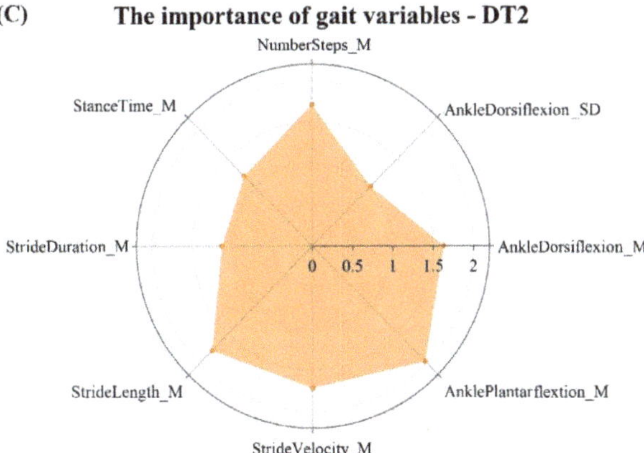

Figure 2. (**A–C**) shows the importance of the gait parameters by orange area (VIP > 1) from ST, DT1 and DT2. M = mean, SD = standard deviation. DT1 = walking and checking boxes on a paper sheet; DT2 = serial 7 s subtraction.

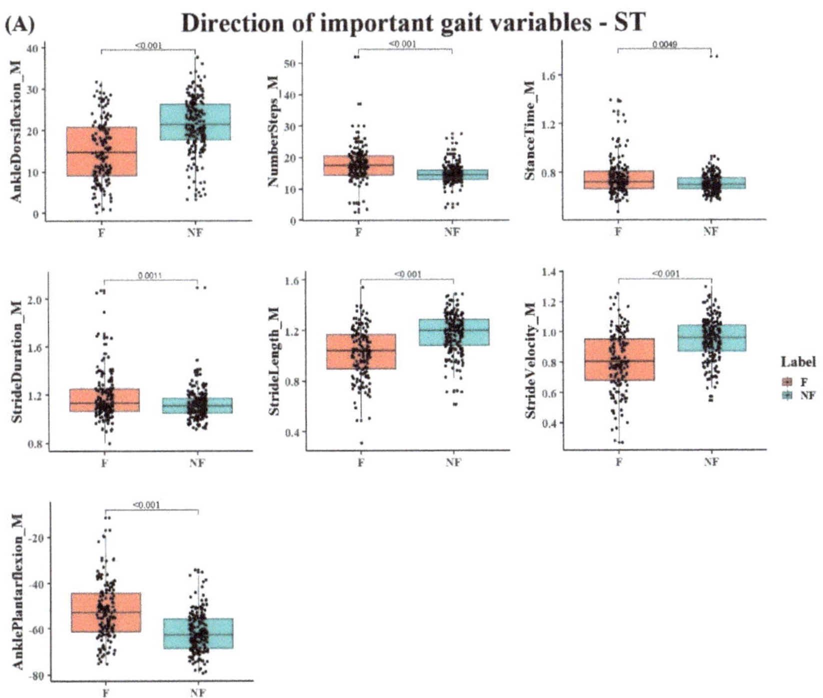

Figure 3. *Cont.*

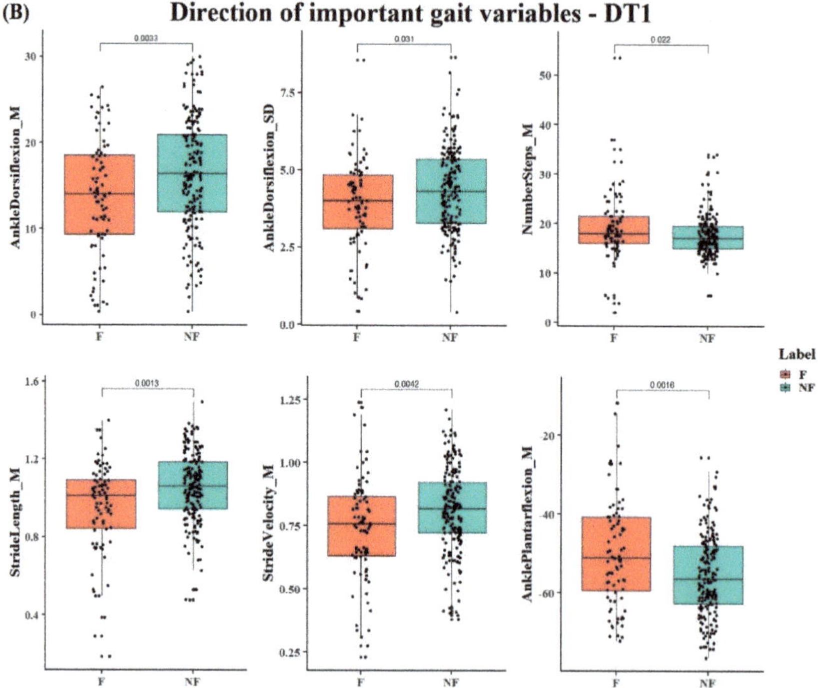

Figure 3. (**A**,**B**) show the direction of variables that contribute more to the PLS-DA model in and DT1. The x-axis represents the groups of fallers and non-fallers, and the y-axis shows the coefficients of each variable in each square. The vertical bars indicate the confidence interval. Dots show the individual data of the participants. DT1 = walking and checking boxes on a paper sheet. Note that the results for DT2 were similar as for DT1.

For DT1 and DT2, fallers were characterized by lower mean stride velocity and stride length, and a lower mean and standard deviation of ankle dorsiflexion at heel strike than non-fallers. On the other hand, non-fallers showed a greater higher number of steps with a larger mean ankle plantar flexion at toe-off than fallers (see Figure 3B). Additionally, for DT2, non-fallers showed a lower mean stance time and stride duration than fallers. The variables from the PLS-DA model with a VIP score >1 were also significantly different between the groups when separately tested (all $p < 0.05$). Figure 3 shows the individual data of the participants in addition to the mean values and confidence intervals of the variables with a VIP score >1 that contribute most to the PLS-DA model during ST and DT1. As can be seen in figure irrespective of the variability between participants within a group, these variables were significantly different between non-fallers and fallers. The results for DT2 were similar to DT1, implying that no major difference was present between walking with an additional cognitive task and walking when performing an additional motor task.

4. Discussion

Many people with neurological deficits have an increased fall risk. The present study aimed to develop a model to classify fallers and non-fallers within a heterogeneous group of older adults with neurological disorders. The model is based on spatial-temporal gait variables derived from IMU during walking at a comfortable speed, walking with an additional motor task (DT1), and walking when also performing a cognitive task (DT2) to identify spatial-temporal gait variables that differentiate fallers from non-fallers. We found that gait differed between fallers and non-fallers, and single task walking resulted in the highest classification accuracy in the neurological patients.

4.1. Classification Performance of Fallers and Non-Fallers by ST, DT1, and DT2

Overall, the results showed that using PLS-DA fallers could be identified from non-fallers with an AUC of 0.7. Adding DT2, the cognitive dual task, to the model, the AUC was still 0.7, but, with the inclusion of DT1, the motor dual task in the model, AUC decreased to 0.6.

Random forest machine learning method classified Parkinson's Disease patients versus controls based on gait with an AUC of 0.76 [30]. Given the heterogeneity of the sample in the present study, identification of fallers vs non-fallers with an AUC of 0.7 seems reasonably accurate (Figure 1). The corresponding true positive rate and true negative rate in the fallers group provide more insights into classification performance. In the fallers group, classification based on ST and DT1 produced a similar true negative rate (specificity) around 0.7, suggesting that less than 30% of the non-fallers were classified as fallers.

The true positive rate (sensitivity), however, was lower during DT walking (DT1: 0.17, DT2: 0.49) compared with ST walking (0.72). This finding was unexpected since we anticipated that dual task walking would enhance the differences between fallers and non-fallers. Fallers are expected to have a significant different gait pattern in particularly during DT compared to non-fallers [31] because DT increases the influence of supraspinal control mechanisms on gait compared to ST [32]. The type of DT might be important in this respect. A motor-related DT (walking with a glass of water in hand) improved the discrimination of fallers from non-fallers in otherwise disease-free older adults based on spatial-temporal gait variables [33]. Similarly, a DT with a cognitive component (walking while talking) slowed gait and shortened stride length compared to ST in neurological patients [34]. In the present study, we anticipated that the DT1, with a cognitive component, would demand more cognitive flexibility than DT2, with a motor component. Contrary to this expectation, over 50% of the patients were assigned to the incorrect group. In other words, gait performance under dual task conditions did not improve classification performance compared to gait performance under a single task. Likewise, counting backwards while walking was poorly (high *p*-value > 0.1) associated with falls [35]. The heterogeneity of neurological patients in the present could contribute to the poor classification performance during DT. To illustrate, while falls might be related purely to motor symptoms (bradykinesia, hypokinesia, rigidity) in PD, cognitive dysfunction is likely to contribute to falls in dementia [36]. Therefore, the prediction accuracy of fallers by including DT1 and DT2 in the model would not necessarily improve due to a selective sensitivity of DT1 and DT2 to gait markers of falls. DT1 focus on motor destitution should affect PD, but, for other patients, such as Dementia patients, DT2 with a higher cognitive load presumably has a larger effect on them.

4.2. Contribution of Gait Variables to the PLS-DA Classification Model

Gait variables related to the domains of pace (stride velocity), rhythm (stride duration, stance/swing time), variability (standard deviation of ankle dorsiflexion at heel strike), and spatial gait variables (stride length, plantar flexion at toe-off, ankle dorsiflexion at heel strike) contributed significantly to the classification model. These variables appear to be sensitive indicators of gait impairments in a heterogeneous group of neurological patients to identify the risk of falling [37]. In line with previous work [6,23,24], fallers versus non-fallers in the present study walked slower, a different rhythm, higher gait variability, and impaired spatial gait (Figure 3) [38,39]. We extend current data demonstrating that spatial-temporal gait measures can discriminate fallers from non-fallers among healthy older adults by showing that some domains of gait are global and not disease-specific. The classification results in this heterogeneous population might be explained by the fact that multiple types of fallers are included in the dataset and sub-clinical mobility limitations (slow gait, low stability, obesity, arthritis) are randomly distributed among older adults [40]. It is possible that, within different cohorts, they have different risks for falls. The model predicts reasonably well falls because the risks for falls measured through the 'global' gait variables are distributed across the cohorts. Therefore, there is a probabilistic chance for a given outcome to predict a fall related to the risk measured with a relatively low error by this outcome. This process is iterated across cohorts, resulting in a relatively accurate fall prediction across cohorts.

4.3. Improving Classification Accuracy of a Heterogeneous Population

Nevertheless, there are strategies that can be adopted in the future to improve the current classification model to increase the accuracy of fall classification performance. Fall classification is a multifaceted problem that involves complex interactions between physiological, behavioral, and environmental factors. Most studies that aim to identify falls or classify fallers and non-fallers focus on the factor of motor behavior, such as gait and balance, but do not include other indicators of falls, such as patients' characteristics [41].

As we know, the combination of intrinsic and extrinsic risk factors contributes to a fall incident. The intrinsic factors include age, fall history, mobility impairments, sleep disturbances, neurological disorders, the presence of co-morbidities, and medication use. Extrinsic factors include slippery surfaces, improper footwear, poor lighting, and clutter [2,42,43]. A comprehensive fall classification should involve the interactions between these risk factors. Clinically, many different test batteries are used that examine gait and balance performance as indicators of fall risk. One of the most well-known and widely-used clinical tests is the Time Up and Go (TUG) [44]. The advantage of TUG is that the test is simple and easy to perform for older adults [45]. Other examples of the clinical test include the Berg Balance Scale (BBS) [46], the Functional Gait Assessment (FGA) [47], and the developed Balance Evaluation Systems Test (BESTest) [48]. However, clinical tests may suffer from ceiling effects, not able to detect relatively small difference, and provide a general score of functioning.

Therefore, more likely, the combined variables from clinical tests and movement measurements could optimize the classification of falls. For example, compared with the fall classification model with only TUG variables, a six-minute walking test equipped with an IMU was added to the TUG to test the model, the classification accuracy of falls in a group of 73 nursing home residents, using a decision tree classifier, increased from 68% to 76% [49].

On the other hand, adding different types of gait variables also could improve the classification performance. In the present study, commonly used spatial-temporal gait variables were calculated from the data collected from wearable sensors during walking, to establish an accurate PLS-DA classification model. However, when combining time/frequency domain and spatial-temporal gait variables together to establish an advanced classification model to discriminate fallers from non-fallers, the accuracy of classification will be enhanced. For example, a Random Forest (RF) classification model classified eleven stroke patients and nine patients with neurological disorders other than stroke (brain concussion, spinal injury, or brain haemorrhage) based on only spatial-temporal gait variables. The classification model performed a moderate testing accuracy of 76.08%. While combining the time domain gait variables and spatial-temporal gait variables and applying a Multilayer Perceptron (MLP) classification model, the classification performance was increased to 84.78% accuracy [21].

The present study results show that in general for fall classification among diverse neurological patients, spatial-temporal gait properties could be used as a biomarker for fallers, irrespective of the specific diagnosis. However, for the identification of pathology specific gait characteristics, e.g., gait features or gait signatures that are unique for a certain diagnostic group (e.g., Parkinson, Stroke, Dementia), other types of gait data, as input to a classification model are needed [22]. The current gait variables are spatial-temporal gait parameters averaged over a number of strides. These parameters do not take into account of time, i.e., fluctuations of walking over a number of strides. Whereas the spatial-temporal gait variables provide more overall information, adding gait variables that include time, the so-called dynamic gait variables, will improve the sensitivity to identify specific diagnostic groups of patients and provide more detailed information for prediction models [50].

Methodologically, gait classification performance can be improved by using a data pre-processing method [51]. Principal Component Analysis (PCA) usually leads to an improvement in classification accuracy [52]. Although PCA is an often-used method for extracting unique features in multidimensional data sets, its assumption (orthogonality) might neglect the significant interactions between the spatial-temporal gait variables. Alternative a less known method for data pre-processing in advance to applying to machine learning methods is the Signature Method. The Signature Method transforms the original data by using path-integral to generate a continuous pathway of the gait data, which results in new feature sets. It not only reduces the redundancy of the data, as well as PCA, but also generates new features that better represent the interactions among these gait variables [53]. The drawback of this new data pre-processing method is that the newly generated features are hard to relate back to the original variables; therefore, it is very difficult to interpret the clinical meanings behind the new features [53].

4.4. Selection of Classification Models for Clinical Applications

In order to make a proper program to mitigate fall risks and to improve the gait and mobility impairments, to finally prevent a future fall, advanced computational models, such as machine learning, are anticipated to have the capacity to automatize this process. They can provide a transparent and accurate classification of fallers and non-fallers to assist early identification of fall risks before the actual occurrence of a fall [41].

Different computational methods, such as machine learning, have been used for gait assessment, in general, to construct a model for the classification of different patients and/or age-based groups [50,51]. These computing algorithms should have the capacity to weight the predictive variables, to illustrate the additional clinical value of fall detection, and to assist clinicians in identifying the unique factors that increase falls in a specific population [41].

However, many clinical gait datasets suffer from the co-linear and highly correlated data features, as well as relatively small sample sizes, that are not appropriate for many of these approaches since the accuracy of the widely used machine learning models is dependent on large sets of training data. The more training data, the more accurate, sensitive, and specific the model built. However, sufficient training data may not always be available for the populations in a clinical test. Besides, more variables with less data samples would complicate the model with the low bias high variance to overfit the classification results [51]. For clinical gait analysis, the multiple classical machine learning methods, such as Support Vector Machine (SVM) or Neural Network (NN) approaches, have the advantage to automatically selected the features that are used for a classification model, without any prior feature selection [50]. Yet, for clinical relevance, the results of the computational model parameters are necessary to be translated into meaningful clinical knowledge, despite the complex interactions among the variables leading to the classification [51,54]. The lack of transparency of the construction process in the machine learning classification models limits the reproducibility and clinical interpretation of these advanced computational technologies. This 'black box' problem hinders clinical application since clinicians need to understand the specific gait variables for diagnosis [55]. Therefore, in the present study, we applied a method that automatically identified gait parameters, using VIP scores from the PLS-DA model, without a clinical diagnosis as a predictor. PLS-DA is two-fold: Firstly, in clinical gait analysis, many of the gait variables are interrelated. PLS-DA is not impeded by collinearity among variables, which will have a negative impact on the normal Linear Discriminant Analysis. In terms of the fall classification model for clinical gait data, we mostly have more variables than the number of subjects in a dataset. So, one advantage of the PLS-DA model is that we can input various standardized gait variables even more than the number of participants, without prior knowledge to select in advance, but the model still can accurately detect falls. In the present study, we used Leave-One-Out cross-validation (LOOCV) to increase the sample size for training and testing PLS-DA model and to minimize the drawbacks of limited sample size and bias of data [56]. Secondly, PLS-DA is transparent, i.e., several statistical parameters can be derived, such as the VIP scores, to identify the weight of contribution by each variable to the classification model and Q^2, representing the predictability and validity of the model [57]. In this case, PLS-DA could provide more information to interpret fallers' gait in clinical. Clinicians using these data science approaches might, at an early stage, improve the identification of patients with fall risk.

When we construct a reliable and accurate computational model, gait patterns can be identified by the model of new patients to identify the at-risk gait on early stage, to diagnose the potential fallers without the prior knowledge of an accurate neurological pathology, and to finally determine the high risk of falling for patients based on their mobility decline [58]. However, human clinical decision-making can be supported and assisted by computational models, such as PLS-DA, but not replace the diagnosis from clinicians. For instance, the identified gait variables could be used for new individuals, to predict the fall risk for potential patients. Moreover, the established computational model might be instrumented in IMU to monitor the interventions in patients' real-world daily lives and to optimize the efficacy of specific rehabilitation protocols [51].

5. Conclusions

The present study classified non-fallers and fallers based on spatial-temporal gait variables derived from IMUs using PLS-DA while walking with or without a DT. The model successfully classified fallers and non-fallers with a satisfactory AUC of 0.69 to 0.77. Thus, differences in gait among neurological patients could be used to identify potential fallers from non-fallers even without DT gait that does not seem to improve classification accuracy among patients with a diverse neurological diagnosis. Number of steps, plantar flexion at toe-off, and ankle dorsiflexion at heel strike, stride length, stride duration, stride velocity, and stance time were sensitive variables to classify fallers and non-fallers. Fallers versus non-fallers have a slow pace and rhythm, high gait variability, and impaired spatial gait pattern. Improving our understanding of how falls risk-related gait impairments in neurological patients could aid the design of tailored fall-prevention interventions to decrease the fall risk for people with neurological deficits.

Author Contributions: Conceptualization, Y.Z., C.H. and C.J.C.L.; methodology, Y.Z. and C.J.C.L.; software, Y.Z.; validation, Y.Z. and C.J.C.L.; formal analysis, Y.Z. and C.J.C.L.; investigation, Y.Z. and C.J.C.L.; resources, W.M. and C.H.; data curation, W.M. and C.H.; writing—original draft preparation, Y.Z.; writing—review and editing, Y.Z., C.J.C.L., R.Z.U.R., S.D.D., C.H., W.M., T.H., L.R.; visualization, Y.Z.; supervision, C.J.C.L. and T.H.; project administration, C.J.C.L. and W.M.; funding acquisition, C.J.C.L. and W.M. All authors have read and agreed to the published version of the manuscript.

Funding: This research was funded by Keep Control project, funding from the European Union's Horizon 2020 research and innovation program under the Marie Skłodowska-Curie grant agreement No 721577.

Conflicts of Interest: The authors declare no conflict of interest.

Appendix A

Table A1 indicates an overview of the PLS-DA fall classification evaluation results.

Table A1. Results of PLS-DA classification for the single task (ST) a motor dual task (DT1) and a cognitive dual task (DT2).

	Evaluation	True Positive Rate %		Ture Negative Rate %	
	AUC	Non-Fallers	Fallers	Non-Fallers	Fallers
ST	0.77	84	60	76	72
DT1	0.69	95	17	58	72
DT2	0.77	88	49	72	73

ST = walking at a comfortable speed without an additional task, DT1 = walking and checking boxes on a paper sheet, DT2 = serial 7 s subtraction.

References

1. OECD/EU. *Health at a Glance: Europe 2018: State of Health in the EU Cycle*; OECD Publishing: Paris, France, 2018. Available online: https://doi.org/10.1787/health_glance_eur-2018-en (accessed on 11 May 2020).
2. Ambrose, A.F.; Paul, G.; Hausdorff, J.M. Risk factors for falls among older adults: A review of the literature. *Maturitas* **2013**, *75*, 51–61. [CrossRef] [PubMed]
3. Nishtala, P.S.; Narayan, S.W.; Wang, T.; Hilmer, S.N. Associations of drug burden index with falls, general practitioner visits, and mortality in older people. *Pharmacoepidemiol. Drug Saf.* **2014**, *23*, 753–758. [CrossRef] [PubMed]
4. Gillespie, L.D.; Gillespie, W.J.; Robertson, M.C.; Lamb, S.E.; Cumming, R.G.; Rowe, B.H. Interventions for preventing falls in elderly people. *Cochrane Database Syst. Rev.* **2003**. [CrossRef]
5. Gleason, C.E.; Gangnon, R.E.; Fischer, B.L.; Mahoney, J.E. Increased risk for falling associated with subtle cognitive impairment: Secondary analysis of a randomized clinical trial. *Dement. Geriatr. Cogn. Disord.* **2009**, *27*, 557–563. [CrossRef] [PubMed]
6. Tinetti, M.E.; Kumar, C. The patient who falls: "It's always a trade-off". *J. Am. Med. Assoc.* **2010**, *303*, 258–266. [CrossRef] [PubMed]
7. Ganz, D.A.; Bao, Y.; Shekelle, P.G.; Rubenstein, L.Z. Will my patient fall? *J. Am. Med. Assoc.* **2007**, *297*, 77–86. [CrossRef]
8. Zampogna, A.; Mileti, I.; Palermo, E.; Celletti, C.; Paoloni, M.; Manoni, A.; Mazzetta, I.; Costa, G.D.; Pérez-López, C.; Camerota, F.; et al. Fifteen years of wireless sensors for balance assessment in neurological disorders. *Sensors* **2020**, *20*, 3247. [CrossRef]
9. Verghese, J.; Ambrose, A.F.; Lipton, R.B.; Wang, C. Neurological gait abnormalities and risk of falls in older adults. *J. Neurol.* **2010**, *257*, 392–398. [CrossRef]
10. Stolze, H.; Klebe, S.; Zechlin, C.; Baecker, C.; Friege, L.; Deuschl, G. Falls in frequent neurological diseases: Prevalence, risk factors and aetiology. *J. Neurol.* **2004**, *251*, 79–84. [CrossRef]
11. Allen, N.E.; Schwarzel, A.K.; Canning, C.G. Recurrent falls in parkinson's disease: A systematic review. *Parkinsons Dis.* **2013**, *2013*, 906274. [CrossRef]
12. Doi, T.; Hirata, S.; Ono, R.; Tsutsumimoto, K.; Misu, S.; Ando, H. The harmonic ratio of trunk acceleration predicts falling among older people: Results of a 1-year prospective study. *J. Neuroeng. Rehabil.* **2013**, *10*, 7. [CrossRef] [PubMed]
13. Hausdorff, J.M.; Edelberg, H.K.; Mitchell, S.L.; Goldberger, A.L.; Wei, J.Y. Increased gait unsteadiness in community-dwelling elderly fallers. *Arch. Phys. Med. Rehabil.* **1997**, *78*, 278–283. [CrossRef]
14. Montero-Odasso, M.; Verghese, J.; Beauchet, O.; Hausdorff, J.M. Gait and cognition: A complementary approach to understanding brain function and the risk of falling. *J. Am. Geriatr. Soc.* **2012**, *60*, 2127–2136. [CrossRef]
15. Kikkert, L.H.J.; Vuillerme, N.; van Campen, J.P.; Hortobágyi, T.; Lamoth, C.J. Walking ability to predict future cognitive decline in old adults: A scoping review. *Ageing Res. Rev.* **2016**, *27*, 1–14. [CrossRef] [PubMed]
16. Morris, R.; Lord, S.; Bunce, J.; Burn, D.; Rochester, L. Gait and cognition: Mapping the global and discrete relationships in ageing and neurodegenerative disease. *Neurosci. Biobehav. Rev.* **2016**, *64*, 326–345. [CrossRef] [PubMed]
17. Tomas-Carus, P.; Biehl-Printes, C.; Pereira, C.; Vieiga, G.; Costa, A.; Collado-Mateo, D. Dual task performance and history of falls in community-dwelling older adults. *Exp. Gerontol.* **2019**, *120*, 35–39. [CrossRef]
18. Chelli, A.; Patzold, M. A machine learning approach for fall detection and daily living activity recognition. *IEEE Access* **2019**, *7*, 38670–38687. [CrossRef]
19. Liang, S.; Ning, Y.; Li, H.; Wang, L.; Mei, Z.; Ma, Y.; Zhao, G. Feature selection and predictors of falls with foot force sensors using KNN-based algorithms. *Sensors* **2015**, *15*, 29393–29407. [CrossRef]
20. Bet, P.; Castro, P.C.; Ponti, M.A. Fall detection and fall risk assessment in older person using wearable sensors: A systematic review. *Int. J. Med. Inform.* **2019**, *130*, 103946. [CrossRef]
21. Hsu, W.-C.; Sugiarto, T.; Lin, Y.-J.; Yang, F.-C.; Lin, Z.-Y.; Sun, C.-T.; Hsu, C.-L.; Chou, K.-N. Multiple-wearable-sensor-based gait classification and analysis in patients with neurological disorders. *Sensors* **2018**, *18*, 3397. [CrossRef]

22. Kikkert, L.H.J.; De Groot, M.H.; Van Campen, J.P.; Beijnen, J.H.; Hortobágyi, T.; Vuillerme, N.; Lamoth, C.C.J. Gait dynamics to optimize fall risk assessment in geriatric patients admitted to an outpatient diagnostic clinic. *PLoS ONE* **2017**, *12*, e0178615. [CrossRef] [PubMed]

23. Weinstein, M.; Booth, J. Preventing falls in older adults: A multifactorial approach. *Home Health Care Manag. Pract.* **2006**, *19*, 45–50. [CrossRef]

24. Bernhard, F.P.; Sartor, J.; Bettecken, K.; Hobert, M.A.; Arnold, C.; Weber, Y.G.; Poli, S.; Margraf, N.G.; Schlenstedt, C.; Hansen, C.; et al. Wearables for gait and balance assessment in the neurological ward—Study design and first results of a prospective cross-sectional feasibility study with 384 inpatients. *BMC Neurol.* **2018**, *18*, 114. [CrossRef] [PubMed]

25. Hobert, M.A.; Meyer, S.I.; Hasmann, S.E.; Metzger, F.G.; Suenkel, U.; Eschweiler, G.W.; Berg, D.; Maetzler, W. Gait is associated with cognitive flexibility: A dual-tasking study in healthy older people. *Front. Aging Neurosci.* **2017**, *9*, 154. [CrossRef]

26. Donath, L.; Faude, O.; Lichtenstein, E.; Nüesch, C.; Mündermann, A. Validity and reliability of a portable gait analysis system for measuring spatiotemporal gait characteristics: Comparison to an instrumented treadmill. *J. Neuroeng. Rehabil.* **2016**, *13*, 6. [CrossRef]

27. König, N.; Taylor, W.R.; Armbrecht, G.; Dietzel, R.; Singh, N.B. Identification of functional parameters for the classification of older female fallers and prediction of 'first-time' fallers. *J. R. Soc. Interface* **2014**, *11*, 20140353. [CrossRef]

28. Mann-Whitney-Wilcoxon Test/R Tutorial. Available online: http://www.r-tutor.com/elementary-statistics/non-parametric-methods/mann-whitney-wilcoxon-test (accessed on 11 May 2020).

29. Westerhuis, J.A.; Hoefsloot, H.C.J.; Smit, S.; Vis, D.J.; Smilde, A.K.; van Velzen, E.J.J.; van Duijnhoven, J.P.M.; van Dorsten, F.A. Assessment of PLSDA cross validation. *Metabolomics* **2008**, *4*, 81–89. [CrossRef]

30. Chang, D.; Alban-Hidalgo, M.; Hsu, K. Diagnosing Parkinson's Disease from Gait. Available online: https://pdfs.semanticscholar.org/2885/6bad53ccd2e81feea72ca5c2511c92a3e84f.pdf (accessed on 11 May 2020).

31. Howcroft, J.; Kofman, J.; Lemaire, E.D.; McIlroy, W.E. Analysis of dual-task elderly gait in fallers and non-fallers using wearable sensors. *J. Biomech.* **2016**, *49*, 992–1001. [CrossRef]

32. Maetzler, W.; Nieuwhof, F.; Hasmann, S.E.; Bloem, B.R. Emerging therapies for gait disability and balance impairment: Promises and pitfalls. *Mov. Disord.* **2013**, *28*, 1576–1586. [CrossRef]

33. Toulotte, C.; Thevenon, A.; Watelain, E.; Fabre, C. Identification of healthy elderly fallers and non-fallers by gait analysis under dual-task conditions. *Clin. Rehabil.* **2006**, *20*, 269–276. [CrossRef]

34. Fritz, N.E.; Cheek, F.M.; Nichols-Larsen, D.S. Motor-cognitive dual-task training in persons with neurologic disorders: A systematic review. *J. Neurol. Phys. Ther.* **2015**, *39*, 142–153. [CrossRef]

35. Beauchet, O.; Allali, G.; Annweiler, C.; Berrut, G.; Maarouf, N.; Herrmann, F.R.; Dubost, V. Does change in gait while counting backward predict the occurrence of a first fall in older adults? *Gerontology* **2008**, *54*, 217–223. [CrossRef] [PubMed]

36. Wood, B.H.; Bilclough, J.A.; Bowron, A.; Walker, R.W. Incidence and prediction of falls in Parkinson's disease: A prospective multidisciplinary study. *J. Neurol. Neurosurg. Psychiatry* **2002**, *72*, 721–725. [CrossRef] [PubMed]

37. Qiu, H.; Rehman, R.Z.U.; Yu, X.; Xiong, S. Application of wearable inertial sensors and a new test battery for distinguishing retrospective fallers from non-fallers among community-dwelling older people. *Sci. Rep.* **2018**, *8*, 16349. [CrossRef] [PubMed]

38. Mbourou, G.A.; Lajoie, Y.; Teasdale, N. Step length variability at gait initiation in elderly fallers and non-fallers, and young adults. *Gerontology* **2003**, *49*, 21–26. [CrossRef]

39. Shimada, H.; Kim, H.; Yoshida, H.; Suzukawa, M.; Makizako, H.; Yoshida, Y.; Saito, K.; Suzuki, T. Relationship between age-associated changes of gait and falls and life-space in elderly people. *J. Phys. Ther. Sci.* **2010**, *22*, 419–424. [CrossRef]

40. Hortobágyi, T.; Lesinski, M.; Gäbler, M.; VanSwearingen, J.M.; Malatesta, D.; Granacher, U. Effects of three types of exercise interventions on healthy old adults' gait speed: A systematic review and meta-analysis. *Sports Med.* **2015**, *45*, 1627–1643.

41. Rajagopalan, R.; Litvan, I.; Jung, T.P. Fall prediction and prevention systems: Recent trends, challenges, and future research directions. *Sensors* **2017**, *17*, 2509. [CrossRef]

42. Rossat, A.; Fantino, B.; Nitenberg, C.; Annweiler, C.; Poujol, L.; Herrmann, F.R.; Beauchet, O. Risk factors for falling in community-dwelling older adults: Which of them are associated with the recurrence of falls? *J. Nutr. Health Aging* **2010**, *14*, 787–791. [CrossRef]

43. Moreland, J.D.; Richardson, J.A.; Goldsmith, C.H.; Clase, C.M. Muscle weakness and falls in older adults: A systematic review and meta-analysis. *J. Am. Geriatr. Soc.* **2004**, *52*, 1121–1129. [CrossRef]

44. Richardson, S. The timed "Up & Go": A test of basic functional mobility for frail elderly persons. *J. Am. Geriatr. Soc.* **1991**, *39*, 142–148.

45. Herman, T.; Giladi, N.; Hausdorff, J.M. Properties of the "Timed Up and Go" test: More than meets the eye. *Gerontology* **2011**, *57*, 203–210. [CrossRef] [PubMed]

46. Berg, K.; Wood-Dauphinee, S.; Williams, J.I.; Gayton, D. Measuring balance in the elderly: Preliminary development of an instrument. *Physiother. Can.* **1989**, *41*, 304–311. [CrossRef]

47. Wrisley, D.M.; Marchetti, G.F.; Kuharsky, D.K.; Whitney, S.L. Reliability, internal consistency, and validity of data obtained with the functional gait assessment/physical therapy/oxford academic. *Phys. Ther.* **2004**, *84*, 906–918. [CrossRef]

48. Horak, F.B.; Wrisley, D.M.; Frank, J. The Balance Evaluation Systems Test (BESTest) to differentiate balance deficits. *Phys. Ther.* **2009**, *89*, 484–498. [CrossRef]

49. Buisseret, F.; Catinus, L.; Grenard, R.; Jojczyk, L.; Fievez, D.; Barvaux, V.; Dierick, F. Timed up and go and six-minute walking tests with wearable inertial sensor: One step further for the prediction of the risk of fall in elderly nursing home people. *Sensors* **2020**, *20*, 3207. [CrossRef]

50. Zhou, Y.; Romijnders, R.; Hansen, C.; van Campen, J.; Maetzler, W.; Hortobágyi, T.; Lamoth, C.J.C. The detection of age groups by dynamic gait outcomes using machine learning approaches. *Sci. Rep.* **2020**, *10*, 1–12. [CrossRef]

51. Halilaj, E.; Rajagopal, A.; Fiterau, M.; Hicks, J.L.; Hastie, T.J.; Delp, S.L. Machine learning in human movement biomechanics: Best practices, common pitfalls, and new opportunities. *J. Biomech.* **2018**, *81*, 1–11. [CrossRef]

52. Daffertshofer, A.; Lamoth, C.J.C.; Meijer, O.G.; Beek, P.J. PCA in studying coordination and variability: A tutorial. *Clin. Biomech.* **2004**, *19*, 415–428. [CrossRef]

53. Chevyrev, I.; Kormilitzin, A. A primer on the signature method in machine learning. *arXiv* **2016**, arXiv:1603.03788.

54. Dinov, I.D. Black box machine-learning methods: Neural networks and support vector machines. In *Data Science and Predictive Analytics*; Springer International Publishing: Cham, Switzerland, 2018; pp. 383–422.

55. Aicha, A.N.; Englebienne, G.; van Schooten, K.S.; Pijnappels, M.; Kröse, B. Deep learning to predict falls in older adults based on daily-life trunk accelerometry. *Sensors* **2018**, *18*, 1654. [CrossRef] [PubMed]

56. Vabalas, A.; Gowen, E.; Poliakoff, E.; Casson, A.J. Machine learning algorithm validation with a limited sample size. *PLoS ONE* **2019**, *14*, e0224365. [CrossRef] [PubMed]

57. Cramer, R.D. Partial Least Squares (PLS): Its strengths and limitations. *Perspect. Drug Discov. Des.* **1993**, *1*, 269–278. [CrossRef]

58. Phinyomark, A.; Petri, G.; Ibáñez-Marcelo, E.; Osis, S.T.; Ferber, R. Analysis of big data in gait biomechanics: Current trends and future directions. *J. Med. Biol. Eng.* **2018**, *38*, 244–260. [CrossRef]

Article

3D Motion Capture May Detect Spatiotemporal Changes in Pre-Reaching Upper Extremity Movements with and without a Real-Time Constraint Condition in Infants with Perinatal Stroke and Cerebral Palsy: A Longitudinal Case Series

Julia Mazzarella [1], Mike McNally [2], Daniel Richie [1], Ajit M. W. Chaudhari [1,3,4], John A. Buford [1], Xueliang Pan [5] and Jill C. Heathcock [1,*]

[1]   Physical Therapy Division, School of Health and Rehabilitation Sciences, College of Medicine, The Ohio State University, 453 W 10th Ave., Columbus, OH 43210, USA; julia.mazzarella@osumc.edu (J.M.); daniel.richie@osumc.edu (D.R.); Ajit.Chaudhari@osumc.edu (A.M.W.C.); john.buford@osumc.edu (J.A.B.)
[2]   Tampa Bay Rays, 1 Tropicana Dr., St. Petersburg, FL 33705, USA; mcnallym08@gmail.com
[3]   Department of Mechanical and Aerospace Engineering, College of Engineering, The Ohio State University, 453 W 10th Ave., Columbus, OH 43210, USA
[4]   Department of Biomedical Engineering, College of Engineering, The Ohio State University, 453 W 10th Ave., Columbus, OH 43210, USA
[5]   Center for Biostatistics, Department of Biomedical Informatics, College of Medicine, The Ohio State University, 1800 Cannon Drive, Columbus, OH 43210, USA; jeff.pan@osumc.edu
*   Correspondence: jill.heathcock@osumc.edu

Received: 3 November 2020; Accepted: 16 December 2020; Published: 19 December 2020

Abstract: Perinatal stroke (PS), occurring between 20 weeks of gestation and 28 days of life, is a leading cause of hemiplegic cerebral palsy (HCP). Hallmarks of HCP are motor and sensory impairments on one side of the body—especially the arm and hand contralateral to the stroke (involved side). HCP is diagnosed months or years after the original brain injury. One effective early intervention for this population is constraint-induced movement therapy (CIMT), where the uninvolved arm is constrained by a mitt or cast, and therapeutic activities are performed with the involved arm. In this preliminary investigation, we used 3D motion capture to measure the spatiotemporal characteristics of pre-reaching upper extremity movements and any changes that occurred when constraint was applied in a real-time laboratory simulation. Participants were N = 14 full-term infants: N = six infants with typical development; and N = eight infants with PS (N = three infants with PS were later diagnosed with cerebral palsy (CP)) followed longitudinally from 2 to 6 months of age. We aimed to evaluate the feasibility of using 3D motion capture to identify the differences in the spatiotemporal characteristics of the pre-reaching upper extremity movements between the diagnosis group, involved versus uninvolved side, and with versus and without constraint applied in real time. This would be an excellent application of wearable sensors, allowing some of these measurements to be taken in a clinical or home setting.

Keywords: perinatal stroke; kinematics; upper extremity; cerebral palsy; hemiplegia; constraint

---

1. Introduction

Perinatal stroke (PS) is caused by interrupted blood flow to the brain between 20 weeks gestation and 28 days of life [1,2]. PS affects ~24.7/100,000 live births in the US annually [3]. Common impairments as a result of the stroke include delays and impairments in motor, sensory, cognitive, speech, and hearing abilities [4]. Many infants with PS do not demonstrate clinical signs right away, unless they develop seizures (resulting in a referral for brain imaging), and are often diagnosed once motor asymmetries appear—months, if not years, after the original brain injury [5]. PS is the leading cause of hemiplegic cerebral palsy (HCP), in which the arm, leg, and trunk on one side of the body are more affected (involved side) by motor and sensory impairments than the other side (uninvolved side). HCP leads to the impaired or delayed development of reach and grasp, affecting the child's upper extremity skills. The early detection of these impairments in at-risk infants is crucial in order to provide appropriate referral and begin rehabilitation interventions.

The diagnosis of cerebral palsy (CP) is occurring at earlier ages, due in part to the recently published guidelines and implementation of early diagnosis criteria [6–8]. Magnetic resonance imaging (MRI) is sometimes used in the diagnostic process, but a diagnosis of CP is not given without an observation of motor impairment. Current tools used to aid in diagnosis are based on reflexes and motor skill performance, such as the Hammersmith Infant Neurological Exam (HINE) and the Test of Infant Motor Performance (TIMP) [6]. The early identification of CP is enhanced using the General Movement Assessment (GMA), which uses a gestalt observation of movement quantity and quality [9,10]. The observations of abnormal movement patterns using the GMA as early as 10 weeks of age have been shown to have strong predictive validity for a later CP diagnosis [11]. Combinations of these tools have been used to improve diagnostic predictive ability, but there remains substantial room for improvement, given the time between the injury to the central nervous system (CNS) and diagnosis of motor impairment. There is a need for objective measures of motor impairment in these infants for earlier, definitive diagnosis and a better understanding of the underlying pattern of impaired coordination which is a result of an injured, developing nervous system. 3D motion capture may offer a technology-based solution to detect the asymmetries and abnormal movement patterns in at-risk infants [12,13]. By measuring kinematic characteristics of pre-reaching movements, we might be able to use 3D motion capture to detect motor impairment prior to the emergence of clinical signs like an early arm preference or delayed onset of reaching. Such a tool would aid in the early identification of the underlying movement impairment and provide impairment-level targets for rehabilitation. Importantly, delivering targeted interventions earlier in development leads to better motor outcomes for children with CP [14].

Previous studies using 3D motion capture have identified the spatial and temporal parameters of reaching and pre-reaching or spontaneous movements in infants with typical development (TD) and in infants at high risk for CP [15–23]. A reaching movement is a movement of the hand that ends in a hand–toy contact, typically in a midline position, whereas a pre-reaching or spontaneous movement is any movement that does not meet the definition of a reach [17]. In research studies, these pre-reaching movements are often defined as movements toward a toy and against gravity [13,17]. Most typically developing infants demonstrate an onset of reaching in a midline position around 3–5 months of age [21]. The developmental trajectory of reaching movements includes an (i) increased frequency of successful hand–toy contacts with a decrease in spontaneous or pre-reaching movements; (ii) straighter hand path measured by a lower straightness ratio, which means the length of the hand path is getting shorter relative to the distance between the start and end point of the reach [23]; and (iii) faster movement speed early in development with a decrease in movement speed with the onset of reaching, in order to perform a more accurate movement [21]. Infants with CP show delays in the onset of reaching, less coordinated reaches (a smaller number of reaches, longer straightness ratio, faster speeds, and an increased number of movement units), and often an early arm preference [16,24,25], whereas TD infants will demonstrate a fluctuating arm preference [15,20,23,26].

Constraint-induced movement therapy (CIMT) is a promising treatment option for infants with HCP. CIMT involves constraining the uninvolved arm, using a cast, a mitt, or by a therapist holding the arm, in order to provide therapy and encourage as much meaningful use and repetition of the involved arm as possible. Pre- and post-CIMT treatment outcomes are commonly measured, and have demonstrated positive treatment effects including improvements in kinematics [27–33]. Recent studies have also shown an increased overall use of the involved limb during CIMT [34]. The effects of constraint in real time on spatial and temporal kinematics of upper extremity movements, however,

extremities were categorized as potentially involved in the analysis. Due to the small sample size, no formal statistical tests were conducted to compare the changes from 2 to 6 months among different diagnosis groups. Exploratory analyses were conducted to estimate the changes of the same measure over time for each side by different groups. In addition, the estimated mean and standard deviation at 2-month and 6-month time points for all dependent variables were summarized for each group to be used for the design of future studies. Variability was represented by the standard deviation of the mean. SAS version 9.4 (The SAS Institute, Cary, NC, USA) [38] was used to conduct the statistical analysis.

3. Results

In this study, we successfully collected longitudinal 3D motion capture data on 14 infants with and without PS and CP, prior to and during the emergence of reaching. A total of 642 trials containing 9784 unique movements were included in the statistical analysis, with at least one usable trial of each type per session per participant. We will first give a description of the overall group differences from the first to the final data collection. In the subsequent sections, we will describe the changes between and within diagnosis groups in terms of the side (involved versus uninvolved), age, and constraint versus no constraint (bilateral) in more detail.

To address our first question, whether 3D motion capture can detect group differences in the timing and coordination of pre-reaching and reaching movements, we observed an overall change in the means from 2 to 6 months for both spatial and temporal kinematic variables, indicative of possible group–age interactions (Table 1). At 2 months of age, all three groups had similar means for all kinematic variables, in that they were all within one standard deviation of the TD group. At 6 months of age, there were visible differences between the group means for many of the variables. There was some noticeable variability in the means from 3 to 5 months, with trends becoming most apparent at 5 and 6 months.

Table 1. Overall group mean (standard deviation) of the kinematic variables at 2–6 months of age.

		2 Months	3 Months	4 months	5 Months	6 Months
Movement Length (mm)	CP	62.4 (16.1)	68.6 (19.5)	58.4 (20)	79.8 (15)	52 (13.1)
	NS	62.7 (13.1)	61.4 (16.1)	62.2 (18.4)	87.2 (14.2)	83.9 (16.8)
	TD	64.8 (14.5)	62.2 (13.4)	73.5 (25.9)	79.9 (15.4)	90.2 (33.3)
Path Length (mm)	CP	119 (41.9)	116 (26)	112 (43.6)	138 (29.6)	103 (26.3)
	NS	116 (20.4)	111 (30.9)	113 (38.7)	142 (22.2)	156 (32.1)
	TD	121 (34.9)	113 (20.8)	129 (37.1)	144 (33.7)	155 (47.2)
Straightness Ratio	CP	1.9 (0.3)	1.74 (0.23)	2.09 (0.86)	1.87 (0.33)	2.12 (0.49)
	NS	1.94 (0.3)	1.88 (0.4)	1.91 (0.47)	1.7 (0.22)	2.04 (0.36)
	TD	1.89 (0.31)	1.91 (0.32)	1.9 (0.44)	1.89 (0.31)	1.9 (0.36)
Movement Speed (mm/s)	CP	116 (40.4)	188 (61.4)	111 (55)	244 (130.6)	154 (74.4)
	NS	151 (49.6)	149 (75.6)	130 (67.9)	207 (117.1)	203 (109.4)
	TD	138 (46.8)	174 (72)	203 (136.6)	290 (148)	272 (158.8)
Movement Frequency (#/min)	CP	13.12 (6.94)	24.22 (16.29)	5.25 (3.99)	10.67 (4.88)	6.28 (3.28)
	NS	24.06 (15.14)	14.77 (12.81)	8.44 (6.8)	16.84 (9.61)	17.23 (5.11)
	TD	17.94 (9.27)	14.5 (6.96)	12.93 (9.97)	18.26 (11.94)	17.99 (11.83)
Reach Frequency (#/min)	CP	0.02 (0.07)	1.18 (1.19)	0.22 (0.46)	1.64 (1.52)	0.08 (0.2)
	NS	0.15 (0.32)	0.58 (0.58)	1.17 (1.7)	1.63 (1.82)	3.46 (2.5)
	TD	0.3 (0.66)	1.19 (1.35)	2.06 (3.06)	2.67 (3)	3.63 (3.48)

CP = cerebral palsy, PS = perinatal stroke, TD = typically developing.

In terms of spatial variables, the infants with PS and TD both demonstrated an increase in the mean movement length and path length from 2 to 6 months but the CP group did not: for the infants with PS, both the movement and path length means were more than one standard deviation longer at 6 months than at 2 months (62.7 ± 13.1 to 83.9 ± 16.8, 116 ± 20.4 to 156 ± 32.1, respectively). Conversely, the infants with CP demonstrated *decreased mean movement and path lengths* at 6 months compared to 2 months. Not surprisingly, since movement and path length both changed in the same direction within groups, the straightness ratio did not change noticeably in any of the groups from 2 to 6 months.

In terms of the temporal variables, the infants with CP showed a slightly slower mean movement speed of 116 mm/s, although still within one standard deviation of the other two at 2 months. At 6 months, the infants with CP increased their mean movement speed to 154 mm/s, comparable to the 2-month means of the other two groups; however, the infants with PS and TD increased their mean movement speeds from 151 and 138 to 203 and 273 mm/s by 6 months, respectively. For movement frequency, all groups had similar mean frequencies at 2 months, and the infants with PS and TD appeared to maintain their mean frequencies around 18 movements/min from 2 to 6 months, while the infants with CP decreased to a mean of just 6.3 movements/min at 6 months, which is within two standard deviations below the means of the other two groups. Finally, all three groups had a mean reach frequency close to 0 reaches/min at 2 months. The infants with CP remained near 0 reaches/min at 6 months, while the infants with PS had a mean of 3.46 reaches/min and the infants with TD a mean of 3.63 reaches/min, indicative of the typical onset of reaching.

To address our second question, whether 3D motion capture detected changes with versus without constraint, we found that this varied based on the diagnosis group, age, and side (involved versus uninvolved). We will describe the differences that were observed for each variable in the proceeding sections.

3.1. Spatial Variables

We measured three spatial variables of pre-reaching movement: movement length (Figure 2), length of hand path (Figure 3), and straightness ratio (Figure 4), as defined in Section 2.2. With regards to our second question, the infants with CP did demonstrate some consistent differences in their involved side with constraint versus bilateral, with a longer mean path length and larger mean straightness ratio in the constraint versus bilateral condition, which was not observed for the PS group or for any group with movement length.

3.1.1. Movement Length

Infants in all three groups (TD, PS, CP) had an average movement length around 60 mm at 2 months for both sides in the constraint and bilateral conditions (Figure 2). The changes with age differed between diagnosis groups. The infants with TD and PS both increased mean movement length over time, while the infants with CP did not, and even showed a trend for shorter movement length in the involved side with age, which addresses our first question.

The infants with TD demonstrated increased movement length with age, more so in the bilateral condition than the constraint condition. In the constraint condition, there was more variability, but still a gross increase in mean movement length from 2 to 6 months.

The infants with PS showed similar changes with age in the bilateral condition as the infants with TD, although the mean movement length for the uninvolved side in the bilateral condition did decrease from 5 to 6 months. As with the infants with TD, infants with PS had more variability with age in the constraint condition, particularly with the uninvolved limb, which showed a decrease at 4 months, an increase at 5 months, followed by another decrease at 6 months. The overall change from 2 to 6 months was an increase in mean movement length for both limbs and both conditions. At 6 months, the infants with PS had mean movement lengths about 20 mm less in the uninvolved limb versus the involved limb in both the constraint and bilateral conditions.

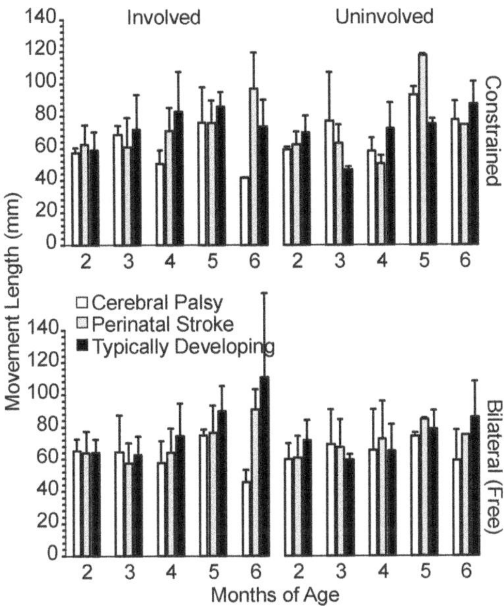

Figure 2. Mean movement length by age, side and condition. For typically developing infants, the left arm is plotted alongside the involved limb and the right arm is plotted alongside the uninvolved limb. Error bars show 1 standard deviation.

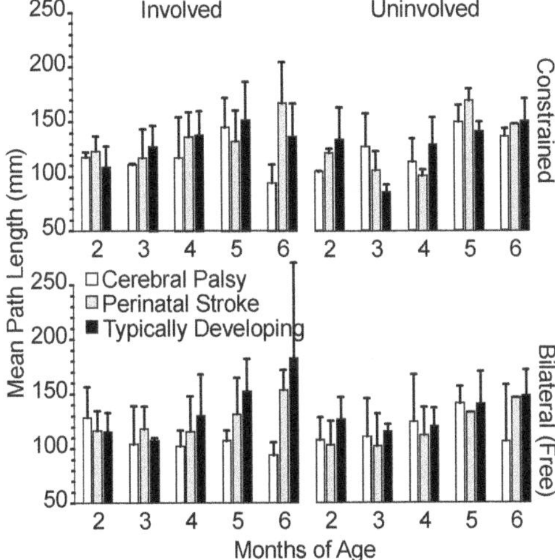

Figure 3. Mean hand path length by age, side, and condition. For typically developing infants, the left arm is plotted alongside the involved limb and the right arm is plotted alongside the uninvolved limb. Error bars show 1 standard deviation.

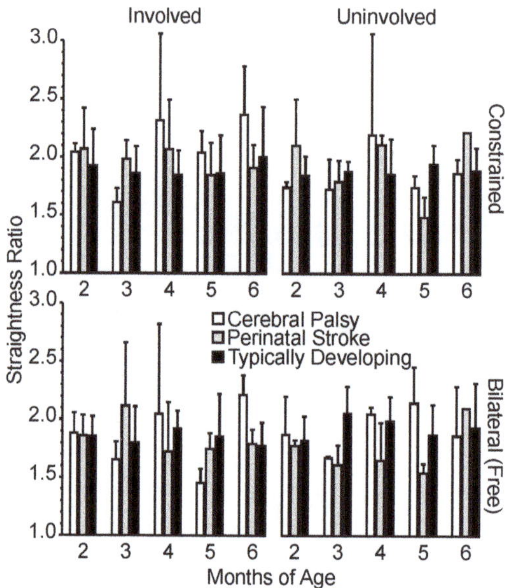

Figure 4. Mean straightness ratio by age, side, and condition. For typically developing infants, the left arm is plotted alongside the involved limb and the right arm is plotted alongside the uninvolved limb. Error bars show 1 standard deviation.

The infants with CP followed a variable trend in both sides and conditions similar to the uninvolved limb of the infants with PS in the constraint condition. There were peaks in the movement length at 3 and 5 months for all conditions for the infants with CP and decreases at 4 and 6 months. The movement length for infants with CP was about the same at 2 months as it was at 6 months for the uninvolved side in the bilateral condition. For the involved side of the infants with CP in both conditions, the movement length was shorter at 4 and 6 months than at 2 months. Only the uninvolved side in the constraint condition for infants with CP showed increased movement length from 2 months to 6 months. In all conditions, the longest movement length for infants with CP occurred at 5 months.

3.1.2. Length of Hand Path

In human movement, the length of the hand path (Figure 3) can never be shorter than the movement length (Figure 2). In this study, all groups had similar mean path lengths at 2 months. The infants with TD and PS overall had increased hand path length with age, corresponding to the increased mean movement lengths they demonstrated. The infants with CP showed a more variable pattern again, with shorter path lengths on the involved side at 6 months versus 2 months. At the age of 6 months, the path lengths for CP, PS and TD were: 103 ± 26.3, 156 ± 32.1 and 155 ± 47.2 mm respectively, demonstrating a difference of more than one standard deviation between the infants with CP and TD.

With constraint, the infants with PS and TD both showed more variation over time compared to bilateral trials. The infants with PS showed a decrease from 2 to 4 months, followed by a sharp spike from 4 to 5 months, followed by a decrease from 5 to 6 months in the uninvolved side constraint condition, which was similar to the trend seen with movement length. The infants with PS showed an inverse pattern with the constraint of the potentially involved limb, increasing from 3 to 4 months, decreasing from 4 to 5 months, then increasing again from 5 to 6 months. The infants with CP did not show clear differences between the constraint and bilateral conditions for mean path length, although they did demonstrate a shorter mean path length in their involved side, particularly in the bilateral condition, after 2 months.

3.1.3. Straightness Ratio

Infants in all three groups started out with a similar mean straightness ratio at 2 months (Figure 4). Noticeably, the infants with CP and PS showed greater variation in the straightness ratio between sides and conditions, whereas the infants with TD maintained a relatively constant straightness ratio for both sides and conditions over time.

The infants with CP showed decreased straightness ratios at 3 and 5 months for both sides and conditions. Interestingly, at 5 months, the infants with CP had a lower straightness on their involved side versus their uninvolved side in the bilateral condition, indicating straighter movements with the involved arm. The infants with CP also had a higher straightness ratio for the involved side in the constraint condition compared to bilateral, while the uninvolved side showed the opposite pattern. At 6 months, the infants with CP had a higher straightness ratio with the involved side in both conditions compared to the uninvolved side, which indicated less straight movements on the involved side.

The infants with PS showed a slight trend for decreased straightness ratio in the involved side from 2 to 6 months, with some variability. The uninvolved side followed a similar trend, although there was a sharp increase for the uninvolved side in both conditions from 5 to 6 months. For the infants with PS, the uninvolved side in the bilateral condition showed the lowest straightness ratio at all time points, except at 6 months, indicating the straightest movements in that condition. Additionally, at most time points the straightness ratio for infants with PS is higher for the constraint condition than the bilateral condition for both involved and uninvolved sides, indicating straighter movements in the bilateral condition.

The infants with TD did not demonstrate notable differences in straightness ratio with condition, side, or age.

3.2. Temporal Variables

We measured three temporal variables of upper extremity movement: movement speed (Figure 5), movement frequency (Figure 6), and reach frequency (Figure 7), as defined in Section 2.2.

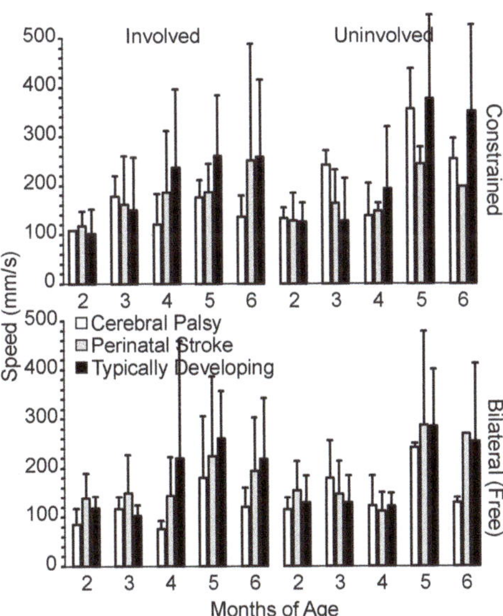

Figure 5. Mean movement speed by age, side, and condition. For typically developing infants, the left arm is plotted alongside the involved limb and the right arm is plotted alongside the uninvolved limb. Error bars show 1 standard deviation.

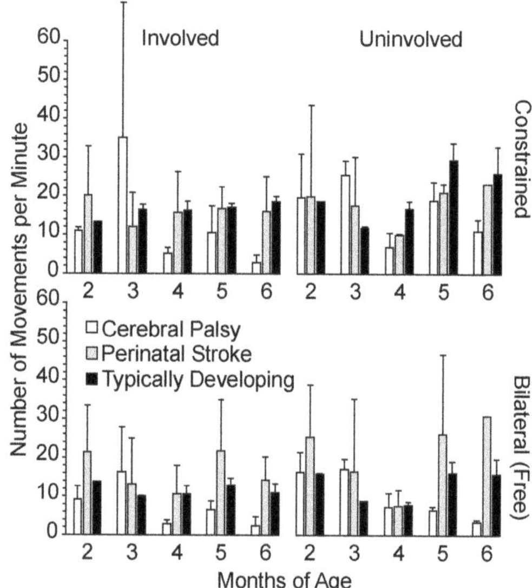

Figure 6. Mean movement frequency by age, side, and condition. For typically developing infants, the left arm is plotted alongside the involved limb and the right arm is plotted alongside the uninvolved limb. Error bars show 1 standard deviation.

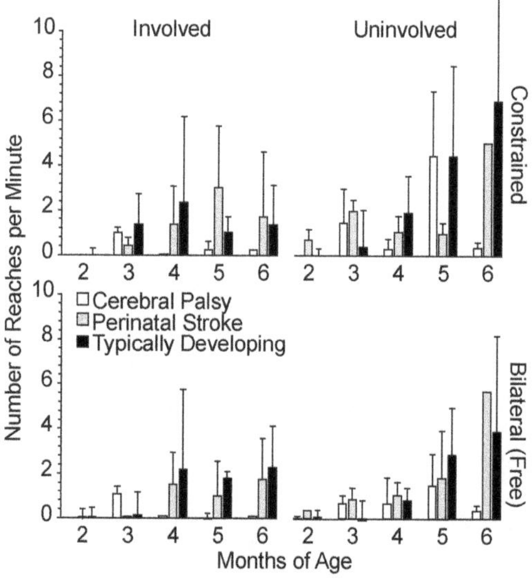

Figure 7. Mean reach frequency by age, side, and condition. For typically developing infants, the left arm is plotted alongside the involved limb and the right arm is plotted alongside the uninvolved limb. Error bars show 1 standard deviation.

3.2.1. Movement Speed

All three groups demonstrated an increase in movement speed in both sides and both conditions from 2 to 6 months (Figure 5). There were potential differences between groups at 2 months; the mean for the involved side of the infants with CP was lower than those for the infants with PS and TD. At 6 months, the mean movement speed for the uninvolved side in the constraint condition for infants with CP was similar to the means of infants with TD and PS, but the means for the involved side in both conditions, and the uninvolved side in the bilateral condition, were substantially lower. Additionally, all three groups showed a decrease in the movement speed from 5 to 6 months for both sides and conditions, with the exception of the involved side of the infants with PS, and the right hand of the infants with TD, in the constraint condition.

The infants with CP showed peaks in their movement speed at 3 and 5 months and decreases at 4 and 6 months for both sides and conditions. The infants with CP appeared to move at faster speeds, on average, with their uninvolved limb compared to the involved limb. Additionally, they moved at faster speeds, on average, in the constraint conditions, compared to the bilateral conditions. The slowest movement speeds were consistently in the involved side in the bilateral condition at all time points for the infants with CP.

The infants with PS demonstrated a general increase in movement speed with age in both conditions and sides, although there was a decrease noted at 4 and 6 months for all except the involved side in the constraint condition. There was no clear difference in movement speed between sides or conditions for the infants with PS, although at 5 and 6 months, the mean movement speeds were fastest with the uninvolved side in the bilateral condition.

The infants with TD demonstrated a sharp increase in mean movement speed from 4 to 5 months with the left side, and from 3 to 5 months with the right side. From 5 to 6 months, we saw a decrease in the mean movement speed in all but the right side in the constraint condition. The infants with TD also had faster mean speeds in the constraint condition compared to the bilateral for the left side at 4 to 6 months, but not with the right side.

3.2.2. Movement Frequency

The infants with CP and TD both had higher mean movement frequencies with constraint compared to the bilateral conditions (Figure 6). The infants with PS showed no clear difference in the mean movement frequency between conditions, although they did have a higher mean movement frequency with the uninvolved side compared to the involved side at 6 months. The infants with TD and PS both showed a minor trend for increased movement frequency between 4 to 6 months, although this increase was more pronounced in the infants with PS, particularly in the uninvolved limb. Meanwhile, the infants with CP had a sharp decrease in the mean movement frequency from 3 to 4 months, which increased slightly at 5 months, then decreased again at 6 months. Interestingly, the infants with CP started out with similar mean movement frequencies as the infants with TD at 2 months, then increased to much higher mean frequencies at 3 months in the constraint condition, then decreased sharply to frequencies below those of the infants with PS and TD at 4 to 6 months. The infants with PS started at higher movement frequencies than the two other groups, but then showed mean movement frequencies similar to the infants with TD for the remaining 4 months.

3.2.3. Reach Frequency

Infants with TD and PS both showed an overall increase in the reach frequency with age, particularly after 3 months (Figure 7). The infants with CP mostly did not reach with their involved side, except at 3 months and 5 months, and only in the constraint condition. The infants with CP did show an increase in the mean reach frequency of the uninvolved side at 3 and 5 months, more so in the constraint condition than bilateral, however, there were no reaches at 6 months. The infants with PS demonstrated a fairly steady increase in the mean reach frequency with age, albeit with some fluctuation. At 6 months, the involved side of the infants with PS had a much lower mean reach frequency than the uninvolved side. The infants with TD showed a steady increase from 3 to 6 months with the left side, with the highest mean reach frequency in the left side constraint condition at 6 months. There was an increase in mean reach frequency from 2 to 4 months on the right side in both conditions for infants with TD, but then a plateau from 4 to 6 months.

4. Discussion

In this study, we demonstrated that the 3D motion capture may feasibly be used to measure objective changes in pre-reaching upper extremity movements in infants with TD and infants with neuromotor impairment, namely PS and CP. We successfully measured three spatial and three temporal variables of upper extremity movement over a 5-month period in 14 infants with TD, PS, and CP. Moreover, we observed trends in the data that generated several hypotheses, which can be used for the design of further investigation. The main trends that we identified were related to differences with diagnosis, changes with the real-time constraint of one arm, and changes over time with age and development. In the subsequent sections, we will discuss the specific hypotheses that we generated from these data, which can be applied to future research, and support for those hypotheses based on previous research.

4.1. Hypothesis 1: 3D Motion Capture Can Be Used to Detect Differences in Kinematic Characteristics of Pre-Reaching Movements between a Diagnosis of CP and TD, Particularly through an Interaction Effect with Age

We observed trends for an age–diagnosis interaction for many of the dependent variables measured. Most noticeable was the variability seen in the infants with CP and PS compared to infants with TD. The infants with TD showed apparently linear trends with age that were consistent with previous research, such as increased reach frequency, increased movement length, and increased movement speed [17,20,21,36]. The infants with PS largely followed the trends of typical development, however, with more fluctuation from month to month than the TD group. In contrast, the infants with CP had trends in the opposite direction of typical development for multiple variables, including decreased movement frequency, decreased movement length, and no change in reach frequency. Furthermore, their month-to-month performance was more variable. These differences might be indicative of motor impairment as a result of CNS injury as these infants all had a later diagnosis of CP. The results from this study suggest that it might be possible to use 3D motion capture to differentiate between CP and no CP in infants with PS and between CP and TD, based on kinematic measures of pre-reaching movements.

4.2. Hypothesis 2: 3D Motion Capture Can Be Used to Detect Asymmetries in Kinematic Characteristics of Pre-Reaching Movements between the Involved and Uninvolved Side in Infants with PS and CP

We found preliminary evidence that 3D motion capture could be used to detect asymmetry in pre-reaching movements in infants with PS and CP for certain spatiotemporal variables. In our analysis, the infants with CP demonstrated largely decreased movement length, path length, movement speed, movement frequency and reach frequency in their involved side versus uninvolved side. Previous research has found that infants with HCP often begin to present with an increased asymmetry in upper extremity use with development, but also that many of them demonstrate some impairment in the uninvolved limb as well, in terms of the speed and accuracy of movement [24]. In this study, the infants with TD and PS showed a steady increase in the movement speed and reach frequency with age, consistent with typical development [21,39]. The infants with CP showed only a slight increase in movement speed and more notably, only reached at 3 and 5 months, predominately with the uninvolved arm. The slower movement speed and lack of reaches in the infants with CP might indicate impairment in the involved upper extremity, and to a lesser extent, the uninvolved upper extremity, as infants with HCP often demonstrate some impairment or delays in both upper extremities [40].

The infants with PS did not show consistent differences between the potentially involved and uninvolved upper extremities, but based on our results, a larger sample size might reveal a larger straightness ratio, slower movement speed, and decreased movement and reach frequencies in the potentially involved limb, particularly at 5 and 6 months of age. The infants with PS who did not later receive a diagnosis of CP might be demonstrating typical upper extremity development, given that their means were similar to those of the infants with TD. We might speculate, however, that there were some signs that the infants with PS were not following the same pattern of development as the infants with TD, based on the increased fluctuation in the means from month to month. The subtle differences between infants with PS and TD for some kinematic variables of pre-reaching movements (movement frequency, reach frequency, movement speed, straightness ratio), particularly with the involved limb, might represent sub-clinical signs of motor impairment, or it might just be due to the small sample size. Further research is needed to make a definitive conclusion about the results of the PS group.

4.3. Hypothesis 3: 3D Motion Capture Can Be Used to Detect and Monitor Changes in Spatial and Temporal Characteristics of Pre-Reaching Upper Extremity Movements with Constraint Versus without Constraint in Infants with TD and CP

Our preliminary analysis indicated that real-time changes with constraint could be observed in the infants with CP and infants with TD for multiple variables of upper extremity movement. These changes were most apparent in the temporal variables. Infants with CP and TD both demonstrated increased movement speed, movement frequency, and reach frequency in the constraint condition compared to the bilateral. For the infants with TD, the difference between the conditions became more apparent at 4–6 months of age, which was consistent with the onset of reach development, usually occurring between 3 and 5 months of age [21]. The infants with CP also demonstrated less precise movement with constraint on their involved side, indicated by a longer hand path length and a larger straightness ratio. Previous studies found the reverse trend for typical development, where the infants demonstrated straighter movements, with a smaller straightness ratio, as they got older, highlighting another possible indication of motor impairment in the infants with CP in this study [23]. Based on our preliminary results, it is possible that constraint increases the use of the free arm in real time in both infants with TD and CP. Our preliminary results also indicate that the infants with CP might show less precise movements in their involved arm with constraint in real time, likely due to poorer coordination and overall development in that arm as a result of both motor impairment and neglect [4,25,41].

4.4. Limitations

This study has limitations due to its small sample size because it is a preliminary study. Insights that the data from this initial exploration provide are valuable for generating hypotheses for future investigation. We attempted to not overstate our results and present them in a way that was consistent with the study design. In order to draw any firm conclusions from the results of this study, it needs to be replicated with a larger cohort and sufficient power. In this study, we generated hypotheses about many independent and dependent variables by observing trends in the means from the data we collected. While we observed many plausible and hypothesis-generating trends, there is a risk with the small sample size that results could be skewed by a single participant. Based on our results, we observed the difference in movement length between infants with CP and TD to be at least one standard deviation with an 85% lower boundary of about 0.7. A sample size of 34 per group in a future study could provide at least 80% power to detect such a difference.

Additionally, in this study, we used a 10-camera, 3D motion capture system (Vicon Motion Systems, Ltd., Oxford, UK). It is possible that the cost and access to a 3D motion capture system could pose a barrier to the replication or widespread application of our results. The ongoing development of more inexpensive and portable wearable sensors that can calculate similar variables as 3D motion capture will hopefully make the collection of kinematic data much more accessible and widely applicable. There has been considerable progress in the methods and technology for wearable use in infants with most of the information on physical activity and 24 h monitoring [42–47].

Another limitation of this study is the relatively large number of biomechanical variables involved and their interdependence with each other. For example, the straightness ratio is a combination of movement length and path length. Movement speed and movement length are inextricably related. Thus, in a future study, care would need to be exercised for appropriate statistical comparisons between the correlated variables. Furthermore, the relatively large number of measurements that is possible increases the risk of false positives.

4.5. Application and Future Directions

A major takeaway from this study is that 3D motion capture may be feasible for the longitudinal tracking of pre-reaching and reaching behavior in infants with TD, PS, and CP. It is possible that further research could identify specific kinematic variables that could be used as biomarkers in infants with neuromotor impairment [12]. The benefit of a biomarker is that it can be monitored with intervention and over time in at-risk infants, in order to measure progress and inform precise rehabilitation to attain optimum health outcomes. 3D motion capture or future wearable sensors could be used for the monitoring of spatial and temporal characteristics of pre-reaching movements. The ability to measure objective changes in the spatial and temporal characteristics of upper extremity movement could be extremely beneficial to a clinician who is using CIMT or another intensive upper extremity intervention for a child with HCP. In the future, we would like to replicate this study with a larger cohort, as well as with wearable sensors.

Wearable sensors, such as inertial measurement unit (IMU) sensors, are becoming increasingly useful for measuring the kinematics of human movement, as they are a small, portable alternative to 3D motion capture. Previous studies in adults have found IMU sensors to perform comparably to the Vicon 3D motion capture system, in terms of measuring joint angles [48,49]. IMU sensors have been successfully applied to measure infant leg movement, mostly in terms of movement frequency [43–46]. Of note, most of these studies collect data on leg movements that are mostly single-plane and sampled over a duration of one or several days. These are simpler than the multi-dimensional upper extremity movements we measured and are sampled from a much larger data set. Despite the greater complexity of tracking infant upper extremity movements, rapidly emerging wearable technologies using IMU sensors show promise in this application. Notably, IMU sensor systems have been evaluated for their feasibility to measure infant limb kinematics, and with the collection of movement data over a shorter duration, like 60 s [42,47,50]. The variables measured in these kinematic studies using IMUs include the frequency, duration, and acceleration of movements in both the upper and lower limbs, similar to some of the variables we measured in this study [42,47,50]. Some of these recent infant IMU studies aimed to develop sensor systems specifically for detecting early motor delay in infants [45,50]. In a future study, we would like to test the feasibility of implementing the protocol in this paper with IMU sensors. The use of these sensors would allow for the ability to measure and track some kinematic variables of infant movement in a home or clinical setting.

Another promising approach for prediction in cases where multiple factors influence diagnosis, and early detection is desirable, is machine learning. These approaches are well suited, often superior to traditional means-based statistics, for handling the relatively large number of measurements possible from biomechanics [51]. Machine learning approaches have proven successful at detecting upper limb movement patterns from electromyography (EMG) with wearable sensors [52]. They have also proven useful to predict outcomes from CIMT rehabilitation in adults after stroke [53]. Although machine learning approaches have been used to predict the eventual diagnosis of CP based on components of the GMA [54], the specificity of predictions from such models is not at the level desired for widespread adoption [55]. We are not aware of attempts to combine data from various domains, including 3D motion analysis like that presented here, along with observations such as the HINE, TIMP, and GMA, into a machine learning approach for the prediction of an eventual CP diagnosis. Given the relationships evident in a 3D biomechanical analysis from this study, a machine learning approach combining biomechanical measurements with other observational data such as the GMA would seem promising.

5. Conclusions

This is a preliminary study with important findings. The exploration of spatiotemporal characteristics of upper extremity movements prior to and during the age of reach onset shows some interesting patterns of change over time, and potential differences between infants with TD and CP. Constraint in real time might increase movement speed and frequency, and reach frequency in infants with CP and TD. Importantly, constraint might increase the movement frequency of an involved arm in real time in infants with CP. Finally, 3D motion capture or wearable sensors might be useful in tracking upper extremity pre-reaching and reaching movements in infants with neuromotor impairment in the upper extremity, such as HCP. Due to the preliminary nature of this study, no recommendations for the immediate application of these results can be made. Further investigation of the hypotheses generated from this study is necessary for the further interpretation of results.

Author Contributions: Conceptualization, J.C.H. and J.M.; methodology, J.C.H., A.M.W.C., D.R., M.M., J.M., and X.P.; software, A.M.W.C., D.R. and M.M.; validation, J.M., A.M.W.C., M.M. and D.R.; formal analysis, X.P., J.M. and D.R.; investigation, J.C.H. and J.M.; resources, J.C.H. and A.M.W.C.; data curation, J.M., D.R., M.M. and A.M.W.C.; writing—original draft preparation, J.M.; writing—review and editing, J.C.H., X.P., A.M.W.C., J.A.B. and D.R.; visualization, J.A.B., X.P. and J.M.; supervision, J.C.H.; project administration, J.C.H. and J.M.; funding acquisition, J.C.H. and J.M. All authors have read and agreed to the published version of the manuscript.

Funding: This research was partially funded by the Foundation for Physical Therapy Research Promotion of Doctoral Studies I grant and Eunice Kennedy Shriver National Institute of Child Health and Human Development (1 P2CHD10191201). The funding bodies had no influence on study design, participant recruitment, methods, data analysis, or manuscript preparation.

Conflicts of Interest: The authors declare no conflict of interest. The funders had no role in the design of the study; in the collection, analyses, or interpretation of data; in the writing of the manuscript, or in the decision to publish the results.

References

1. Dunbar, M.; Kirton, A. Review Perinatal stroke: Mechanisms, management, and outcomes of early cerebrovascular brain injury. *Lancet Child Adolesc. Heal.* **2018**, *2*, 666–676. [CrossRef]
2. Dunbar, M.; Kirton, A. Perinatal Stroke. *Semin. Pediatr. Neurol.* **2019**, *32*, 100767. [CrossRef] [PubMed]
3. Lynch, J.K.; Nelson, K.B. Epidemiology of perinatal stroke. *Curr. Opin. Pediatr.* **2001**, *13*, 499–505. [CrossRef] [PubMed]
4. Golomb, M.R. Outcomes of perinatal arterial ischemic stroke and cerebral sinovenous thrombosis. *Semin. Fetal Neonatal Med.* **2009**, *14*, 318–322. [CrossRef] [PubMed]
5. Lynch, J.K. Epidemiology and classification of perinatal stroke. *Semin. Fetal Neonatal Med.* **2009**, *14*, 245–249. [CrossRef] [PubMed]
6. Novak, I.; Morgan, C.; Adde, L.; Blackman, J.; Boyd, R.N.; Brunstrom-Hernandez, J.; Cioni, G.; Damiano, D.; Darrah, J.; Eliasson, A.-C.; et al. Early, Accurate Diagnosis and Early Intervention in Cerebral Palsy: Advances in Diagnosis and Treatment. *JAMA Pediatr.* **2017**, *171*, 897–907. [CrossRef]
7. Maitre, N.L.; Burton, V.J.; Duncan, A.F.; Iyer, S.; Ostrander, B.; Winter, S.; Ayala, L.; Burkhardt, S.; Gerner, G.; Getachew, R.; et al. Network implementation of guideline for early detection decreases age at cerebral palsy diagnosis. *Pediatrics* **2020**, *145*, e20192126. [CrossRef]
8. Byrne, R.; Noritz, G.; Maitre, N.L. Implementation of Early Diagnosis and Intervention Guidelines for Cerebral Palsy in a High-Risk Infant Follow-Up Clinic. *Pediatr. Neurol.* **2017**, *76*, 66–71. [CrossRef]
9. Guzzetta, A.; Mercuri, E.; Rapisardi, G.; Ferrari, F.; Roversi, M.F.; Cowan, F.; Rutherford, M.; Paolicelli, P.B.; Einspieler, C.; Boldrini, A.; et al. General Movements Detect Early Signs of Hemiplegia in Term Infants with Neonatal Cerebral Infarction. *Neuropediatrics* **2003**, *34*, 61–66.
10. Hay, K.; Nelin, M.A.; Carey, H.; Chorna, O.; Moore-Clingenpeel, M.A.; Mas, M.; Maitre, N. Hammersmith Infant Neurological Examination Asymmetry Score Distinguishes Hemiplegic Cerebral Palsy From Typical Development. *Pediatr. Neurol.* **2018**, *87*, 70–74. [CrossRef]
11. Kwong, A.K.L.; Fitzgerald, T.L.; Doyle, L.W.; Cheong, J.L.Y.; Spittle, A.J. Predictive validity of spontaneous early infant movement for later cerebral palsy: A systematic review. *Dev. Med. Child Neurol.* **2018**, *60*, 480–489. [CrossRef] [PubMed]
12. Marchi, V.; Belmonti, V.; Cecchi, F.; Coluccini, M.; Ghirri, P.; Grassi, A.; Sabatini, A.M.; Guzzetta, A. Movement analysis in early infancy: Towards a motion biomarker of age. *Early Hum. Dev.* **2020**, *142*, 104942. [CrossRef] [PubMed]
13. Mazzarella, J.; McNally, M.; Chaudhari, A.; Pan, J.; Heathcock, J. Kinematic differences in spontaneous arm movements may be an indicator of cerebral palsy in infants with stroke: An initial investigation. *Exp. Brain Res.* **2020**, submitted.
14. Mcintyre, S.; Morgan, C.; Walker, K.; Novak, I. Cerebral Palsy - Don't Delay. *Dev. Disabil.* **2011**, *17*, 114–129. [CrossRef] [PubMed]

15. Corbetta, D.; Thelen, E. Lateral Biases and Fluctuations in Infants' Spontaneous Arm Movements and Reaching. *Dev. Psychobiol.* **1999**, *34*, 237–255. [CrossRef]

16. Van der Heide, J.C.; Fock, J.M.; Otten, B.; Stremmelaar, E.; Hadders-Algra, M. Kinematic Characteristics of Postural Control during Reaching in Preterm Children with Cerebral Palsy. *Pediatr. Res.* **2005**, *58*, 586–593. [CrossRef]

17. Bhat, A.N.; Galloway, J.C. Toy-oriented changes during early arm movements: Hand kinematics. *Infant Behav. Dev.* **2006**, *29*, 358–372. [CrossRef]

18. Bhat, A.N.; Lee, H.M.; Galloway, J.C. Toy-oriented changes in early arm movements II — Joint kinematics. *Infant Behav. Dev.* **2007**, *30*, 307–324. [CrossRef]

19. Lee, H.M.; Bhat, A.; Scholz, J.P.; Galloway, J.C. Toy-oriented changes during early arm movements IV: Shoulder – elbow coordination. *Infant Behav. Dev.* **2008**, *31*, 447–469. [CrossRef]

20. Lynch, A.; Lee, H.M.; Bhat, A.; Galloway, J.C. No stable arm preference during the pre-reaching period: A comparison of right and left hand kinematics with and without a toy present. *Dev. Psychobiol.* **2008**, *50*, 390–398. [CrossRef]

21. Thelen, E.; Corbetta, D.; Spencer, J.P. Development of reaching during the first year: Role of movement speed. *J. Exp. Psychol. Hum. Percept. Perform.* **1996**, *22*, 1059–1076. [CrossRef] [PubMed]

22. Karch, D.; Kim, K.; Wochner, K.; Pietz, J.; Dickhaus, H.; Philippi, H. Quantification of the segmental kinematics of spontaneous infant movements. *J. Biomech.* **2008**, *41*, 2860–2867. [CrossRef] [PubMed]

23. Nelson, E.L.; Konidaris, G.D.; Berthier, N.E. Hand preference status and reach kinematics in infants. *Infant Behav. Dev.* **2014**, *37*, 615–623. [CrossRef] [PubMed]

24. Van der Heide, J.C.; Fock, J.M.; Otten, B.; Stremmelaar, E.; Hadders-algra, M. Kinematic Characteristics of Reaching Movements in Preterm Children with Cerebral Palsy. *Pediatr. Res.* **2005**, *57*, 883–889. [CrossRef] [PubMed]

25. Chen, C.-Y.; Tafone, S.; Lo, W.; Heathcock, J.C. Perinatal stroke causes abnormal trajectory and laterality in reaching during early infancy. *Res. Dev. Disabil.* **2015**, *38*, 301–308. [CrossRef]

26. Rönnqvist, L.; Domellöf, E. Quantitative Assessment of Right and Left Reaching Movements in Infants: A Longitudinal Study from 6 to 36 Months. *Dev. Psychobiol.* **2006**, *48*, 444–459. [CrossRef] [PubMed]

27. Eliasson, A.; Nordstrand, L.; Ek, L.; Lennartsson, F.; Sjöstrand, L.; Tedro, K.; Krumlinde-sundholm, L. The effectiveness of Baby-CIMT in infants younger than 12 months with clinical signs of unilateral-cerebral palsy; an explorative study with randomized design. *Res. Dev. Disabil.* **2018**, *72*, 191–201. [CrossRef]

28. Coker, P.; Lebkicher, C.; Harris, L.; Snape, J. The effects of constraint-induced movement therapy for a child less than one year of age. *NeuroRehabilitation* **2009**, *24*, 199–208. [CrossRef]

29. Chorna, O.; Heathcock, J.; Key, A.; Noritz, G.; Carey, H.; Hamm, E.; Nelin, M.A.; Murray, M.; Needham, A.; Slaughter, J.C.; et al. Early childhood constraint therapy for sensory / motor impairment in cerebral palsy: A randomised clinical trial protocol. *BMJ Open* **2015**, *5*, 33–35. [CrossRef]

30. Novak, I.; Morgan, C.; Fahey, M.; Finch-Edmondson, M.; Galea, C.; Hines, A.; Langdon, K.; Namara, M.M.; Paton, M.C.; Popat, H.; et al. State of the Evidence Traffic Lights 2019: Systematic Review of Interventions for Preventing and Treating Children with Cerebral Palsy. *Curr. Neurol. Neurosci. Rep.* **2020**, *20*, 1–21. [CrossRef]

31. Maitre, N.L.; Jeanvoine, A.; Yoder, P.J.; Key, A.P.; Slaughter, J.C.; Carey, H.; Needham, A.; Murray, M.M.; Heathcock, J.; Burkhardt, S.; et al. Kinematic and Somatosensory Gains in Infants with Cerebral Palsy After a Multi-Component Upper-Extremity Intervention: A Randomized Controlled Trial. *Brain Topogr.* **2020**, *33*, 751–766. [CrossRef] [PubMed]

32. Hoare, B.; Wallen, M.; Thorley, M.; Jackman, M.; Carey, L.; Imms, C. Constraint-induced movement therapy in children with unilateral cerebral palsy. *Cochrane Database Syst. Rev.* **2019**. [CrossRef] [PubMed]

33. Chen, H.; Kang, L.; Chen, C.; Lin, K.; Chen, F.; Wu, K.P.H. Younger Children with Cerebral Palsy Respond Better Than Older Ones to Therapist-Based Constraint-Induced Therapy at Home on Functional Outcomes and Motor Control. *Phys. Occup. Ther. Pediatr.* **2016**, *36*, 171–185. [CrossRef] [PubMed]

34. Goodwin, B.M.; Sabelhaus, E.K.; Pan, Y.C.; Bjornson, K.F.; Pham, K.L.D.; Walker, W.O.; Steele, K.M. Accelerometer Measurements Indicate That Arm Movements of Children With Cerebral Palsy Do Not Increase After Constraint-Induced Movement Therapy (CIMT). *Am. J. Occup. Ther.* **2020**, *74*, 1–9. [CrossRef]

35. *Vicon Nexus 1.8.5*; Vicon Motion Systems Ltd.: Oxford, UK, 2013.

36. Lee, M.H.; Ranganathan, R.; Newell, K.M. Changes in object-oriented arm movements that precede the transition to goal-directed reaching in infancy. *Dev. Psychobiol.* **2011**, *53*, 685–693. [CrossRef] [PubMed]

37. The Mathworks Inc. MATLAB (R2019b). Available online: https://www.mathworks.com/products/new_products/release2019b.html (accessed on 19 December 2020).

38. The SAS Institute SAS Version 9.4M6. Available online: https://documentation.sas.com/?cdcId=pgmsascdc&cdcVersion=9.4_3.5&docsetId=whatsdiff&docsetTarget=p0owjfzx15uhnan1rye9smrllx19.htm&locale=en (accessed on 19 December 2020).

39. Piek, J.P.; Carman, R. Developmental profiles of spontaneous movements in infants. *Early Hum. Dev.* **1994**, *39*, 109–126. [CrossRef]

40. De Campos, A.C.; Kukke, S.N.; Hallett, M.; Alter, K.E.; Damiano, D.L. Characteristics of Bilateral Hand Function in Individuals With Unilateral Dystonia Due to Perinatal Stroke: Sensory and Motor Aspects. *J. Child Neurol.* **2014**, *29*, 623–632. [CrossRef]

41. Kirton, A.; DeVeber, G. Life After Perinatal Stroke. *Stroke* **2013**, 3265–3271. [CrossRef]

42. Trujillo-Priego, I.A.; Smith, B.A. Kinematic characteristics of infant leg movements produced across a full day. *J. Rehabil. Assist. Technol. Eng.* **2017**, *4*, 205566831771746. [CrossRef]

43. Deng, W.; Trujillo-Priego, I.A.; Smith, B.A. How Many Days Are Necessary to Represent an Infant's Typical Daily Leg Movement Behavior Using Wearable Sensors? *Phys. Ther.* **2019**, *99*, 730–738. [CrossRef]

44. Jiang, C.; Lane, C.J.; Perkins, E.; Schiesel, D.; Smith, B.A. Determining if wearable sensors affect infant leg movement frequency. *Dev. Neurorehabil.* **2018**, *21*, 133–136. [CrossRef] [PubMed]

45. Abrishami, M.S.; Nocera, L.; Mert, M.; Trujillo-Priego, I.A.; Purushotham, S.; Shahabi, C.; Smith, B.A. Identification of developmental delay in infants using wearable sensors: Full-day leg movement statistical feature analysis. *IEEE J. Transl. Eng. Heal. Med.* **2019**, *7*, 1–7. [CrossRef] [PubMed]

46. Smith, B.A.; Trujillo-Priego, I.A.; Lane, C.J.; Finley, J.M.; Horak, F.B. Daily quantity of infant leg movement: Wearable sensor algorithm and relationship to walking onset. *Sensors* **2015**, *15*, 19006–19020. [CrossRef] [PubMed]

47. Trujillo-Priego, I.; Lane, C.; Vanderbilt, D.; Deng, W.; Loeb, G.; Shida, J.; Smith, B. Development of a Wearable Sensor Algorithm to Detect the Quantity and Kinematic Characteristics of Infant Arm Movement Bouts Produced across a Full Day in the Natural Environment. *Technologies* **2017**, *5*, 39. [CrossRef] [PubMed]

48. Sers, R.; Forrester, S.; Moss, E.; Ward, S.; Ma, J.; Zecca, M. Validity of the Perception Neuron inertial motion capture system for upper body motion analysis. *Meas. J. Int. Meas. Confed.* **2020**, *149*, 107024. [CrossRef]

49. Chaudhari, A.M.W.; McKenzie, C.S.; Pan, X.; Oñate, J.A. Lumbopelvic control and days missed because of injury in professional baseball pitchers. *Am. J. Sports Med.* **2014**, *42*, 2734–2740. [CrossRef]

50. Redd, C.B.; Barber, L.A.; Boyd, R.N.; Varnfield, M.; Karunanithi, M.K. Development of a Wearable Sensor Network for Quantification of Infant General Movements for the Diagnosis of Cerebral Palsy. In Proceedings of the Annual International Conference of the IEEE Engineering in Medicine and Biology Society, Berlin, Germany, 23–27 July 2019; pp. 7134–7139. [CrossRef]

51. Li, F.; Shirahama, K.; Nisar, M.A.; Huang, X.; Grzegorzek, M. Deep Transfer Learning for Time Series Data Based on Sensor Modality Classification. *Sensors* **2020**, *20*, 4271. [CrossRef]

52. Burns, A.; Adeli, H.; Buford, J.A. Upper Limb Movement Classification Via Electromyographic Signals and an Enhanced Probabilistic Network. *J. Med. Syst.* **2020**, *44*. [CrossRef]

53. Rafiei, M.; Kelly, K.; Borstad, A.; Adeli, H.; Gauthier, L. Predicting Improved Daily Use of the More Affected Arm Poststroke Following Constraint-Induced Movement Therapy. *Phys. Ther.* **2019**, *99*, 1667–1678. [CrossRef]

54. Ihlen, E.A.F.; Støen, R.; Boswell, L.; de Regnier, R.-A.; Fjørtoft, T.; Gaebler-Spira, D.; Labori, C.; Loennecken, M.C.; Msall, M.E.; Möinichen, U.I.; et al. Machine Learning of Infant Spontaneous Movements for the Early Prediction of Cerebral Palsy: A Multi-Site Cohort Study. *J. Clin. Med.* **2019**, *9*, 5. [CrossRef]

55. Schmidt, W.; Regan, M.; Fahey, M.; Paplinski, A. General movement assessment by machine learning: Why is it so difficult? *J. Med. Artif. Intell.* **2019**, *2*. [CrossRef]

Publisher's Note: MDPI stays neutral with regard to jurisdictional claims in published maps and institutional affiliations.

Communication

Quantifying Circadian Aspects of Mobility-Related Behavior in Older Adults by Body-Worn Sensors—An "Active Period Analysis"

Tim Fleiner [1,2,†], Rieke Trumpf [1,2,†], Anna Hollinger [1], Peter Haussermann [2] and Wiebren Zijlstra [1,*]

[1] Institute of Movement and Sport Gerontology, German Sport University Cologne, 50933 Cologne, Germany; t.fleiner@dshs-koeln.de (T.F.); rieke.trumpf@lvr.de (R.T.); A.Hollinger@dshs-koeln.de (A.H.)

[2] Department of Geriatric Psychiatry & Psychotherapy, LVR Hospital Cologne, 51109 Cologne, Germany; peter.haeussermann@lvr.de

* Correspondence: zijlstra@dshs-koeln.de

† These authors contributed equally to this work.

Abstract: Disruptions of circadian motor behavior cause a significant burden for older adults as well as their caregivers and often lead to institutionalization. This cross-sectional study investigates the association between mobility-related behavior and subjectively rated circadian chronotypes in healthy older adults. The physical activity of 81 community-dwelling older adults was measured over seven consecutive days and nights using lower-back-worn hybrid motion sensors (MM+) and wrist-worn actigraphs (MW8). A 30-min and 120-min active period for the highest number of steps (MM+) and activity counts (MW8) was derived for each day, respectively. Subjective chronotypes were classified by the Morningness-Eveningness Questionnaire into 40 (50%) morning types, 35 (43%) intermediate and six (7%) evening types. Analysis revealed significantly earlier starts for the 30-min active period (steps) in the morning types compared to the intermediate types ($p \leq 0.01$) and the evening types ($p \leq 0.01$). The 120-min active period (steps) showed significantly earlier starts in the morning types compared to the intermediate types ($p \leq 0.01$) and the evening types ($p = 0.02$). The starting times of active periods determined from wrist-activity counts (MW8) did not reveal differences between the three chronotypes ($p = 0.36$ for the 30-min and $p = 0.12$ for the 120-min active period). The timing of mobility-related activity, i.e., periods with the highest number of steps measured by hybrid motion sensors, is associated to subjectively rated chronotypes in healthy older adults. The analysis of individual active periods may provide an innovative approach for early detecting and individually tailoring the treatment of circadian disruptions in aging and geriatric healthcare.

Keywords: circadian motor behavior; body-worn sensors; older adults

Citation: Fleiner, T.; Trumpf, R.; Hollinger, A.; Haussermann, P.; Zijlstra, W. Quantifying Circadian Aspects of Mobility-Related Behavior in Older Adults by Body-Worn Sensors—An "Active Period Analysis". *Sensors* **2021**, *21*, 2121. https://doi.org/10.3390/s21062121

Academic Editor: Giuseppe Vannozzi

Received: 14 February 2021
Accepted: 16 March 2021
Published: 18 March 2021

1. Introduction

Morning lark or night owl—what is your preferred time of the day? The growing knowledge of and interest in the impact of circadian rhythms in daily life refers to circadian medicine [1], where individual chronotypes and circadian characteristics play a key role in society and health care [2].

Physiological processes and behaviors synchronized to a 24 h structure are defined as circadian (lat. circa = approximately, -dian = day) [3,4]. The stability of circadian behaviors is especially relevant in older adults and geriatric health care, where aspects of circadian behavior may show deviations ranging from age-associated changes in subjective chronotypes [5] to clinical syndromes [6]. Disease-related changes of the circadian system occur, for example, as sleep disturbances with reversed day-night rhythms [7], or sundowning phenomena with increased levels of physical activity (PA) and behavioral disturbances in the afternoon and evening hours [8,9]. Disturbed circadian rhythms cause a significant burden for both the patients themselves as well as their caregivers [10] and often lead to institutionalization, especially in home-dwelling dementia care [11].

Within chronobiological research and geriatric sleep medicine, subjective or proxy-based psychopathometric instruments [12,13] and objective approaches like polysomnography or body-worn motion sensors are usually applied to assess circadian characteristics [14]. Most commonly, uni- and multi-axial accelerometers attached to the non-dominant wrist, so called actigraphs [15], are used as ambulatory assessment to quantify circadian motor behavior. The accumulated raw activity counts of wrist movements are analyzed by non-parametric methods—e.g., by deriving an acrophase that refers to the timing of the peak activity within an day [16], or analyzing the intradaily variability, and the interdaily stability of the counts per minute [17]. As these actigraphs only record wrist movements, these measurements and analyses provide a general assumption of temporal aspects of PA and do not enable to detect specific motor behavior patterns. Studies in geriatric care and investigations in community-dwelling older adults indicate the wrist activity to be independent of the distribution of the step count [18,19]. Therefore, such actigraphic measurements do not allow to derive personalized interventions, e.g., like physical activity programs scheduled as circadian zeitgebers [20,21].

Hybrid motion sensors attached to the lower back can detect the patients' body postures over several days, allowing to analyze individual mobility patterns concerning mobility-related behavior [22]. First studies conducted with older adults have investigated inter-daily walking duration and step count with sensors attached to the participants' trunk or thigh [23,24]. Up to now, only a few approaches have been developed and applied to investigate the temporal distribution 65 of mobility-related PA in older adults. For example, the investigation led by Lim [18] analyzed the gait activity during the day using the number of active minutes ($\geq$4 steps per minute), and the study reported by Paraschiv-Ionescu [25] analyzed the complexity of motor behavior by barcoding the participants' motor behavior during the day. Both studies used sensor-based approaches to monitor mobility-related physical activity but did not address chronotypes and circadian aspects of motor behavior. As these mobility-related measurements promise an added value over wrist-worn actigraphs for use in diagnostics and treatment, the primary aim of this study is to investigate the association between the timing of mobility-related active periods and subjectively rated chronotypes in healthy older adults.

2. Materials and Methods

2.1. Study Design

This investigation was part of the ChronoSense project—a cross-sectional study to investigate the use of body-worn motion sensors to quantify chronotypes in older adults (DRKS00015069, German clinical trial register). The study protocol was approved by the Ethics Committee of the Medical Association North Rhine (registration number 2018192) and the Ethics Committee of the German Sport University Cologne (registration number 156/2017).

2.2. Participants

Participants were recruited by sending out emails with information brochures to local senior citizens' networks and to employees of a large municipal association in the Rhineland region in Germany, and through word-of-mouth referrals. Furthermore, invitation letters were sent to persons who expressed interest in participating in studies of the Institute of Movement and Sport Gerontology in the past. These persons had not participated in any studies in the previous year.

Inclusion criteria for participation in the project were as follows: age of 65 years or older, community-dwelling, a score on the Mini-Mental Status Examination (MMSE) $\geq$24 [26,27], subjective health (self-reported), no full-time employment and written informed consent regarding the study procedures. Any acute or severe mobility impairment, cardiovascular disorder, cognitive disorder or neurological disease (assessed with the

Functional Comorbidity Index (FCMI)) [28], which can interfere with functional mobility, led to exclusion from the project.

2.3. Instruments

The self-estimation of chronotype was assessed using the Morning-Eveningness Questionnaire (MEQ) [29]. The MEQ is a self-administered questionnaire, determining the circadian chronotype based on 19 questions concerning the participant's usual daytime preferences. Five chronotypes are distinguished based on the total score of the MEQ: definite evening type (16–30 points), moderate evening type (31–41 points), intermediate type (42–58 points), moderate morning type (59–69 points) and definite morning type (70–86 points). For further analysis, the moderate and definite evening type as well as the moderate and definite morning type were each grouped together. In order to determine the waking time during the day, the participants were asked to log their get up and got to bed times in a sleep diary [30].

The wrist-worn MotionWatch 8 (Camntech, Cambridge, UK) was used for the actigraphy-assessments. It integrates a triaxial accelerometer, a light sensor and an event marker button. The MotionWatch 8 (MW8) was attached to the wrist of the participants' non-dominant hand. The participants were asked to push the event marker button when getting up and going to bed. The sample frequency of the accelerometer was 50 Hz. The raw acceleration measurements were processed by the on-board software of the MW8 to produce a quantitative measure of the activity during each epoch. For this, the X, Y and Z-axes of the accelerometer were sampled with filtering at 3–11 Hz. The peak $X^2 + Y^2 + Z^2$ value was tracked. At the end of each second, the square root of the peak value from that second was calculated. This was compared to a threshold of 0.1 g. Values below this threshold were ignored to simplify the final activity graph. Activity that caused the acceleration signal to exceed the threshold was counted as activity. At the end of each epoch of 60 s, the number of activity counts were accumulated. This value was recorded as the 'Tri-Axial count' for the epoch.

The mobility-related measurements were conducted using the Dynaport Move Monitor + (MM+; McRoberts, The Hague, NL). The MM+ consists of a triaxial accelerometer, a triaxial gyroscope (sample frequency for both sensors: 100 Hz), a triaxial magnometer, a barometer and a temperature sensor. Data can be collected for up to seven consecutive days. In order to enable a consistent recording of PA, waterproof self-adhesive foil (Opsite Flexifix, Smith and Nephew, London, UK) was used to attach the MM+ to the participants' lower back, approximately 3 cm right to the fifth vertebra of the lumbar spine (L5). The participants were asked not to remove either sensor during the measurement period. To ensure an assessment of habitual awake and rest phases, only sensor data of participants with four or more complete measurement days were included.

2.4. Data Collection

Data collection covered the period of one week. During an individual appointment in the laboratory, the MEQ was administered and participants were equipped with the two sensors and received the sleep diary. Furthermore, the participants' living situation (e.g., marital status, income) as well as their health status (e.g., number and kind of chronic diseases) were assessed using a custom-made questionnaire.

At the end of the measurement period, the sensors were removed from the participants' body. The participants were asked to specify whether or not they had removed one or both sensors during the measurement and to indicate the period if this was the case. In order to ensure that the measurement period represented a habitual week in terms of the participant's PA and wake and rest periods, special events (e.g., acute illness) were noted.

2.5. Data Processing and Statistical Analysis

The MW8 raw data were processed using the validated proprietary MotionWare software (CamNtech, Fenstanton, UK). Total counts per 60 s epochs as well as the getting up and bedtimes based on the event markers set by the participants were included in the output. Average counts per minute were calculated for 24 h. The duration of wakefulness (time from getting up to bedtime) for each day was calculated. In case a participant forgot to set the marker, the corresponding time from the sleep diary was used instead.

The MM+ raw data were processed using the validated manufacturer's own algorithm (MoveMonitor, McRoberts, The Hague, NL). PA category (walking, stair walking, cycling, shuffling, standing, sitting and lying) as well as the categories not-worn, activity duration and number of steps per 60 s epoch were provided within the output. For the description of this study sample, the average durations of PA types and total number of steps were calculated for 24 h.

In order to quantify circadian aspects of mobility-related behavior, we determined an active period for each day. The active period was defined as the time interval in which the highest PA was measured during wakefulness (from getting up to bedtime). For the MW8, the active period was determined based on activity counts, and the active period of the MM+ was determined based on the number of steps. According to the recommendations of the American College of Sports Medicine to be active for a minimum of 30 min per day on five days per week [31], we chose to determine the active period for an interval of 30 min. Referring to the MEQ, rating the best time of the day for performing two hours of physically hard work, we chose to determine a 120 min active period [29]. Matlab R2020a (The Mathworks, Natick, MA, USA) was used to detect the time of the beginning of each active period. To this purpose, the total number of steps or counts over a time interval of 30 or 120 min was calculated repeatedly, starting with the first available data (when participants got up) and repeated by shifting the start of the time intervals to each next minute. This was repeated until the very last interval (30 or 120 min before the participant went to bed). Subsequently, all intervals were sorted in ascending order and the interval with the highest value (number of steps or number of counts) was determined as the active period. As the sensor measurements were started at 8 pm on day one and ended at 8 pm on day 8, the active periods were analyzed for day two (getting up) to day seven (going to bed). Finally, the mean start times of the 30-min and 120-min active phases were determined for each participant.

IBM SPSS Statistics 26.0 for Windows (International Business Machines, Armonk, NY, USA) was used for statistical analysis. Boxplots were used to identify extreme values. Values of more than three times the interquartile distance were excluded from further analysis. Subsequently, the Kolmogorov Smirnov test was used to examine data for normal distribution. One-way analyses of variance (ANOVAs) or Kruskal–Wallis tests were performed to assess differences in the start times of the active period between the three groups. Bonferroni post-hoc tests were used to examine significant differences. An alpha < 0.05 was considered to be statistically significant.

3. Results

3.1. Participants

A total of 118 persons were screened with regard to the inclusion criteria. Twenty-three persons declined to participate; 10 persons did not fit to the inclusion criteria. Eighty-five persons agreed to participate. Two participants became ill during the measurement period and were excluded from data analysis. One participant indicated that he was less active than usual during the measurement period, and one participant did not wear the sensors according to the instructions. The data of these two participants were excluded from analysis. Finally, the data of 81 participants were analyzed. Table 1 shows their characteristics.

Table 1. Sample characteristics.

		N (%)	Mean	SD	Min	Max
Sample		81				
	Female	40 (49.4)				
Age (years)			71.5	5.0	65	84
Mass (kg)			76.9	15.4	54	119
Height (cm)			170.3	8.3	154	188
MEQ score			57.7	9.9	31	77
Number of diseases			2.0	1.4	0	7
Duration of wakefulness (h)			15.9	0.8	13.5	18.1
Move monitor+		75				
Activity/posture [hh:mm]/24 h						
lying			09:01	01:38	06:18	14:29
sitting			09:13	01:53	05:08	14:13
standing			03:02	00:48	01:02	04:48
shuffling			00:29	00:10	00:11	01:11
walking			01:57	00:36	00:44	03:51
other activities *			00:05	00:11	00:00	00:59
not worn			00:13	00:25	00:00	03:07
steps/24 h			9860.1	3279.9	3278.0	17,319.2
MotionWatch 8		66				
counts/min [24 h]			317.9	87.3	121.6	556.9

MEQ—Morningness-Eveningness Questionnaire (assessment of subjective chronotypes; scores can range from 16–86. Scores of 41 and below indicate "evening types". Scores of 59 and above indicate "morning types". Scores between 42 and 58 indicate "intermediate types"); * summation of total activity durations for cycling and stair walking.

MM+ data were available for 75 (93%) participants. The MM+ data of six participants (7%) were missing due to technical problems. Six complete measurement days were available for 67 participants (83%). Seven participants (9%) completed five measurement days. One participant (1%) completed the minimum requirement of four measurement days. MW8 data were available for 66 (82%) participants. MW8 data of 15 (18%) participants were missing due to technical problems. Six complete measurement days were available for all 66 participants.

The distribution of self-estimated chronotypes and the corresponding characteristics of subgroups is shown in Table 2. Statistical analysis revealed no significant differences between groups in the sample characteristics and their general level of PA.

Table 2. Sample characteristics of self-estimated chronotype subgroups.

		Morning Type			Intermediate Type			Evening Type			*p*
		N (%)	M	SD	N (%)	M	SD	N (%)	M	SD	
Sample		40 (49.4)			35 (43.2)			6 (7.4)			
	female	15 (37.5)			22 (62.9)			3 (50.0)			
Age (years)			72.1	5.2		71.4	5.1		67.8	3.1	0.10
Wakefulness (h/day)			16.1	0.6		15.8	0.8		15.9	0.8	0.55
MM+ Steps (number/24 h)		37	9791.2	3157.1	32	9823.9	3101.1	6	10,478.4	4310.3	0.98
MW8 Counts/min (24 h)		30	316.8	73.9	30	319.1	103.2	6	317.9	40.0	0.98

MM+ Move Monitor+, MW8 MotionWatch 8.

3.2. Active Period Analysis

Figure 1 shows the comparison of the summed 30 min time intervals of the number of steps per minute for one subject from each chronotype group. The time of day at the peak of each curve indicates the beginning of the 30 min active period based on the step count for one participant of each chronotype group.

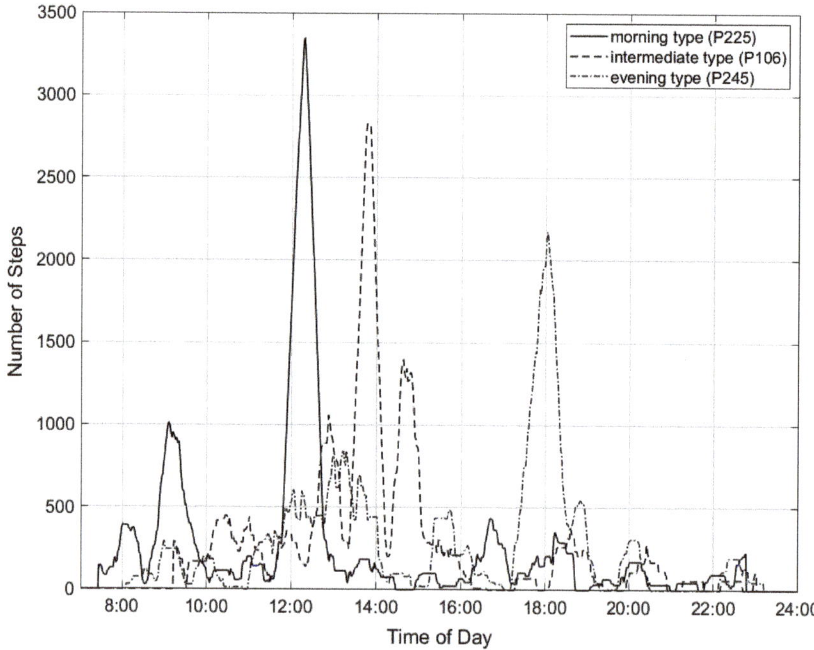

Figure 1. Exemplary analysis of 30-min active period.

Significant differences within the circadian aspects of the step count (MM+) between the three groups were detected (Figures 2 and 3).

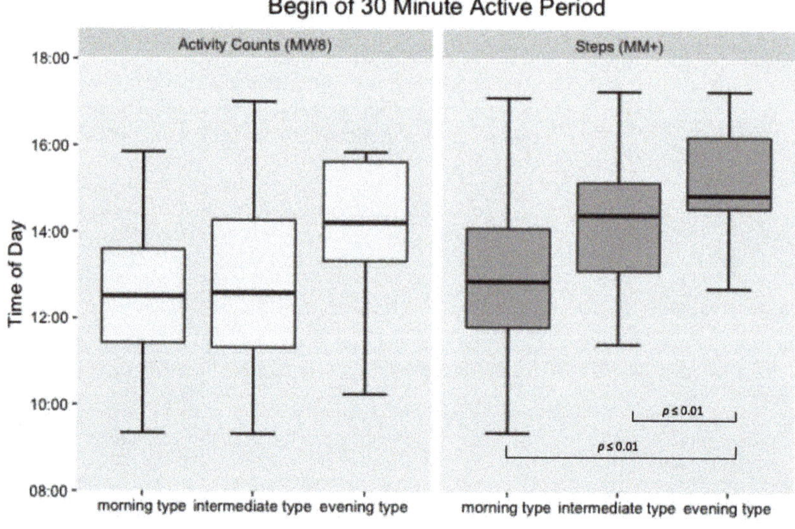

Figure 2. Box-plot illustration of 30-min active periods beginnings in relation to the subjectively rated chronotypes.

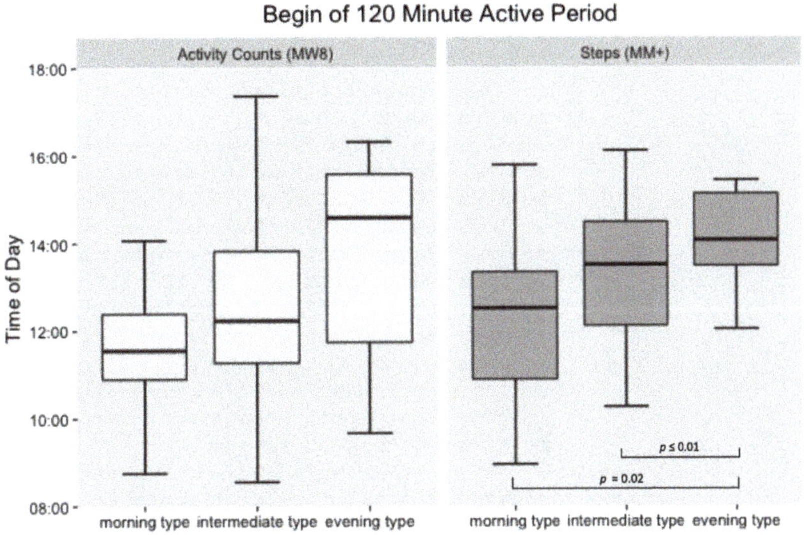

Figure 3. Box-plot illustration of 120-min active periods beginnings in relation to the subjectively rated chronotypes.

Compared to the morning type group, the intermediate type group showed a delay in their 30-min active period of approximately one hour ($p \leq 0.01$) and the evening type group a delay of approximately two hours ($p \leq 0.01$), respectively. These results were also found for the 120-min active period. Compared to the morning type group, the intermediate type group showed a delay of approximately one hour ($p \leq 0.01$) and the evening type group a delay of approximately two hours ($p = 0.02$), respectively. No differences within the circadian aspects of the activity counts (MW8) were detected regarding the three chronotype groups for both the 30- and 120-min active periods.

4. Discussion

The primary aim of this study was to investigate the association between the timing of mobility-related active periods and subjectively rated chronotypes in healthy older adults. The analysis revealed significant differences in the starting times for the 30-min and 120-min active period (steps) between the chronotypes. The starting times of the active periods regarding the wrist-activity counts did not reveal differences between the three chronotype groups.

The "active-period" analysis is a novel approach in this field of research. Whereas the usually applied wrist-worn actigraph approach showed no differences in activity-related behaviors over the three chronotype groups, this study's results reveal the hybrid motion sensor to be able to quantify circadian aspects of mobility-related behavior, i.e., regarding the number of steps. The timing of the peak wrist activity within a day, usually reported as acrophase for wrist-worn actigraphy [16], seems to be independent of the timing of the peak gait activity, reported as active period. These differences between objectively measured wrist-based activity (counts) and mobility-related behavior (steps, postures) are comparable to previous results [19]. Additionally, studies applied in an acute geriatric care setting reported three peaks in the wrist-measured physical activity at 9 am, 12 pm and 5 pm, referring to the patients' meal times and showing no relation to the distribution of the step count within the patients' day [18]. As compared to the usually used wrist-worn actigraphy approach in chronobiological research and geriatric sleep medicine, analyzing the active period of mobility-related behavior seems to provide more clinically relevant and essential information. Such circadian aspects of mobility-related behavior could be applied to assess circadian disruptions based on the temporal distribution of the step count within a day and subsequently derive, individually tailor and evaluate interventions to treat circadian disruptions, e.g., exercise approaches based on step counts [20,21,32].

The participants included in this study were healthy, community-dwelling older adults, on average 72 years old (SD 5), with a daily step count ranging from 3278 to 17,320 with on average 9860 steps per day (SD 3280). These activity measures reveal a general active lifestyle, as the endorsed level of 7000–10,000 steps per day was almost achieved in this group [33]. With a mean of 317 activity counts per minute [24 h] (SD 87), the study sample shows comparable levels to other studies using the same actigraphy approach [34]. The included participants subjectively rated themselves mainly as morning type ($n = 40$, 49.4%) and intermediate type ($n = 35$, 43.2%), but only six participants rated themselves as evening type (7.4%). This distribution of chronotypes is comparable to previous studies, which reported more morning types in association with higher age [5].

An analysis and interpretation of this study and its results should consider the following methodological limitations: established by Horne and Ostberg [29], the Morningness-Eveningness Questionnaire usually categorizes five chronotypes. The definite and moderate morning- and evening types were accumulated in order to analyze differences between the three main chronotypes. The current analysis did not reflect inter-daily consistency of active periods. Future analysis should address these aspects, e.g., via coefficient of variance. A potential selection bias should be taken into account, as the sample has been shown to be highly active with approximately 10,000 steps per 24 h.

The results of this study contribute to the growing knowledge and interest on the impact of circadian rhythms in daily life and healthcare [1,2]. Analyzing the starting times of the active periods for mobility-related behavior, e.g., by the number of steps measured by hybrid motion sensors (MM+), seems to be a clinically relevant approach to quantify circadian aspects in healthy older adults. Analyzing circadian aspects of mobility-related activity, and potentially also temporal patterns of inactivity, could play a key role in aging research and geriatric healthcare, especially in the assessment and treatment of circadian disruptions. In addition to the presented results of not showing differences in the assessment of active periods, the wrist-worn actigraphy approach (here MW8) does not allow to derive, apply and evaluate individually tailored interventions. This is essential for its clinical application, and therefore limits its use in general and especially in geriatric healthcare [19]. The presented "active period analysis" provides an innovative and clinically relevant approach to quantify circadian aspects of mobility-related behavior with body-worn sensors in older adults. Especially in patients suffering from circadian disruptions, an individual (in)active period could be used to derive, apply and evaluate step-based interventions [35] potentially combined with day-structuring approaches. The individual active period analysis may improve the early detection and individual tailoring in the treatment of circadian disruptions in aging and geriatric healthcare that may have promising effects for patients, caregivers and geriatric healthcare [1,2].

Author Contributions: Conceptualization, T.F., R.T., P.H. and W.Z.; Data assessment: R.T. and A.H.; Data processing: R.T. and A.H.; Analysis and interpretation: T.F., R.T. and A.H.; Writing—Original Draft Preparation, T.F. and R.T. Writing—Review and Editing, T.F., R.T., A.H., P.H. and W.Z.; Supervision, T.F., P.H., and W.Z. Funding Acquisition, T.F. All authors have read and agreed to the published version of the manuscript.

Funding: This analysis is part of the Chronosense project, supported by the Paul-Kuth-Stiftung (Wuppertal). This funding had no influence on the study design, collection, analysis and interpretation of the data.

Institutional Review Board Statement: The study was conducted according to the guidelines of the Declaration of Helsinki and has been approved by the Ethics Committee of the German Sport University Cologne (registration number 156/2017) and the Ethics Committee of the Medical Chamber Northrhine (registration number 2018192).

Informed Consent Statement: Informed consent was obtained from all subjects involved in the study.

Data Availability Statement: The data presented in this study are available on request from the corresponding author, upon reasonable request. The data are not publicly available due to privacy/ethical restrictions.

Acknowledgments: The authors would like to thank all participants.

Conflicts of Interest: The authors declare no conflict of interest.

References

1. Colwell, C.S. *Circadian Medicine*; John Wiley & Sons Inc.: Hoboken, NJ, USA, 2015.
2. Van Someren, E.J.W.; Riemersma-Van Der Lek, R.F. Live to the Rhythm, Slave to the Rhythm. *Sleep Med. Rev.* **2007**, *11*, 465–484. [CrossRef]
3. Van Someren, E.J.W. Circadian and Sleep Disturbances in the Elderly. *Exp. Gerontol.* **2000**, *35*, 1229–1237. [CrossRef]
4. Roenneberg, T.; Merrow, M. The Circadian Clock and Human Health. *Curr. Biol.* **2016**, *26*, R432–R443. [CrossRef]
5. Huang, Y.-L.; Liu, R.-Y.; Wang, Q.-S.; Van Someren, E.J.; Xu, H.; Zhou, J.-N. Age-Associated Difference in Circadian Sleep–Wake and Rest–Activity Rhythms. *Physiol. Behav.* **2002**, *76*, 597–603. [CrossRef]
6. Van Someren, E.J.W. Circadian Rhythms and Sleep in Human Aging. *Chronobiol. Int.* **2000**, *17*, 233–243. [CrossRef]
7. Harper, D.G. Sleep and Circadian Rhythm Disturbances in Alzheimer's Disease. In *Principles and Practice of Geriatric Sleep Medicine*; Pandi-Perumal, S.R., Monti, J.M., Monjan, A.A., Eds.; Cambridge University Press: Cambridge, UK, 2010; pp. 214–226.
8. Coogan, A.N.; Schutová, B.; Husung, S.; Furczyk, K.; Baune, B.T.; Kropp, P.; Häßler, F.; Thome, J. The Circadian System in Alzheimer's Disease: Disturbances, Mechanisms, and Opportunities. *Biol. Psychiatry* **2013**, *74*, 333–339. [CrossRef]
9. Boronat, A.C.; Ferreira-Maia, A.P.; Wang, Y.P. Sundown Syndrome in Older Persons: A Scoping Review. *J. Am. Med. Dir. Assoc.* **2019**, 664–671.e5. [CrossRef] [PubMed]
10. Landry, G.J.; Liu-Ambrose, T. Buying Time: A Rationale for Examining the Use of Circadian Rhythm and Sleep Interventions to Delay Progression of Mild Cognitive Impairment to Alzheimer's Disease. *Front. Aging Neurosci.* **2014**, *6*, 325. [CrossRef] [PubMed]
11. Bianchetti, A.; Scuratti, A.; Zanetti, O.; Binetti, G.; Frisoni, G.B.; Magni, E.; Trabucchi, M. Predictors of Mortality and Institutionalization in Alzheimer Disease Patients 1 Year after Discharge from an Alzheimer Dementia Unit. *Dement. Geriatr. Cogn. Disord.* **1995**, *6*, 108–112. [CrossRef] [PubMed]
12. Di Milia, L.; Adan, A.; Natale, V.; Randler, C. Reviewing the Psychometric Properties of Contemporary Circadian Typology Measures. *Chronobiol. Int.* **2013**, *30*, 1261–1271. [CrossRef]
13. Cummings, J.L.; Mega, M.; Gray, K.; Rosenberg-Thompson, S.; Carusi, D.A.; Gornbein, J. The Neuropsychiatric Inventory: Comprehensive Assessment of Psychopathology in Dementia. *Neurology* **1994**, *44*, 2308–2314. [CrossRef]
14. Ancoli-Israel, S.; Cole, R.; Alessi, C.A.; Chambers, M.; Moorcroft, W.; Pollak, C.P. The Role of Actigraphy in the Study of Sleep and Circadian Rhythms. *Sleep* **2003**, *26*, 342–392. [CrossRef] [PubMed]
15. Morgenthaler, T.; Alessi, C.; Friedman, L.; Owens, J.; Kapur, V.; Boehlecke, B.; Brown, T.; Chesson, A.; Coleman, J.; Lee-Chiong, T.; et al. Practice Parameters for the Use of Actigraphy in the Assessment of Sleep and Sleep Disorders: An Update for 2007. *Sleep* **2007**, *30*, 519–529. [CrossRef]
16. Van Someren, E.J.W. Actigraphic Monitoring of Sleep and Circadian Rhythms. In *Handbook of Clinical Neurology*; Montagna, P., Chokroverty, S., Eds.; Elsevier: Amsterdam, Netherlands, 2011; Volume 98, pp. 55–63. [CrossRef]
17. Witting, W.; Kwa, I.H.; Eikelenboom, P.; Mirmiran, M.; Swaab, D.F. Alterations in the Circadian Rest-Activity Rhythm in Aging and Alzheimer's Disease. *Biol. Psychiatry* **1990**, *27*, 563–572. [CrossRef]
18. Lim, S.E.R.; Dodds, R.; Bacon, D.; Sayer, A.A.; Roberts, H.C. Physical Activity among Hospitalised Older People: Insights from Upper and Lower Limb Accelerometry. *Aging Clin. Exp. Res.* **2018**, *30*, 1363–1369. [CrossRef] [PubMed]
19. Trumpf, R.; Zijlstra, W.; Haussermann, P.; Fleiner, T. Quantifying Habitual Physical Activity and Sedentariness in Older Adults—Different Outcomes of Two Simultaneously Body-Worn Motion Sensor Approaches and a Self-Estimation. *Sensors* **2020**, *20*, 1877. [CrossRef]
20. Baldacchino, F.V.; Pedrinolla, A.; Venturelli, M. Exercise-Induced Adaptations in Patients with Alzheimer's Disease: The Role of Circadian Scheduling. *Sport Sci. Health* **2018**, *14*, 227–234. [CrossRef]
21. Fleiner, T.; Zijlstra, W.; Dauth, H.; Haussermann, P. Evaluation of a Hospital-Based Day-Structuring Exercise Programme on Exacerbated Behavioural and Psychological Symptoms in Dementia—The Exercise Carrousel: Study Protocol for a Randomised Controlled Trial. *Trials* **2015**, *16*, 228. [CrossRef]
22. Zijlstra, W.; Becker, C.; Pfeiffer, K. Wearable Systems for Monitoring Mobility Related Activities. In *E-Health, Assistive Technologies and Applications for Assisted Living*; Röcker, C., Ziefle, M., Eds.; IGI Global: Hershey, PA, USA, 2011; pp. 244–267. [CrossRef]
23. Klenk, J.; Peter, R.S.; Rapp, K.; Dallmeier, D.; Rothenbacher, D.; Denkinger, M.; Büchele, G.; Group, S.; Becker, T.; Böhm, B.; et al. Lazy Sundays: Role of Day of the Week and Reactivity on Objectively Measured Physical Activity in Older People. *Eur. Rev. Aging Phys. Act.* **2019**, *1*, 16–19.
24. Abel, B.; Pomiersky, R.; Werner, C.; Lacroix, A.; Schäufele, M.; Hauer, K. Day-to-Day Variability of Multiple Sensor-Based Physical Activity Parameters in Older Persons with Dementia. *Arch. Gerontol. Geriatr.* **2019**, *85*, 103911. [CrossRef] [PubMed]

25. Paraschiv-Ionescu, A.; Mellone, S.; Colpo, M.; Bourke, A.; Ihlen, E.A.F.; el Achkar, C.M.; Chiari, L.; Becker, C.; Aminian, K. Patterns of Human Activity Behavior. In *Proceedings of the 2016 ACM International Joint Conference on Pervasive and Ubiquitous Computing Adjunct—UbiComp '16*; Lukowicz, P., Krüger, A., Bulling, A., Lim, Y.-K., Patel, S.N., Eds.; ACM Press: New York, NY, USA, 2016; pp. 841–845. [CrossRef]
26. Folstein, M.F.; Folstein, S.E.; McHugh, P.R. "Mini-Mental State". A Practical Method for Grading the Cognitive State of Patients for the Clinician. *J. Psychiatr. Res.* **1975**, *12*, 189–198. [CrossRef]
27. Creavin, S.T.; Wisniewski, S.; Noel-Storr, A.H.; Trevelyan, C.M.; Hampton, T.; Rayment, D.; Thom, V.M.; Nash, K.J.E.; Elhamoui, H.; Milligan, R.; et al. Mini-Mental State Examination (MMSE) for the Detection of Dementia in Clinically Unevaluated People Aged 65 and over in Community and Primary Care Populations. *Cochrane Database Syst. Rev.* **2016**, *2016*, CD011145. [CrossRef]
28. Groll, D.; To, T.; Bombardier, C.; Wright, J. The Development of a Comorbidity Index with Physical Function as the Outcome. *J. Clin. Epidemiol.* **2005**, *58*, 595–602. [CrossRef] [PubMed]
29. Horne, J.A.; Ostberg, O. A Self-Assessment Questionnaire to Determine Morningness-Eveningness in Human Circadian Rhythms. *Int. J. Chronobiol.* **1976**, *4*, 97–110. [PubMed]
30. Hoffmann, R.M.; Müller, T.; Hajak, G.; Cassel, W. Abend-Morgenprotokolle in Schlafforschung Und Schlafmedizin—Ein Standardinstrument Für Den Deutschsprachigen Raum. *Somnologie* **1997**, *1*, 103–109. [CrossRef]
31. Chodzko-Zajko, W.J.; Proctor, D.N.; Fiatarone Singh, M.A.; Minson, C.T.; Nigg, C.R.; Salem, G.J.; Skinner, J.S. Exercise and Physical Activity for Older Adults. *Med. Sci. Sport. Exerc.* **2009**, *41*, 1510–1530. [CrossRef]
32. Venturelli, M.; Sollima, A.; Cè, E.; Limonta, E.; Bisconti, A.V.; Brasioli, A.; Muti, E.; Esposito, F. Effectiveness of Exercise- and Cognitive-Based Treatments on Salivary Cortisol Levels and Sundowning Syndrome Symptoms in Patients with Alzheimer's Disease. *J. Alzheimer's Dis.* **2016**, *53*, 1631–1640. [CrossRef]
33. Tudor-Locke, C.; Craig, C.L.; Aoyagi, Y.; Bell, R.C.; Croteau, K.A.; De Bourdeaudhuij, I.; Ewald, B.; Gardner, A.W.; Hatano, Y.; Lutes, L.D.; et al. How Many Steps/Day Are Enough? For Older Adults and Special Populations. *Int. J. Behav. Nutr. Phys. Act.* **2011**, *8*, 80. [CrossRef]
34. Landry, G.J.; Falck, R.S.; Beets, M.W.; Liu-Ambrose, T. Measuring Physical Activity in Older Adults: Calibrating Cut-Points for the MotionWatch 8©. *Front. Aging Neurosci.* **2015**, *7*, 165. [CrossRef]
35. Larsen, R.T.; Wagner, V.; Keller, C.; Juhl, C.B.; Langberg, H.; Christensen, J. Physical Activity Monitors to Enhance Amount of Physical Activity in Older Adults—A Systematic Review and Meta-Analysis. *Eur. Rev. Aging Phys. Act.* **2019**, *16*, 53. [CrossRef]

Letter

Maximal Walking Distance in Persons with a Lower Limb Amputation

Cheriel J. Hofstad [1,*], Kim T.J. Bongers [1], Mark Didden [1,2], René F. van Ee [3] and Noël L.W. Keijsers [1,4]

[1] Department of Research & Innovation, Sint Maartenskliniek, 6500 GM Nijmegen, The Netherlands; k.bongers@interzorgthuiszorg.nl (K.T.J.B.); m.didden@tolbrug.nl (M.D.); n.keijsers@maartenskliniek.nl (N.L.W.K.)
[2] Department of Rehabilitation, Tolbrug, 5223 GZ Den Bosch, The Netherlands
[3] Department of Rehabilitation, Sint Maartenskliniek, 6500 GM Nijmegen, The Netherlands; r.vanee@maartenskliniek.nl
[4] Department of Rehabilitation, Donders Institute for Brain, Cognition and Behaviour, Radboud University Medical Center, 6525 AJ Nijmegen, The Netherlands
* Correspondence: c.hofstad@maartenskliniek.nl; Tel.: +31-24-3659329

Received: 30 October 2020; Accepted: 24 November 2020; Published: 26 November 2020

Abstract: The distance one can walk at a time could be considered an important functional outcome in people with a lower limb amputation. In clinical practice, walking distance in daily life is based on self-report (SIGAM mobility grade (Special Interest Group in Amputee Medicine)), which is known to overestimate physical activity. The aim of this study was to assess the number of consecutive steps and walking bouts in persons with a lower limb amputation, using an accelerometer sensor. The number of consecutive steps was related to their SIGAM mobility grade and to the consecutive steps of age-matched controls in daily life. Twenty subjects with a lower limb amputation and ten age-matched controls participated in the experiment for two consecutive days, in their own environment. Maximal number of consecutive steps and walking bouts were obtained by two accelerometers in the left and right trouser pocket, and one accelerometer on the sternum. In addition, the SIGAM mobility grade was determined and the 10 m walking test (10 MWT) was performed. The maximal number of consecutive steps and walking bouts were significantly smaller in persons with a lower limb amputation, compared to the control group ($p < 0.001$). Only 4 of the 20 persons with a lower limb amputation had a maximal number of consecutive steps in the range of the control group. Although the maximal covered distance was moderately correlated with the SIGAM mobility grade in participants with an amputation ($r = 0.61$), for 6 of them, the SIGAM mobility grade did not match with the maximal covered distance. The current study indicated that mobility was highly affected in most persons with an amputation and that the SIGAM mobility grade did not reflect what persons with a lower limb amputation actually do in daily life. Therefore, objective assessment of the maximal number of consecutive steps of maximal covered distance is recommended for clinical treatment.

Keywords: inertial measurement units; walking distance; lower limb amputation; rehabilitation; gait

---

1. Introduction

The two primary concerns for people with a lower limb amputation are mobility [1,2] and wearing comfort of the prosthesis, in which mobility is most relevant for their quality of life [3,4]. However, many persons with a lower limb amputation report that they are unable to use their prosthesis to the extent they desire [2] and, moreover, they lose their independence [5,6]. To function independently, one should be able to walk sufficient bouts. Therefore, in the context of independency, the walking distance one can walk consecutively could be considered as an important outcome in persons with a lower limb amputation. In clinical practice, the self-reporting SIGAM (Special Interest Group in Amputee Medicine) mobility grades [4] are often used to classify prosthetic users. The SIGAM mobility grades describe a single-item scale comprising six clinical grades (A–F) of amputee mobility, and the scale consists of 21 'yes'/'no' items. The SIGAM mobility grades include a walking distance item; a threshold of 50 m at a time is used as a benchmark to denote an improvement of mobility [4] and reflects sufficient independency. It is known, however, that people tend to overestimate their physical activity when self-report measures are used [7–9]. As the SIGAM mobility grade is a self-reporting questionnaire, it is very likely that the activity levels are overestimated in the SIGAM mobility grades. This results in false positive outcomes, also known as bias towards independency. Since clinical interventions, like prosthetic fitting, are partially based on questionnaires assessing functional level [10], it is conceivable that clinical care might be subject to bias or subjectivity.

In contrast to self-reported measures (diaries and questionnaires [11–14]), there are also technical approaches that were used to assess prosthetic mobility. All techniques differ in the type and number of mobility aspects they measure, ranging from categories of ambulation to prosthetic use over a variety of ambulation activities [12] and performance tests in laboratory settings [15]. Another more objective way to measure mobility is the use of activity monitors [16–24]. The advantage of activity monitors is that they can measure long-term and continuously in a person's own environment, and assess what persons with a lower limb amputation actually do, in a reliable and valid way [25]. Although it was demonstrated that persons with a lower limb amputation are significantly less physically active compared to the age-matched controls [17,18], none of the studies focused on the length of walking bouts and the number of consecutive steps in these bouts.

The aim of this study was to assess the number of consecutive steps and walking bouts in persons with a lower limb amputation and age-matched controls in daily life, using an accelerometer sensor. We hypothesized that the maximal number of consecutive steps was correlated to the level of the SIGAM mobility grades. We were particularly interested in whether physically active or independent persons with a lower limb amputation (SIGAM mobility grade D or higher) covered longer distances than 50 m during walking bouts, which is an important benchmark for mobility, as stated by Ryall [4]. We also assessed the relationship between the SIGAM mobility grade, maximal covered distance and preferred walking velocity, to indicate the effect of gait capacity on physical functioning. Age-matched subjects were included for comparison.

2. Materials and Methods

2.1. Subjects

Patients were recruited from the Prosthetics and Orthotics Centre in Nijmegen and from the prosthetic training group at the rehabilitation clinic Sint Maartenskliniek in Nijmegen, The Netherlands. Persons with a lower limb amputation were included when they had a unilateral transfemoral or transtibial amputation or knee exarticulation, were at least 18 years old, and had no cognitive disorders. They had to be free from neurological and clinical orthopedic problems (other than the amputation), without stump pain, stump wounds, and foot pathology, which could influence their daily activities. A control group of age-matched subjects without an amputation also participated in this study. All participants gave written informed consent in accordance with the Declaration of Helsinki. The study was approved by the internal review board of the Sint Maartenskliniek. The study was carried out in the Netherlands, in accordance with the applicable rules concerning the review of research ethics committees, and did not fall within the remit of Medical Research Involving Human Subjects Act.

2.2. Accelerometers

The accelerometers (62 mm (length) × 41 mm (width) × 18 mm (height)) used in this study were tri-axial piezo-capacitive MiniMods from Dynaport (McRoberts BV, The Hague, The Netherlands).

The sample rate of the accelerometers was 100 Hz and data were stored on secure digital (SD) memory cards. Three accelerometers were used during the measurement. Two were placed in the left and right trouser pocket and one on the lower part of the sternum of the subject. The accelerometer that was placed on the sternum was attached with means of a 10 cm wide elastic band around the chest, to prevent irritation of the skin.

2.3. Protocol

The accelerometers were worn over two consecutive days in the participant's own environment. The researcher explained the measurement protocol, instructed the patient on how to attach and detach the accelerometers and administered the SIGAM mobility grade. The participants were instructed to perform their normal daily life activities during the two measurement days. At the end of both measurement days, the participant had to fill out a short questionnaire on whether the activities performed were representative for someone's usual daily activities. The accelerometers were not worn overnight.

To be able to calculate the maximal walking distance, step length was estimated by data form a 10-m walk test (10 MWT). In addition, the 10 MWT was also used to assess the preferred walking velocity, which is an excellent indicator of gait capacity. After attaching the accelerometers on the first day, the participant performed a 10 MWT. The participant was instructed to walk 10 m at his own comfortable pace. The start and finish of the 10 MWT were marked in the accelerometer data by pushing a remote button, which was connected to the accelerometers. The researcher timed the 10 MWT and counted the number of steps.

2.4. Data Analysis and Outcome Measures

The main outcome measure of this study was the number of steps a subject walked consecutively during the two measuring days. Walking could be well detected by accelerometers on a thigh [26] or trunk [27]. Two custom written algorithms were used (MATLAB 7.1, The Mathworks Inc, Natick, MA, USA). The first algorithm identified walking bouts, in which a subject was walking. A subject was considered as walking when the orientation of all three accelerometers was upright and there was sufficient movement of the sensors. As a measure for the movement of the sensor, we took the square root of the sum of squares of the derivative of the three orthogonal accelerometer signals [28]. Finally, the signals of the accelerometers should have a repetitive character, which was determined by the autocorrelation of the accelerometer signals. The second algorithm counted the number of consecutive steps within each walking bout. The number of steps for each walking bout was calculated by dividing the time of the walking bout by the step frequency of the walking bout, which was the dominant frequency in the auto correlation of the accelerometer signals. Subsequent walking bouts with an interval within 1 s were seen as a single walking bout.

All walking bouts were visually checked on time and steps, and the remaining data were visually screened for walking bouts missed by the algorithm. Walking bouts missed by the algorithm were added.

In addition to the number of consecutive steps within each walking bout, we were interested in the frequency of walking bouts per hour. Therefore, categories of walking bouts were created in bins of 5 steps (for the walking bouts in which 0 to 50 consecutive steps were walked), bins of 25 steps (from 50–100 consecutive steps), and bins of 100 steps (from 100–400 consecutive steps). These frequencies were determined for both the persons with a lower limb amputation and the elderly control group.

To estimate the maximal walking distance in the persons with a lower limb amputation, the maximal number of consecutive steps was multiplied by the individual step length, based on the 10 MWT. The individual step length was 10 m divided by the number of steps needed to accomplish the 10 MWT. This estimated maximal walking distance was compared with the specific answer on the walking distance questions of the SIGAM mobility grades ("Do you usually manage to walk more than 50 m (55 yards) at a time?").

2.5. Statistical Analysis

Differences in the group characteristics and results of the 10 MWT were calculated with a nonparametric independent samples test (Mann–Whitney test). To calculate the difference of the frequency per hour between the groups (persons with a lower limb amputation vs. the elderly control group) and walking bouts, a mixed model ANOVA was performed with persons with a lower limb amputation or the elderly control group as between-factor, and walking bout bins as the within-group factor. Spearman's rank correlation coefficients were calculated between 10 MWT, the SIGAM mobility grade, and the maximal covered walking distance, to indicate the relationship between gait capacity and physical functioning. Statistics were performed in SPSS 12.0.1 (SPSS Inc. Chicago, IL, USA). Differences were considered significant when $p < 0.05$.

3. Results

3.1. Participants

Twenty subjects with a lower limb amputation and ten age-matched controls participated in this study. See Table 1 for characteristics of both groups. Nineteen subjects had data on two complete measurement days. Eleven subjects generated data on only one complete day, because some participants failed to start or recharge the accelerometers adequately or due to technical problems. Mean measurement time for the complete days was 9:45 h ± 2:37 (SD) for the persons with a lower limb amputation and 11:20 h ± 1:40 (SD) for the elderly control group. All subjects, except one control subject who was sick during the measurement days, indicated that the measurement days were normal with regards to their standard daily activities.

Table 1. Characteristics of persons with a lower limb amputation and elderly controls, median (interquartile range).

Characteristic	Persons with A Lower Limb Amputation	Elderly Control Group	*p*-Value
Gender (M:F)	13:7	5:5	
Age (years)	68 (60–74)	76 (69–81)	0.43
Height (cm)	171 (165–179)	172 (168–175)	0.93
Weight (kg)	77 (67–85) *	78 (75–78)	0.83
Amputation level	TT $n = 9$ KE $n = 4$ TF $n = 7$	n.a.	
Reason for amputation	Traumatic $n = 6$ Vascular $n = 10$ Oncological $n = 2$ VOther $n = 2$	n.a.	
SIGAM mobility grade	B $n = 1$ C $n = 5$ D $n = 6$ E $n = 2$ F $n = 6$	n.a.	

TT = Transtibial amputation; KE = Knee exarticulation; TF = Transfemoral amputation; SIGAM mobility grades: B = Therapeutic use only for transfers, C = Walks on level ground less than or equal to 50 m with/without aids, D = Walks outdoor on level ground only, in good weather, more than 50 m with/without walking aid, and E = Walks more than 50 m. No aids, except in adverse terrain or weather, F = normal or near normal gait. * Body weight including the prosthesis. n.a. = not applicable.

3.2. Maximal Number of Consecutive Steps and 10 MWT

Table 2 shows the median and interquartile range of the maximal number of consecutive steps and the 10 MWT. For both the maximal consecutive steps and the 10 MWT, the elderly control group performed better than the persons with a lower limb amputation ($p < 0.001$ for the Mann–Whitney test).

Table 2. Median and IQR (interquartile range) of the outcome measures for the persons with a lower limb amputation and the elderly control group.

Variable	Persons with A Lower Limb Amputation $n = 20$	Elderly Control Group $n = 10$	*p*-Value
Maximal consecutive steps	141 (60–217)	883 (362–1168)	<0.001
10 MWT (s)	17.4 (10.3–25.5)	9.4 (9.1–10.6)	<0.001

Maximal Number of Consecutive Steps per Individual

The maximal number of consecutive steps was significantly larger in the elderly control group ($p < 0.001$, Table 2). However, some active persons with a lower limb amputation achieved similar maximal consecutive steps. Figure 1 shows the maximal number of consecutive steps for each subject. All elderly controls achieved more than 250 consecutive steps, except one. This elderly control subject reached a maximal of 94 steps, but reported on the activities questionnaire that she walked less than normal, due to illness. In contrast, only 4 of the 20 persons with a lower limb amputation achieved the 250 consecutive steps. Furthermore, eight persons with a lower limb amputation had even less than 100 consecutive steps. It was remarkable that one person with a lower limb amputation (SIGAM mobility grade F) revealed the highest maximal number of consecutive steps of almost 2500.

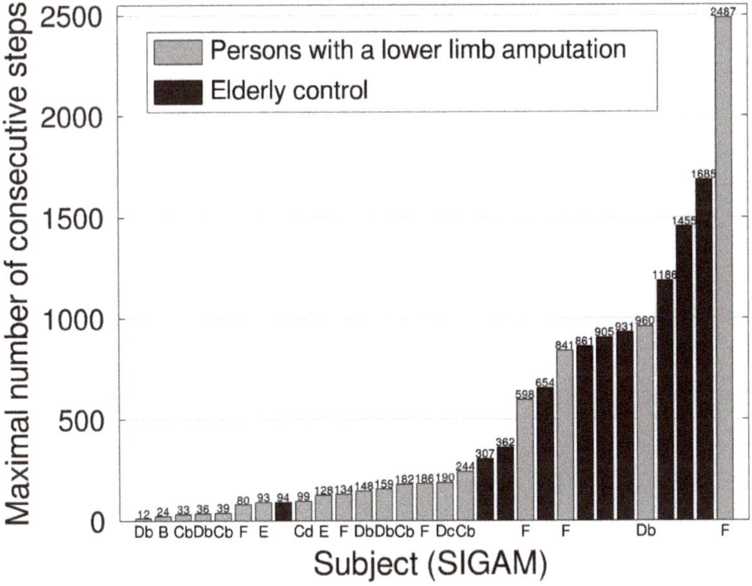

Figure 1. Maximal numbers of consecutive steps of each subject. The SIGAM mobility grade is given for every subject of the persons with a lower limb amputation. B = Therapeutic use only for transfers, C = Walks on level ground less than or equal to 50 m with (Cb)/without aids (Cd), D = Walks outdoor on level ground only, in good weather, more than 50 m with 2 crutches/sticks (Db) or 1 crutch/stick (Dc), and E = Walks more than 50 m, no aids, except in adverse terrain or weather, F = normal or near normal gait.

3.3. Frequency of Number of Steps per Hour

Figure 2 shows the frequency per hour per bin (number of consecutive steps). The mixed ANOVA revealed an interaction effect ($F_{14,392} = 2.41$, $p = 0.003$) and a significant main effect for the number of consecutive steps ($F_{14,29} = 56.1$, $p < 0.001$) and no significant main effect for group ($F_{1,28} = 4.13$, $p = 0.052$). Post-hoc analysis showed that the elderly controls had significantly more walking bouts with 10–25 consecutive steps and more than 100 consecutive steps (as indicated by the * in Figure 2).

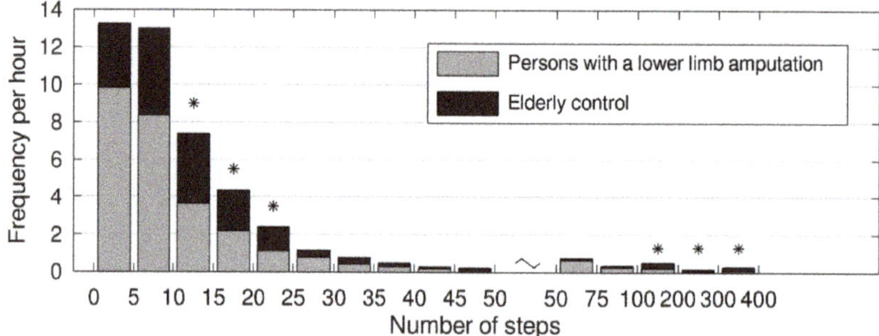

Figure 2. The mean frequency per hour of the number of steps per bin. Grey bars are persons with a lower limb amputation, black bars are the elderly controls. * Post-hoc difference between the persons with a lower limb amputation and the elderly control group.

3.4. 10 MWT and Maximal Covered Walking Distance

The left panel of Figure 3 shows the performance on the 10 MWT for the SIGAM mobility grades for all persons with a lower limb amputation and the elderly controls (EC). Spearman's rank correlation between the SIGAM mobility grade and the 10 MWT was −0.78 (p = 0.0001). Based on the 10 MWT, the median estimated maximal covered distance in the persons with a lower limb amputation was 67 m (with an interquartile range of 22–93). The maximal covered distance is shown for the SIGAM mobility grades in the center panel of Figure 3. Obviously, the higher the SIGAM mobility grade, the higher the maximal covered distance (r = 0.61, p = 0.006). Nevertheless, a closer look showed that even some persons with a lower limb amputation with a SIGAM mobility grade higher than C did not reach the 50 m. The maximal covered distance was also significantly correlated with the 10 MWT (r = −0.66, p = 0.002).

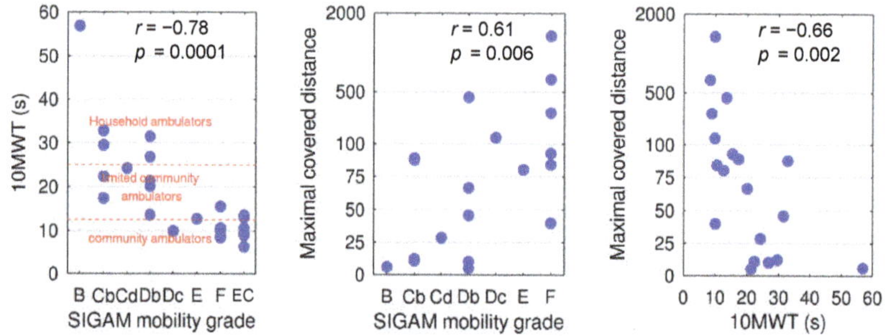

Figure 3. Relation between the SIGAM mobility grades, the 10 MWT, and the maximal covered distance. B = Therapeutic use only for transfers, C = Walks on level ground less than or equal to 50 m with (Cb)/without aids (Cd), D = Walks outdoor on level ground only, in good weather, more than 50 m with 2 crutches/sticks (Db) or 1 crutch/stick (Dc), and E = Walks more than 50 m, no aids, except in adverse terrain or weather, F = normal or near normal gait. EC = elderly controls.

4. Discussion

The goal of this study was to assess the maximal covered walking distance and walking bouts in a wide range of persons with a lower limb amputation in daily life. Forty percent of the persons with a lower limb amputation (8 out of 20) did not reach walking distances of 50 m during a single walking bout, which was indicated as an important benchmark for mobility and, therefore, important for independent living and social participation. There was a significant positive correlation between the maximal covered distance and the SIGAM mobility grades (Figure 3). In contrast to the persons with a lower limb amputation, the elderly control group, except for the sick subject, covered a walking distance of at least 150 m, based on the maximal number of consecutive steps of at least 300 (Figure 1). These results imply that the current SIGAM mobility grades do not sufficiently reflect what a person with lower limb amputation actually does in daily life, but more what a person is able to do.

Several studies performed activity measurements in persons with a lower limb amputation with daily duration of dynamic activities or daily number of steps as the main outcome measure [16–24]. The lower number of walking bouts, especially in the long walking bouts, compared to the age-matched control subjects, supports the finding that persons with a lower limb amputation are less active. However, none of these studies investigated walking bouts and the related maximal number of consecutive steps. For persons with a lower limb amputation, maximal walking distance is an important measure for social mobility and ADL independence. Since SIGAM mobility grades uses the 50 m walking distance as a limit for indoor and outdoor walking, this 50 m limit should correspond with independence, and the level at which a person can participate in society [4]. Forty percent of the persons with a lower limb amputation did not cover a walking distance of more than 50 m. Except for the sick subjects, the elderly control group had a maximal number of steps of at least 300, which was at least 150 m with a 0.5 m step length. Therefore, walking bouts of at least 300 steps seemed to be the lower bound for walking mobility in the elderly control. In contrast, only 4 of 12 persons with a lower limb amputation with normal or near normal gait (3 persons with a lower limb amputation with SIGAM mobility scale grade F and one person with a lower limb amputation with grade D) took more than 300 steps consecutively. A minimal walking distance of approximately 300–350 m is required for community walking tasks, such as walking from the parking lot to the grocery shop or visiting a health care practitioner [29–31]. In our study, 4 out of 20 persons with a lower limb amputation and 7 out of the 10 elderly control had walking bouts of more than 600 steps, which indicated at least community walking. Hence, walking mobility was affected in most persons with a lower limb amputation who were defined as normal or near normal walkers.

The limited walking distance at a time, for persons with a lower limb amputation, could be compensated by walking consecutive short distances more frequently, with rest periods in between. However, persons with a lower limb amputation had significantly smaller short walking bouts compared to the elderly control. Furthermore, detailed analysis revealed that in the persons with a lower limb amputation data, consecutive short walking bouts with rest periods were not present, making it impossible to reach similar long walking distances as the elderly control. There might be several reasons why persons with a lower limb amputation avoid walking long distances. One explanation might be that the persons with a lower limb amputation adapt their walking distance to keep their heart rate response within a normal range [18]. Another explanation might be that persons with a lower limb amputation had a poorer joint coordination, and thus might be easier to get fatigued, feel discomfort, and have an unstable gait [32,33]. It seems that walking is already a maximum effort for a great part of the persons with a lower limb amputation. Beside the physical limitation, outdoor gait performance of the persons with a lower limb amputation is of course also dependent on a variety of other factors, including personal interest, weather, terrain, comorbidities, prosthetic fit, social interactions, etc. [24].

Evaluation of daily functioning of persons with a lower limb amputation is highly based on questionnaires such as the SIGAM mobility grade, which are based on self-report and estimates of the physician. Several studies found that one of the risks of self-report activity questionnaires is an overestimation of activity levels when using self-reported measures [7–9]. Bootsma-van der Wiel et al. [34] found that discrepancies between what the elderly (>85 years) can do and actually do in activities of daily living had important consequences when estimating disability in old people. As a consequence, incorrect assessment of daily functioning might influence the care given. The clear positive correlation between maximal covered distance and the SIGAM mobility grades and maximal covered distance and 10 MWT implies a high association between gait mobility and gait capacity, which justifies the SIGAM mobility grade as an evaluation for daily functioning. However, the limit of 50 m walking at a time as a threshold for the SIGAM mobility grades of D and higher was not established by all persons with a lower limb amputation, with a SIGAM mobility of D or higher. Furthermore, 2 of the 6 persons with a lower limb amputation with a SIGAM mobility grade of B or C, covered a larger distance than 50 m. Therefore, a discrepancy exists between the SIGAM mobility grades and performance in daily life, which corresponds to the results of Albert et al. [35]. This finding implies that the SIGAM mobility grade of persons with a lower limb amputation is more dependent on the type of activities one can perform, than purely on walking distance. Therefore, daily functioning should not only question what a person with a lower limb amputation can do but should also monitor the amputees actual daily activities and walking distance. For the assessment of daily functioning, more information can be obtained than only maximal number of consecutive steps and gait bouts. For example, the distribution of walking bouts across the day, use of walking aids, how long the prosthesis was worn during the day, and specific activities for a person with a lower limb amputation.

Limitations

A limitation of the current study is that data collection was limited to two days and the maximum covered distance was estimated by multiplying the number of steps, with the step length measured with the 10 MWT. Although a larger number of measurement days than 2 would have resulted in a more accurate estimates of the maximal number of consecutive steps, the walking bouts of at least 300 steps in the elderly control group indicated that 1–2 days was sufficient to indicate their mobility. The estimated covered distance was most likely an overestimate since daily life walking was less regular than walking during a 10 MWT. Inertial measurement units attached to the shoe or ankle would be a better alternative as it estimates gait velocity and step length in a valid and reliable way [36,37]. We chose the most convenient and easy way, by focusing on the number of steps, which could also be simply assessed by using, for example, a smart watch or a smart phone [38–41]. The relatively small sample size did not allow us to perform a sub-analysis within the persons with a lower limb amputation. We expect that the level of amputation and reason for amputation group would affect the maximal covered distance. Persons with a transfemoral amputation would most likely have a reduced walking distance compared to the persons with a transtibial amputation.

5. Conclusions

The current study indicates that mobility is highly affected in most persons with a lower limb amputation and that the SIGAM mobility grade does not reflect what persons with a lower limb amputation actually do in daily life. Therefore, objective assessment of the maximal number of consecutive steps of the maximal covered distance, is recommended for clinical treatment.

Author Contributions: Conceptualization, C.J.H. and N.L.W.K.; methodology, C.J.H. and N.L.W.K.; software, N.L.W.K.; validation, C.J.H. and N.L.W.K.; formal analysis, K.T.J.B., M.D. and N.L.W.K.; investigation, C.J.H., K.T.J.B. and M.D.; resources, C.J.H. and N.L.W.K.; data curation, C.J.H. and N.L.W.K.; writing—original draft preparation, C.J.H., K.T.J.B. and M.D.; writing—review and editing, C.J.H., R.F.v.E. and N.L.W.K.; visualization, C.J.H. and N.L.W.K.; supervision, N.L.W.K.; project administration, C.J.H.; funding acquisition, N.L.W.K. All authors have read and agreed to the published version of the manuscript.

Funding: This research received no external funding.

Acknowledgments: We thank all our participants for their time and effort. We would also like to thank Ivo Koekkoek, OIM Orthopedie Nijmegen, for inclusion and recruitment of the lower limb amputees.

Conflicts of Interest: The authors declare no conflict of interest.

References

1. Legro, M.W.; Reiber, G.; del Aguila, M.; Ajax, M.J.; Boone, D.A.; Larsen, J.A.; Smith, D.G.; Sangeorzan, B. Issues of importance reported by persons with lower limb amputations and prostheses. *J. Rehabil. Res. Dev.* **1999**, *36*, 155–163.

2. Christensen, B.; Ellegaard, B.; Bretler, U.; Ostrup, E.L. The effect of prosthetic rehabilitation in lower limb amputees. *Prosthet. Orthot. Int.* **1995**, *19*, 46–52. [CrossRef]

3. Pell, J.P.; Donnan, P.T.; Fowkes, F.G.; Ruckley, C.V. Quality of life following lower limb amputation for peripheral arterial disease. *Eur. J. Vasc. Surg.* **1993**, *7*, 448–451. [CrossRef]
4. Ryall, N.H.; Eyres, S.B.; Neumann, V.C.; Bhakta, B.B.; Tennant, A. The SIGAM mobility grades: A new population-specific measure for lower limb amputees. *Disabil. Rehabil.* **2003**, *25*, 833–844. [CrossRef]
5. Pohjolainen, T.; Alaranta, H.W. Primary survival and prosthetic fitting of lower limb amputees. *J. Prosthet. Orthot. Int.* **1989**, *13*, 63–69. [CrossRef]
6. Inderbitzi, R.; Buettiker, M.; Enzler, M. The long-term mobility and mortality of patients with peripheral arterial disease following bilateral amputation. *Eur. J. Vasc. Endovasc. Surg.* **2003**, *26*, 59–64. [CrossRef]
7. Klesges, R.C.; Eck, L.H.; Mellon, M.W.; Fulliton, W.; Somes, G.W.; Hanson, C.L. The accuracy of self-reports of physical activity. *Med. Sci. Sports Exerc.* **1990**, *22*, 690–697. [CrossRef]
8. Lagerros, Y.T.; Lagiou, P. Assessment of physical activity and energy expenditure in epidemiological research of chronic diseases. *Eur. J. Epidemiol.* **2007**, *22*, 353–362. [CrossRef]
9. Cuperus, N.; Hoogeboom, T.J.; Neijland, Y.; van den Ende, C.H.; Keijsers, N.L. Are people with rheumatoid arthritis who undertake activity pacing at risk of being too physically inactive? *Clin. Rehabil.* **2012**, *26*, 1048–1052. [CrossRef]
10. Orendurff, M.S.; Kobayashi, T.; Villarosa, C.Q.; Coleman, K.L.; Boone, D.A. Comparison of a computerized algorithm and prosthetists' judgment in rating functional levels based on daily step activity in transtibial amputees. *J. Rehabil. Assist. Technol. Eng.* **2016**, *3*. [CrossRef]
11. Burger, H.; Marincek, C.; Isakov, E. Mobility of persons after traumatic lower limb amputation. *Disabil. Rehabil.* **1997**, *19*, 272–277. [CrossRef] [PubMed]
12. Legro, M.W.; Reiber, G.D.; Smith, D.G.; del Aguila, M.; Larsen, J.; Boone, D. Prosthesis evaluation questionnaire for persons with lower limb amputations: Assessing prosthesis-related quality of life. *Arch. Phys. Med. Rehabil.* **1998**, *79*, 931–938. [CrossRef]
13. Miller, W.C.; Deathe, A.B.; Speechley, M. Lower extremity prosthetic mobility: A comparison of 3 self-report scales. *Arch. Phys. Med. Rehabil.* **2001**, *82*, 1432–1440. [CrossRef]
14. Hagberg, B.R.; Hägg, O. Questionnaire for Persons with a Transfemoral Amputation (Q-TFA): Initial validity and reliability of a new outcome measure. *J. Rehabil. Res. Dev.* **2004**, *41*, 695–706. [CrossRef]
15. Raschke, S.U.; Orendurff, M.S.; Mattie, J.L.; Kenyon, D.E.; Jones, O.Y.; Moe, D.; Winder, L.; Wong, A.S.; Moreno-Hernández, A.; Highsmith, M.J.; et al. Biomechanical characteristics, patient preference and activity level with different prosthetic feet: A randomized double blind trial with laboratory and community testing. *J. Biomech.* **2015**, *48*, 146–152. [CrossRef]
16. Anastasiades, P.; Johnston, D.W. A simple activity measure for use with ambulatory subjects. *Psychophysiology* **1990**, *27*, 87–93. [CrossRef]
17. Bussmann, J.B.; Grootscholten, E.A.; Stam, H.J. Daily physical activity and heart rate response in people with a unilateral transtibial amputation for vascular disease. *Arch. Phys. Med. Rehabil.* **2004**, *85*, 240–244. [CrossRef]
18. Bussmann, J.B.; Schrauwen, H.J.; Stam, H.J. Daily physical activity and heart rate response in people with a unilateral traumatic transtibial amputation. *Arch. Phys. Med. Rehabil.* **2008**, *89*, 430–434. [CrossRef] [PubMed]
19. Selles, R.W.; Janssens, P.J.; Jongenengel, C.D.; Bussmann, J.B. A randomized controlled trial comparing functional outcome and cost efficiency of a total surface-bearing socket versus a conventional patellar tendon-bearing socket in transtibial amputees. *Arch. Phys. Med. Rehabil.* **2005**, *86*, 154–161. [CrossRef] [PubMed]
20. Houdijk, H.; Appelman, F.M.; Van Velzen, J.M.; Van der Woude, L.H.; Van Bennekom, C.A. Validity of DynaPort GaitMonitor for assessment of spatiotemporal parameters in amputee gait. *J. Rehabil. Res. Dev.* **2008**, *45*, 1335–1342. [CrossRef] [PubMed]
21. Theeven, P.J.; Hemmen, B.; Geers, R.P.; Smeets, R.J.; Brink, P.R.; Seelen, H.A. Influence of advanced prosthetic knee joints on perceived performance and everyday life activity level of low-functional persons with a transfemoral amputation or knee disartidulation. *J. Rehabil. Med.* **2012**, *44*, 454–461. [CrossRef] [PubMed]
22. Redfield, M.T.; Cagle, J.C.; Hafner, B.J.; Sanders, J.E. Classifying prosthetic use via accelerometry in persons with transtibial amputations. *J. Rehabil. Res. Dev.* **2013**, *50*, 1201–1212. [CrossRef] [PubMed]

23. Samuelsen, B.T.; Andrews, K.L.; Houdek, M.T.; Terry, M.; Shives, T.C.; Sim, F.H. The Impact of the Immediate Postoperative Prosthesis on Patient Mobility and Quality of Life after Transtibial Amputation. *Am. J. Phys. Med. Rehabil.* **2017**, *96*, 116–119. [CrossRef]

24. Orendurff, M.S.; Raschke, S.U.; Winder, L.; Moe, D.; Boone, D.A.; Kobayashi, T. Functional level assessment of individuals with transtibial limb loss: Evaluation in the clinical setting versus objective community ambulatory activity. *J. Rehabil. Assist. Technol. Eng.* **2016**, *3*. [CrossRef] [PubMed]

25. Van Dam, M.S.; Kok, G.J.; Munneke, M.; Vogelaar, F.J.; Vliet Vlieland, T.P.; Taminiau, A.H. Measuring physical activity in patients after surgery for a malignant tumour in the leg. The reliability and validity of a continuous ambulatory activity monitor. *J. Bone Joint Surg. Br.* **2001**, *83*, 1015–1019. [CrossRef] [PubMed]

26. Bussmann, J.B.; Damen, L.; Stam, H.J. Analysis and decomposition of signals obtained by thigh-fixed accelerometry during walking. *Med. Biol. Eng. Comput.* **2000**, *38*, 632–638. [CrossRef]

27. Zijlstra, W.; Hof, A.L. Assessment of spatio-temporal gait parameters from trunk accelerations during human walking. *Gait Posture* **2003**, *18*, 1–10. [CrossRef]

28. Keijsers, N.L.; Horstink, M.W.; Gielen, S.C. Ambulatory motor assessment in Parkinson's disease. *Mov. Disord.* **2006**, *21*, 34–44. [CrossRef]

29. Shumway-Cook, A.; Patla, A.E.; Stewart, A.; Ferrucci, L.; Ciol, M.A.; Guralnik, J.M. Environmental demands associated with community mobility in older adults with and without mobility disabilities. *Phys. Ther.* **2002**, *82*, 670–681. [CrossRef]

30. Robinett, C.S.; Vondran, M.A. Functional ambulation velocity and distance requirements in rural and urban communities. A clinical report. *Phys. Ther.* **1988**, *68*, 1371–1373. [CrossRef]

31. Lapointe, R.; Lajoie, Y.; Serresse, O.; Barbeau, H. Functional community ambulation requirements in incomplete spinal cord injured subjects. *Spinal Cord* **2001**, *39*, 327–335. [CrossRef] [PubMed]

32. Wong, D.W.C.; Lam, W.K.; Yeung, L.F.; Lee, W.C. Does long-distance walking improve or deteriorate walking stability of transtibial amputees? *Clin. Biomech.* **2015**, *30*, 867–873. [CrossRef]

33. Yeung, L.F.; Leung, A.K.; Zhang, M.; Lee, W.C. Effects of long-distance walking on socket-limb interface pressure, tactile sensitivity and subjective perceptions of trans-tibial amputees. *Disabil. Rehabil.* **2013**, *35*, 888–893. [CrossRef] [PubMed]

34. Bootsma-van der Wiel, A.; Gussekloo, J.; de Craen, A.J.; van Exel, E.; Knook, D.L.; Lagaay, A.M.; Westendorp, R.G. Disability in the oldest old: "can do" or "do do"? *J. Am. Geriatr. Soc.* **2001**, *49*, 909–914. [CrossRef] [PubMed]

35. Albert, M.V.; McCarthy, C.; Valentin, J.; Herrmann, M.; Kording, K.; Jayaraman, A. Monitoring functional capability of individuals with lower limb amputations using mobile phones. *PLoS ONE* **2013**, *8*, e65340. [CrossRef]

36. Washabaugh, E.P.; Kalyanaraman, T.; Adamczyk, P.G.; Claflin, E.S.; Krishnan, C. Validity and repeatability of inertial measurement units for measuring gait parameters. *Gait Posture* **2017**, *55*, 87–93. [CrossRef]

37. Hung, T.N.; Suh, Y.S. Inertial Sensor-Based Two Feet Motion Tracking for Gait Analysis. *Sensors* **2013**, *13*, 5614–5629. [CrossRef]

38. Bort-Roig, J.; Gilson, N.D.; Puig-Ribera, A.; Contreras, R.S.; Trost, S.G. Measuring and influencing physical activity with smartphone technology: A systematic review. *Sports Med.* **2014**, *44*, 671–686. [CrossRef]

39. Gonze, B.B.; Padovani, R.D.C.; Simoes, M.D.S.; Lauria, V.; Proença, N.L.; Sperandio, E.F.; Ostolin, T.L.V.D.P.; Gomes, G.A.O.; Castro, P.C.; Romiti, M.; et al. Use of a Smartphone App to Increase Physical Activity Levels in Insufficiently Active Adults: Feasibility Sequential Multiple Assignment Randomized Trial (SMART). *JMIR Res. Protoc.* **2020**, *9*. [CrossRef]

40. Silsupadol, P.; Teja, K.; Lugade, V. Reliability and validity of a smartphone-based assessment of gait parameters across walking speed and smartphone locations: Body, bag, belt, hand, and pocket. *Gait Posture* **2017**, *58*, 516–522. [CrossRef]

41. Kang, X.; Huang, B.; Qi, G. A Novel Walking Detection and Step Counting Algorithm Using Unconstrained Smartphones. *Sensors* **2018**, *18*, 297. [CrossRef] [PubMed]

Publisher's Note: MDPI stays neutral with regard to jurisdictional claims in published maps and institutional affiliations.

Perspective

Implementation of Wearable Sensing Technology for Movement: Pushing Forward into the Routine Physical Rehabilitation Care Field

Catherine E. Lang [1,2,3,*], Jessica Barth [1], Carey L. Holleran [1,3], Jeff D. Konrad [1] and Marghuretta D. Bland [1,2,3]

[1] Program in Physical Therapy, Washington University School of Medicine, St. Louis, MO 63122, USA; jessica.barth@wustl.edu (J.B.); cholleran@wustl.edu (C.L.H.); jdkonrad@wustl.edu (J.D.K.); blandm@wustl.edu (M.D.B.)
[2] Program in Occupational Therapy, Washington University School of Medicine, St. Louis, MO 63122, USA
[3] Department of Neurology, Washington University School of Medicine, St. Louis, MO 63122, USA
* Correspondence: langc@wustl.edu

Received: 14 September 2020; Accepted: 8 October 2020; Published: 10 October 2020

Abstract: While the promise of wearable sensor technology to transform physical rehabilitation has been around for a number of years, the reality is that wearable sensor technology for the measurement of human movement has remained largely confined to rehabilitation research labs with limited ventures into clinical practice. The purposes of this paper are to: (1) discuss the major barriers in clinical practice and available wearable sensing technology; (2) propose benchmarks for wearable device systems that would make it feasible to implement them in clinical practice across the world and (3) evaluate a current wearable device system against the benchmarks as an example. If we can overcome the barriers and achieve the benchmarks collectively, the field of rehabilitation will move forward towards better movement interventions that produce improved function not just in the clinic or lab, but out in peoples' homes and communities.

Keywords: rehabilitation; motor function; wearable sensors; outcomes; measurement; implementation

---

1. Introduction

Wearable sensor technology to measure human movement is rapidly evolving. Motion-sensing wearable devices continue to get smaller, lighter and have more data storage space, with even better products anticipated in the future. These wearable devices are ubiquitous in the general public in the form of a variety of commercially available, consumer-grade products. Here, we use the term 'device' to refer to the wearable unit, 'sensor' to refer to the sensors within the device, and 'wearable device system' to refer to the collective hardware and software package (see Box 1 for other operational definitions used in this paper). This paper intentionally excludes cell phones as a wearable because cell phones are used for a multitude of purposes and are not routinely 'worn' in the same location within and across people. Because of the focus on movement, this paper also excludes wearable sensors designed to measure physiological signals such as heart rate, oxygen saturation, and respiratory rate. Wearable devices include one or more specific sensors, with accelerometers being the most common sensor for quantifying human movement. Many devices also include magnometers, inclinometers, gyroscopes, and light sensors.

Box 1. Operational Definitions.

> *Commercially-available:* device systems that can be purchased through companies or organizations.
> *Laboratory-available:* device systems developed in one or more labs and not available for purchase by the general public.
> *Consumer-grade:* device systems that are marketed directly to consumers, intended to be used by anyone.
> *Research-grade:* device systems that are marketed to researchers and healthcare professionals, intended for use to be managed by someone with specialized training.
> *Activity:* Execution of a task or action, such as walking, dressing, or bathing [1].
> *Capacity:* Activity that is assessed in a structured setting, usually with a standardized tool, such as the 10 m walk test or the Box and Block test. Alternate terms are function and functional capacity [1].
> *Performance:* Activity that is assessed in an unstructured, free living setting. Performance can be measured directly via wearable devices, or via self-report with questionnaires [1].

The potential benefit of wearable sensor technology for physical rehabilitation has been discussed in the literature since the early 2000s [2]. The real promise for rehabilitation is that wearable device systems can measure human movement in real world settings, not just in the structured environment of the clinic or laboratory. People seek physical rehabilitation services to improve their movement in daily life, so rehabilitation professionals need a way to quantify that movement in order to best help their patients. In the published literature to date, two overlapping scientific cohorts are responsible for the majority of the progress with wearable sensor technology. The first cohort includes the engineering groups, who are developing and testing sensors, software, and algorithms. The second cohort includes clinician scientists who are validating sensor-derived metrics in clinical populations and deploying these metrics to answer questions and measure outcomes in rehabilitation studies. While wearable device systems to measure human movement have become more commonplace in research studies, they have not yet been widely implemented in routine clinical practice [3]. Multiple barriers arising both from clinical rehabilitation practice and from the current state of commercially-available devices need to be overcome before there is widespread adoption into routine care. We note that these barriers and their relative importance may vary somewhat around the globe.

The purposes of this paper are to: (1) discuss the major barriers in clinical practice and available wearable sensing technology; (2) propose benchmarks for wearable device systems that would make it feasible to implement them into clinical practice across the world; and (3) evaluate a current wearable device system against the benchmarks as an example. This field is in its early stages. The evolution of the telephone is a good analogy for our field. Early telephones look and function nothing like the sleek, smart phones we use today, with advancements in telephone technology spread out over more than 100 years. This paper is intended as a small step, specifically focused on helping the collective advancement of wearable sensor technology from research groups into routine clinical practice. Our larger goal is to move the field of physical rehabilitation forward towards better interventions that produce improved movement performance in homes and communities, not just improved capacity in our clinics and labs.

2. The Current Situation in Clinical Care

There are multiple factors influencing physical rehabilitation that may impact the implementation of wearable device systems into routine clinical rehabilitation practice. Many influencing factors are generic, i.e. they influence any type of change in clinical practice. This paper intentionally focuses on two key factors most strongly influencing clinical uptake of wearable sensor technology. These key factors are the time constraints present in everyday clinical practice and the salience of information from wearable device systems by physical rehabilitation professionals. Note that medical- or research-grade devices that are deployed into clinical care must meet all the clinical standards, defined by the International Organization of Standards (ISO/TC 173 for Assistive Products and ISO 10667-1:2011 for Assessment Service Delivery), and evaluated by regulatory agencies. Regulatory approval is a long, challenging process, varying somewhat by geographic location; a discussion of this process is outside the scope of this review.

2.1. Busy Clinical Practice Affords Little Time for Anything Else

Across the world, patients are seen in a variety of rehabilitation settings including acute and long-term hospitals, specialized facilities (skilled nursing, inpatient or acute rehabilitation), outpatient clinics, virtual visits, or at home, each with different requirements for productivity. In the United States for example, productivity standards for physical therapists can be upwards of 85% with reimbursement only for billable units [4]. This means that therapists are working under the expectation that upwards of 6 h 48 min of their 8-h workday are devoted to direct patient care, with patients scheduled back to back and little or no time in between. Meeting these standards can be challenging as direct patient care time is often limited by the patients themselves, both within the hospital or inpatient setting and outside the hospitals in outpatient, ambulatory care settings. Common disruptions to direct patient care time in the inpatient setting are delays or interruptions due to hygiene or dietary concerns, and conflicting medical needs. Similar disruptions in the outpatient setting most frequently arise from transportation delays. Figure 1 provides two examples of typical rehabilitation sessions (Figure 1A: inpatient, physical therapy, Figure 1B: outpatient occupational therapy) to illustrate how planned, scheduled direct patient care time does not often equate to actual care time [5,6].

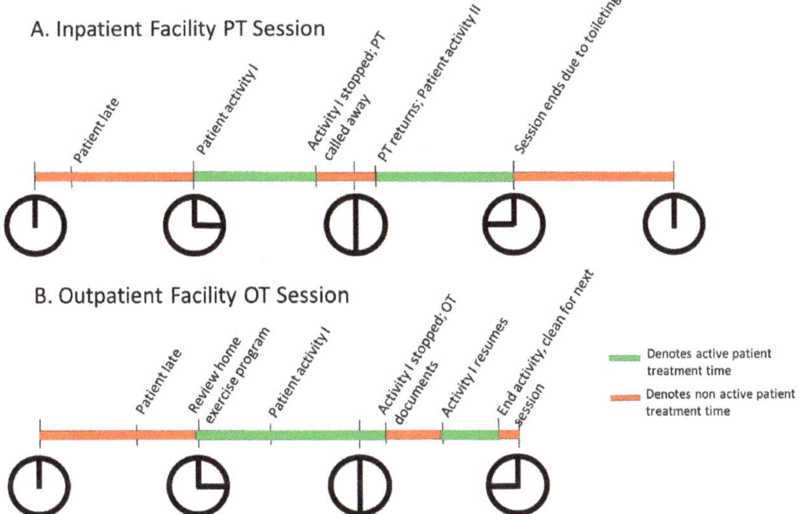

Figure 1. Clinicians encounter many barriers to spending time in direct patient care. Examples of inpatient physical therapy (PT) 1 h session (**A**) and outpatient occupational therapy (OT) 45 min session (**B**).

In addition, regardless of the clinical setting or geographic location, physical rehabilitation professionals (physicians, physical and occupational therapists, etc.) in today's busy clinical environment have many competing priorities above and beyond direct patient care that limit available time [3,7–9]. Rehabilitation clinicians have administrative demands related to documentation, addressing patient and family concerns, and coordination of care with other healthcare team members [10,11]. Competing demands on time are often balanced by the rehabilitation professional multi-tasking, providing group treatments (allowable in some settings or countries, but not allowable in others), and completing administrative tasks while providing treatment during a session. Collectively, these pressures limit time and energy to trial new technology, including wearable device systems.

Figure 2A shows an example of current time costs for using a wearable sensor system within clinical practice, with time estimates based on our laboratory protocol [12]. Asking rehabilitation professionals to charge, program, don/doff, process, and share results (both with the patient and other health professionals through an electronic medical record) of wearable device systems is unrealistic if it cannot be completed quickly during a patient treatment session. Figure 2B shows a clinically-feasible time cost that would foster implementation of wearable device systems into routine clinical rehabilitation practice.

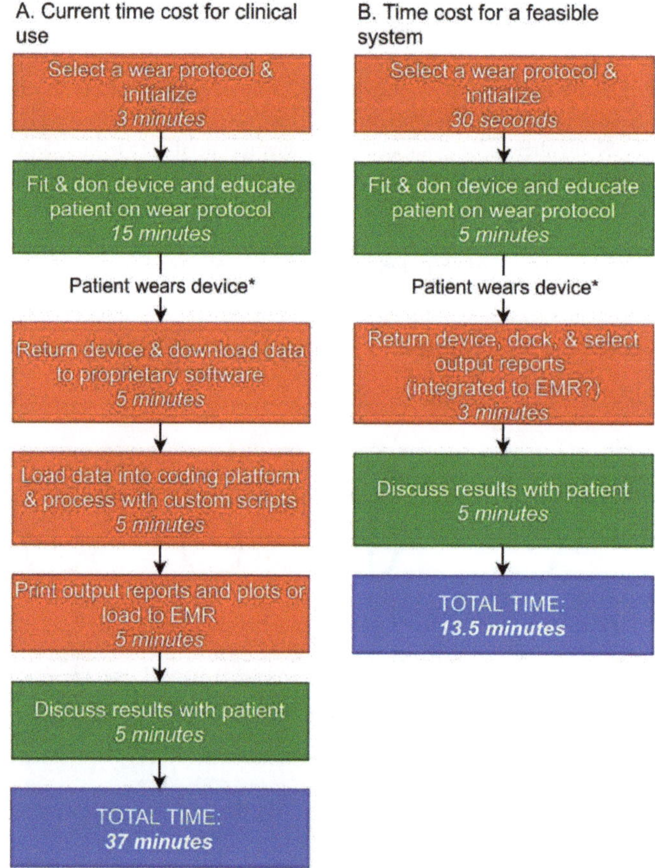

Figure 2. Time Cost for Clinical Implementation of Wearable Device Systems. **A:** Current time cost in clinical practice. **B:** Estimated time cost for realistic implementation into clinical practice. The first two boxes in **A** & **B** are estimates for an experienced clinician using a wearable device system with a new patient. Red boxes indicate non-billable time while green boxes indicate billable time. EMR: electronic medical record. * Wear time may vary from hours to several days.

The above examples are drawn from the healthcare system in the United States. While other countries might not currently face such severe time constraints, all counties face at least some. Additional challenges faced across multiple continents include under-sourced physical rehabilitation services (i.e. a limited number of professionals available), which may lead to self-imposed pressures for professionals to treat as many people as quickly as possible. Regardless of the unique situations in each country, it is unlikely that these time constraints and competing priorities are going to change in the foreseeable future. For wearable device systems to be widely implemented into rehabilitation care around the globe, the technology needs to fit in seamlessly, minimizing disruption to busy, clinical settings.

2.2. Clinicians Are Still Building towards Understanding the Added Value of Wearable Sensor Data for Clinical Rehabilitation Practice

The World Health Organization's International Classification of Function, Disability, and Health (WHO-ICF) describes three levels of classification for any health condition (disorder or disease), including body function and structure, activity, and participation [1]. Activity is considered the ability to execute tasks or actions and can be subdivided into the capacity for activity, i.e. what a person can do in a structured environment, and performance of activity in daily life, i.e. what a person actually does (see Box 1). Established in the field of physical rehabilitation is the importance of standardized outcome assessments to evaluate change over the course of clinical care [13]. Recommended activity level measures included in rehabilitation clinical practice guidelines around the world nearly always assess capacity, not performance [14]. But as mentioned in the Introduction, individuals who engage in physical rehabilitation services typically seek improvement in movement performance within their daily lives [15]. Assessing the impact of rehabilitation interventions in the context of an individual's life could, therefore, serve as a primary indicator of effectiveness of rehabilitation interventions. Examples of using performance-level tracking in physical rehabilitation are the use of a single wearable sensor worn at the ankle to track daily walking (e.g., steps/day), or the use of two wearable sensors worn on the left and right wrists to track upper limb activity (e.g., use ratio, which is the relative duration of activity in one limb vs. the other). Objective, performance-level tracking via wearable devices across clinical episodes of care for decision-making, however, is an emerging, yet not established practice.

While there is promise for the adoption of performance-level tracking within clinical populations, several clinical assumptions stand in the way of wide-spread adoption. First, there is the widespread assumption that capacity measures taken in the clinic reflect performance measures in daily life. Capacity is a snapshot of ability at a singular time point in a structured environment, whereas performance measures capture real-world activity that includes the ecological validity of an individual's free-living condition. Published data on gait speed, an index of walking capacity, and steps per day, an index of walking performance, illustrate this problem. Gait speed, measured in the clinic, generally accounted for 30–45% of the variance in steps/day, leaving up to 70% of the variance in daily stepping unexplained [16,17]. For example, individuals in a recent study with self-selected gait speeds around 0.8 m/s ranged from about 750 steps/day to over 6000 steps/day (see Figure 2B of [18]). Without a direct measure such as those provided by wearable device systems, a clinician has limited insight about walking or other movement performance in daily life. Up to the present time, physical rehabilitation clinicians have had to rely primarily on self-report measures to quantify the amount, frequency and duration of movement outside of clinical services. Unfortunately, self-report measures have been shown to lack consistency with more direct assessments [19,20]. Thus, wide-spread adoption of wearable device systems into routine clinical rehabilitation practice will provide new, important information for clinical management.

The second clinical assumption is that a change in an individual's capacity is equivalent to a change in that individual performance in daily life. Over the past few years, multiple reports have now demonstrated discrepant outcomes in capacity and performance over the course of research and clinical interventions [21–24]. Each of these aforementioned belief barriers will require educational strategies tailored towards rehabilitation professionals to improve adoption of performance tracking with wearable device systems within clinical populations.

The great news is that emerging data exist demonstrating the utility of performance-level tracking at improving a variety of outcomes and monitoring the effects of disease processes on movement longitudinally [25,26]. Performance monitoring as part of research interventions has been effective at increasing daily stepping, improving daily physical activity, and reducing sedentary time in healthy populations (see [27] for review). In clinical populations, performance tracking has been effective at improving functional outcomes including walking endurance [28,29] and daily walking activity [30,31] in the lower limb, and has proven to be sensitive to changes in real-world upper limb use across both research and clinical interventions [22,32]. Further, tracking has also been sensitive to the degradation of movement in daily life in both static [33] and progressive [34] disease processes. Advantages of having performance-level measurements from electronic, internet-connected wearable device systems include: (1) the storage of data in secure, remote databases; (2) the ability to analyze these measurements across large data sets; and the subsequent ability to recommend specific actions based on the results of these measurements. Increased adoption of performance monitoring in the clinical environment could provide the big data required to elucidate the full potential of wearable sensing technology.

3. The Current Situation with Wearable Device Systems

3.1. Commerically-Available, Consumer-Grade Device Algorithms Have Limited Accuracy in Disabled Patient Populations

A major barrier for widespread clinical adoption of wearable sensor technology is that the most accessible wearable device systems, those marketed directly to consumers, have questionable accuracy in rehabilitation populations. Using terminology set out in the V3 framework [35], the problem is not in the verification of the sensor itself, but rather lies in the analytic and clinical validation of the algorithm. People seeking rehabilitation services often do not move normally, such that the algorithms programmed into consumer-grade devices are inaccurate in identifying or quantifying their movement [36–41]. Continuing with our mainstream metric of walking performance, steps/day has been evaluated across many consumer-grade devices. Studies have evaluated the accuracy of these devices across a variety of functional activities and environmental settings, at different placements on the body, and across a wide range of abilities. There was wide variability in the accuracy of these devices in individuals with normal gait speed [42,43]. Furthermore, in individuals who utilize assistive devices (e.g., cane, walker) [44–46], walk with slower speeds (e.g., < 0.8 m/s) [36–41], or have interruptions in continuous walking [47,48], even higher levels of inaccuracy have been identified. The Fit-Bit Ultra (Fitbit Inc., San Francisco, CA, USA) consumer-level device, for example, has been shown to systematically under-estimate steps for individuals with a diagnosis of stroke and traumatic brain injury over a 2-min walk test, with greater inaccuracy for those who took less steps per minute and those that walked ≤ 0.58 m/s [41]. As many individuals who seek physical rehabilitation walk at slower speeds, this poses a major barrier for accurate, objective monitoring. In contrast to the consumer-grade devices, commercially-available, research-grade sensors such as the Step Activity Monitor (Modus Inc., Edmonds, WA, USA) have demonstrated strong reliability and accuracy across varied levels of abilities, including differing medical diagnosis, variable gait speeds, and use of assistive devices [41,49–54] with limited data across differing environmental conditions [40,54–56]. Though this research grade device is accurate and reliable in individuals with physical impairments, the device may lack key features that would be essential for widespread clinical adoption (see Section 4 below). Unfortunately, this discussion is an example of only one variable derived from a wearable device. When additional variables are examined, the depth of work to integrate these devices into clinical practice grows exponentially.

3.2. Research-Grade Device Systems Are Expensive and Not yet Clinician- and Patient-Friendly

Most research-grade devices work in conjunction with software systems that require a separate computer program to set-up, download and examine the data. To be feasible in the clinic, every rehabilitation professional would need the necessary computer program loaded onto their laptop or other computer they use for clinical care, which would be expensive. In addition, cost of devices for multiple patients and multiple limbs (e.g., wear one device on each wrist to measure upper limb performance) could quickly make wearable sensor technology unreachable for most clinics.

Beyond the cost, current output from research-grade wearable device systems is not easily accessible in a timely manner for rehabilitation professionals or for their patients. Current research-grade systems require training to use. Training involves both how to use the device (e.g., turning on/off, charging, specific requirements for sensor placement wear) and how to use the system (e.g., programming the device, uploading/storing data, translating device output into clinically relevant values). For wearable device systems with proprietary software algorithms, only the default outputs (or variables) are available. If the clinician needs variables beyond the defaults, they would need a research colleague to write code and extract them. Research-grade system outputs (default or otherwise, see Table 1 for variables summary) are difficult to provide quickly during or outside of a treatment session. Furthermore, if the rehabilitation clinician has taken the effort to extract the data and variables, the output is not yet patient-friendly. Researchers have tried to bridge this gap by transforming outputs into patient-friendly graphs, but the process is cumbersome and has not yet yielded improved results [57]. For the clinically-important information derived from the sensors to be widely utilized by patients and clinicians, wearable device systems will likely need to be less expensive, continuously streamed [58], and on an accessible consumer-based platform [9]. Wearable device systems will need to be compatible with and seamlessly integrated into the electronic medical record to be readily adopted [59] and to contribute to the quality and effectiveness of rehabilitation services.

3.3. Standardization of Output Variables in Research Is Limited to Date, with Much Work to Do

The pathway to routine clinical rehabilitation implementation for wearable sensor technology is not just impeded by current clinical care and available devices, but also by the state of the science about the output variables. While it is clear that measurement of outcomes is essential in both research and clinical practice, the pathway to an established outcome measure or sensor-based variable is a long, hard one [60,61]. It is critical that measures and variables are thoroughly vetted, since scores obtained may be used to make diagnostic and rehabilitation management decisions. Prior to any measure being routinely implemented into clinical rehabilitation care, multiple studies evaluating the psychometric properties and clinical utility of the measure are necessary [60–63]. Beyond the verification of sensor signals [35], variables must first demonstrate reliability, or consistency of results obtained, indicating that one can trust that the obtained value is stable. Validity is the second hurdle, with multiple layers of validity. These layers, in hierarchical order, include: (1) face validity (does the variable appear to capture the underlying construct); (2) content validity (does the variable adequately capture/sample the underlying construct); (3) criterion-related validity (how does the variable agree with other measures or the gold-standard measure of the same construct); and (4) construct validity (how well does the variable measure the construct). If a sensor-based variable were to be used to make diagnostic decisions, discriminitive validity (how does the variable distinguish between those with and without a specific condition) and predictive validity (how well does the variable predict future outcome) would also need to be demonstrated in the population of interest. Responsiveness is the third and final hurdle and includes the ability of the variable to capture the range of the construct and its sensitivity to change over time.

Table 1 is a sample list of variables proposed in the literature to measure important rehabilitation constructs that might be used for diagnostic or clinical decision-making. These variables have largely been derived from data captured on research-grade devices and calculated with custom software developed in laboratories. The variables are intended to capture different movement constructs that may be of interest to physical rehabilitation clinicians. We have excluded most measures of general physical activity (e.g., caloric expenditure), except steps/day, since general physical activity measures are an extensive topic in their own right. Variables are presented based on their intention to quantify movement at either the lower (generally captured via one sensor on one leg) or upper limb (generally captured with one sensor on each arm). Based on the available data, variables are marked as to their exploratory status and how quality and quantity of data are related to reliability, validity, and responsiveness.

Table 1. A sample of wearable-derived variables proposed in the literature. This list is not exhaustive, with new variables proposed all the time. Published variable data have been judged as: Green = sufficient data for implementation into clinical practice; Yellow = some data available to-date, but not yet sufficient for implementation; and Red = no data yet.

Variable Name	Explored In:		Evaluation in Health Condition:		
	Absence of Health Condition	Health Condition	Reliability	Validity	Responsiveness
Lower Limb [16,18,21,30,34,40,41,49,53,57,64–70]					
Time-based variables					
% time inactive	Green	Green	Yellow	Yellow	Red
Walking duration	Green	Green	Yellow	Yellow	Red
Amount-based variables					
Steps/day	Green	Green	Green	Green	Green
Bouts/day	Green	Green	Yellow	Yellow	Red
Steps/bout	Green	Green	Yellow	Yellow	Red
Intensity-based variables					
Stepping intensity	Green	Green	Yellow	Yellow	Yellow
Maximum output	Green	Green	Yellow	Yellow	Yellow
Mod. intensity minutes	Green	Green	Yellow	Yellow	Yellow
Peak activity index	Green	Green	Yellow	Yellow	Yellow
Other variables					
Step length variability	Green	Green	Red	Red	Red
Upper Limb [21,32,71–96]					
Time-based variables					
Hours/duration of use	Green	Green	Green	Yellow	Yellow
Use/activity ratio	Green	Green	Green	Green	Yellow
Amount-based variables					
Acceleration area	Yellow	Red	Red	Red	Red
Activity counts	Green	Green	Yellow	Red	Red
Mono-arm use index	Green	Green	Yellow	Yellow	Red
Intensity-based variables					
Acceleration variability	Yellow	Green	Yellow	Yellow	Red
Acceleration magnitude	Green	Green	Yellow	Yellow	Red
Acceleration asymmetry	Green	Green	Red	Yellow	Red
Laterality index	Yellow	Yellow	Red	Yellow	Red
Magnitude ratio	Green	Green	Yellow	Yellow	Red
Bilateral magnitude	Green	Green	Yellow	Yellow	Red
Other variables					
Variation ratio	Yellow	Green	Red	Yellow	Red
Jerk asymmetry	Yellow	Yellow	Red	Red	Red
Spectral arc length	Yellow	Red	Red	Red	Red

There are three key points to take away from this table. First in looking at the left-hand column, there are many variables with different names and often different formulae that may be capturing similar or related constructs of movement. This is both good and bad. Similarity is good because it signifies that the movement construct is considered important by multiple groups. Different names and different formulae are bad for future progress because they make comparisons across studies, samples, and populations difficult. Second in looking at the next two columns, exploratory data are available for many variables. This highlights the creativity that will be needed to really understand movement constructs, variables, and relationships with clinical practice. Third and most importantly in looking at the right-hand columns, there is a tremendous amount of work needed for most of these variables to demonstrate the reliability, validity, and responsiveness necessary for adoption into routine rehabilitation clinical practice. Only one variable, steps/day has sufficient established psychometric and clinical utility to make it worthy of recommendation to clinicians, i.e. evidence of strong, stable psychometric properties across multiple studies with large samples. Another variable, the use ratio for the upper limb, is making good progress towards this goal, after its initial proposal some 20 years ago [81]. Since many of the other variables are only recently proposed, it may be a long time before needed data are available. We are hopeful that once one or a few variables are deployed into routine care, the process for deploying other variables may be accelerated.

To achieve standardization of variables derived from wearable devices, our field will need to communicate and collaborate on issues such as: (1) sensor placement on the limbs; (2) filtering algorithms for raw data; and (3) sensor-independent reporting of variables (e.g., gravitational units and not activity counts, which vary by manufacturer). Comparisons across studies are often impossible due to these issues [97]. For example, it is unfortunately not possible to compare upper limb activity levels in a small sample of children with Duchene Muscular Dystrophy [98] to a larger, referent sample of typically developing children [83]. The two studies evaluated conceptually-similar variables, but used different calculations, different sensors, and different filtering algorithms. Going forward, it would be of great benefit to develop open-access databases for normative or referent data, in order to compare those with and without specific health conditions. Further progress might be made by pooling or sharing currently available datasets for meta-analyses.

3.4. Different Clinical Populations Will Need Different Metrics for Clinical Decision-Making

Much of the variable development summarized above has occurred in the stroke rehabilitation population. While individuals with stroke represent a substantial portion of the world-wide physical rehabilitation population, there are many other clinical populations that could benefit from the ability to capture motor performance in daily life. Given the heterogeneity of physical rehabilitation populations, it is highly likely that different clinical populations will need different wearable-derived variables for clinical decision-making [99]. Important sensor variables developed for one population may not be clinically relevant for another population. For example, the use ratio is an upper limb variable reflecting the relative time one limb is active compared to the other [81,100]. The use ratio has clear clinical relevance in rehabilitation populations with asymmetric effects on the limbs, such as stroke, hemiparetic cerebral palsy, limb amputation/prosthetic use, and recovery from unilateral upper limb surgery. It has little clinical relevance, however, for those with very mild or no asymmetries in motor abilities, such as children with Duchene's Muscular Dystrophy or hyperactivity disorders, or for adults with some brain injuries, and many spinal cord injuries. The ultimate goal of any clinical assessment, including variables derived from wearable devices is that a value or score on the variable informs clinical-decision making, such that without this value or score, a different clincial decision would have been made and a worse outcome might have occurred. Thus, our challenge going forward extends beyond establishing reliability, validity, and responsiveness for wearable-derived variables towards demonstration of how the score or value can change clinical practice.

3.5. Special Considerations for Complexity in Some Populations

3.5.1. Children

Wearable device systems can be a powerful assessment tool for children as objective assessment of their motor activity is difficult. Direct behavioral observation can be used to objectively assess children in their typical environments but that is costly, time-consuming, and only feasible in research. Clinic-based assessments are time consuming, require trained experts, have subjective components, and require the children to be assessed outside of their familiar environment. Furthermore, children under 10 or children with special needs cannot always follow commands or accurately report on their performance [101]. Proxy-reports from parents or teachers have limited accuracy [101,102]. Wearable devices, therefore, can provide pediatric rehabilitation clinicians with real-world, objective performance data on their patients [103–106]. Wearable devices have been used to research the behavior of children with attention deficit-hyperactivity disorder (ADHD) in the classroom [107]. In autism spectrum disorder research, wearable devices have been used to study sleep patterns, stereotyped behaviors, and metabolic disease risk [108–114]. Researchers have also used wearables to identify infants with developmental delays and infants with movement disorders associated with cerebral palsy [115,116] and children with ADHD [105]. While their value for pediatric physical rehabilitation is clear, two special considerations for pediatric wearable device system use are the data collection protocol and the physical design of the device.

Particularly in pediatric studies, there has been wide variation in data collection procedures with respect to device placement and wearing duration. Devices are placed on the waist, ankle, one or both wrists, or a combination of these sites. The issue of device placement is further complicated by the different, and often more intense, movement activity in children (e.g., swinging, climbing, and sliding on playgrounds) compared to adults. While multiple sensors can quantify movement in ways a single sensor cannot, multiple sensors create a greater burden for the wearer, particularly on infants and small children. Implementation of wearable sensor technology into pediatric clinical practice will require thoroughly vetted protocols that can maximize sensor information and minimize the number of sensors and duration of wear. Thoroughly vetted protocols will also be an important part of convincing parents to allow and promote device wearing in their children.

Routine clinical implementation of wearable devices and systems for pediatric rehabilitation populations will also depend on the physical design of the device and how it is secured to the individual. Size, weight, aesthetics, and ease of donning/doffing, are all key considerations. An ideal wearable device for pediatric patients would be small and light enough for infant wrists and ankles but robust enough to handle the stresses encountered while being worn by a child or teen for prolonged periods. The device must not impede or alter the movements that are being recorded [117]. The strap should be strong, non-absorbent, easy to remove and clean. For young children, playful or colorful aesthetics may increase compliance [118]. Simple hook-and-loop or buckle closures make donning and doffing devices easy for caregivers. Some pediatric populations will have sensory sensitivities and, for them, strapping that is soft and compliant but secure needs to be available. Devices are ideally waterproof, which can prevent recording interruptions during hand washing or bathing, and data capture can continue during swimming or other water-based activities. If these special considerations to implementation are overcome, wearable device systems can bring uniquely powerful diagnostic and prognostic information into pediatric care, a setting where real-world, objective assessment is otherwise very difficult.

3.5.2. Individuals with Cognitive Deficits

Great potential exists for the use of wearable devices in the routine clinical care of individuals with cognitive impairment. For example, ankle-mounted accelerometers have been used to distinguish between healthy controls and participants with Alzheimer's disease prior to the onset of major clinical behavioral impairments [119]. Cognitively impaired individuals cannot always accurately report on their performance in daily life and proxy-reports from caregivers may lack accuracy [120]. Wearable devices can avoid these issues by providing clinicians with objective, real-world data.

Several special considerations arise from the individual's cognitive impairment. Those individuals with more severe impairment may not understand the purpose of a wearable device. Devices that are small, light, and comfortable are more unobtrusive and will help avoid resistance to wear [3]. Device discomfort has led to as much as 25% withdrawal from studies of a neurological population [121]. The level of caregiver support available will be critical for success, as individuals with severe impairment may not be able to communicate discomfort (e.g., excess tightness, skin irritation) during device use. Device design could facilitate comfortable wearing by employing soft but strong materials and avoiding hard edges near areas of fragile skin. There will be a trade-off between the simplicity to don and doff by the patient themselves [3,121] versus minimizing the proclivity to remove the devices when it is not appropriate to do so. An option of a more secure clasp that requires caregivers or clinicians to remove may be appropriate. When possible, wearable devices will need to be attached to the patient and not their clothing, to avoid inadvertently laundering or discarding the device [121]. Wearable devices with long battery life could remove the need for cognitively-impaired individuals and/or caregivers to remember to monitor and charge the battery. Devices that are aesthetically pleasing may facilitate compliance [3], such as the emerging devices designed to appear as watches or jewelry (e.g., Motionlogger Micro Watch, AMI, Ardsley, NY, USA; Motiv Ring, Motiv, San Francisco, CA, USA). Lastly, cognitive impairment heightens the need for clear caregiver education related to the device protocol and interpretation of results, preferably with simple graphical instructions and data visualizations.

4. Benchmarks for Future Development

4.1. Proposed Benchmarks

While there are serious barriers to implementation of wearable device systems into routine physical rehabilitation practice, there is also substantial opportunity for growth and development in the nascent wearable sensor technology field. Here, we propose benchmarks for wearable device systems that might be readily adopted into clinical practice. Figure 3 is a visual illustration of the barriers and benchmarks. Each barrier is represented by a circle, with barriers that will be more difficult to overcome indicated by larger circles and hotter colors. Once barriers are overcome (circles shrink enough to fit through the funnel), then wide-spread clinical implementation (green rectangle) will be possible. The specific benchmarks (Table 2) are intended to serve as a guide for engineers, software developers, clinician-scientists, and clinicians alike as we pursue this important goal. They arise from the 10+ year history of using wearable device systems in our laboratory with more than 400 research participants, our clinical practice experience, and discussions with a range of rehabilitation clinicians.

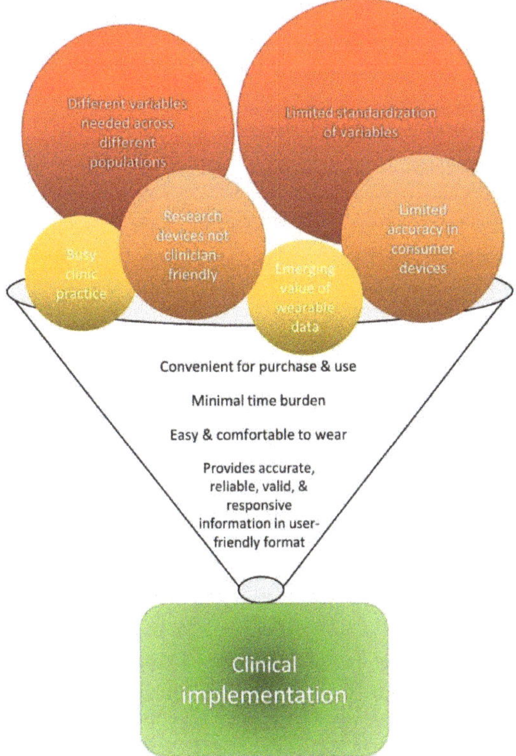

Figure 3. Barriers and benchmarks for implementation of wearable sensor technology into routine clinical rehabilitation practice. The warmest colors and largest circles indicate barriers that will take more work to overcome. Once the barriers are sufficiently reduced, then widespread clinical implementation (green rectangle) will be possible.

Table 2. Benchmarks for wearable device systems that will facilitate implementation into routine physical rehabilitation clinical practice.

	Benchmark
Convenience for purchase and use	Commercially-available, consumer-grade device system that can be easily used by clinicians and consumers; comprehensive, accessible tech support.
Initial set-up time for clinician	5–6 min for first time with new patient.
Routine set-up time for clinician	≤1 min in subsequent times with same patient.
Time to extract data and generate output or report	≤5 min
Ease of donning/doffing for patient	≤2 min; without assistance from another person if intended for home use.
Comfort for extended wear	Soft plastic or other flexible strapping that can be tolerated 12–24 h/day; no hard edges on device that push into skin; water resistant so does not have to be removed for bathing, dishwashing, etc.
Device operations	≥95% of the time, device collects, stores, and/or uploads data as programmed and does not malfunction.
Algorithms for extracting data and generating variables of interest	≥90% accuracy to measure intended construct; must be accurate across a broad range of movement abilities typically seen in physical rehabilitation clinics.
Standardization of variables of interest	Reliability: consistently captures construct with reliability coefficients of ≥0.80.
	Validity: comprehensively captures construct that has known relevance to clinical decision-making and management.
	Responsiveness: detects changes of ≥5%; changes of 5–10% or higher provide relevant information for clinical decision-making and management.
	Values can be computed & reported in sensor-independent units.
Report to clinician and patient	Consumer friendly, targeting audience with ≤ secondary school education; 1–3 key outcome variables presented; simple graphics with colors to make accessible across languages and language and/or cognitive deficits; ability to integrate into electronic medical record.

As with any technology, wearable device systems that are commercially-available with comprehensive, accessible technology support for end-users (rehabilitation professionals and patients) are necessary for routine implementation into clinical practice [122]. Moving a system from laboratory development into commercial production therefore becomes an important goal for system developers. System developers and their manufacturing partners will need to ensure that any systems sold as medical- (or research-grade) devices will need to meet the clinical safety standards defined by the International Organization of Standards (ISO/TC 173 for Assistive Products and ISO 10667-1:2011 for Assessment Service Delivery) and evaluated by national or multi-national regulatory agencies. Ideally, a system would be marketable to both clinicians and direct to consumers, as that would facilitate uptake [59,123]. While it is possible that research-grade device systems that are compliant with clinical regulatory standards can be adopted into clinical care, systems that are sold directly to patients and other consumers will better enable the wide-spread, ubiquitous adoption that would be most desirable. Time benchmarks flow from the previously discussed need to maximize patient care time and minimize undue burden for busy clinicians (Figures 1 and 2). While many, varying protocols for wearing duration may be used going forward, a device that can be comfortably worn for 12–24 h at time would be an excellent staring point and would facilitate adherence to wearing. A device system that works as intended/programmed for every patient is a prerequisite for wide-spread clinical adoption [59]. If devices are fickle and do not work (i.e. fail to record, store, or upload data) even a small percentage of the time, rehabilitation clinicians and patients will become frustrated and discontinue use. The most challenging and time-consuming benchmarks to achieve will be accurate algorithms and standardization of variables of interest that are reliable, valid, and responsive in the heterogeneous physical rehabilitation clinic populations. Since clinicians typically carry patient caseloads that include more than one patient population (e.g., Monday morning caseload includes a person with stroke, a person with multiple sclerosis, and a person with Parkinson disease), algorithms that work across populations and the ability to select specific variables of interest will make it easier to implement a wearable device system into daily practice. Finally, consumer-friendly reports that are quickly understandable to clinicians, patients, and families and integrated into the electronic medical record will allow the movement performance data collected from daily life to be used for shared clinical decision making and motivation.

4.2. Example Application of Benchmarks to A Currently-Available System

This section applies the benchmarks to a currently-available wearable device system as an example. The example system is the Actigraph wGT3X-BT device paired with the ActiLife software (Actigraph Inc, Pensacola, FL, USA). This system is one of the most commonly used in North America. Figure 4 is a picture of the wearable devices on a participant (Figure 4A) and a screen-shot of the software interface (Figure 4B). The wGT3X-BT contains a solid state, triaxial, microelectromechanical system (MEMS) accelerometer, an ambient light sensor, and a capacitive proximity sensor to detect wear time. It measures $4.6 \times 3.3 \times 1.5$ cm and weighs 19 g.

A.

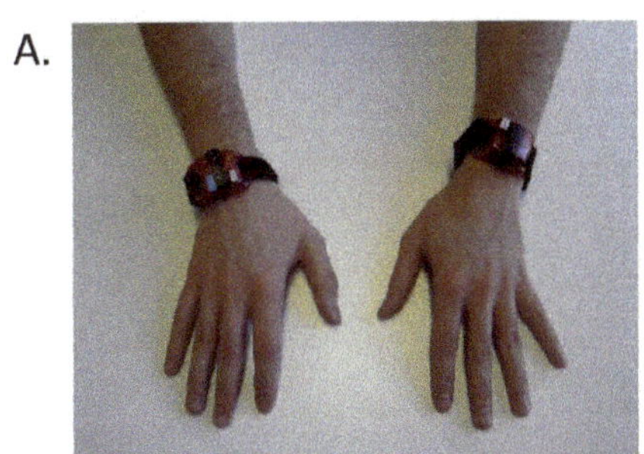

B.

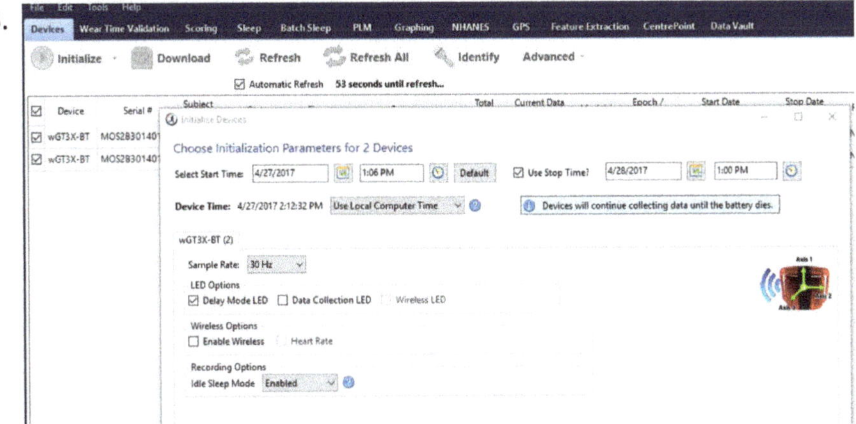

Figure 4. (**A**): Photo of a participant wearing the Actigraph wGT3X-BT devices on each wrist. (**B**): A screen shot of the ActiLife software interface used to interact with the wearable device. The software provides options to change various recording parameters (start and stop times, sampling rate, etc.) prior to recording/collecting data.

Sampling rates can be set within a 30–100 Hz range, with slower sampling rates enabling longer data collection periods. The accelerometer has a dynamic range of $+/-$ 8 g. Memory is 4 GB, with a battery life of 25 days and data storage possible for up to 180 days. The device has Bluetooth communication capability, which when activated, reduces battery life.

We benchmark this wearable device system's use for capturing upper limb performance in daily life, indexed by the use ratio variable in Table 3. As briefly mentioned in the previous section, the use ratio is a variable quantifying the relative duration of movement in one limb to the other. Determination of benchmark achievement was decided based on ours and others experience with the wearable device system (convenience, time, ease, comfort, device operations, and reporting benchmarks) and the references included in the prior section (algorithm and standardization benchmarks).

Table 3. Application of benchmarks to a current wearable device system, the Actigraph wGT3X-BT and ActiLife software for measuring upper limb performance in daily life, indexed by the use ratio variable.

	Progress toward Benchmark
Convenience for purchase and use	*Commercial-availability achieved.* Can be easily purchased. *Consumer-grade not achieved.* Marketed and sold as research-grade device. Technology support helpful for researchers but would be too difficult for clinician or patient consumers.
Initial set-up time for clinician	*Not achieved.* Current time estimate is 18 min.
Routine set-up time for clinician	*Not achieved.* Current time estimate is 8–10 min.
Time to extract data and generate output or report	*Not achieved.* Current time estimate, using ActiLife + custom-written software in MATLAB or R is 15 min
Ease of donning/doffing for patient	*Achieved.* Can be done at home for most patients without assistance from another person.
Comfort for extended wear	*Achieved.* Allows for variety of strapping options and has been worn 12–24 h by hundreds of patients, with many wearing it for 24 hrs 1x/wk or 1x/month. Water resistant.
Device operations	*Achieved.* Have lost data <2% of the time.
Algorithms for extracting data and generating variables of interest	*Achieved for use ratio.* Algorithm is stable across a range of movement abilities in typical adults and children, and persons with stroke.
Standardization of variable of interest: Use ratio	*Reliability achieved.* Test-retest reliability coefficient = 0.86 [79]
	Validity achieved for adult stroke population, but not other populations [99]. Captures relative use of the upper limbs, which is stable and narrowly distributed in referent populations [83,88], but wide-ranging post stroke.
	Responsive to change achieved. Can detect changes of ≤5% [32,79]. *Clinical relevance of change not achieved.* Currently unknown how much change is clinically meaningful.
	Sensor-independent units achieved. Values are a ratio, making differences across sensors irrelevant.
Report to clinician and patient	*Not achieved.* Current output can be consumed by trained researchers but is not clinician-, patient-, or family friendly. Output is not integrated with electronic medical record.

As can be seen in the right column of Table 3, around half of the proposed benchmarks have been achieved to date for this wearable device system and variable of interest. We respectfully note that this system is targeted for researchers and not for rehabilitation clinicians and patients. The conclusion drawn from Table 3 is that this wearable device system and its corresponding variable of interest does not yet meet all the benchmarks proposed here for widespread implementation into routine clinical rehabilitation practice. This variable of interest first appeared in published literature in the year 2000. Similarly, the wearable device system has been commercially available for more than 10 years and is the device system that is most commonly used in rehabilitation research studies. Thus, even a system that is commonly used by researchers is not yet ready for clinicians and patients.

Many new commercially- and laboratory-available wearable device systems are introduced each year. We are unable to benchmark a newer device system for this review because new systems, by default, have limited publications about how they operate and the variables that are derived from them. Publications from commercial entities are often not available at all, or come in the form of white papers that are not peer-reviewed. Publications from research laboratories about brand new device systems often report information about the device system and testing in a few healthy, young participants and/or a few patients, but do not yet have sufficient information/detail for assessing progress towards achievement of the proposed benchmarks for adoption into clinical care. As can be seen in this benchmarking exercise, it takes a long time for new devices to be picked up by the clinical research community and the generation of subsequent publications documenting clinical feasibility and standardization of variables. Nonetheless, the benchmarking exercise serves as a useful example of the collective progress that needs occur.

5. Conclusions

In the foreseeable future, the promise of wearable sensor technology for improving physical rehabilitation practice will be realized (Figure 3). A key goal for the field is to work to move wearable device systems that measure human movement into routine clinical practice. Major barriers to implementation arise from both current clinical practice and from the wearable device systems themselves. Clinical practice barriers include the busy clinical environment and a not-yet full realization of the value of the critical information that can be obtained. Wearable device system barriers include: (1) consumer-grade devices that are not accurate for many physical rehabilitation patient populations; (2) research-grade devices that are not user-friendly for clinicians or patients; (3) insufficient published data regarding reliability, validity, and responsiveness of output variables that can inform clinical decisions; and (4) the need to have these data on a range of output variables so that clinicians can select the most appropriate ones for specific patients.

As noted in the Introduction, we are at the beginning of this effort. Next generation wearable sensors and device systems will be smaller, faster, and allow enormous flexibility for clinicians and patients to gather data via wireless sensors networks and secure cloud technology, facilitating widespread adoption. Comparing the current status of the field to the development of telephones through smart phones, we are entering the early 1900s, when telephones were present and available in select locations, and switchboard operators were required to make connections between parties. Similarly, wearable devices systems have become increasing common in the research-world and require research-trained personnel to make them work. The collective efforts of engineers, computer scientists, clinician-scientists, and clinicians can push us all towards a future with sleek, ubiquitous, and easy to use wearable device systems in physical rehabilitation practice.

Author Contributions: All authors participated in conceptualization, writing, editing, reviewing, and approving this paper. All authors have read and agreed to the published version of the manuscript.

Funding: This work was funded by NIH R01HD068290.

Acknowledgments: We thank our current and former lab members and all of the research participants who have contributed to our understanding of how, when, and why to deploy wearable sensor systems for rehabilitation research and clinical practice.

Conflicts of Interest: The authors declare no conflict of interest. The funders had no role in the design of the study; in the collection, analyses, or interpretation of data; in the writing of the manuscript, or in the decision to publish the results.

References

1. WHO. *Towards A Common Language for Functioning, Disability, and Health: ICF*; World Health Organization: Geneva, Switzerland, 2002.
2. Dobkin, B.H.; Martinez, C. Wearable Sensors to Monitor, Enable Feedback, and Measure Outcomes of Activity and Practice. *Curr. Neurol. Neurosci. Rep.* **2018**, *18*, 87. [CrossRef] [PubMed]
3. Louie, D.R.; Bird, M.L.; Menon, C.; Eng, J.J. Perspectives on the prospective development of stroke-specific lower extremity wearable monitoring technology: A qualitative focus group study with physical therapists and individuals with stroke. *J. Neuroeng. Rehabil.* **2020**, *17*, 31. [CrossRef] [PubMed]
4. Hayhurst, C. Measuring by Value Not Volume. Available online: https://www.apta.org/apta-magazine/2015/07/01/measuring-by-value-not-volume (accessed on 15 July 2020).
5. Host, H.H.; Lang, C.E.; Hildebrand, M.W.; Zou, D.; Binder, E.F.; Baum, C.M.; Freedland, K.E.; Morrow-Howell, N.; Lenze, E.J. Patient Active Time During Therapy Sessions in Postacute Rehabilitation: Development and Validation of a New Measure. *Phys. Occup. Ther. Geriatr.* **2014**, *32*, 169–178. [CrossRef] [PubMed]

6.   Talkowski, J.B.; Lenze, E.J.; Munin, M.C.; Harrison, C.; Brach, J.S. Patient participation and physical activity during rehabilitation and future functional outcomes in patients after hip fracture. *Arch. Phys. Med. Rehabil.* **2009**, *90*, 618–622. [CrossRef] [PubMed]

7.   Bayley, M.T.; Hurdowar, A.; Richards, C.L.; Korner-Bitensky, N.; Wood-Dauphinee, S.; Eng, J.J.; McKay-Lyons, M.; Harrison, E.; Teasell, R.; Harrison, M.; et al. Barriers to implementation of stroke rehabilitation evidence: Findings from a multi-site pilot project. *Disabil. Rehabil.* **2012**, *34*, 1633–1638. [CrossRef] [PubMed]

8.   Noonan, V.K.; Moore, J.L. Knowledge Translation: The Catalyst for Innovation of Neurologic Physical Therapy. *J. Neurol. Phys. Ther.* **2016**, *40*, 67–70. [CrossRef]

9.   Wright, S.P.; Hall Brown, T.S.; Collier, S.R.; Sandberg, K. How consumer physical activity monitors could transform human physiology research. *Am. J. Physiol. Regul. Integr. Comp. Physiol.* **2017**, *312*, R358–R367. [CrossRef]

10. Standards of Practice for Physical Therapy. Available online: https://www.apta.org/apta-and-you/leadership-and-governance/policies/standards-of-practice-pt (accessed on 15 July 2020).

11. Blau, R.; Bolus, S.; Carolan, T.; Kramer, D.; Mahoney, E.; Jette, D.U.; Beal, J.A. The experience of providing physical therapy in a changing health care environment. *Phys. Ther.* **2002**, *82*, 648–657. [CrossRef]

12. Lang, C.E.; Waddell, K.J.; Klaesner, J.W.; Bland, M.D. A Method for Quantifying Upper Limb Performance in Daily Life Using Accelerometers. *J. Vis. Exp.* **2017**, *122*, 55673. [CrossRef]

13. Physical Therapy Outcomes Registry. Available online: http://www.ptoutcomes.com/home.aspx (accessed on 4 September 2020).

14. Moore, J.L.; Potter, K.; Blankshain, K.; Kaplan, S.L.; O'Dwyer, L.C.; Sullivan, J.E. A Core Set of Outcome Measures for Adults with Neurologic Conditions Undergoing Rehabilitation: A CLINICAL PRACTICE GUIDELINE. *J. Neurol. Phys. Ther.* **2018**, *42*, 174–220. [CrossRef]

15. Waddell, K.J.; Birkenmeier, R.L.; Bland, M.D.; Lang, C.E. An exploratory analysis of the self-reported goals of individuals with chronic upper-extremity paresis following stroke. *Disabil. Rehabil.* **2016**, *38*, 853–857. [CrossRef] [PubMed]

16. Danks, K.A.; Pohlig, R.T.; Roos, M.; Wright, T.R.; Reisman, D.S. Relationship Between Walking Capacity, Biopsychosocial Factors, Self-efficacy, and Walking Activity in Persons Poststroke. *J. Neurol. Phys. Ther.* **2016**, *40*, 232–238. [CrossRef] [PubMed]

17. Thilarajah, S.; Mentiplay, B.F.; Bower, K.J.; Tan, D.; Pua, Y.H.; Williams, G.; Koh, G.; Clark, R.A. Factors Associated with Post-Stroke Physical Activity: A Systematic Review and Meta-Analysis. *Arch. Phys. Med. Rehabil.* **2018**, *99*, 1876–1889. [CrossRef] [PubMed]

18. Holleran, C.L.; Bland, M.D.; Reisman, D.S.; Ellis, T.D.; Earhart, G.M.; Lang, C.E. Day-to-Day Variability of Walking Performance Measures in Individuals Poststroke and Individuals with Parkinson Disease. *J. Neurol. Phys. Ther.* **2020**, *44*, 241–247. [CrossRef] [PubMed]

19. Prince, S.A.; Adamo, K.B.; Hamel, M.E.; Hardt, J.; Connor Gorber, S.; Tremblay, M. A comparison of direct versus self-report measures for assessing physical activity in adults: A systematic review. *Int. J. Behav. Nutr. Phys. Act.* **2008**, *5*, 56. [CrossRef] [PubMed]

20. Waddell, K.J.; Lang, C.E. Comparison of Self-Report Versus Sensor-Based Methods for Measuring the Amount of Upper Limb Activity Outside the Clinic. *Arch. Phys. Med. Rehabil.* **2018**, *99*, 1913–1916. [CrossRef]

21. Rand, D.; Eng, J.J. Disparity between functional recovery and daily use of the upper and lower extremities during subacute stroke rehabilitation. *Neurorehabil. Neural. Repair.* **2012**, *26*, 76–84. [CrossRef]

22. Doman, C.A.; Waddell, K.J.; Bailey, R.R.; Moore, J.L.; Lang, C.E. Changes in Upper-Extremity Functional Capacity and Daily Performance During Outpatient Occupational Therapy for People with Stroke. *Am. J. Occup. Ther.* **2016**, *70*, 7003290040p1–7003290040p11. [CrossRef]

23. Waddell, K.J.; Strube, M.J.; Bailey, R.R.; Klaesner, J.W.; Birkenmeier, R.L.; Dromerick, A.W.; Lang, C.E. Does Task-Specific Training Improve Upper Limb Performance in Daily Life Poststroke? *Neurorehabil. Neural. Repair.* **2017**, *31*, 290–300. [CrossRef]

24. Ardestani, M.M.; Henderson, C.E.; Hornby, T.G. Improved walking function in laboratory does not guarantee increased community walking in stroke survivors: Potential role of gait biomechanics. *J. Biomech.* **2019**, *91*, 151–159. [CrossRef]

25. Benson, L.C.; Clermont, C.A.; Bosnjak, E.; Ferber, R. The use of wearable devices for walking and running gait analysis outside of the lab: A systematic review. *Gait Posture* **2018**, *63*, 124–138. [CrossRef]

26. Braito, I.; Maselli, M.; Sgandurra, G.; Inguaggiato, E.; Beani, E.; Cecchi, F.; Cioni, G.; Boyd, R. Assessment of upper limb use in children with typical development and neurodevelopmental disorders by inertial sensors: A systematic review. *J. Neuroeng. Rehabil.* **2018**, *15*, 94. [CrossRef] [PubMed]

27. Brickwood, K.J.; Watson, G.; O'Brien, J.; Williams, A.D. Consumer-Based Wearable Activity Trackers Increase Physical Activity Participation: Systematic Review and Meta-Analysis. *JMIR Mhealth Uhealth* **2019**, *7*, e11819. [CrossRef] [PubMed]

28. Danks, K.A.; Pohlig, R.; Reisman, D.S. Combining Fast-Walking Training and a Step Activity Monitoring Program to Improve Daily Walking Activity After Stroke: A Preliminary Study. *Arch. Phys. Med. Rehabil.* **2016**, *97* (Suppl. 9), S185–S193. [CrossRef]

29. Dobkin, B.H.; Dorsch, A. The promise of mHealth: Daily activity monitoring and outcome assessments by wearable sensors. *Neurorehabil. Neural. Repair.* **2011**, *25*, 788–798. [CrossRef] [PubMed]

30. Danks, K.A.; Roos, M.A.; McCoy, D.; Reisman, D.S. A step activity monitoring program improves real world walking activity post stroke. *Disabil. Rehabil.* **2014**, *36*, 2233–2236. [CrossRef]

31. Moore, J.L.; Roth, E.J.; Killian, C.; Hornby, T.G. Locomotor training improves daily stepping activity and gait efficiency in individuals poststroke who have reached a "plateau" in recovery. *Stroke* **2010**, *41*, 129–135. [CrossRef]

32. Urbin, M.A.; Waddell, K.J.; Lang, C.E. Acceleration Metrics Are Responsive to Change in Upper Extremity Function of Stroke Survivors. *Arch. Phys. Med. Rehabil.* **2015**, *96*, 854–861. [CrossRef]

33. Moore, S.A.; Hallsworth, K.; Plotz, T.; Ford, G.A.; Rochester, L.; Trenell, M.I. Physical activity, sedentary behaviour and metabolic control following stroke: A cross-sectional and longitudinal study. *PLoS ONE* **2013**, *8*, e55263. [CrossRef]

34. Cavanaugh, J.T.; Ellis, T.D.; Earhart, G.M.; Ford, M.P.; Foreman, K.B.; Dibble, L.E. Capturing ambulatory activity decline in Parkinson's disease. *J. Neurol. Phys. Ther.* **2012**, *36*, 51–57. [CrossRef]

35. Goldsack, J.C.; Coravos, A.; Bakker, J.P.; Bent, B.; Dowling, A.V.; Fitzer-Attas, C.; Godfrey, A.; Godino, J.G.; Gujar, N.; Izmailova, E.; et al. Verification, analytical validation, and clinical validation (V3): The foundation of determining fit-for-purpose for Biometric Monitoring Technologies (BioMeTs). *NPJ Digit. Med.* **2020**, *3*, 55. [CrossRef] [PubMed]

36. Evenson, K.R.; Goto, M.M.; Furberg, R.D. Systematic review of the validity and reliability of consumer-wearable activity trackers. *Int. J. Behav. Nutr. Phys. Act.* **2015**, *12*, 159. [CrossRef] [PubMed]

37. Fokkema, T.; Kooiman, T.J.; Krijnen, W.P.; Van der Schans, C.P.; De Groot, M. Reliability and Validity of Ten Consumer Activity Trackers Depend on Walking Speed. *Med. Sci. Sports Exerc.* **2017**, *49*, 793–800. [CrossRef] [PubMed]

38. Tophoj, K.H.; Petersen, M.G.; Saebye, C.; Baad-Hansen, T.; Wagner, S. Validity and Reliability Evaluation of Four Commercial Activity Trackers' Step Counting Performance. *Telemed. J. E Health* **2018**, *24*, 669–677. [CrossRef] [PubMed]

39. Duclos, N.C.; Aguiar, L.T.; Aissaoui, R.; Faria, C.; Nadeau, S.; Duclos, C. Activity Monitor Placed at the Nonparetic Ankle Is Accurate in Measuring Step Counts During Community Walking in Poststroke Individuals: A Validation Study. *PM R* **2019**, *11*, 963–971. [CrossRef]

40. Macko, R.F.; Haeuber, E.; Shaughnessy, M.; Coleman, K.L.; Boone, D.A.; Smith, G.V.; Silver, K.H. Microprocessor-based ambulatory activity monitoring in stroke patients. *Med. Sci. Sports Exerc.* **2002**, *34*, 394–399. [CrossRef]

41. Fulk, G.D.; Combs, S.A.; Danks, K.A.; Nirider, C.D.; Raja, B.; Reisman, D.S. Accuracy of 2 activity monitors in detecting steps in people with stroke and traumatic brain injury. *Phys. Ther.* **2014**, *94*, 222–229. [CrossRef]

42. Tedesco, S.; Sica, M.; Ancillao, A.; Timmons, S.; Barton, J.; O'Flynn, B. Accuracy of consumer-level and research-grade activity trackers in ambulatory settings in older adults. *PLoS ONE* **2019**, *14*, e0216891. [CrossRef]

43. Feehan, L.M.; Geldman, J.; Sayre, E.C.; Park, C.; Ezzat, A.M.; Yoo, J.Y.; Hamilton, C.B.; Li, L.C. Accuracy of Fitbit Devices: Systematic Review and Narrative Syntheses of Quantitative Data. *JMIR Mhealth Uhealth* **2018**, *6*, e10527. [CrossRef]

44. Larsen, R.T.; Korfitsen, C.B.; Juhl, C.B.; Andersen, H.B.; Langberg, H.; Christensen, J. Criterion validity for step counting in four consumer-grade physical activity monitors among older adults with and without rollators. *Eur. Rev. Aging Phys. Act.* **2020**, *17*, 1. [CrossRef]

45. Rozanski, G.M.; Aqui, A.; Sivakumaran, S.; Mansfield, A. Consumer Wearable Devices for Activity Monitoring Among Individuals After a Stroke: A Prospective Comparison. *JMIR Cardio* **2018**, *2*, e1. [CrossRef] [PubMed]

46. Clarke, C.L.; Taylor, J.; Crighton, L.J.; Goodbrand, J.A.; McMurdo, M.E.T.; Witham, M.D. Validation of the AX3 triaxial accelerometer in older functionally impaired people. *Aging Clin. Exp. Res.* **2017**, *29*, 451–457. [CrossRef] [PubMed]

47. Wendel, N.; Macpherson, C.E.; Webber, K.; Hendron, K.; DeAngelis, T.; Colon-Semenza, C.; Ellis, T. Accuracy of Activity Trackers in Parkinson Disease: Should We Prescribe Them? *Phys. Ther.* **2018**, *98*, 705–714. [CrossRef] [PubMed]

48. Maganja, S.A.; Clarke, D.C.; Lear, S.A.; Mackey, D.C. Formative Evaluation of Consumer-Grade Activity Monitors Worn by Older Adults: Test-Retest Reliability and Criterion Validity of Step Counts. *JMIR Form. Res.* **2020**, *4*, e16537. [CrossRef] [PubMed]

49. Sandroff, B.M.; Motl, R.W.; Pilutti, L.A.; Learmonth, Y.C.; Ensari, I.; Dlugonski, D.; Klaren, R.E.; Balantrapu, S.; Riskin, B.J. Accuracy of StepWatch and ActiGraph accelerometers for measuring steps taken among persons with multiple sclerosis. *PLoS ONE* **2014**, *9*, e93511. [CrossRef] [PubMed]

50. Wendland, D.M.; Sprigle, S.H. Activity monitor accuracy in persons using canes. *J. Rehabil. Res. Dev.* **2012**, *49*, 1261–1268. [CrossRef] [PubMed]

51. Mudge, S.; Stott, N.S. Test–retest reliability of the StepWatch Activity Monitor outputs in individuals with chronic stroke. *Clin. Rehabil.* **2008**, *22*, 871–877. [CrossRef]

52. Bowden, M.G.; Behrman, A.L. Step Activity Monitor: Accuracy and test-retest reliability in persons with incomplete spinal cord injury. *J. Rehabil. Res. Dev.* **2007**, *44*, 355–362. [CrossRef]

53. Haeuber, E.; Shaughnessy, M.; Forrester, L.W.; Coleman, K.L.; Macko, R.F. Accelerometer monitoring of home- and community-based ambulatory activity after stroke. *Arch. Phys. Med. Rehabil.* **2004**, *85*, 1997–2001. [CrossRef]

54. Schmidt, A.L.; Pennypacker, M.L.; Thrush, A.H.; Leiper, C.I.; Craik, R.L. Validity of the StepWatch Step Activity Monitor: Preliminary findings for use in persons with Parkinson disease and multiple sclerosis. *J. Geriatr. Phys. Ther.* **2011**, *34*, 41–45. [CrossRef]

55. Rand, D.; Eng, J.J.; Tang, P.F.; Jeng, J.S.; Hung, C. How active are people with stroke? Use of accelerometers to assess physical activity. *Stroke* **2009**, *40*, 163–168. [CrossRef] [PubMed]

56. Paul, S.S.; Ellis, T.D.; Dibble, L.E.; Earhart, G.M.; Ford, M.P.; Foreman, K.B.; Cavanaugh, J.T. Obtaining Reliable Estimates of Ambulatory Physical Activity in People with Parkinson's Disease. *J. Parkinsons. Dis.* **2016**, *6*, 301–305. [CrossRef] [PubMed]

57. Dorsch, A.K.; Thomas, S.; Xu, X.; Kaiser, W.; Dobkin, B.H.; Investigators, S. SIRRACT: An International Randomized Clinical Trial of Activity Feedback During Inpatient Stroke Rehabilitation Enabled by Wireless Sensing. *Neurorehabil. Neural. Repair.* **2015**, *29*, 407–415. [CrossRef] [PubMed]

58. Appelboom, G.; Yang, A.H.; Christophe, B.R.; Bruce, E.M.; Slomian, J.; Bruyere, O.; Bruce, S.S.; Zacharia, B.E.; Reginster, J.Y.; Connolly, E.S., Jr. The promise of wearable activity sensors to define patient recovery. *J. Clin. Neurosci.* **2014**, *21*, 1089–1093. [CrossRef] [PubMed]

59. Morris, J.; Jones, M.; Thompson, N.; Wallace, T.; DeRuyter, F. Clinician Perspectives on mRehab Interventions and Technologies for People with Disabilities in the United States: A National Survey. *Int. J. Environ. Res. Public Health* **2019**, *16*, 4220. [CrossRef] [PubMed]

60. Jerosch-Herold, C. An evidenced-based approach to choosing outcome measures: A checklist for the critical appraisal of validity, reliability, and responsiveness studies. *Brit. J. Occup. Ther.* **2005**, *68*, 347–353. [CrossRef]

61. Mokkink, L.B.; Terwee, C.B.; Patrick, D.L.; Alonso, J.; Stratford, P.W.; Knol, D.L.; Bouter, L.M.; de Vet, H.C. The COSMIN study reached international consensus on taxonomy, terminology, and definitions of measurement properties for health-related patient-reported outcomes. *J. Clin. Epidemiol.* **2010**, *63*, 737–745. [CrossRef]

62. Portney, L.G.; Watkins, M.P. *Foundations of Clinical Research: Applications to Clinical Practice*, 1st ed.; Appleton & Lange: Norwalk, CT, USA, 1993; p. 722.

63. *Guidance for Industry: Patient-Reported Outcome Measures Use in Medical Product Development to Support Labeling Claims*; Center for Biologics Evaluation and Research: Washington DC, USA, 2009.

64. Cavanaugh, J.T.; Ellis, T.D.; Earhart, G.M.; Ford, M.P.; Foreman, K.B.; Dibble, L.E. Toward Understanding Ambulatory Activity Decline in Parkinson Disease. *Phys. Ther.* **2015**, *95*, 1142–1150. [CrossRef]

65. Roos, M.A.; Rudolph, K.S.; Reisman, D.S. The structure of walking activity in people after stroke compared with older adults without disability: A cross-sectional study. *Phys. Ther.* **2012**, *92*, 1141–1147. [CrossRef]

66. Orendurff, M.S.; Schoen, J.A.; Bernatz, G.C.; Segal, A.D.; Klute, G.K. How humans walk: Bout duration, steps per bout, and rest duration. *J. Rehabil. Res. Dev.* **2008**, *45*, 1077–1089. [CrossRef]

67. Fulk, G.D.; Reynolds, C.; Mondal, S.; Deutsch, J.E. Predicting home and community walking activity in people with stroke. *Arch. Phys. Med. Rehabil.* **2010**, *91*, 1582–1586. [CrossRef]

68. Mudge, S.; Stott, N.S.; Walt, S.E. Criterion validity of the StepWatch Activity Monitor as a measure of walking activity in patients after stroke. *Arch. Phys. Med. Rehabil.* **2007**, *88*, 1710–1715. [CrossRef]

69. Jarchi, D.; Pope, J.; Lee, T.K.M.; Tamjidi, L.; Mirzaei, A.; Sanei, S. A Review on Accelerometry-Based Gait Analysis and Emerging Clinical Applications. *IEEE Rev. Biomed. Eng.* **2018**, *11*, 177–194. [CrossRef]

70. Bregou Bourgeois, A.; Mariani, B.; Aminian, K.; Zambelli, P.Y.; Newman, C.J. Spatio-temporal gait analysis in children with cerebral palsy using, foot-worn inertial sensors. *Gait Posture* **2014**, *39*, 436–442. [CrossRef] [PubMed]

71. Chadwell, A.; Kenney, L.; Granat, M.; Thies, S.; Head, J.S.; Galpin, A. Visualisation of upper limb activity using spirals: A new approach to the assessment of daily prosthesis usage. *Prosthet. Orthot. Int.* **2018**, *42*, 37–44. [CrossRef] [PubMed]

72. Chadwell, A.; Kenney, L.; Granat, M.H.; Thies, S.; Head, J.; Galpin, A.; Baker, R.; Kulkarni, J. Upper limb activity in myoelectric prosthesis users is biased towards the intact limb and appears unrelated to goal-directed task performance. *Sci. Rep.* **2018**, *8*, 11084. [CrossRef]

73. Gebruers, N.; Vanroy, C.; Truijen, S.; Engelborghs, S.; De Deyn, P.P. Monitoring of physical activity after stroke: A systematic review of accelerometry-based measures. *Arch. Phys. Med. Rehabil.* **2010**, *91*, 288–297. [CrossRef]

74. Trujillo-Priego, I.A.; Lane, C.J.; Vanderbilt, D.L.; Deng, W.; Loeb, G.E.; Shida, J.; Smith, B.A. Development of a Wearable Sensor Algorithm to Detect the Quantity and Kinematic Characteristics of Infant Arm Movement Bouts Produced across a Full Day in the Natural Environment. *Technologies* **2017**, *5*, 39. [CrossRef]

75. de Lucena, D.S.; Stoller, O.; Rowe, J.B.; Chan, V.; Reinkensmeyer, D.J. Wearable sensing for rehabilitation after stroke: Bimanual jerk asymmetry encodes unique information about the variability of upper extremity recovery. *IEEE Int. Conf. Rehabil. Robot.* **2017**, *2017*, 1603–1608.

76. Urbin, M.A.; Bailey, R.R.; Lang, C.E. Validity of body-worn sensor acceleration metrics to index upper extremity function in hemiparetic stroke. *J. Neurol. Phys. Ther.* **2015**, *39*, 111–118. [CrossRef]

77. Urbin, M.A.; Hong, X.; Lang, C.E.; Carter, A.R. Resting-State Functional Connectivity and Its Association with Multiple Domains of Upper-Extremity Function in Chronic Stroke. *Neurorehabil. Neural. Repair.* **2014**, *28*, 761–769. [CrossRef] [PubMed]

78. van der Pas, S.C.; Verbunt, J.A.; Breukelaar, D.E.; van Woerden, R.; Seelen, H.A. Assessment of arm activity using triaxial accelerometry in patients with a stroke. *Arch. Phys. Med. Rehabil.* **2011**, *92*, 1437–1442. [CrossRef] [PubMed]

79. Uswatte, G.; Foo, W.L.; Olmstead, H.; Lopez, K.; Holand, A.; Simms, L.B. Ambulatory monitoring of arm movement using accelerometry: An objective measure of upper-extremity rehabilitation in persons with chronic stroke. *Arch. Phys. Med. Rehabil.* **2005**, *86*, 1498–1501. [CrossRef] [PubMed]

80. Uswatte, G.; Giuliani, C.; Winstein, C.; Zeringue, A.; Hobbs, L.; Wolf, S.L. Validity of accelerometry for monitoring real-world arm activity in patients with subacute stroke: Evidence from the extremity constraint-induced therapy evaluation trial. *Arch. Phys. Med. Rehabil.* **2006**, *87*, 1340–1345. [CrossRef] [PubMed]

81. Uswatte, G.; Miltner, W.H.; Foo, B.; Varma, M.; Moran, S.; Taub, E. Objective measurement of functional upper-extremity movement using accelerometer recordings transformed with a threshold filter. *Stroke* **2000**, *31*, 662–667. [CrossRef]

82. Hoyt, C.R.; Brown, S.K.; Sherman, S.K.; Wood-Smith, M.; Van, A.N.; Ortega, M.; Nguyen, A.L.; Lang, C.E.; Schlaggar, B.L.; Dosenbach, N.U.F. Using accelerometry for measurement of motor behavior in children: Relationship of real-world movement to standardized evaluation. *Res. Dev. Disabil.* **2020**, *96*, 103546. [CrossRef]

83. Hoyt, C.R.; Van, A.N.; Ortega, M.; Koller, J.M.; Everett, E.A.; Nguyen, A.L.; Lang, C.E.; Schlaggar, B.L.; Dosenbach, N.U.F. Detection of Pediatric Upper Extremity Motor Activity and Deficits with Accelerometry. *JAMA Netw. Open* **2019**, *2*, e192970. [CrossRef]

84. Bailey, R.R. *Assessment of Real-World Upper Limb Activity in Adults with Chronic Stroke*; Washington University: St. Louis, MO, USA, 2015.

85. Bailey, R.R.; Birkenmeier, R.L.; Lang, C.E. Real-world affected upper limb activity in chronic stroke: An examination of potential modifying factors. *Top. Stroke Rehabil.* **2015**, *22*, 26–33. [CrossRef]

86. Bailey, R.R.; Klaesner, J.W.; Lang, C.E. An accelerometry-based methodology for assessment of real-world bilateral upper extremity activity. *PLoS ONE* **2014**, *9*, e103135. [CrossRef]

87. Bailey, R.R.; Klaesner, J.W.; Lang, C.E. Quantifying Real-World Upper-Limb Activity in Nondisabled Adults and Adults with Chronic Stroke. *Neurorehabilit. Neural Repair* **2015**, *29*, 969–978. [CrossRef]

88. Bailey, R.R.; Lang, C.E. Upper-limb activity in adults: Referent values using accelerometry. *J. Rehabil. Res. Dev.* **2013**, *50*, 1213–1222. [CrossRef] [PubMed]

89. Lemmens, R.J.; Timmermans, A.A.; Janssen-Potten, Y.J.; Pulles, S.A.; Geers, R.P.; Bakx, W.G.; Smeets, R.J.; Seelen, H.A. Accelerometry measuring the outcome of robot-supported upper limb training in chronic stroke: A randomized controlled trial. *PLoS ONE* **2014**, *9*, e96414. [CrossRef] [PubMed]

90. Lang, C.E.; Wagner, J.M.; Edwards, D.F.; Dromerick, A.W. Upper Extremity Use in People with Hemiparesis in the First Few Weeks After Stroke. *J. Neurol. Phys. Ther.* **2007**, *31*, 56–63. [CrossRef]

91. Lang, C.E.; Edwards, D.F.; Birkenmeier, R.L.; Dromerick, A.W. Estimating minimal clinically important differences of upper-extremity measures early after stroke. *Arch. Phys. Med. Rehabil.* **2008**, *89*, 1693–1700. [CrossRef] [PubMed]

92. Sokal, B.; Uswatte, G.; Vogtle, L.; Byrom, E.; Barman, J. Everyday movement and use of the arms: Relationship in children with hemiparesis differs from adults. *J. Pediatr. Rehabil. Med.* **2015**, *8*, 197–206. [CrossRef]

93. Seitz, R.J.; Hildebold, T.; Simeria, K. Spontaneous arm movement activity assessed by accelerometry is a marker for early recovery after stroke. *J. Neurol.* **2011**, *258*, 457–463. [CrossRef]

94. Balasubramanian, S.; Melendez-Calderon, A.; Burdet, E. A robust and sensitive metric for quantifying movement smoothness. *IEEE Trans. Biomed. Eng.* **2012**, *59*, 2126–2136. [CrossRef] [PubMed]

95. Balasubramanian, S.; Melendez-Calderon, A.; Roby-Brami, A.; Burdet, E. On the analysis of movement smoothness. *J. Neuroeng. Rehabil.* **2015**, *12*, 112. [CrossRef]

96. Thrane, G.; Emaus, N.; Askim, T.; Anke, A. Arm use in patients with subacute stroke monitored by accelerometry: Association with motor impairment and influence on self-dependence. *J. Rehabil. Med.* **2011**, *43*, 299–304. [CrossRef]

97. Lang, C.E.; Cade, W.T. A step toward the future of seamless measurement with wearable sensors in pediatric populations with neuromuscular diseases. *Muscle Nerve* **2020**, *61*, 265–267. [CrossRef] [PubMed]

98. van der Geest, A.; Essers, J.M.N.; Bergsma, A.; Jansen, M.; de Groot, I.J.M. Monitoring daily physical activity of upper extremity in young and adolescent boys with Duchenne muscular dystrophy: A pilot study. *Muscle Nerve* **2020**, *61*, 293–300. [CrossRef]

99. Smith, B.A.; Lang, C.E. Sensor Measures of Symmetry Quantify Upper Limb Movement in the Natural Environment Across the Lifespan. *Arch. Phys. Med. Rehabil.* **2019**, *100*, 1176–1183. [CrossRef]

100. Lang, C.E.; Bland, M.D.; Bailey, R.R.; Schaefer, S.Y.; Birkenmeier, R.L. Assessment of upper extremity impairment, function, and activity after stroke: Foundations for clinical decision making. *J. Hand. Ther.* **2013**, *26*, 104–115. [CrossRef]

101. Loprinzi, P.D.; Cardinal, B.J. Measuring children's physical activity and sedentary behaviors. *J. Exerc. Sci. Fit.* **2011**, *9*, 15–23. [CrossRef]

102. Rice, K.R.; Joschtel, B.; Trost, S.G. Validity of family child care providers' proxy reports on children's physical activity. *Child. Obes.* **2013**, *9*, 393–398. [CrossRef]

103. Lobenius-Palmer, K.; Sjoqvist, B.; Hurtig-Wennlof, A.; Lundqvist, L.O. Accelerometer-Assessed Physical Activity and Sedentary Time in Youth with Disabilities. *Adapt. Phys. Activ. Q.* **2018**, *35*, 1–19. [CrossRef]

104. King-Dowling, S.; Rodriguez, C.; Missiuna, C.; Timmons, B.W.; Cairney, J. Health-related Fitness in Preschool Children with and without Motor Delays. *Med. Sci. Sports Exerc.* **2018**, *50*, 1442–1448. [CrossRef]

105. Wood, A.C.; Asherson, P.; Rijsdijk, F.; Kuntsi, J. Is overactivity a core feature in ADHD? Familial and receiver operating characteristic curve analysis of mechanically assessed activity level. *J. Am. Acad. Child. Adolesc. Psychiatry* **2009**, *48*, 1023–1030. [CrossRef]

106. Pan, C.Y.; Tsai, C.L.; Chu, C.H.; Sung, M.C.; Ma, W.Y.; Huang, C.Y. Objectively Measured Physical Activity and Health-Related Physical Fitness in Secondary School-Aged Male Students with Autism Spectrum Disorders. *Phys. Ther.* **2016**, *96*, 511–520. [CrossRef]

107. Brewis, A. Social and biological measures of hyperactivity and inattention: Are they describing similar underlying constructs of child behavior? *Soc. Biol.* **2002**, *49*, 99–115. [CrossRef]

108. Uebel, H.; Albrecht, B.; Kirov, R.; Heise, A.; Dopfner, M.; Freisleder, F.J.; Gerber, W.D.; Gunter, M.; Hassler, F.; Ose, C.; et al. What can actigraphy add to the concept of labschool design in clinical trials? *Curr. Pharm. Des.* **2010**, *16*, 2434–2442. [CrossRef]

109. Lea, S.E.; Matt Alderson, R.; Patros, C.H.G.; Tarle, S.J.; Arrington, E.F.; Grant, D.M. Working Memory and Motor Activity: A Comparison Across Attention-Deficit/Hyperactivity Disorder, Generalized Anxiety Disorder, and Healthy Control Groups. *Behav. Ther.* **2018**, *49*, 419–434. [CrossRef]

110. Gapin, J.; Etnier, J.L. The relationship between physical activity and executive function performance in children with attention-deficit hyperactivity disorder. *J. Sport Exerc. Psychol.* **2010**, *32*, 753–763. [CrossRef]

111. Heathers, J.A.J.; Gilchrist, K.H.; Hegarty-Craver, M.; Grego, S.; Goodwin, M.S. An analysis of stereotypical motor movements and cardiovascular coupling in individuals on the autism spectrum. *Biol. Psychol.* **2019**, *142*, 90–99. [CrossRef]

112. Garcia-Pastor, T.; Salinero, J.J.; Theirs, C.I.; Ruiz-Vicente, D. Obesity Status and Physical Activity Level in Children and Adults with Autism Spectrum Disorders: A Pilot Study. *J. Autism. Dev. Disord.* **2019**, *49*, 165–172. [CrossRef]

113. Benson, S.; Bender, A.M.; Wickenheiser, H.; Naylor, A.; Clarke, M.; Samuels, C.H.; Werthner, P. Differences in sleep patterns, sleepiness, and physical activity levels between young adults with autism spectrum disorder and typically developing controls. *Dev. Neurorehabil.* **2019**, *22*, 164–173. [CrossRef]

114. Goldman, S.E.; Alder, M.L.; Burgess, H.J.; Corbett, B.A.; Hundley, R.; Wofford, D.; Fawkes, D.B.; Wang, L.; Laudenslager, M.L.; Malow, B.A. Characterizing Sleep in Adolescents and Adults with Autism Spectrum Disorders. *J. Autism. Dev. Disord.* **2017**, *47*, 1682–1695. [CrossRef]

115. Abrishami, M.S.; Nocera, L.; Mert, M.; Trujillo-Priego, I.A.; Purushotham, S.; Shahabi, C.; Smith, B.A. Identification of Developmental Delay in Infants Using Wearable Sensors: Full-Day Leg Movement Statistical Feature Analysis. *IEEE J. Transl. Eng. Health Med.* **2019**, *7*, 2800207. [CrossRef]

116. Heinze, F.; Hesels, K.; Breitbach-Faller, N.; Schmitz-Rode, T.; Disselhorst-Klug, C. Movement analysis by accelerometry of newborns and infants for the early detection of movement disorders due to infantile cerebral palsy. *Med. Biol. Eng. Comput.* **2010**, *48*, 765–772. [CrossRef]

117. Jiang, C.; Lane, C.J.; Perkins, E.; Schiesel, D.; Smith, B.A. Determining if wearable sensors affect infant leg movement frequency. *Dev. Neurorehabil.* **2018**, *21*, 133–136. [CrossRef]

118. Zhu, Z.; Liu, T.; Li, G.; Li, T.; Inoue, Y. Wearable sensor systems for infants. *Sensors* **2015**, *15*, 3721–3749. [CrossRef] [PubMed]

119. Kirste, T.; Hoffmeyer, A.; Koldrack, P.; Bauer, A.; Schubert, S.; Schroder, S.; Teipel, S. Detecting the effect of Alzheimer's disease on everyday motion behavior. *J. Alzheimers. Dis.* **2014**, *38*, 121–132. [CrossRef] [PubMed]

120. Matthews, L.; Hankey, C.; Penpraze, V.; Boyle, S.; Macmillan, S.; Miller, S.; Murray, H.; Pert, C.; Spanos, D.; Robinson, N.; et al. Agreement of accelerometer and a physical activity questionnaire in adults with intellectual disabilities. *Prev. Med.* **2011**, *52*, 361–364. [CrossRef] [PubMed]

121. Block, V.A.; Pitsch, E.; Tahir, P.; Cree, B.A.; Allen, D.D.; Gelfand, J.M. Remote Physical Activity Monitoring in Neurological Disease: A Systematic Review. *PLoS ONE* **2016**, *11*, e0154335. [CrossRef] [PubMed]

122. Jacob, C.; Sanchez-Vazquez, A.; Ivory, C. Social, Organizational, and Technological Factors Impacting Clinicians' Adoption of Mobile Health Tools: Systematic Literature Review. *JMIR Mhealth Uhealth* **2020**, *8*, e15935. [CrossRef] [PubMed]

123. Nussbaum, R.; Kelly, C.; Quinby, E.; Mac, A.; Parmanto, B.; Dicianno, B.E. Systematic Review of Mobile Health Applications in Rehabilitation. *Arch. Phys. Med. Rehabil.* **2019**, *100*, 115–127. [CrossRef]

Communication

Wearable Activity Monitoring in Day-to-Day Stroke Care: A Promising Tool but Not Widely Used

Hanneke E. M. Braakhuis [1,2,3,*], **Johannes B. J. Bussmann** [1,3], **Gerard M. Ribbers** [1,3] **and Monique A. M. Berger** [2]

[1] Department of Rehabilitation Medicine, Erasmus MC University Medical Center, P.O. Box 2040, 3000 CA Rotterdam, The Netherlands; j.b.j.bussmann@erasmusmc.nl (J.B.J.B.); g.ribbers@erasmusmc.nl (G.M.R.)

[2] Faculty of Health, Nutrition and Sport, The Hague University of Applied Sciences, 2521 EN The Hague, The Netherlands; m.a.m.berger@hhs.nl

[3] Rijndam Rehabilitation, 3015 LJ Rotterdam, The Netherlands

* Correspondence: h.braakhuis@erasmusmc.nl

Abstract: Physical activity monitoring with wearable technology has the potential to support stroke rehabilitation. Little is known about how physical therapists use and value the use of wearable activity monitors. This cross-sectional study explores the use, perspectives, and barriers to wearable activity monitoring in day-to-day stroke care routines amongst physical therapists. Over 300 physical therapists in primary and geriatric care and rehabilitation centers in the Netherlands were invited to fill in an online survey that was developed based on previous studies and interviews with experts. In total, 103 complete surveys were analyzed. Out of the 103 surveys, 27% of the respondents were already using activity monitoring. Of the suggested treatment purposes of activity monitoring, 86% were perceived as useful by more than 55% of the therapists. The most recognized barriers to clinical implementation were lack of skills and knowledge of patients (65%) and not knowing what brand and type of monitor to choose (54%). Of the non-users, 79% were willing to use it in the future. In conclusion, although the concept of remote activity monitoring was perceived as useful, it was not widely adopted by physical therapists involved in stroke care. To date, skills, beliefs, and attitudes of individual therapists determine the current use of wearable technology.

Keywords: wearable technology; rehabilitation; stroke; implementation; physical therapy

Citation: Braakhuis, H.E.M.; Bussmann, J.B.J.; Ribbers, G.M.; Berger, M.A.M. Wearable Activity Monitoring in Day-to-Day Stroke Care: A Promising Tool but Not Widely Used. *Sensors* **2021**, *21*, 4066. https://doi.org/10.3390/s21124066

Academic Editor: Mario Munoz-Organero

Received: 7 April 2021
Accepted: 9 June 2021
Published: 12 June 2021

Publisher's Note: MDPI stays neutral with regard to jurisdictional claims in published maps and institutional affiliations.

1. Introduction

Stroke is a major cause of disability and is an age-dependent problem [1]. With an aging society and improved acute care, the number of stroke survivors living with long-term stroke consequences is increasing beyond the level of increase of professional capacity [2,3]. Many stroke survivors show deteriorated levels of functioning, with low levels of physical activity [4,5]. Being physically active is an important determinant of social participation and is a major target of stroke rehabilitation [6]. Furthermore, being physically active is related to physical and psychosocial functioning, quality of life, and reduction of cardiovascular risk factors [7–10].

Physical activity is one of the components of physical behavior, that covers all movements, postures, and activities of a person's during their daily life [11]. Another component is sedentary behavior, which is associated with cardiovascular disease incidence and mortality and depressive symptoms too [12,13]. Targeting stroke rehabilitation by increasing physical activity and decreasing sedentary behaviors may help to suppress the burden of stroke.

Stroke rehabilitation could benefit from remote monitoring of physical behavior with wearable sensor technology [14]. The development of wearable activity monitors has rapidly evolved over the last decades in academic research and the consumer market [15,16]. They provide an objective insight into behavior in a non-invasive and continuous way and can be applied in the home environments as well as in in- and outpatient settings to patients and therapists [17]. In addition, increased patient involvement by providing feedback on physical activity may enhance compliance and stimulate self-management [18]. The objective insights also allow therapists to set tailored therapy goals, guide patients towards them, and evaluate progress [19,20].

Although the body of evidence of remote monitoring of physical activity is growing in academic research, its clinical implementation lags behind [21,22]. Adopting technologies in day-to-day care routines seems challenging for therapists, who are key players in adopting remote monitoring of physical activity [22], since it requires careful attention, precious time, sufficient organizational and technical infrastructure, and knowledge [23–27]. Studies indicate that physical therapists acknowledge the potential benefits and practical purposes of wearable activity monitoring in rehabilitation therapy [28–30]. However, so far these studies have applied individual interviews and small focus groups. To provide an extensive insight into the current uses and clues on how to push the clinical implementation of this technology in stroke care forward, a study with a wide group of physical therapists involved in stroke care is needed. Therefore, the current study aimed to explore the use, perspectives, and barriers to potential applications of wearable activity monitoring in day-to-day stroke care amongst physical therapists in the Netherlands.

2. Materials and Methods

2.1. Participants and Data Collection

This cross-sectional study used an online survey (LimeSurvey®) among physical therapists in the Netherlands involved in post-stroke rehabilitation. Therapists were included if they were involved in the treatment of at least one stroke patient in the last month in a rehabilitation center, geriatric care center, or in primary care in the Netherlands. Participants were invited by e-mail with a web link via contact persons of seven primary care stroke networks in the Netherlands and ten Dutch rehabilitation centers and via a newsletter of the special interest group "rehabilitation" of the Royal Dutch Society of Physical therapy (KNGF: Koninklijk Nederlands Genootschap voor Fysiotherapie). After three weeks, a reminder for filling in the questionnaire was sent. Surveys were filled in anonymously.

2.2. Survey Development

A research team of physical therapists, human movement scientists, and researchers developed the survey based on literature and interviews. The survey included questions on demographic and occupational characteristics. Literature was used to formulate questions on the following topics: innovativeness (multiple choice answers to the question on innovativeness were based on the descriptions of the adoption categories of Rogers [31]), health care technology, activity monitoring outcome measures, perceived usefulness, barriers, and willingness to use it in the future [15,16,27,29,32] (See Supplementary Materials for the complete survey). To measure the attitudes of the participants regarding these questions, a 5-point Likert scale was used [33]. Participants were also asked if they were familiar with activity monitoring, if they use it for tracking their own activities, and if they already use it in stroke care. If a participant answered "yes" to the question concerning use in stroke care, they were defined as a user, and otherwise as a non-user. The users received additional questions about the use in day-to-day practice. They were asked how long they have been applying it, for how many patients per week, for what purpose, and what outcome of physical behavior they were interested in. Additionally, with an open-ended question, the reason for use was questioned. At the end of the survey, all participants were asked by an open-ended question if they wanted to share anything else on activity monitoring in stroke care.

To ensure common understanding, definitions were explained in between the questions (see Supplementary Materials). Experts and physical therapists checked the initial survey for face validity, comprehensibility, vocabulary, and layout. The survey was pilot-tested by five physical therapists in primary care before distribution.

2.3. Data Analysis

RStudio (version 1.2.50001, RStudio, Inc., Boston, MA, USA) was used for the data analyses. Descriptive analysis was provided for all questions with means (SD), frequencies, and percentages. The Likert package [34] was used to visualize the questions answered with a Likert scale. Differences between users and non-users were carried out with Chi^2 and Mann–Whitney U tests. The significance was set at $\alpha = 0.05$.

All individual answers to the open-ended question were collected in Microsoft Excel for qualitative analysis. All answers were divided into emergent themes. The most frequent, remarkable, or important issues that were relevant to this study were extracted and reported in the results.

3. Results

3.1. Participants

Over 300 physical therapists received the e-mail with the invitation to fill in the online questionnaire. Of them, approximately 100 therapists were recruited via a primary care stroke network and approximately 200 therapists were recruited via a contact person within their rehabilitation center. The survey was available from 1 March till 1 June 2020. n = 132 started the survey via the web link and n = 103 completed the questionnaire (78%). Only complete surveys were used for further analysis.

Table 1 shows the demographic characteristics of the participants. The mean age of the study sample was 42.2 (SD 12.1) years. Most of the participants worked in a rehabilitation center as a physical therapist (n = 58). Nine participants were employed in two or three different settings. All therapists were involved in the treatment of stroke patients. Other patient groups treated by the therapists were congenital and acquired brain injuries, (inactive) elderly, chronic diseases, orthopedic conditions, and sports injuries.

Table 1. Demographic characteristics of respondents.

		Total (n = 103)	Users (n = 28) (27%)	Non-Users (n = 75) (73%)	*p*-Value
Age, mean (SD)		42.2 (12.06)	41.70 (13.24)	45.30 (12.11)	0.212
Gender (m/f)		26/76	8/20	18/56	0.420
	<5	9 (8.7%)	2 (7.1%)	7 (9.3%)	0.331
Years of work	5–10	18 (17.5%)	8 (28.6%)	10 (13.3%)	
Experience, n	10–15	22 (21.4%)	7 (25.0%)	15 (20.0%)	
(%)	15–20	7 (6.8%)	2 (7.1%)	5 (6.7%)	
	>20 years	47 (45.6%)	9 (32.1%)	38 (50.7%)	
Setting [a] (n)	Primary care	34	7	27	
	Rehabilitation	59	21	38	
	Geriatric care	20	2	18	

[a] = participants were allowed to fill in multiple answers; user is defined by answering "yes" on the question if they already use activity monitoring during their work as a physical therapist.

Twenty-seven percent used activity monitoring in the treatment of stroke patients and were defined as users. Characteristics of both groups and differences between them are presented in Table 1.

More than half of the non-users (59%) were familiar with activity monitoring before filling in this questionnaire. Similar percentages of users (54%) and non-users (53%) used a smartphone app or consumer-grade activity tracker for monitoring their own lifestyle and sports activities. Two participants (1.9%) considered themselves as people who were initially reluctant to use new healthcare technology and innovations. Most of the therapists in the total study sample described themselves as a person who had no problem going along with pioneers in healthcare technology and innovation but who did not initiate it themselves (60%). Only one (0.9%) of the therapists described himself as someone who invented and designed new healthcare technology and innovations and 18% of the total sample said they were someone who followed the latest developments in healthcare technology and innovation and looked for applications in practice.

The most often used health care technologies in the total study sample, other than activity monitoring, were applications and websites supporting the patient with practicing (21% often, 5% very often). The least often used was technology that supported diagnostics (15% often, 0% very often). Users of activity monitoring used significantly more other health care technologies (apps/websites, $p = 0.036$; online consulting (expert) colleagues, $p = 0.023$; technology that supports diagnostics, $p = 0.009$; and technology that supports treatment, $p = 0.026$) compared to non-users (Figure 1).

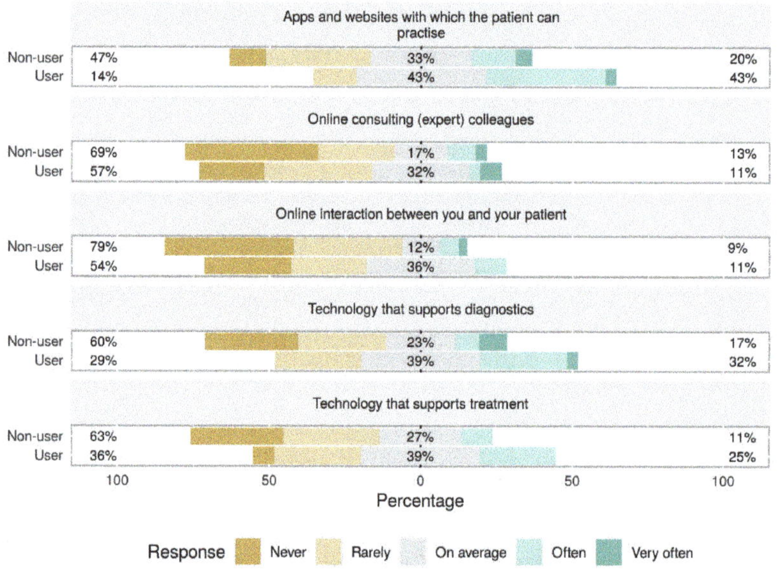

Figure 1. Other health technology used by participants, with differences between users and non-users.

3.2. Users

Most users (54%) have been applying activity monitoring between six months and two years. Thirty-six percent have been applying activity monitoring shorter than six months, and eleven percent longer than two years. Most of the users applied activity monitoring between one and five patients per month (61%). Thirty-two percent applied activity monitoring in one patient per month or less, and seven percent in more than five patients per month.

Figure 2A shows the treatment purposes of activity monitoring of the users. Almost all therapists used the monitor to create awareness for the patient with regard to their physical behavior (96%). Giving feedback about their physical behavior (82%) was also often recognized as a useful activity monitoring purpose. Figure 2B shows the activity monitor outcomes of interest during treatment. Most of them were interested in the number of steps. Additional outcomes of interest reported by users were heart rate and demands vs. capability, or in other words, the relation between what a patient did compared to what the patient was capable of.

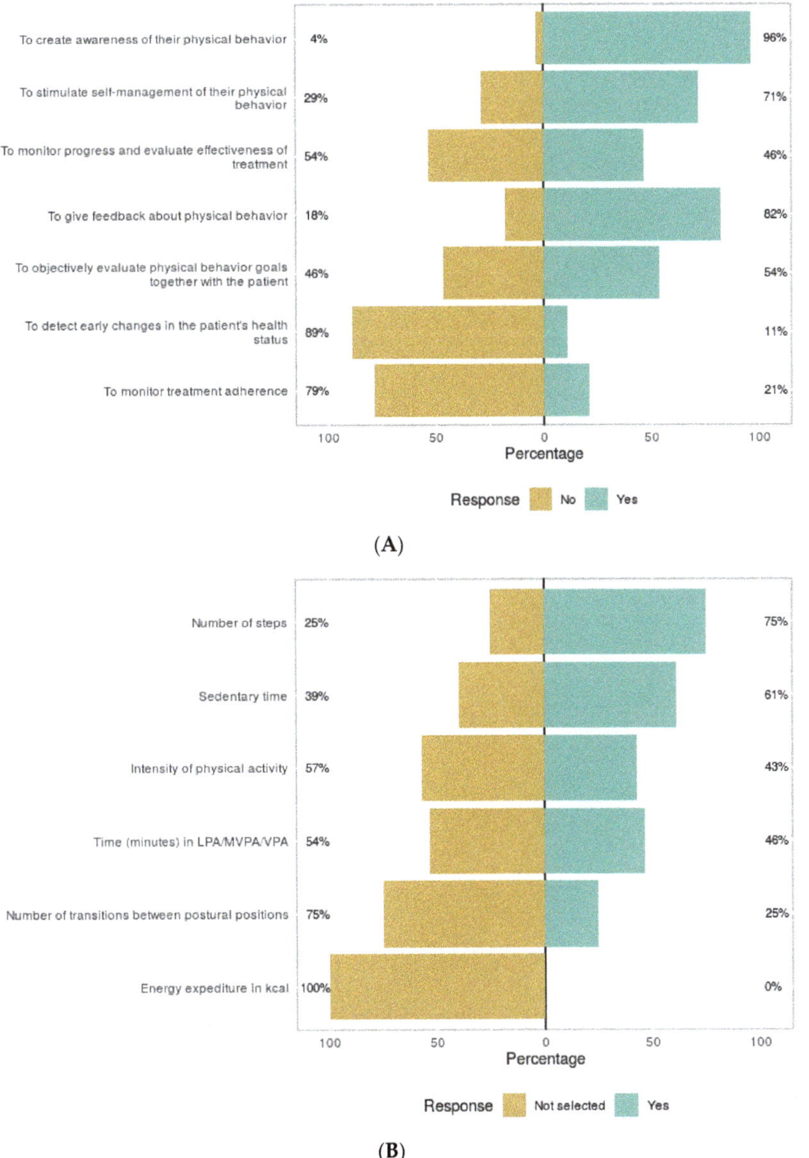

Figure 2. Treatment purposes (**A**) evaluated by users and outcome of interest of users (**B**). LPA= low physical activity, MVPA = moderate to vigorous physical activity, VPA = vigorous physical activity.

In addition to the purposes in Figure 2A, users filled in for what reason they applied activity monitoring. Some of them reported new purposes compared to the ones provided in the answers; that they were instructed or motivated by external factors such as other colleagues who were already working with activity monitors or research/projects initiated by their organizations.

3.3. Perceived Usefulness

All participants (users and non-users) were asked for their opinion about the usefulness of activity monitoring for stroke patients. Six out of seven suggested purposes were considered useful by more than half of the study sample (Figure 3).

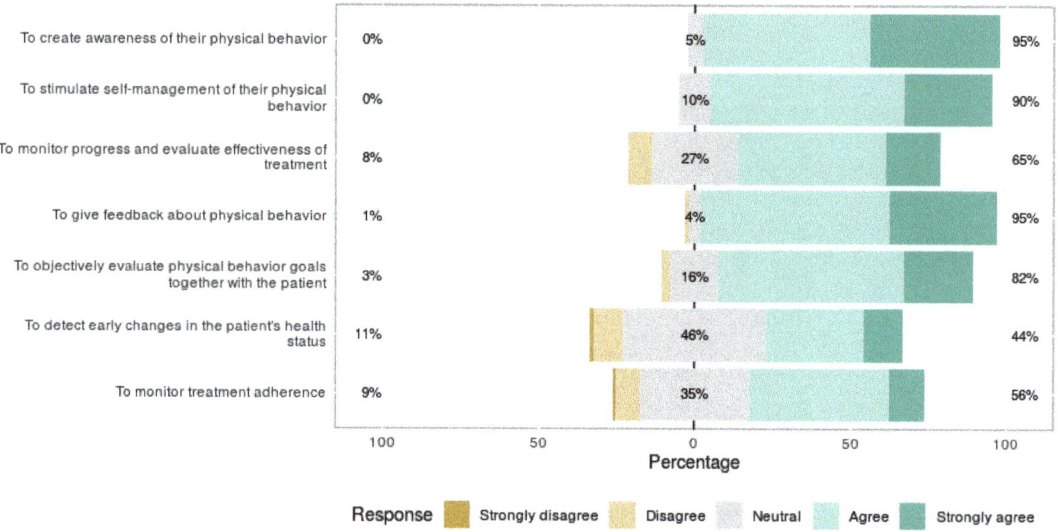

Figure 3. Perceived usefulness for eight different activity monitor purposes.

One significant difference was found between users and non-users: the users perceived creating awareness as more useful than non-users ($p = 0.031$). The participants were asked if they could come up with useful purposes other than noted in the question. Sixteen participants (16%) filled in the open-ended question on useful purposes other than mentioned in the question (Figure 3). Providing insight into a patients' demands vs. their capabilities (n = 6) was the most common purpose. Two mentioned heart rate and one mentioned arm/hand use.

3.4. Barriers

The most present barriers reported by the whole sample were lack of skills and knowledge of patients (65%), not being sure what monitor to purchase (54%), finding it too expensive (47%), and taking too much time (27%). Overall, seeing no added value for their patients and their work as physical therapists was not recognized as a barrier by participants (Figure 4).

Non-users agreed more strongly with the following barriers compared to users: not knowing much about the effectiveness ($p = 0.015$), lacking knowledge and ability to apply the technology themselves ($p = 0.013$), finding it too expensive ($p = 0.043$), and not being sure what monitor to purchase ($p = 0.035$). Other barriers did not show significant differences between users and non-users.

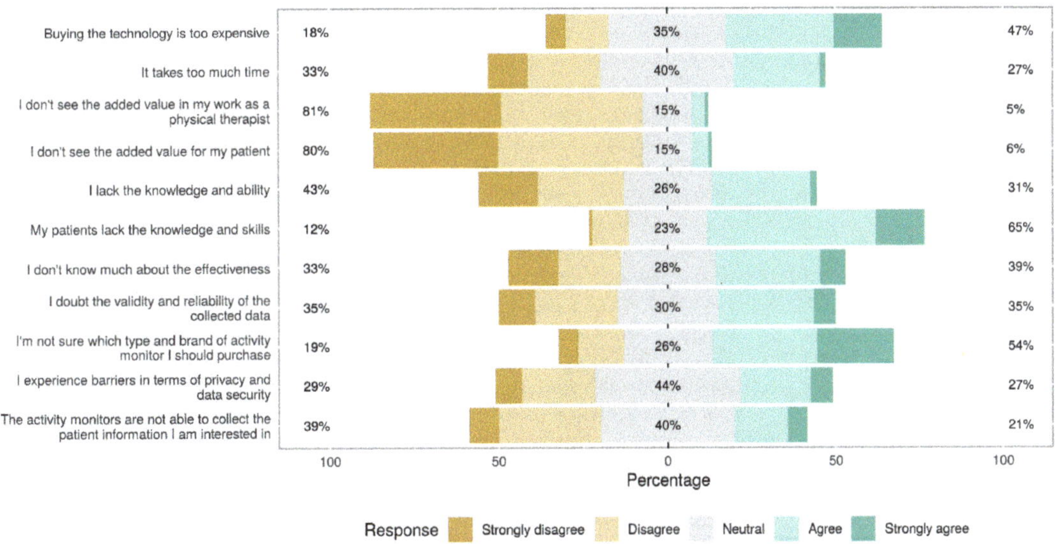

Figure 4. Barriers of using activity monitoring as a physical therapist.

3.5. Future Thoughts of Non-Users

Seventy-nine percent of the therapists who were not currently using activity monitoring were willing to use it in the future. Nineteen percent were neutral to this question, and two percent did not want to use activity monitoring in the future. In addition, participants were asked whether they would likely use activity monitoring in the next five years. Fifty-five percent of the non-users considered the change big or very big. Eight percent considered the change small or very small.

3.6. Additional Thoughts

The survey's last question asked all participants if they wanted to share anything else on activity monitoring. Thirty-two participants (31%) filled in this question. Several positive and enthusiastic thoughts on activity monitoring were provided. Participants report that activity monitoring offers valuable insight into a patients' behavior. About half of the 32 participants added some critical notes; they had doubts about the added value to the standard care relative to the effort. A few stated that applying technology was not always a holy grail and could not define therapy. Multiple participants mentioned that the usefulness was highly dependent on the age and stroke severity of the population.

4. Discussion

This study showed that, although physical therapists perceived wearable monitoring as potentially useful in stroke rehabilitation, only a minority of 25% actually used it in clinical care. Therapists that already used activity monitoring during treatment of stroke patients used it more often than other health care technologies and described themselves as being more innovative compared to non-users. The most recognized barriers were lack of knowledge and skills of patients, financial constraints, and not being sure what monitor to purchase.

The vast majority of our sample had not yet adopted the use of activity monitoring in day-to-day stroke care. The low numbers of technology used in treatment amongst physical therapists were in accordance with other studies that focused on technology use in rehabilitation practice [21,22]. A majority of 80% of therapists not using remote monitoring technology (non-users) did see value in the concept of objective physical behavior measurements with wearable technology, such as raising the patients' awareness of their behavior and the ability of providing objective feedback in order to promote physical activity and were willing to use it in the future. Correspondingly, a majority disagreed with seeing no added value for their work as a therapist and for their patients as a barrier. Other studies also found positive attitudes and excitement of therapists towards the concept of objective physical behavior data collection in clinical practice [28,35].

The discrepancy between the levels of adoption of activity monitoring and its perceived potential value suggests the presence of barriers. Potential barriers to adoption were indeed identified. The most frequently recognized barrier (65%) was perceived lack of skills and knowledge to use wearable monitoring technology in patients. Obviously, cognitive problems and generally older age might complicate the use of technological devices in daily life in stroke patients [36]. Especially for this group of patients, a user-friendly design of technology is desirable [14,28]. Issues with older and more severely affected patients were also explicitly stressed by the therapists in the open-ended questions. It should be noted that these results represent a perception of the therapists and are not confirmed by the patients themselves. Mercer et al. [37] found that older patients with chronic conditions also saw meaningful potential for wearable activity trackers but acknowledged that help from health professionals was desired to integrate the use in their daily life. In addition, caregivers who know the patient and his circumstances can play a crucial role in successful adoption [38,39]. Their support and encouragement might help patients to learn how to use wearable technology in their daily lives. To further improve the adoption of remote monitoring of physical behavior, collaboration with end-users, both therapists, patients, and their caregivers is to be recommended [28]. Whether the device matches the needs of end-users seems a critical factor for successful use [40].

Another frequently recognized barrier, especially by the non-users, is the lack of skills in selecting and using the appropriate wearable activity monitor suitable for the patient. This might be aggravated by the increasing amount of available consumer and research-grade wearable monitors and their different specifications [23,41]. Research-grade devices are generally accurate and reliable but are not easy to use in clinical practice, whereas consumer-grade devices have limited accuracy in rehabilitation populations [23]. A clear overview of best practices and skill training for therapists may help to overcome this barrier. The non-users also expressed significant doubts about the effectiveness of wearable monitoring for stroke patients' treatment. The field of research on the effectiveness for stroke patients is still evolving, more high-quality evidence might be a positive stimulus for use in the future [40,42]. Another critical concern physical therapists shared in the open-ended questions was that using technology can not define the course of therapy. Using technology should address the clinical need and the interaction between a patient and professional should not be forgotten [40].

Next to the individual skills and knowledge, successful, sustainable, and widespread adoption of technology is likely to be dependent on beliefs and attitudes of health care professionals [25,43,44]. Only one percent of the therapists in our study explicitly indicated being a person designing health care technologies and only 18% indicate that they are up-to-date and are looking for ways to adopt technology in daily practice. This low or absent innovative attitude might hamper the wide adoption in clinical practice. Therefore, if it is not widely accepted and fully integrated within organizations or the health care system, the use of wearable monitors will depend on the individual professional. Other stakeholders that have the potential to support and facilitate wider adoption of wearable technology are, for example, the policymakers of health care organizations, activity monitor companies, educational programs, and post-graduate training of professionals.

Our study has some limitations. As common in electronic surveys [21,27], non-response bias might have influenced our results. Respondents were probably more interested in contributing to a study on innovative technology than non-respondents, which may have overestimated the results. Since our respondents were selected based on being a physical therapist involved in stroke care, caution against generalizing our results to other health care occupations and patient populations is at its place. In addition, generalizability to other countries is limited since health care can be organized in a different way. We do not expect that geographical differences within the Netherlands have influenced our results since we tried to attempt diverse regions. No validated questionnaire that met our study purpose was available in the literature, and therefore to the best of our knowledge, we developed a survey with experts from the field and based on sufficient previous literature. The survey was pilot-tested amongst therapists and showed to be understandable and feasible. In addition, due to our study's narrative and exploratory nature, we could not establish in-depth and underlying thoughts regarding the use of wearable technology for stroke patients. From our results, no extensive requirements or (sensor) features of wearable monitors for clinical practice could be derived. Future studies should provoke a more profound discussion with therapists about the need and requirements for wearable monitors and relevant datasets for clinical use. However, together with qualitative studies [28,29], our study contributed to a comprehensive understanding of physical therapists' perspectives who, in the present years, are key stakeholders in adopting wearable technology in stroke care.

5. Conclusions

Our explorative study showed that despite physical behavior monitoring with wearable technology becoming commonplace in the consumer market and in academic research, it is not widely used by physical therapists involved in treatment of stroke patients. The concept of quantifying physical behavior with wearable monitors was perceived as useful by therapists, however, several barriers were identified. In current stroke care, physical therapists' skills, beliefs, and attitudes determine the current use of wearable technology.

Supplementary Materials: The following are available online at https://www.mdpi.com/article/10.3390/s21124066/s1, Questionnaire wearable activity monitoring for physical therapists involved in stroke care.

Author Contributions: Conceptualization, H.E.M.B., J.B.J.B., M.A.M.B., G.M.R.; methodology, H.E.M.B., J.B.J.B.; formal analysis, H.E.M.B., J.B.J.B., M.A.M.B.; investigation, H.E.M.B., M.A.M.B.; resources, H.E.M.B.; data curation, H.E.M.B.; writing—original draft preparation, H.E.M.B., M.A.M.B.; writing—review and editing, H.E.M.B., J.B.J.B., M.A.M.B., G.M.R.; visualization, H.E.M.B., J.B.J.B.; supervision, J.B.J.B., G.M.R., M.A.M.B.; project administration, H.E.M.B. All authors have read and agreed to the published version of the manuscript.

Funding: This research received no external funding.

Institutional Review Board Statement: Not applicable.

Informed Consent Statement: Not applicable.

Data Availability Statement: The data presented in this study are available on request from the corresponding author. The data are not publicly available.

Conflicts of Interest: The authors declare no conflict of interest.

References

1. WHO. *Global Health Estimates: Deaths by Cause, Age, Sex and Country, 2000–2012*; WHO: Geneva, Switzerland, 2014; Volume 9.
2. Rajsic, S.; Gothe, H.; Borba, H.H.; Sroczynski, G.; Vujicic, J.; Toell, T.; Siebert, U. Economic burden of stroke: A systematic review on post-stroke care. *Eur. J. Health Econ.* **2019**, *20*, 107–134. [CrossRef]
3. Collaborators, G.S. Global, regional, and national burden of stroke, 1990-2016: A systematic analysis for the global burden of disease study 2016. *Lancet Neurol.* **2019**, *18*, 439–458.
4. Billinger, S.A.; Arena, R.; Bernhardt, J.; Eng, J.J.; Franklin, B.A.; Johnson, C.M.; MacKay-Lyons, M.; Macko, R.F.; Mead, G.E.; Roth, E.J.; et al. Physical activity and exercise recommendations for stroke survivors: A statement for healthcare professionals from the american heart association/american stroke association. *Stroke* **2014**, *45*, 2532–2553. [CrossRef] [PubMed]
5. Shaughnessy, M.; Michael, K.M.; Sorkin, J.D.; Macko, R.F. Steps after stroke: Capturing ambulatory recovery. *Stroke* **2005**, *36*, 1305–1307. [CrossRef] [PubMed]
6. Langhorne, P.; Bernhardt, J.; Kwakkel, G. Stroke rehabilitation. *Lancet* **2011**, *377*, 1693–1702. [CrossRef]

7. Thilarajah, S.; Mentiplay, B.F.; Bower, K.J.; Tan, D.; Pua, Y.H.; Williams, G.; Koh, G.; Clark, R.A. Factors associated with post-stroke physical activity: A systematic review and meta-analysis. *Arch. Phys. Med. Rehabil.* **2018**, *99*, 1876–1889. [CrossRef] [PubMed]

8. Saunders, D.H.; Greig, C.A.; Mead, G.E. Physical activity and exercise after stroke: Review of multiple meaningful benefits. *Stroke* **2014**, *45*, 3742–3747. [CrossRef]

9. Lawrence, M.; Pringle, J.; Kerr, S.; Booth, J.; Govan, L.; Roberts, N.J. Multimodal secondary prevention behavioral interventions for tia and stroke: A systematic review and meta-analysis. *PLoS ONE* **2015**, *10*, e0120902. [CrossRef]

10. Han, P.; Zhang, W.; Kang, L.; Ma, Y.; Fu, L.; Jia, L.; Yu, H.; Chen, X.; Hou, L.; Wang, L.; et al. Clinical evidence of exercise benefits for stroke. *Adv. Exp. Med. Biol.* **2017**, *1000*, 131–151.

11. Bussmann, J.B.; van den Berg-Emons, R.J. To total amount of activity..... And beyond: Perspectives on measuring physical behavior. *Front. Psychol.* **2013**, *4*, 463. [CrossRef]

12. Teychenne, M.; Ball, K.; Salmon, J. Sedentary behavior and depression among adults: A review. *Int. J. Behav. Med.* **2010**, *17*, 246–254. [CrossRef]

13. Biswas, A.; Alter, D.A. Sedentary time and risk for mortality. *Ann. Intern. Med.* **2015**, *162*, 875–876. [CrossRef]

14. Marwaa, M.N.; Kristensen, H.K.; Guidetti, S.; Ytterberg, C. Physiotherapists' and occupational therapists' perspectives on information and communication technology in stroke rehabilitation. *PLoS ONE* **2020**, *15*, e0236831. [CrossRef] [PubMed]

15. Wright, S.P.; Hall Brown, T.S.; Collier, S.R.; Sandberg, K. How consumer physical activity monitors could transform human physiology research. *Am. J. Physiol. Regul. Integr. Comp. Physiol.* **2017**, *312*, R358–R367. [CrossRef]

16. Piwek, L.; Ellis, D.A.; Andrews, S.; Joinson, A. The rise of consumer health wearables: Promises and barriers. *PLoS Med.* **2016**, *13*, e1001953. [CrossRef] [PubMed]

17. Vegesna, A.; Tran, M.; Angelaccio, M.; Arcona, S. Remote patient monitoring via non-invasive digital technologies: A systematic review. *Telemed. J. E Health* **2017**, *23*, 3–17. [CrossRef] [PubMed]

18. Hailey, D.; Roine, R.; Ohinmaa, A.; Dennett, L. Evidence of benefit from telerehabilitation in routine care: A systematic review. *J. Telemed. Telecare* **2011**, *17*, 281–287. [CrossRef] [PubMed]

19. Braakhuis, H.E.M.; Berger, M.A.M.; Bussmann, J.B.J. Effectiveness of healthcare interventions using objective feedback on physical activity: A systematic review and meta-analysis. *J. Rehabil. Med.* **2019**, *51*, 151–159. [CrossRef]

20. Dobkin, B.H.; Martinez, C. Wearable sensors to monitor, enable feedback, and measure outcomes of activity and practice. *Curr. Neurol. Neurosci. Rep.* **2018**, *18*, 87. [CrossRef]

21. Langan, J.; Subryan, H.; Nwogu, I.; Cavuoto, L. Reported use of technology in stroke rehabilitation by physical and occupational therapists. *Disabil. Rehabil. Assist. Technol.* **2018**, *13*, 641–647. [CrossRef] [PubMed]

22. Allouch, S.B.; van Velsen, L. Fit by bits: An explorative study of sports physiotherapists' perception of quantified self technologies. *Stud. Health Technol. Inform.* **2018**, *247*, 296–300.

23. Lang, C.E.; Barth, J.; Holleran, C.L.; Konrad, J.D.; Bland, M.D. Implementation of wearable sensing technology for movement: Pushing forward into the routine physical rehabilitation care field. *Sensors* **2020**, *20*, 5744. [CrossRef]

24. Hoy, M.B. Personal activity trackers and the quantified self. *Med. Ref. Serv. Q.* **2016**, *35*, 94–100. [CrossRef]

25. Sabus, C.; Spake, E. Innovative physical therapy practice: A qualitative verification of factors that support diffusion of innovation in outpatient physical therapy practice. *J. Healthc. Leadersh.* **2016**, *8*, 107–120. [CrossRef]

26. Brouns, B.; Meesters, J.J.L.; Wentink, M.M.; de Kloet, A.J.; Arwert, H.J.; Boyce, L.W.; Vliet Vlieland, T.P.M.; van Bodegom-Vos, L. Factors associated with willingness to use erehabilitation after stroke: A cross-sectional study among patients, informal caregivers and healthcare professionals. *J. Rehabil. Med.* **2019**, *51*, 665–674. [CrossRef] [PubMed]

27. Liu, L.; Miguel Cruz, A.; Rios Rincon, A.; Buttar, V.; Ranson, Q.; Goertzen, D. What factors determine therapists' acceptance of new technologies for rehabilitation–a study using the unified theory of acceptance and use of technology (utaut). *Disabil. Rehabil.* **2015**, *37*, 447–455. [CrossRef] [PubMed]

28. Louie, D.R.; Bird, M.L.; Menon, C.; Eng, J.J. Perspectives on the prospective development of stroke-specific lower extremity wearable monitoring technology: A qualitative focus group study with physical therapists and individuals with stroke. *J. Neuroeng. Rehabil.* **2020**, *17*, 31. [CrossRef]

29. Papi, E.; Murtagh, G.M.; McGregor, A.H. Wearable technologies in osteoarthritis: A qualitative study of clinicians' preferences. *BMJ Open* **2016**, *6*, e009544. [CrossRef]

30. Hamilton, C.; McCluskey, A.; Hassett, L.; Killington, M.; Lovarini, M. Patient and therapist experiences of using affordable feedback-based technology in rehabilitation: A qualitative study nested in a randomized controlled trial. *Clin. Rehabil.* **2018**, *32*, 1258–1270. [CrossRef]

31. Rogers, E.M. Diffusion of innovations: Modifications of a model for telecommunications. In *Die Diffusion von Innovationen in der Telekommunikation*; Springer: Berlin/Heidelberg, Germany, 1995; pp. 25–38.

32. Kononova, A.; Li, L.; Kamp, K.; Bowen, M.; Rikard, R.V.; Cotten, S.; Peng, W. The use of wearable activity trackers among older adults: Focus group study of tracker perceptions, motivators, and barriers in the maintenance stage of behavior change. *JMIR Mhealth Uhealth* **2019**, *7*, e9832. [CrossRef] [PubMed]

33. Likert, R. A technique for the measurement of attitudes. *Arch. Psychol.* **1932**, *22*, 44–53.

34. Bryer, J.; Speerschneider, K.; Bryer, M.J. Package 'likert'. In *Analysis and Visualization Likert Items. CRAN*; Institute for Statistics and Mathematics; Vienna University of Economics and Business: Vienna, Austria, 2016.

35. Pak, P.; Jawed, H.; Tirone, C.; Lamb, B.; Cott, C.; Brunton, K.; Mansfield, A.; Inness, E.L. Incorporating research technology into the clinical assessment of balance and mobility: Perspectives of physiotherapists and people with stroke. *Physiother. Can.* **2015**, *67*, 1–8. [CrossRef]

36. Farivar, S.; Abouzahra, M.; Ghasemaghaei, M. Wearable device adoption among older adults: A mixed-methods study. *Int. J. Inf. Manag.* **2020**, *55*, 102209. [CrossRef]

37. Mercer, K.; Giangregorio, L.; Schneider, E.; Chilana, P.; Li, M.; Grindrod, K. Acceptance of commercially available wearable activity trackers among adults aged over 50 and with chronic illness: A mixed-methods evaluation. *JMIR Mhealth Uhealth* **2016**, *4*, e7. [CrossRef]

38. Brouns, B.; Meesters, J.J.L.; Wentink, M.M.; de Kloet, A.J.; Arwert, H.J.; Vliet Vlieland, T.P.M.; Boyce, L.W.; van Bodegom-Vos, L. Why the uptake of erehabilitation programs in stroke care is so difficult-a focus group study in the netherlands. *Implement. Sci.* **2018**, *13*, 133. [CrossRef]

39. Saywell, N.; Taylor, D. Focus group insights assist trial design for stroke telerehabilitation: A qualitative study. *Physiother. Theory Pract.* **2015**, *31*, 160–165. [CrossRef]

40. Chen, C.C.; Bode, R.K. Factors influencing therapists' decision-making in the acceptance of new technology devices in stroke rehabilitation. *Am. J. Phys. Med. Rehabil.* **2011**, *90*, 415–425. [CrossRef]

41. Shin, G.; Jarrahi, M.H.; Fei, Y.; Karami, A.; Gafinowitz, N.; Byun, A.; Lu, X. Wearable activity trackers, accuracy, adoption, acceptance and health impact: A systematic literature review. *J. Biomed Inform.* **2019**, *93*, 103153. [CrossRef] [PubMed]

42. Lynch, E.A.; Jones, T.M.; Simpson, D.B.; Fini, N.A.; Kuys, S.S.; Borschmann, K.; Kramer, S.; Johnson, L.; Callisaya, M.L.; Mahendran, N.; et al. Activity monitors for increasing physical activity in adult stroke survivors. *Cochrane Database Syst. Rev.* **2018**, *7*, Cd012543. [CrossRef] [PubMed]

43. Bagot, K.L.; Moloczij, N.; Barclay-Moss, K.; Vu, M.; Bladin, C.F.; Cadilhac, D.A. Sustainable implementation of innovative, technology-based health care practices: A qualitative case study from stroke telemedicine. *J. Telemed. Telecare* **2020**, *26*, 79–91. [CrossRef] [PubMed]

44. Urquhart, R.; Kendell, C.; Cornelissen, E.; Madden, L.L.; Powell, B.J.; Kissmann, G.; Richmond, S.A.; Willis, C.; Bender, J.L. Defining sustainability in practice: Views from implementing real-world innovations in health care. *BMC Health Serv. Res.* **2020**, *20*, 87. [CrossRef] [PubMed]

MDPI

St. Alban-Anlage 66

4052 Basel

Switzerland

Tel. +41 61 683 77 34

Fax +41 61 302 89 18

www.mdpi.com

Sensors Editorial Office

E-mail: sensors@mdpi.com

www.mdpi.com/journal/sensors

9 783036 520636